Adaptive High-Order Methods in Computational Fluid Dynamics

Advances in Computational Fluid Dynamics

Editors-in-Chief: Chi-Wang Shu *(Brown University, USA)* and
Chang Shu *(National University of Singapore, Singapore)*

Published

Vol. 2 Adaptive High-Order Methods in Computational Fluid Dynamics
edited by Z. J. Wang (Iowa State University, USA)

Forthcoming

Vol. 1 Computational Methods for Two-Phase Flows
by Peter D. M. Spelt (Imperial College London, UK), Stephen J. Shaw (X'ian Jiaotong – University of Liverpool, Suzhou, China) & Hang Ding (University of California, Santa Barbara, USA)

Advances in
Computational Fluid Dynamics

Vol.
2

Adaptive High-Order Methods in Computational Fluid Dynamics

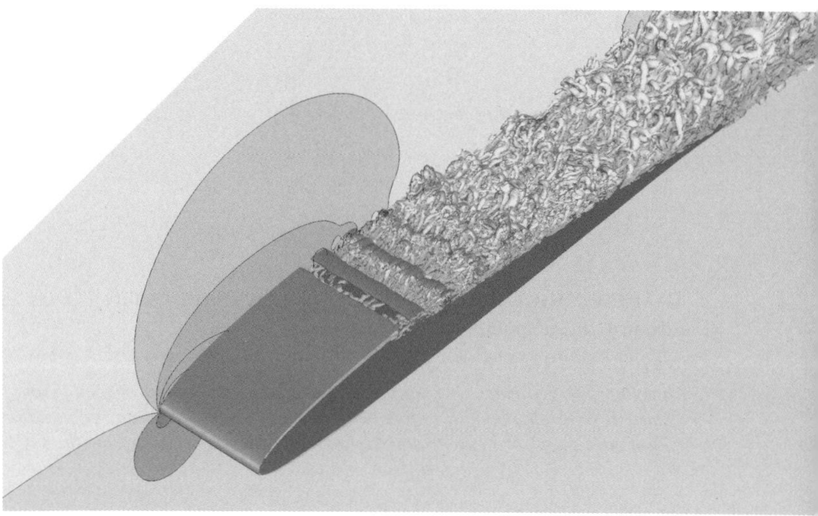

Editor

Z J Wang

Iowa State University, USA

World Scientific

NEW JERSEY • LONDON • SINGAPORE • BEIJING • SHANGHAI • HONG KONG • TAIPEI • CHENNAI

Published by

World Scientific Publishing Co. Pte. Ltd.

5 Toh Tuck Link, Singapore 596224

USA office: 27 Warren Street, Suite 401-402, Hackensack, NJ 07601

UK office: 57 Shelton Street, Covent Garden, London WC2H 9HE

British Library Cataloguing-in-Publication Data
A catalogue record for this book is available from the British Library.

ADAPTIVE HIGH-ORDER METHODS IN COMPUTATIONAL FLUID DYNAMICS
Advances in Computational Fluid Dynamics — Vol. 2

ISBN-13 978-981-4313-18-6
ISBN-10 981-4313-18-1

Printed in Singapore by B & Jo Enterprise Pte Ltd

To My Family

PREFACE

This book contains invited chapters written by leading international experts on adaptive high-order methods in computational fluid dynamics (CFD). It covers several widely used, and still intensively researched methods, including the discontinuous Galerkin (DG), residual distribution, differential quadrature, k-exact finite volume, spectral volume/spectral difference, P_NP_M, and correction procedure via reconstruction methods. The reasons for including such a wide coverage of methods are to: (1) provide a single source of reference, (2) present a snapshot of the state-of-the-art, and (3) facilitate the observation of similarities and differences as well as pros and cons of these methods.

In the present context, adaptive high-order methods refer to numerical methods that are capable of handling unstructured adaptive meshes with accuracy higher than second-order. These methods are compact, scalable, capable of handling both complex physics and geometry, and suitable for modern parallel supercomputers and graphics processing units (GPUs). They are widely considered the next major breakthrough in CFD, and have already found applications in computational aeroacoustics, computational electromagnetics, vortex dominated flows, and large eddy simulation and direct numerical simulation of turbulent flows.

A concerted effort was made to minimize overlaps among the chapters. For example, the first 7 chapters describe different aspects of the DG methods, while the last 8 chapters are devoted to other high-order methods. Main topics covered include innovative formulations, analyses, efficient solution and time marching algorithms, parallel implementation, turbulence modeling, discontinuity-capturing techniques, error estimates, hp-adaptations, and dynamic mesh techniques, etc.

The book requires a graduate student level of understanding. It should serve as an excellent source of information for CFD developers, educators, researchers, users, and students who are interested in the state-of-the-art and the remaining challenges in adaptive high-order methods.

I am grateful to Dr. Chang Shu, a close friend and the Co-Editor-in-Chief of the book series, *Advances in Computational Fluid Dynamics* in World Scientific, for suggesting the book. Heartfelt thanks are due to all the contributors of this volume. Needless to say, the book would not exist without their hard work. Finally, I'd like to thank Ying Zhou for producing the color cover graphic, and Varun Vikas for help with Latex.

Z.J. Wang
Ames, Iowa
June 30, 2011

CONTENTS

CHAPTER 1

DISCONTINUOUS GALERKIN FOR TURBULENT FLOWS

Francesco Bassi*, Lorenzo Botti† and Alessandro Colombo‡

Dipartimento di Ingegneria Industriale,
Università degli studi di Bergamo,
Viale Marconi 5, 24044 Dalmine (BG), Italy
**francesco.bassi@unibg.it*
†lorenzo.botti@unibg.it
‡alessandro.colombo@unibg.it

Antonio Ghidoni§ and Stefano Rebay¶

Dipartimento di Ingegneria Meccanica e Industriale,
Università degli Studi di Brescia,
Via Branze 38, 25123 Brescia, Italy
§antonio.ghidoni@ing.unibs.it
¶stefano.rebay@ing.unibs.it

The purpose of this chapter is to present all the relevant features of a high-order DG method developed over the years for the numerical solution of the RANS and k-ω equations. The method has been implemented using orthogonal and hierarchical modal shape functions defined in the real space. The code can handle hybrid grids consisting of tetrahedra, prisms, pyramids and hexahedra. Implicit time integration is applied to the fully coupled RANS and k-ω equations, both for steady and unsteady computations. A directional shock-capturing term, proportional to the inviscid residual, is employed to control oscillations around shocks. Most of the numerical results presented in this chapter have been computed within the EU-funded ADIGMA project to investigate the capability of the method for aeronautical applications.

1. Introduction

In recent years several high-order methods have been emerging as practical tools to go beyond the second-order accuracy of standard finite volume discretizations of PDEs on general unstructured grids. For aerospace applications this is of particular importance to further increase the impact of Computational Fluid Dynamics (CFD) on the aerodynamic design of new generation aircraft. The Discontinuous Galerkin (DG) method, in particular, has been gaining popularity as one of the most promising approaches to the accurate and robust numerical solution of ever more complex physical models and has attracted great efforts of many research groups into its development.

In this context, the purpose of this chapter is to describe several developments of the DG method implemented over the years in a fully parallel DG code, named MIGALE, that we have used for the numerical solution of the Euler, Navier-Stokes and the coupled RANS and k-ω turbulence model equations. These developments include: i) a proposal for adapting the smooth-wall treatment of the variable ω to the degree of the polynomial approximation, ii) the adoption of orthonormal and hierarchical modal basis functions defined in the real space for arbitrary shape elements, iii) a shock-capturing technique based on the inviscid residual and applied in the direction of the pressure gradient, iv) an implicit time integration technique suited both for steady and unsteady problems.

The capabilities of the present version of the code will be demonstrated by computing several fairly complex problems taken from the suites of test cases proposed within the EU-funded project ADIGMA. In the conclusions we will give a brief account of other recent implementations which are already quite mature and will outline future directions of development.

2. DG Solution of the RANS and k-ω Equations

This section describes relevant implementation aspects of the DG discretization applied to the coupled set of RANS and k-ω equations, including surface boundary conditions for $\widetilde{\omega}$, choice of shape functions, shock-capturing approach and implicit time integration.

2.1. *Governing equations*

The governing equations can be written as

$$\frac{\partial \rho}{\partial t} + \frac{\partial}{\partial x_j}(\rho u_j) = 0, \tag{1}$$

$$\frac{\partial}{\partial t}(\rho u_i) + \frac{\partial}{\partial x_j}(\rho u_j u_i) = -\frac{\partial p}{\partial x_i} + \frac{\partial \widehat{\tau}_{ji}}{\partial x_j}, \tag{2}$$

$$\frac{\partial}{\partial t}(\rho e_0) + \frac{\partial}{\partial x_j}(\rho u_j h_0) = \frac{\partial}{\partial x_j}\left[u_i \widehat{\tau}_{ij} - q_j\right] - \tau_{ij}\frac{\partial u_i}{\partial x_j} + \beta^* \rho \overline{k} e^{\widetilde{\omega}_r}, \tag{3}$$

$$\frac{\partial}{\partial t}(\rho k) + \frac{\partial}{\partial x_j}(\rho u_j k) = \frac{\partial}{\partial x_j}\left[(\mu + \sigma^* \overline{\mu}_t)\frac{\partial k}{\partial x_j}\right] + \tau_{ij}\frac{\partial u_i}{\partial x_j} - \beta^* \rho \overline{k} e^{\widetilde{\omega}_r}, \tag{4}$$

$$\begin{aligned}\tfrac{\partial}{\partial t}(\rho\widetilde{\omega}) + \tfrac{\partial}{\partial x_j}(\rho u_j \widetilde{\omega}) = \tfrac{\partial}{\partial x_j}\left[(\mu + \sigma\overline{\mu}_t)\tfrac{\partial \widetilde{\omega}}{\partial x_j}\right] + \tfrac{\alpha}{\overline{k}}\tau_{ij}\tfrac{\partial u_i}{\partial x_j} - \beta\rho e^{\widetilde{\omega}_r} \\ + (\mu + \sigma\overline{\mu}_t)\tfrac{\partial \widetilde{\omega}}{\partial x_k}\tfrac{\partial \widetilde{\omega}}{\partial x_k},\end{aligned} \tag{5}$$

where the pressure, the turbulent and total stress tensors, the heat flux vector and the eddy viscosity are given by

$$p = (\gamma - 1)\rho\left(e_0 - u_k u_k/2\right), \tag{6}$$

$$\tau_{ij} = 2\overline{\mu}_t\left[S_{ij} - \frac{1}{3}\frac{\partial u_k}{\partial x_k}\delta_{ij}\right] - \frac{2}{3}\rho\overline{k}\delta_{ij}, \tag{7}$$

$$\widehat{\tau}_{ij} = 2\mu\left[S_{ij} - \frac{1}{3}\frac{\partial u_k}{\partial x_k}\delta_{ij}\right] + \tau_{ij}, \tag{8}$$

$$q_j = -\left(\frac{\mu}{\mathrm{Pr}} + \frac{\overline{\mu}_t}{\mathrm{Pr}_t}\right)\frac{\partial h}{\partial x_j}, \tag{9}$$

$$\overline{\mu}_t = \alpha^* \rho \overline{k} e^{-\widetilde{\omega}_r}, \quad \overline{k} = \max(0, k). \tag{10}$$

Here γ is the ratio of gas specific heats, Pr and Pr_t are the molecular and turbulent Prandtl numbers and

$$S_{ij} = \frac{1}{2}\left(\frac{\partial u_i}{\partial x_j} + \frac{\partial u_j}{\partial x_i}\right), \tag{11}$$

is the mean strain-rate tensor. The closure parameters α, α^*, β, β^*, σ, σ^* are those of the high- or low-Reynolds number k-ω model of Wilcox.[1]

Notice that the RANS and k-ω equations here employed are not in standard form since the variable $\widetilde{\omega} = \log\omega$, instead of ω itself, is used in Eqs. (3), (4), (5). Motivations for this choice have been discussed in Ref. 2. The variable $\widetilde{\omega}_r$ in the source terms of Eqs. (3), (4), (5) and in the eddy viscosity defined by Eq. (10) is introduced to indicate that $\widetilde{\omega}_r$ fulfills suitably defined "realizability" conditions, which set a lower bound on $\widetilde{\omega}$ in such equations. This limitation substantially improves the stability and robustness of turbulent flow computations because there is numerical evidence that too small, though positive, values of $\omega = e^{\widetilde{\omega}}$ can lead to sudden breakdown of computations.

Realizability conditions, which guarantee that the turbulence model predicts positive normal turbulent stresses and satisfies the Schwarz inequality for shear turbulent stresses, lead to the following inequalities

$$\frac{e^{\widetilde{\omega}}}{\alpha^*} - 3\left(S_{ii} - \frac{1}{3}\frac{\partial u_k}{\partial x_k}\right) \geq 0, \quad i = 1, 2, 3, \tag{12}$$

$$\begin{aligned}\left(\frac{e^{\widetilde{\omega}}}{\alpha^*}\right)^2 - 3\left(S_{ii} + S_{jj} - \frac{1}{3}\frac{\partial u_k}{\partial x_k}\right)\frac{e^{\widetilde{\omega}}}{\alpha^*}& \\ + 9\left[\left(S_{ii} - \frac{1}{3}\frac{\partial u_k}{\partial x_k}\right)\left(S_{jj} - \frac{1}{3}\frac{\partial u_k}{\partial x_k}\right) - S_{ij}^2\right] &\geq 0, \\ i, j = 1, 2, 3, \quad i \neq j.&\end{aligned} \tag{13}$$

Denoting with a the maximum value of the unknown $e^{\widetilde{\omega}}/\alpha^*$ corresponding to the zeros of Eqs. (12) and (13), the lower bound $\widetilde{\omega}_{r0}$ that guarantees realizable turbulent stresses is given by

$$\frac{e^{\widetilde{\omega}_{r0}}}{\alpha^*} = a. \tag{14}$$

The solution of Eq. (14) is trivial for the high-Reynolds number k-ω model because in this case α^* is constant. For the low-Reynolds number k-ω model α^* depends on the turbulent Reynolds number according to the equation

$$\alpha^* = \alpha_t^* \frac{\alpha_0^* + \mathrm{Re}_t/R_k}{1 + \mathrm{Re}_t/R_k}, \tag{15}$$

where α_t^*, α_0^* and R_k are model constants and Re_t is the turbulent Reynolds number given by $\mathrm{Re}_t = k/(e^{\widetilde{\omega}}\nu)$. Combining Eqs. (14) and (15) we find $\widetilde{\omega}_{r0}$ from the following second degree equation for the unknown $e^{\widetilde{\omega}_{r0}}$

$$e^{2\widetilde{\omega}_{r0}} - \left(\alpha_t^* \alpha_0^* a - \frac{\overline{k}}{R_k \nu}\right) e^{\widetilde{\omega}_{r0}} - \alpha_t^* \frac{\overline{k}}{R_k \nu} a = 0, \tag{16}$$

and, finally, we set $\widetilde{\omega}_r$ in Eqs. (3), (4), (5) and (10) as

$$\widetilde{\omega}_r = \max\left(\widetilde{\omega}, \widetilde{\omega}_{r0}\right). \tag{17}$$

2.1.1. *Surface boundary condition for $\widetilde{\omega}$*

In the viscous sublayer the $\widetilde{\omega}$ equation reduces to the form

$$\nu \frac{d^2\widetilde{\omega}}{dy^2} + \nu \left(\frac{d\widetilde{\omega}}{dy}\right)^2 = \beta e^{\widetilde{\omega}}, \tag{18}$$

where y is the local coordinate normal to the wall. Eq. (18) admits the following near wall solutions

$$\widetilde{\omega} = \log\left(\frac{6\nu}{\beta}\right) - 2\log\left(y + \sqrt{\frac{6\nu}{\beta e^{\widetilde{\omega}_w}}}\right), \tag{19}$$

where $\widetilde{\omega}_w$ is the value of $\widetilde{\omega}$ at the wall $(y = 0)$. Of course, these solutions are nothing but the logarithm of the viscous sublayer solutions for ω reported in Ref. 1. For $\widetilde{\omega}_w \to \infty$ the solution is singular and is considered the appropriate solution for smooth walls, whereas non-singular solutions are those corresponding to finite values of $\widetilde{\omega}_w$ and provide a way to include effects of surface roughness through surface boundary conditions.

It has been shown in Ref. 1 that singular and non-singular solutions for ω_w can be encompassed in the so-called rough-wall method whereby the smooth wall solution is recovered when the surface roughness tends to zero. In the rough-wall method surface values of ω can be simply set by means of the correlation

$$\omega_w = S_r \frac{u_\tau^2}{\nu_w}, \tag{20}$$

where $u_\tau = \sqrt{\tau_w/\rho_w}$ is the friction velocity and τ_w, ρ_w and ν_w are the shear stress, the density and the kinematic viscosity at the wall. The non-dimensional function S_r given by Wilcox is defined as

$$S_r = \begin{cases} (50/k_r^+)^2 & \text{if} \quad k_r^+ < 25 \\ 100/k_r^+ & \text{if} \quad k_r^+ \geq 25 \end{cases} \tag{21}$$

where $k_r^+ = k_r u_\tau/\nu_w$ denotes the non-dimensional equivalent sand-roughness height. For rough surfaces with prescribed values of k_r, Eq. (20) allows to compute the values ω_w to be set at the wall surface. Of course, the grid density or the degree of polynomial approximation should be high enough to provide accurate solutions.

On the other hand, the implementation of the smooth wall boundary condition for ω requires special care in the numerical treatment of the singularity. Two popular approaches have been proposed by Wilcox[1] and Menter.[3]

Relying on the rough-wall method, the approach recommended by Wilcox is simply to skip the issue of the numerical treatment of the singularity by replacing the perfectly smooth surface with an hydraulically smooth surface. In this so-called "slightly-rough-wall" boundary condition, using again Eqs. (20) and (21) with $k_r^+ < 25$, one obtains

$$\omega_w = 2500\frac{\nu_w}{k_r^2}, \tag{22}$$

where, according to Wilcox, k_r should be low enough to guarantee that $k_r^+ < 5$, *i.e.*, it should ensure that the surface is hydraulically smooth with roughness peaks lying within the viscous sublayer.

The approach proposed by Menter consists in setting at the wall a finite value ω_w given by

$$\omega_w = 10\frac{6\nu_w}{\beta y_1^2}, \tag{23}$$

where y_1 is the distance to the next grid point away from the wall. This means setting at the wall the analytical solution computed at y_1 multiplied by the factor 10, or, put another way, the analytical solution computed at $y = \alpha_M y_1$ where $\alpha_M = 1/\sqrt{10}$.

A comparison of Eqs. (22) and (23) suggests that if k_r is made proportional to y_1 then the two equations have the same form of functional dependence on the length. This observation, reported by Hellsten,[4] allows to find the equivalent slightly-rough-wall roughness implied by Menter's formula as a function of y_1. More important, Hellsten proposed to optimize the factor 10 of Menter's formula by means of an accurate near-wall numerical study of the ω solution and by comparing skin friction distributions of flat plate flows computed on differently refined grids. The value of the factor proposed by Hellsten is 1.25 instead of 10.

In the framework of the DG method, an approach like that of Hellsten was presented in Ref. 2 where it was found that a good agreement between experimental and numerical skin friction distributions of flat plate flows could be obtained by replacing the value of the coefficient α_M used by Menter with a lower value given by $\alpha = 0.3\sqrt{6/\beta}/50$. As the solutions presented in Ref. 2 were computed only up to $\mathbb{P}^2$ polynomial approximation, keeping α not dependent on the polynomial degree of the solution did work accurately. However, as higher degree polynomials can follow closer and closer the exact near wall distribution of ω, it seems reasonable to make α dependent on the degree k of the polynomial approximation.

A possible approach in this direction has been outlined in Ref. 5, where the basic idea was to replace the smooth wall limit $\widetilde{\omega}_w \to \infty$ of Eq. (19) with the value at the wall of the L^2 projection of the singular solution onto the basis of the polynomial approximation.

Here we propose an alternative way which consists of setting at the wall the value $\widetilde{\omega}_w^k$ of the Taylor series expansion of Eq. (19) (with $\widetilde{\omega}_w \to \infty$) around $y = h$, truncated to k terms, *i.e.*,

$$\begin{aligned}\widetilde{\omega}_w^k &= \widetilde{\omega}_h - \left.\frac{\partial \widetilde{\omega}}{\partial y}\right|_h \frac{h}{1!} + \left.\frac{\partial^2 \widetilde{\omega}}{\partial y^2}\right|_h \frac{h^2}{2!} - \cdots \\ &= \widetilde{\omega}_h + \sum_{n=1}^{k} \frac{1}{n}. \end{aligned} \tag{24}$$

The finite values $\widetilde{\omega}_w^k$ at the wall can again be related to the exact solution computed at $y = \alpha^k h$ by setting

$$\widetilde{\omega}_w^k = \log\left(\frac{6\nu_w}{\beta\left(\alpha^k h\right)^2}\right), \tag{25}$$

and thus finding, by comparing Eqs. (24) and (25),

$$\alpha^k = e^{-\sum_{n=1}^{k}\frac{1}{n}}. \tag{26}$$

To actually apply Eq. (25), h needs to be specified. In the flat plate computations presented below h has been set equal to the distance from the wall of the centroid of the elements next to the wall. As Eq. (25) holds for hydraulically smooth surfaces, we remark that the slightly-rough-wall roughness implied by Eq. (25) should satisfy the condition $k_r^+ < 5$. If locally this condition is not satisfied, then $\widetilde{\omega}_w$ is computed by means of Eq. (20) with $k_r^+ = 5$. This could be the case, for instance, of low-degree polynomial solutions computed on relatively coarse grids targeted at high-degree polynomial approximations.

The boundary condition for ω has been tested on the flat plate flow reported by Wieghardt.[6] The computational grid, taken from the NPARC Alliance Validation Archive,[7] is the coarsest one used for the validation of the WIND code and corresponds to $y^+ = 30$ for the first grid point off the wall. The Figure 1 displays the skin-friction distribution along the plate and the profiles of u velocity component and of turbulence quantities at $x/L = 0.923$ of the $\mathbb{P}^{3\to 6}$ solutions. The difference of near wall behavior of k^+ between DG results and "average" experimental data is an effect produced by the high-Reynolds number k-ω model here employed, that disappears

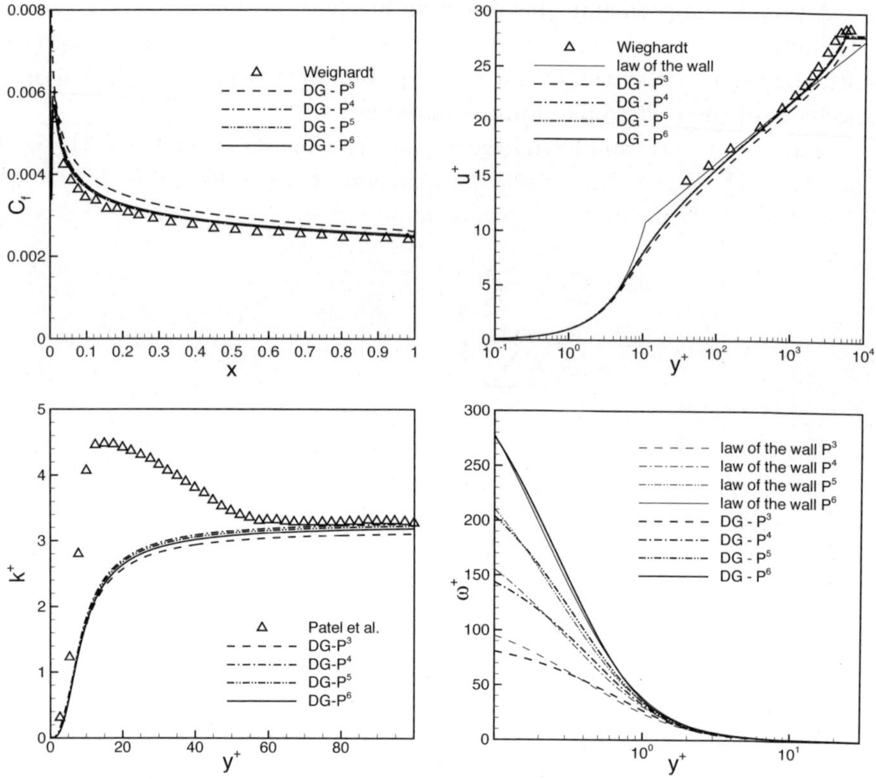

Fig. 1. Flat plate: skin-friction, velocity profiles and turbulence quantities.

using the modified coefficients of the low-Reynolds number version of the model.

2.2. *DG space discretization*

The governing equations can be written in compact form as

$$\frac{\partial \mathbf{u}}{\partial t} + \boldsymbol{\nabla} \cdot \mathbf{F}_c(\mathbf{u}) + \boldsymbol{\nabla} \cdot \mathbf{F}_v(\mathbf{u}, \boldsymbol{\nabla}\mathbf{u}) + \mathbf{s}(\mathbf{u}, \boldsymbol{\nabla}\mathbf{u}) = \mathbf{0}, \tag{27}$$

where $\mathbf{u}, \mathbf{s} \in \mathbb{R}^M$ denote the vectors of the M conservative variables and source terms, $\mathbf{F}_c, \mathbf{F}_v \in \mathbb{R}^M \otimes \mathbb{R}^N$ denote the inviscid and viscous flux functions, respectively, and N is the space dimension.

The weak form of Eq. (27) reads

$$\int_{\Omega} \phi \frac{\partial \mathbf{u}}{\partial t} \, \mathrm{d}\mathbf{x} - \int_{\Omega} \boldsymbol{\nabla}\phi \cdot \mathbf{F}(\mathbf{u}, \boldsymbol{\nabla}\mathbf{u}) \, \mathrm{d}\mathbf{x} + \int_{\partial\Omega} \phi \mathbf{F}(\mathbf{u}, \boldsymbol{\nabla}\mathbf{u}) \cdot \mathbf{n} \, \mathrm{d}\sigma + \int_{\Omega} \phi \mathbf{s}(\mathbf{u}, \boldsymbol{\nabla}\mathbf{u}) \, \mathrm{d}\mathbf{x} = \mathbf{0}, \quad (28)$$

where ϕ denotes any arbitrary, sufficiently smooth, test function and $\mathbf{F}$ is the sum of the inviscid and viscous fluxes. The DG discretization of Eq. (28) is defined on a triangulation $\mathcal{T}_h = \{K\}$ of an approximation Ω_h of Ω, consisting of a set of non-overlapping hybrid-type elements. The following space setting of discontinuous piecewise polynomial functions for each component $u_{h_i} = u_{h_1}, \ldots, u_{h_M}$ of the numerical solution $\mathbf{u}_h$ is assumed:

$$u_{h_i} \in \Phi_h \stackrel{\text{def}}{=} \left\{ \phi_h \in L^2(\Omega) : \phi_h|_K \in \mathrm{P}^k(K) \ \forall K \in \mathcal{T}_h \right\}, \quad (29)$$

for some polynomial degree $k \geq 0$, being $\mathrm{P}^k(K)$ the space of polynomials of global degree at most k on the element K.

The discontinuous approximation of the numerical solution requires introducing a special treatment of the inviscid interface flux and of the viscous flux. For the former it is common practice to use suitably defined numerical flux functions which ensure conservation and account for wave propagation. For the latter we employ the BR2 scheme, presented in Refs. 8, 9 and theoretically analyzed in Refs. 10, 11 (where it is referred to as BRMPS), to obtain a consistent, stable and accurate discretization of the viscous flux. Accounting for these aspects, the DG formulation of problem (28) then requires to find $u_{h_1}, \ldots, u_{h_M} \in \Phi_h$ such that

$$\int_{\Omega_h} \phi_h \frac{\partial \mathbf{u}_h}{\partial t} \, \mathrm{d}\mathbf{x} - \int_{\Omega_h} \boldsymbol{\nabla}_h \phi_h \cdot \mathbf{F}\left(\mathbf{u}_h, \boldsymbol{\nabla}_h \mathbf{u}_h + \mathbf{r}([\![\mathbf{u}_h]\!])\right) \, \mathrm{d}\mathbf{x} + \int_{\Gamma_h} [\![\phi_h]\!] \cdot \widehat{\mathbf{f}}\left(\mathbf{u}_h^{\pm}, (\boldsymbol{\nabla}_h \mathbf{u}_h + \eta_e \mathbf{r}_e([\![\mathbf{u}_h]\!]))^{\pm}\right) \, \mathrm{d}\sigma + \int_{\Omega_h} \phi_h \mathbf{s}\left(\mathbf{u}_h, \boldsymbol{\nabla}_h \mathbf{u}_h + \mathbf{r}([\![\mathbf{u}_h]\!])\right) \, \mathrm{d}\mathbf{x} = \mathbf{0}, \quad (30)$$

for all $\phi_h \in \Phi_h$. In Eq. (30) we have introduced the following jump and average trace operators

$$[\![q]\!] \stackrel{\text{def}}{=} q^+ \mathbf{n}^+ + q^- \mathbf{n}^-, \qquad \{\cdot\} \stackrel{\text{def}}{=} \frac{(\cdot)^+ + (\cdot)^-}{2}, \quad (31)$$

where q denotes a generic scalar quantity and the average operator applies to scalars and vector quantities. By definition, $[\![q]\!]$ is a vector quantity. These definitions can be suitably extended to faces intersecting $\partial\Omega$ accounting for the weak imposition of boundary conditions. The local lifting operator $\mathbf{r}_e$, which is assumed to act on the jumps of $\mathbf{u}_h$ componentwise, is defined as the solution of the following problem

$$\int_{\Omega_h} \boldsymbol{\phi}_h \cdot \mathbf{r}_e(\mathbf{v})\ \mathrm{d}\mathbf{x} = -\int_e \{\boldsymbol{\phi}_h\} \cdot \mathbf{v}\ \mathrm{d}\sigma, \forall \boldsymbol{\phi}_h \in [\Phi_h]^N,\ \mathbf{v} \in \left[L^1(e)\right]^N, \quad (32)$$

and the global lifting operator $\mathbf{r}$ is related to $\mathbf{r}_e$ by the equation

$$\mathbf{r}(\mathbf{v}) \stackrel{\text{def}}{=} \sum_{e \in \mathcal{E}_h} \mathbf{r}_e(\mathbf{v}), \quad (33)$$

where $\mathcal{E}_h$ denotes the set of edges of $\mathcal{T}_h$. We remark that the local lifting operators on the two sides of any edge e have support on the two elements sharing the edge e. Hence, the global lifting operator for any element $K \in \mathcal{T}_h$ has support on the element K itself and on its neighbors.

The inviscid and viscous parts of the numerical flux $\widehat{\mathbf{f}}$ are treated independently. For the former we usually employ the Godunov flux or, alternatively, the van Leer-Hänel[12] flux-splitting scheme. The numerical viscous flux is given by

$$\widehat{\mathbf{f}}_v\left(\mathbf{u}_h^{\pm}, (\boldsymbol{\nabla}_h \mathbf{u}_h + \eta_e \mathbf{r}_e([\![\mathbf{u}_h]\!]))^{\pm}\right) \stackrel{\text{def}}{=} \{\mathbf{F}_v(\mathbf{u}_h, \boldsymbol{\nabla}_h \mathbf{u}_h + \eta_e \mathbf{r}_e([\![\mathbf{u}_h]\!]))\}, \quad (34)$$

where, according to Refs. 10, 11, the stability parameter η_e must be greater than the number of faces of the elements. The BR2 viscous flux discretization is as compact as possible because, for each element K, it only couples the nearest neighbor elements. This feature is obviously very attractive for the implicit implementation of the method.

2.2.1. *Orthonormal and hierarchical basis functions*

The actual implementation of Eq. (30) requires specifying the test and trial functions within each element $K \in \mathcal{T}_h$. The choice of basis functions affects several aspects of the DG discretization, namely, i) numerical efficiency, ii) conditioning of the DG discrete operators, iii) capability of easily handling complex-shape elements.

Modal expansion bases defined in the physical space can be used for irregular and polyhedral elements in a very straightforward manner. Furthermore, it is quite easy to construct hierarchical and orthonormal sets of shape functions that overcome the ill-conditioning of element mass matrices that becomes evident for high-degree polynomial approximations on highly stretched and curved elements. Complex applications presented in the following are in fact based on this type of approximation. The main drawback of such modal polynomial approximations is the cost of numerical integration for elements with non-constant Jacobian mapping.

To avoid cumbersome notation we shall assume that in this section ϕ_h and $\mathbf{u}_h$ denote functions defined within the generic element $K \in \mathcal{T}_h$, *i.e.* $\phi_h = \phi_h^K$ and $\mathbf{u}_h = \mathbf{u}_h^K$. Defining on K a set $\{\varphi_i\}$, $i = 1, \ldots, N_{DOF}$, of linearly independent polynomial basis functions of degree at most k, ϕ_h and $\mathbf{u}_h$ can be simply expressed as

$$\phi_h = \varphi_i, \qquad i = 1, \ldots, N_{DOF},$$
$$\mathbf{u}_h = \sum_{j=1}^{N_{DOF}} \varphi_j \mathbf{U}_j,$$

where $\{\mathbf{U}_j\}$, $j = 1, \ldots, N_{DOF}$, is the set of degrees of freedom of $\mathbf{u}_h$ in K, and

$$N_{DOF} = \dim\{\varphi_i\} = \frac{\prod_{l=1}^{N}(k+l)}{N!}$$

is the number of degrees of freedom of complete polynomials of degree k. Simple choices for $\{\varphi_i\}$, such as the set of monomials $\{x^l y^m z^n : l+m+n \leq k\}$, are not advisable in general and for the sake of improving stability and efficiency a set of orthogonal polynomial basis functions is highly preferable. The procedure to produce a set of orthonormal basis functions on a generic element K relies on the modified Gram-Schmidt (MGS) orthogonalization algorithm. The sole requirement of this procedure is the capability to compute the integral of polynomial functions on the desired element shapes. Let us denote with $\{\varphi_i\}$ and $\{b_i\}$, $i = 1, \ldots, N_{DOF}$, the set of orthonormal basis functions we wish to construct and a starting set of linearly independent basis functions defined on K, respectively. The MGS procedure with re-orthogonalization can be simply setup as shown in the following pseudo-code:

MGS ALGORITHM WITH RE-ORTHOGONALIZATION

for $i \leftarrow 1$ **to** N_{DOF}
    do for $n \leftarrow 1$ **to** 2
        do for $j \leftarrow 1$ **to** $i-1$
            do $r_{ij}^{(n)} \leftarrow (b_i, \varphi_j)_K$
                $b_i \leftarrow b_i - r_{ij}^{(n)} \varphi_j$
        $r_{ii}^{(n)} \leftarrow \sqrt{(b_i, b_i)_K}$
        $b_i \leftarrow b_i / r_{ii}^{(n)}$
    $\varphi_i \leftarrow b_i$

Line 2 indicates that orthogonalization is applied twice. As reported in Ref. 13, this is enough to get a set of basis functions which are orthonormal up to machine precision. It can be shown that the above MGS algorithm amounts to constructing the set of basis functions $\{\varphi_i\}$ according to the following system of equations

$$\varphi_i = \sum_{j=1}^{i-1} a_{ij}\varphi_j + a_{ii} b_i, \qquad i = 1, \ldots, N_{DOF}, \tag{35}$$

where the coefficients a_{ij} are determined by enforcing each new φ_i to be orthogonal to the $i-1$ already orthonormalized basis functions, whereas the coefficient a_{ii} is the normalizing factor for the L^2 norm of the newly created φ_i. For $i = 1, \ldots, N_{DOF}$ these coefficients are given by

$$\frac{a_{ij}}{a_{ii}} = -\left(b_i, \varphi_j\right)_K, \qquad j = 1, \ldots, i-1,$$

$$\frac{1}{a_{ii}} = \sqrt{\int_K \left(b_i - \sum_{j=1}^{i-1} \left(b_i, \varphi_j\right)_K\right)^2 \mathrm{d}\mathbf{x}},$$

and are related to r_{ij} and r_{ii} in the MGS algorithm by

$$r_{ij} = -\frac{a_{ij}}{a_{ii}},$$

$$r_{ii} = \frac{1}{a_{ii}}.$$

From Eq. (35) it is then clear that the orthonormal set $\{\varphi_i\}$ is also hierarchical. In fact, increasing the degree of polynomial approximation entails adding to the existing set of basis functions as many φ_i of the form of Eq. (35) as the number of new b_i up to the required degree, without changing the already existing orthonormal basis functions.

As regards the starting set of basis functions $\{b_i\}$ for a generic element K, we have found that a simple and effective choice is the set of monomials, up to the prescribed degree, expressed in a local frame of reference having its origin in the centroid of K and the coordinate axes coincident with the principal axes of the element.

Finally we remark that the MGS algorithm outlined above is also used to compute the values of basis functions (and of their spatial derivatives, if necessary) at any location other than those needed to compute the integrals of lines 4 and 6. In such cases the symbols b and φ at lines 5, 7 and 8 denote values of starting and orthonormalized basis functions (or of their derivatives) at the desired location.

2.3. *Time integration*

The DG space discretization of Eq. (30) results in the following system of (nonlinear) ODEs in time

$$\mathbf{M}\frac{\mathrm{d}\mathbf{U}}{\mathrm{d}t} + \mathbf{R}(\mathbf{U}) = \mathbf{0}, \tag{36}$$

where $\mathbf{U}$ is the global vector of unknown degrees of freedom, $\mathbf{M}$ is a global block diagonal matrix and $\mathbf{R}(\mathbf{U})$ is the vector of "residuals", *i.e.*, the vector of nonlinear functions of $\mathbf{U}$ resulting from the integrals of the DG discretized space differential operators in Eq. (30). We remark that using the DOFs of the conservative variables $\mathbf{u}_h$ as unknowns of the polynomial approximation, then the matrix $\mathbf{M}$ represents the global block diagonal mass matrix, which, using orthonormal basis functions, reduces to the identity matrix. If, on the other hand, we choose the DOFs of another set of variables $\mathbf{w}_h$ as unknowns of the polynomial expansion of the solution, then the block $\mathbf{M}^K$ of $\mathbf{M}$ for the element K will be given by

$$\mathbf{M}^K = \left\{\int_K \varphi_i^K \frac{\partial u_l}{\partial w_m}\left(\mathbf{w}_h^K\right)\varphi_j^K \,\mathrm{d}\mathbf{x}\right\}, \quad i,j = 1,\ldots,N_{DOF}, \quad l,m = 1,\ldots,M.$$

2.3.1. *Linearly implicit Runge-Kutta schemes*

Implicit time integration of Eq. (36) can be efficiently performed by means of linearly implicit Rosenbrock-type Runge-Kutta schemes. The class of methods here considered can be compactly written as

$$\begin{aligned} \mathbf{U}^{n+1} &= \mathbf{U}^n + \sum_{j=1}^{s} b_j \mathbf{K}_j, \\ \left(\frac{\mathbf{M}}{\Delta t} + \gamma\mathbf{J}\right)\mathbf{K}_i &= -\mathbf{R}\left(\mathbf{U}^n + \sum_{j=1}^{i-1}\alpha_{ij}\mathbf{K}_j\right) - \mathbf{J}\sum_{j=1}^{i-1}\gamma_{ij}\mathbf{K}_j, \quad i = 1,\ldots,s, \end{aligned} \tag{37}$$

where s is the number of stages, b_i, α_{ij}, γ_{ij} are real coefficients and $\mathbf{J} = \partial\mathbf{R}(\mathbf{U}^n)/\partial\mathbf{U}$ is the Jacobian matrix of the residual. The coefficients for the Euler scheme and for the schemes proposed in Refs. 14 and 15 are summarized in Table 1.

Table 1. Coefficients of some linearly implicit Runge-Kutta schemes.

Scheme	s	γ	b_i	α_{ij}	γ_{ij}
Euler	1	1	1	0	γ
Iannelli-Baker	2	$1-\frac{\sqrt{2}}{2}$	$1-\frac{1}{8\gamma}$ $\frac{1}{8\gamma}$	0 $8\gamma\left(\frac{1}{2}-\gamma\right)\ 0$	γ $0\ \gamma$
Lang-Verwer	3	$\frac{1}{2}+\frac{\sqrt{3}}{6}$	$\frac{2}{3}$ 0 $\frac{1}{3}$	0 $1\ 0$ $1\ 0\ 0$	γ $-1\ \ \gamma$ $-\gamma\ \frac{1}{2}-2\gamma\ \gamma$

An implementation of Eq. (37) that saves at each stage the cost of the matrix-vector multiplication $\mathbf{J}\sum_{j=1}^{i-1}\gamma_{ij}\mathbf{K}_j$ can be written as follows

$$\begin{aligned}\mathbf{U}^{n+1} &= \mathbf{U}^n + \sum_{j=1}^{s} m_j \mathbf{W}_j,\\ \left(\frac{\mathbf{M}}{\gamma\Delta t}+\mathbf{J}\right)\mathbf{W}_i &= -\mathbf{R}\left(\mathbf{U}^n + \sum_{j=1}^{i-1} a_{ij}\mathbf{W}_j\right) - \frac{\mathbf{M}}{\gamma\Delta t}\sum_{j=1}^{i-1} c_{ij}\mathbf{W}_j,\\ i &= 1,\ldots,s,\end{aligned} \tag{38}$$

where, for $i = 1,\ldots,s$,

$$\mathbf{K}_i = \frac{1}{\gamma}\mathbf{W}_i - \sum_{j=1}^{i-1} c_{ij}\mathbf{W}_j.$$

The coefficients of the transformed scheme are given by

$$\begin{aligned}&(m_1,\ldots,m_s) = (b_1,\ldots,b_s)\,\Gamma^{-1}, \quad (a_{ij}) = (\alpha_{ij})\,\Gamma^{-1},\\ &C = \operatorname{diag}\left(\gamma^{-1},\ldots,\gamma^{-1}\right) - \Gamma^{-1},\end{aligned}$$

where $\Gamma^{-1} \stackrel{\text{def}}{=} (\gamma_{ij})^{-1}$ denotes the inverse of the coefficient matrices of Table 1. The entries of Γ^{-1} for the schemes of Table 1 are given in Table 2.

The matrix-explicit or the matrix-free GMRES algorithm can be used to actually solve Eq. (38) at each time step. In both cases system preconditioning is required to make the convergence of the GMRES solver acceptable in problems of practical interest. The Jacobian matrix implemented in our code has been derived analytically and takes full account of the dependence of the residual on the unknown vector and on its derivatives, including the

Table 2. Inverse of the (γ_{ij}) matrices of Table 1.

Scheme	s	$\Gamma^{-1} = (\gamma_{ij})^{-1}$
Euler	1	γ^{-1}
Iannelli-Baker	2	$\begin{matrix} \gamma^{-1} & \\ 0 & \gamma^{-1} \end{matrix}$
Lang-Verwer	3	$\begin{matrix} \gamma^{-1} & & \\ \gamma^{-2} & \gamma^{-1} & \\ -\frac{1}{2}\gamma^{-3} + 2\gamma^{-2} + \gamma^{-1} & -\frac{1}{2}\gamma^{-2} + 2\gamma^{-1} & \gamma^{-1} \end{matrix}$

implicit treatment of the lifting operators and of the boundary conditions. Using a suitably accurate time integration scheme, this allows to employ the implicit solver also for accurate unsteady computations.

The choice of the time step can significantly affect both the efficiency and the robustness of the method. For steady computations we have implemented the pseudo-transient continuation strategy with the local time step given by

$$\Delta t_K = CFL \frac{h_K}{c+d},$$

where

$$c = |\mathbf{v}| + a, \qquad d = 2\frac{\mu_e + \lambda_e}{h_K}, \qquad h_K = N\frac{\Omega_K}{S_K},$$

define convective and diffusive velocities and the reference dimension of the generic element K, respectively. The coefficients μ_e and λ_e are the effective dynamic viscosity and conductivity, while Ω_K and S_K denote the volume and the surface of K. All quantities depending on $\mathbf{u}_h$ in the above relations are computed from mean values of $\mathbf{u}_h$. Devising an effective and robust strategy to increase the CFL number as the residual decreases is not an easy task, especially for turbulent computations. The rule here proposed is essentially the result of intensive numerical experimentation and aims at controlling the evolution of CFL number on the basis of both the L^∞ and the L^2 norms of the residual. Denoting with y the CFL number, the rule is as follows

$$\begin{cases} y = \frac{y_0}{x^\alpha} & \text{if } x \le 1 \\ y = y_e + (y_0 - y_e)e^{\alpha\frac{y_0}{y_0 - y_e}(1-x)} & \text{if } x > 1 \end{cases} \tag{39}$$

where, denoting by $x_{L^2} = \max(|R_i|_{L^2}/|R_{i0}|_{L^2})$ and $x_{L^\infty} = \max(|R_i|_{L^\infty}/|R_{i0}|_{L^\infty})$ for $i = 1, \ldots, M$,

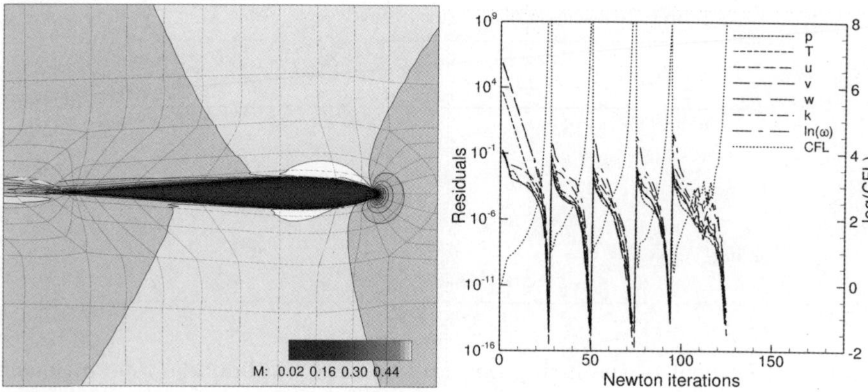

Fig. 2. Streamlined body: Mach number contours of $\mathbb{P}^4$ solution and residuals convergence history of $\mathbb{P}^{0\to4}$ solutions.

$$\begin{cases} x = \min\left(x_{L^2}, 1\right) & \text{if } x_{L^\infty} \leq 1 \\ x = x_{L^\infty} & \text{if } x_{L^\infty} > 1, \end{cases}$$

and $y_0 = \mathrm{CFL}_{min}$, $y_e = \mathrm{CFL}_{exp}$ and α are the user-defined minimum CFL number, the maximum CFL number of explicit schemes and the exponent (usually ≤ 1) governing the growth rate of the CFL number, respectively. The strong CFL number control based on the L^∞ norm of residual has been found useful to prevent sudden breakdown of computations once the CFL number has already reached quite high values. For relatively simple steady test cases, such as the flow around a streamlined body (Figure 2, 232969 DOFs), the implicit time integration combined with the above CFL number evolution rule provides quadratic Newton convergence to machine accuracy.

2.4. *Shock-capturing approach*

The shock-capturing approach consists of adding to the DG discretized equations an artificial viscosity term that aims at controlling the high-order modes of the numerical solution within elements while preserving as much as possible the spatial resolution of discontinuities. The shock-capturing term is local and active in every element, but the amount of artificial viscosity is proportional to the (inviscid) residual of the DG space discretization and thus it is almost negligible except than at locations of flow discontinuities.

The shock-capturing term added to Eq. (30) reads

$$\sum_K \int_K \epsilon_p(\mathbf{u}_h^{\pm}, \mathbf{u}_h) \left(\boldsymbol{\nabla}_h \phi_h \cdot \mathbf{b}\right) \left(\boldsymbol{\nabla}_h \mathbf{u}_h \cdot \mathbf{b}\right) \, \mathrm{d}\mathbf{x}, \tag{40}$$

with the shock sensor and the pressure gradient unit vector defined by

$$\begin{aligned} \epsilon_p(\mathbf{u}_h^{\pm}, \mathbf{u}_h) &= C h_K^2 \frac{|s_p(\mathbf{u}_h^{\pm}, \mathbf{u}_h)| + |d_p(\mathbf{u}_h)|}{p(\mathbf{u}_h)} f_p(\mathbf{u}_h), \\ \mathbf{b}(\mathbf{u}_h) &= \frac{\boldsymbol{\nabla}_h p(\mathbf{u}_h)}{|\boldsymbol{\nabla}_h p(\mathbf{u}_h)| + \varepsilon}, \end{aligned} \tag{41}$$

and

$$s_p(\mathbf{u}_h^{\pm}, \mathbf{u}_h) = \sum_{i=1}^{M} \frac{\partial p(\mathbf{u}_h)}{\partial u_{h_i}} s_i(\mathbf{u}_h^{\pm}), \qquad d_p(\mathbf{u}_h) = \sum_{i=1}^{M} \frac{\partial p(\mathbf{u}_h)}{\partial u_{h_i}} \left(\boldsymbol{\nabla}_h \cdot \mathbf{F}_c(\mathbf{u}_h)\right)_i . \tag{42}$$

The components s_i of the function $\mathbf{s}$, defined by the solution of the problem

$$\int_{\Omega_h} \phi_h \mathbf{s}(\mathbf{u}_h^{\pm}) \, \mathrm{d}\mathbf{x} = \int_{\Gamma_h} [\![\phi_h]\!] \cdot \left(\widehat{\mathbf{f}}_c(\mathbf{u}_h^{\pm}) - \mathbf{F}_c(\mathbf{u}_h)\right)^{\pm} \mathrm{d}\sigma, \tag{43}$$

are actually the lifting of the interface jump in normal direction between the numerical and internal inviscid flux components. The further factor $f_p(\mathbf{u}_h)$ in Eq. (41) is a pressure sensor defined by

$$f_p(\mathbf{u}_h) = \frac{|\boldsymbol{\nabla}_h p(\mathbf{u}_h)|}{p(\mathbf{u}_h)} \left(\frac{h_K}{k}\right), \tag{44}$$

which improves the accuracy of solutions in regions with high but otherwise smooth gradients and allows using the same value of the user-defined parameter C (typically $C = 0.2$) for different degrees of polynomial approximation. Finally, the element dimension h_K is defined as

$$h_K = \frac{1}{\sqrt{\frac{1}{(\Delta x)^2} + \frac{1}{(\Delta y)^2} + \frac{1}{(\Delta z)^2}}}, \tag{45}$$

where $\Delta x, \Delta y$ and Δz are the dimensions of the hexahedral enclosing K, scaled in such a way that their product matches the volume of K. The shock-capturing technique outlined above is highly non-linear and residuals convergence of steady state solutions can be quite difficult, even implementing a fully (linearized) implicit discretization of the shock-capturing term (40). This is in fact the case for the solution of the transonic flow around the RAE 2822 airfoil ($\mathrm{M}_\infty = 0.730$, $\alpha = 3.19°$, $\mathrm{Re}_\infty = 6.5\times10^6$,

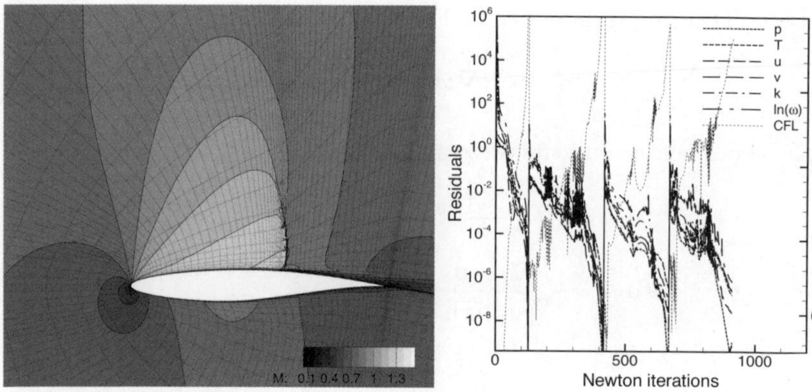

Fig. 3. RAE2822: Mach number contours of $\mathbb{P}^3$ solution and residuals convergence history of $\mathbb{P}^{0\to3}$ solutions.

80860 DOFs), shown in Figure 3, that requires quite a large number of Newton iterations for convergence.

3. Numerical Results

In this section we present the results of high-order DG solutions of several complex turbulent flows of aeronautical interest. All the computations have been run in parallel, initializing the $\mathbb{P}^0$ solution from uniform flow at freestream conditions and the higher-order solutions from the lower-order ones. Solutions have been advanced in time by means of the linearly implicit backward Euler method and the linear system (38) has been solved using the default iterative solver available in PETSc, *i.e.*, the restarted GMRES algorithm preconditioned with the block Jacobi method with one block per process, each of which is solved with ILU(0).

3.1. *L1T2 3-element airfoil*

The flow around the three elements airfoil has been computed with a farfield Mach number $M_\infty = 0.197$, angle of attack $\alpha = 20.18°$ and chord-based Reynolds number $Re_\infty = 3.52\times10^6$. This test case has been computed up to $\mathbb{P}^6$ polynomial approximation on a grid consisting of 4740 quadrilateral elements with curved, four-node edges, see Figure 4. The main difficulties of this test are due to highly distorted elements shapes and to the flow complexity of strongly interacting wakes, see Figure 5. Figure 6 displays

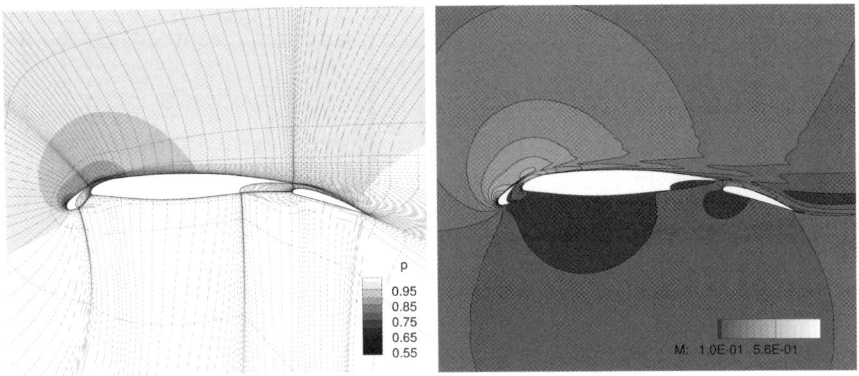

Fig. 4. L1T2: pressure and Mach contours of $\mathbb{P}^6$ solution.

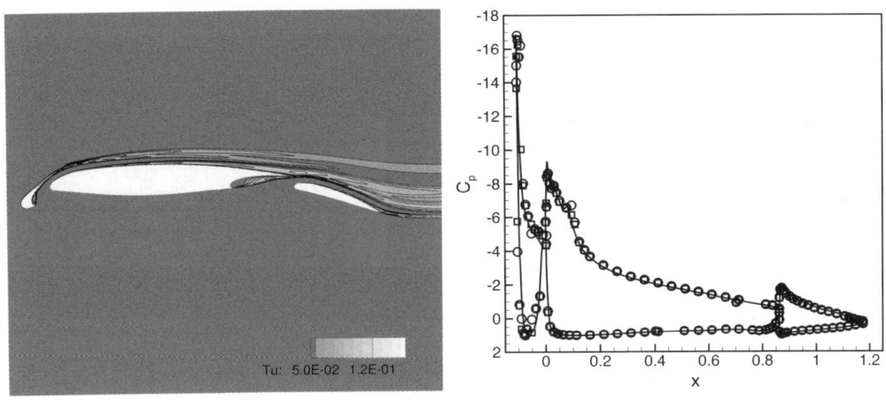

Fig. 5. L1T2: turbulence intensity contours and pressure coefficient distribution of $\mathbb{P}^6$ solution.

the residuals convergence history of $\mathbb{P}^{0\to 6}$ solutions both in terms of Newton iterations and performance index units, which is a relative measure of CPU time established within EU project ADIGMA.[16]

3.2. *ONERA M6 wing*

The flow around the ONERA M6[17] wing is a classical CFD validation case for external flows that combines a simple geometry with complexities of transonic flow, *i.e.*, local supersonic flow, shocks, and turbulent boundary layers separation. The flow conditions are those of Test 2308, *i.e.*, $M_\infty = 0.8395$, $\alpha = 3.06°$ and $Re_\infty = 11.72\times 10^6$ based on the mean aerody-

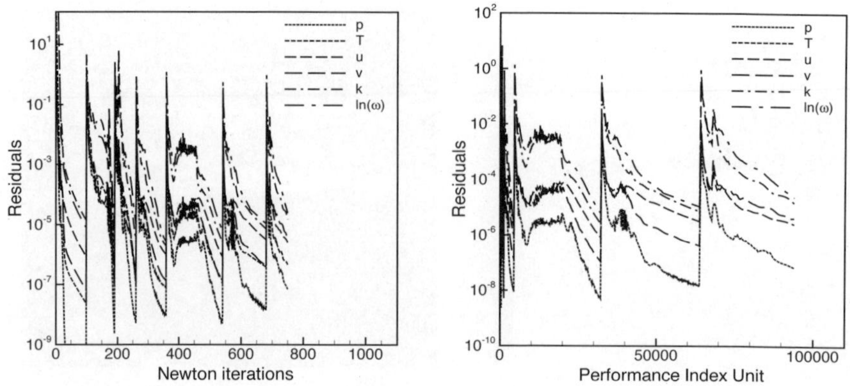

Fig. 6. L1T2: residuals convergence history of $\mathbb{P}^{0\to6}$ solutions.

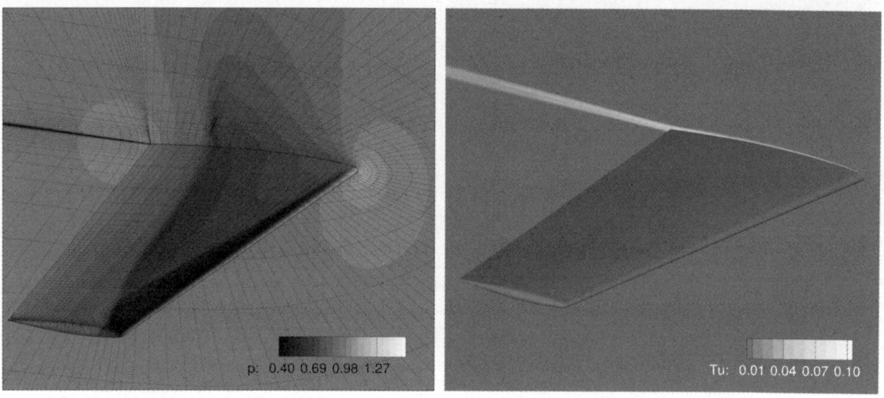

Fig. 7. ONERA M6: pressure and turbulence intensity contours of $\mathbb{P}^2$ solution.

namic chord. The grid consists of 215632 hexahedral elements with curved, eight-node faces, shown in Figure 7 superimposed to the pressure contours. The $\mathbb{P}^0$ solution has been computed using the restarted GMRES algorithm with 60 Krylov subspace vectors and 120 maximum iterations. These parameters have been increased up to 120 and 240 for $\mathbb{P}^1$ and $\mathbb{P}^2$ polynomial approximations. All the computations have been run in parallel using 512 cores of the CINECA BCX/5120 cluster.

In Figure 8 the pressure coefficient distributions of $\mathbb{P}^2$ solution are compared with the experimental data at seven sections along the span of the wing. The shock-capturing technique proves capable of providing accurate

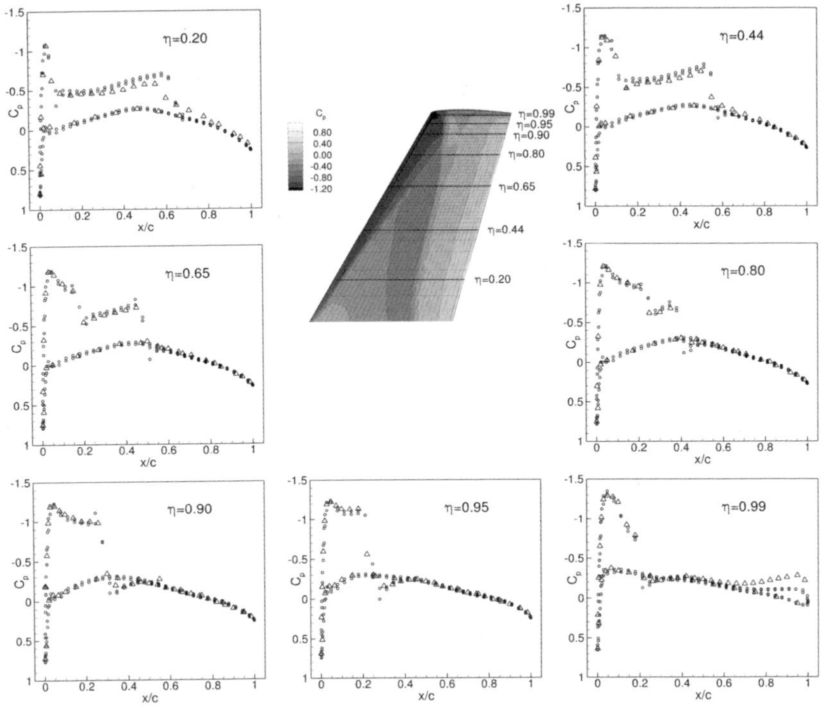

Fig. 8. ONERA M6: pressure coefficient of $\mathbb{P}^2$ solution (∘ 2156320 DOFs) compared with the experimental data (△).

Table 3. ONERA M6: lift and drag coefficients of DG solutions.

	C_l	C_d	C_{d_p}	C_{d_f}
$\mathbb{P}^0$	0.231900	0.0555007	0.0502764	0.00522416
$\mathbb{P}^1$	0.274433	0.0184980	0.0133475	0.00515066
$\mathbb{P}^2$	0.275279	0.0180224	0.0123096	0.00571281

resolution of the lambda shock structure all along the suction surface of the wing and, unlike many results presented in the literature, the shocks can still be clearly distinguished at section $y/b = 0.8$. Despite the quite coarse grid resolution, the $\mathbb{P}^2$ solution is also capable of capturing the flow separation near the wing tip, as shown in Figure 9. Table 3 reports the force coefficients of $\mathbb{P}^{0\to 2}$ solutions, showing that at least a one-degree higher $\mathbb{P}^3$ solution would be useful to assess the convergence of force coefficients.

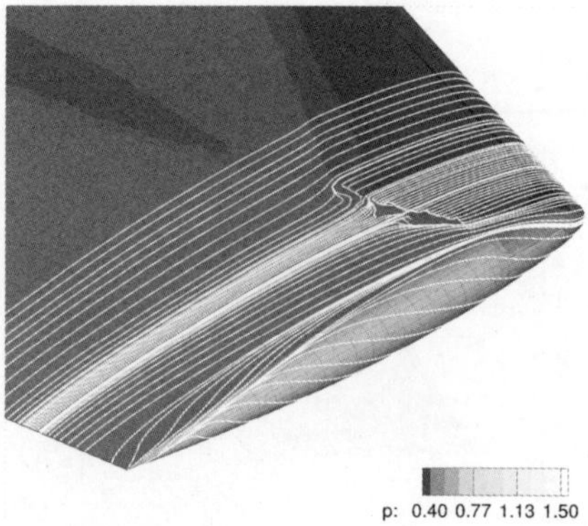

Fig. 9. ONERA M6: flow separation near the wing tip of $\mathbb{P}^2$ solution.

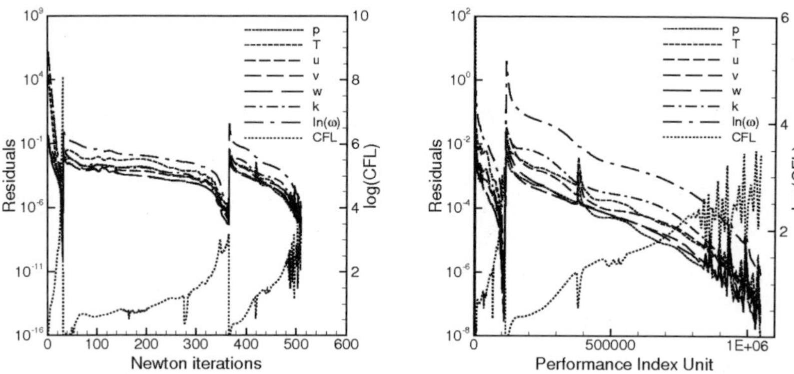

Fig. 10. ONERA M6: residuals convergence history of $\mathbb{P}^{0\to2}$ solutions.

3.3. *DPWIII W1 wing*

This test case has been proposed within the 3rd AIAA CFD Drag Prediction Workshop.[18] The turbulent flow for $M_\infty = 0.76$, $\alpha = 0.5°$ and a chord Reynolds $Re_\infty = 1\times10^7$ has been computed up to $\mathbb{P}^2$ polynomial approximation on a grid of 188928 hexahedral elements with curved, eight-node, faces shown in Figure 11(a). The same Figure 11 displays the pressure and turbulence intensity contours of the $\mathbb{P}^2$ solution. Figure 12 shows the convergence history in terms of Newton iterations and performance index units. The $\mathbb{P}^{0\to2}$ solutions have been computed using the restarted GMRES algorithm with 60 Krylov subspace vectors and 120 maximum iterations. All the computations have been run in parallel using 512 cores of the DLR

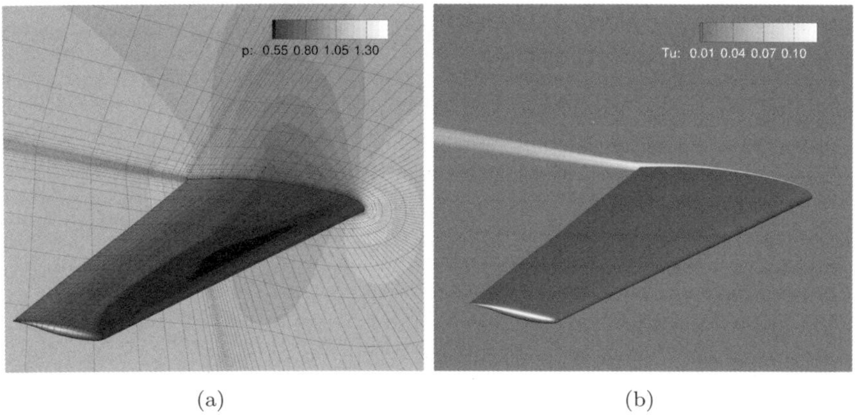

Fig. 11. DPWIII W1: pressure and turbulence intensity contours of $\mathbb{P}^2$ solution.

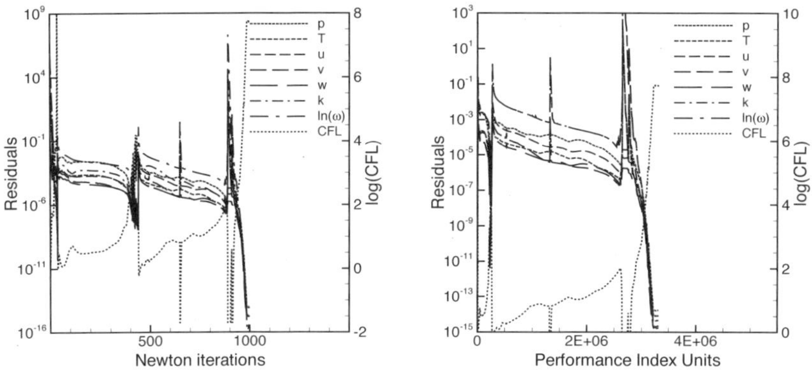

Fig. 12. DPWIII W1: residuals convergence history of $\mathbb{P}^{0\to2}$ solutions.

CASE cluster facility. Figure 13 compares the pressure coefficient distributions of the $\mathbb{P}^2$ solution at eight sections along the span of the wing with those computed by the TAU and FUN3D codes . The FUN3D and TAU solutions have been computed on a grid with 11459041 nodes and on an adapted grid with 17053510 nodes, respectively. The $\mathbb{P}^2$ DG discretization employs 1889280 DOFs.

3.4. *DLR-F6 wing-body configuration*

The DLR-F6 wing-body transport configuration has been the object of several wind-tunnel tests and computational studies, see Ref. 19, and also

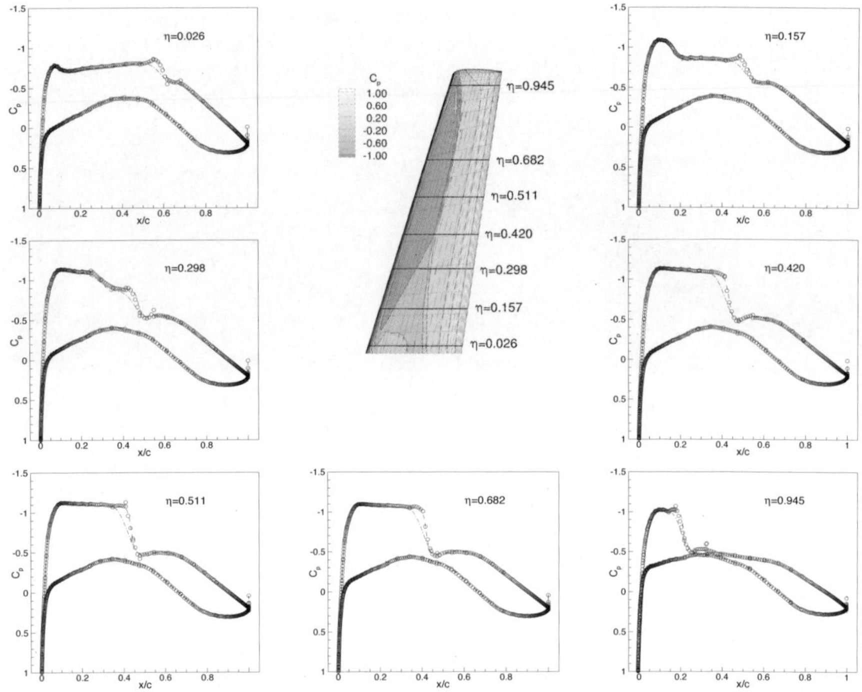

Fig. 13. DPWIII W1: pressure coefficient of $\mathbb{P}^2$ solution (○ 1889280 DOFs) compared with TAU (– – – 17053510 DOFs) and FUN3D (– · – 11459041 DOFs).

deeply investigated within the AIAA CFD Drag Prediction Workshop[18] series with the aim of assessing the state of the art of computational methods as practical aerodynamic tools for aircraft force and moment prediction. In this test case the freestream conditions have been set to $\mathrm{M}_\infty = 0.75$, chord-based $\mathrm{Re}_\infty = 5\times10^6$ and angle of attack capable of achieving a given lift coefficient $C_L = 0.5$. The computations have been carried out on two nested grids with 50618 and 404944 hexahedral elements with curved, eight-node faces, see Figure 14. DG solutions have been computed up to $\mathbb{P}^3$ and up to $\mathbb{P}^2$ polynomial approximation on the coarse and fine grids, respectively. The parameters of the restarted GMRES solver have been set to 60 Krylov subspace vectors and 120 maximum iterations for the coarse grid solutions, and to 120 vectors and 480 iterations for the fine grid solutions. Figure 16 shows the residuals convergence history of the coarse-grid solutions in terms of Newton iterations and performance index units. The coarse and fine grid computations have been run in parallel using respectively 128 and 512 cores

Table 4. DLR F6: force and pitching moment coefficients of DG solutions.

(a) coarse grid

	$\mathbb{P}^0$	$\mathbb{P}^1$	$\mathbb{P}^2$	$\mathbb{P}^3$
DOFs	50618	202472	506180	1012360
α	2.34000	0.22500	-0.07000	-0.07000
C_l	0.49973	0.49996	0.49998	0.49986
C_d	0.16745	0.04232	0.02822	0.02832
C_{d_p}	0.15201	0.02905	0.01672	0.01531
C_{d_f}	0.01544	0.01327	0.01151	0.01301
C_m	0.03812	-0.12468	-0.14526	-0.14642

(b) fine grid

	$\mathbb{P}^0$	$\mathbb{P}^1$	$\mathbb{P}^2$
DOFs	404944	1619776	4049440
α	1.34600	0.10600	0.35700
C_l	0.50002	0.50005	0.49994
C_d	0.11738	0.03045	0.02890
C_{d_p}	0.10306	0.01874	0.01727
C_{d_f}	0.01432	0.01171	0.01163
C_m	-0.03781	-0.13714	-0.12528

of the DLR CASE cluster facility. Figure 15 highlights the capability of the $\mathbb{P}^3$ solution to capture the detail of flow separation at the wing-root junction on the given coarse-grid. Table 4 reports the force and pitching moment coefficients computed on the coarse and fine grids. There is a discrepancy between the more accurate results on the two grids that needs to be further investigated and no clear conclusion can be drawn about the asymptotic values of the aerodynamic coefficients. One issue could be the poor geometrical approximation of surfaces when using only quadratic mappings for the faces of very coarse meshes. Finally, Figures 17, 18 and 19 give an overview of the pressure coefficient and skin friction distributions of DG solutions compared with reference results of the TAU and CFL3D codes taken from Ref. 18.

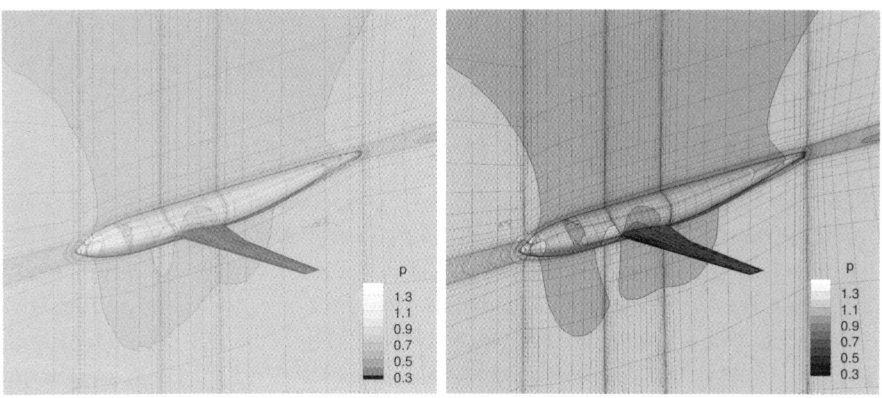

Fig. 14. DLR F6: pressure contours of coarse-grid $\mathbb{P}^3$ and fine-grid $\mathbb{P}^2$ solutions.

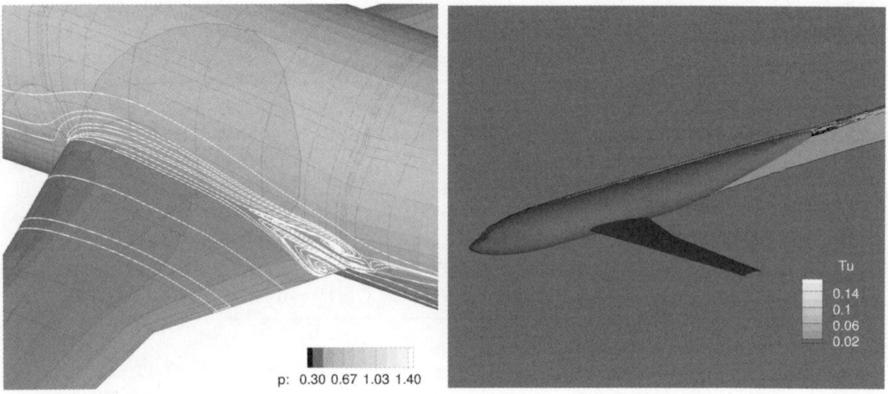

Fig. 15. DLR F6: wing-root juncture flow separation and turbulence intensity contours of coarse-grid $\mathbb{P}^3$ solution.

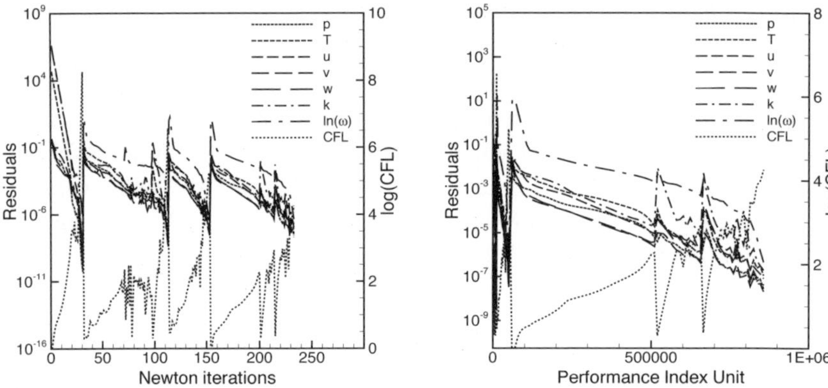

Fig. 16. DLR F6: residuals convergence history of coarse-grid $\mathbb{P}^{0\to3}$ solutions.

3.5. *NASA 65° sweep delta wing*

The NASA 65° sweep delta wing has been proposed and investigated experimentally within the second international Vortex Flow Experiment (VFE-2). The geometry here considered is the delta wing with large-radius leading edge, for which experimental pressure data are available in Ref. 20. The farfield conditions of this test case are $M_\infty = 0.869$, $\alpha = 24.7°$ and $Re_\infty = 19.83 \times 10^6$. Solutions have been computed on a hybrid grid with 614770 tetrahedral and prismatic elements. The prisms fill a few layers of elements within boundary layers on the delta wing and on the sting.

The available grid points allowed to define only a linear mapping of element faces and this resulted in inaccurate pressure distributions on the wing and sting surfaces. All the computations have been run in parallel using 512 cores of the CINECA HPC cluster facility. The convergence of residuals for this test case was quite difficult and, more importantly, neither the $\mathbb{P}^1$ nor the $\mathbb{P}^2$ solutions were able to capture the experimentally observed vortex breakdown on the wing suction surface. This issue could be related to the poor representation of the wing and sting surfaces and needs to be further investigated. Figure 20 shows the pressure coefficient and turbulence intensity contours of $\mathbb{P}^1$ and $\mathbb{P}^2$ solutions. Both parts of Figure 20 clearly highlight the better resolution of vortices provided by the $\mathbb{P}^2$ solution.

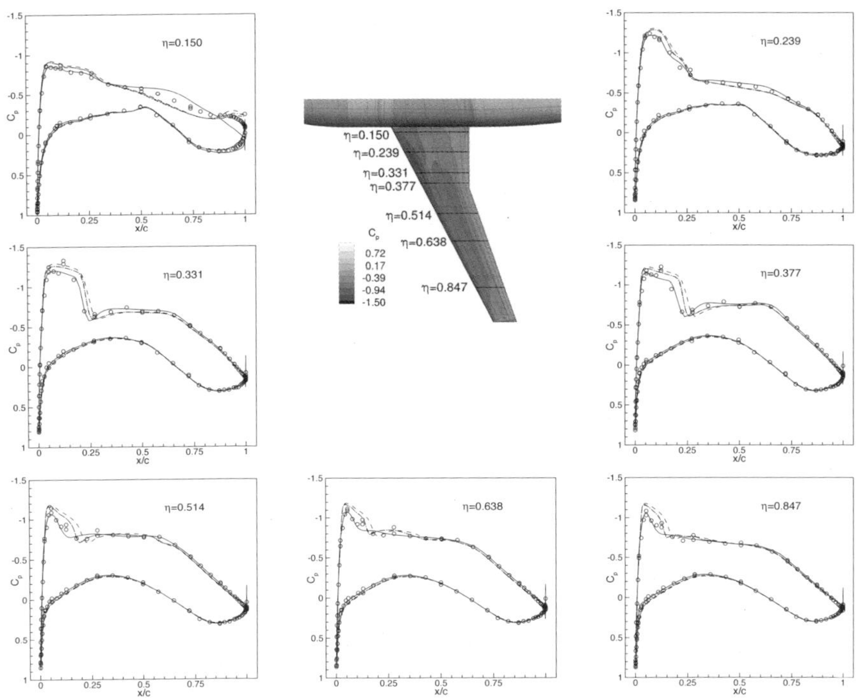

Fig. 17. DLR F6: pressure coefficient of $\mathbb{P}^3$ solution (○ 1012360 DOFs) compared with TAU (—— 5102446 DOFs) and CFL3D (– – – 2256896 DOFs, – · – 7689088 DOFs, – ·· – 26224640 DOFs).

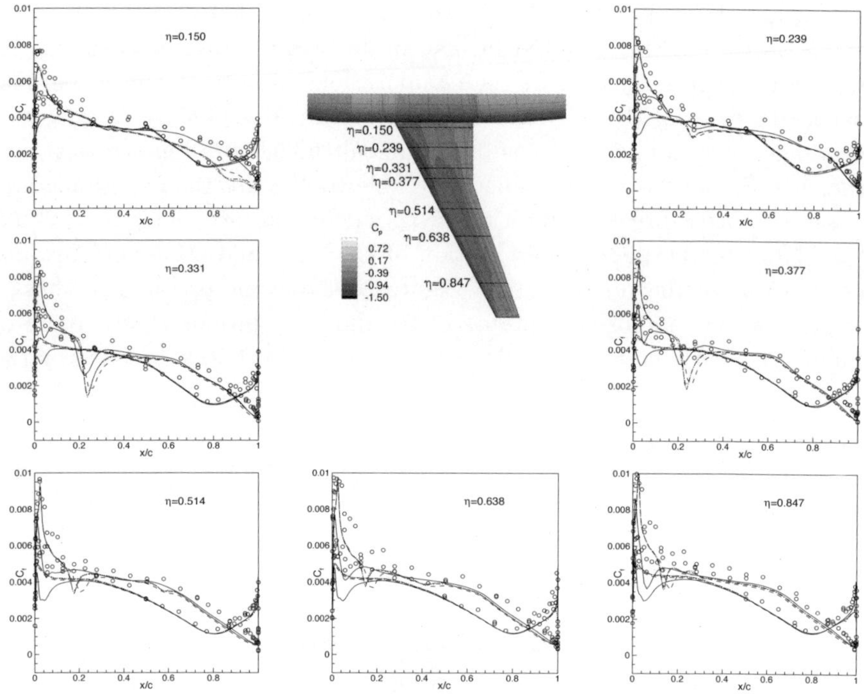

Fig. 18. DLR F6: skin friction coefficient of $\mathbb{P}^3$ solution (○ 1012360 DOFs) compared with TAU (—— 5102446 DOFs) and CFL3D (– – – 2256896 DOFs, – · – 7689088 DOFs, – ·· – 26224640 DOFs).

4. Final Remarks

In this chapter we have presented and demonstrated several well-tried features of the DG code MIGALE, that has been developed over the years for the numerical solution of the coupled RANS and k-ω turbulence model equations.

Open issues of the proposed DG method are mainly related to its computational cost and this has motivated our most recent research efforts in two directions.

On the one hand, we have developed a spectral DG method, with a couple of choices for the sets of collocation and integration points, to improve the computational efficiency and a p-multigrid strategy to reduce the RAM required by the fully coupled implicit solver. The p-multigrid algorithm has been analyzed in Ref. 21 and applied to the solution of the compressible

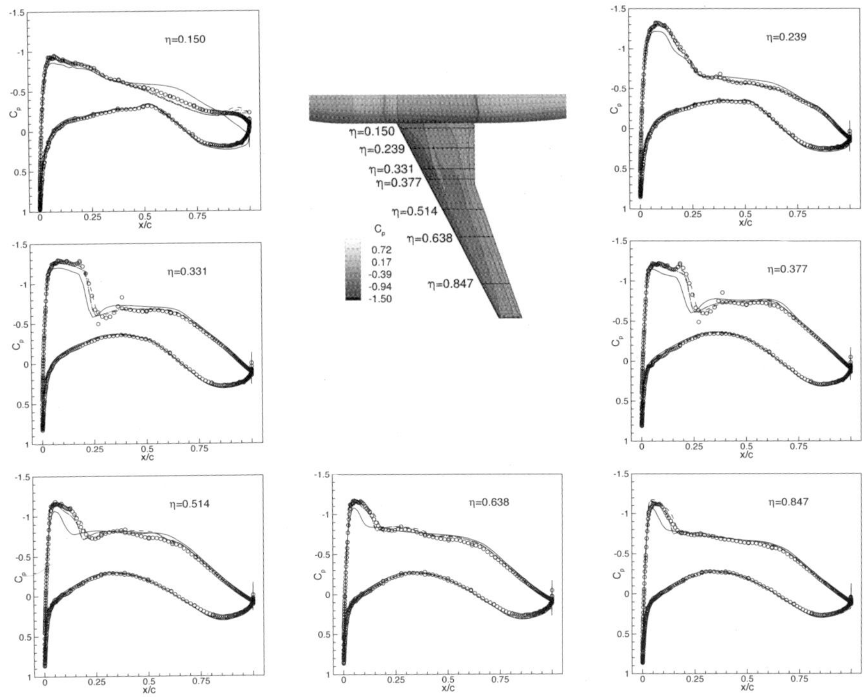

Fig. 19. DLR F6: pressure coefficient of $\mathbb{P}^2$ solution (∘ 4049440 DOFs) compared with TAU (—— 5102446 DOFs) and CFL3D (- - - 2256896 DOFs, – · – 7689088 DOFs, – ·· – 26224640 DOFs).

Euler and Navier-Stokes equations in Refs. 22 and 23. First applications of the p-multigrid strategy to shockless turbulent flows around complex 3D geometries have already provided encouraging results.

On the other hand, we are working on exploiting the flexibility of the modal DG discretization, with shape functions defined in the real space, to improve the computational efficiency by means of agglomeration strategies. The agglomeration technique provides also the natural setting for the development of h-multigrid solution strategies for high-order DG discretizations. First results of this research activity have already been reported in Ref. 24.

Finally, even if the shock-capturing approach turned out to be robust and accurate, further research is needed to make its formulation fully consistent with a residual-based artificial viscosity. Moreover, the adverse impact on the regularity of convergence of residuals needs to be further investigated.

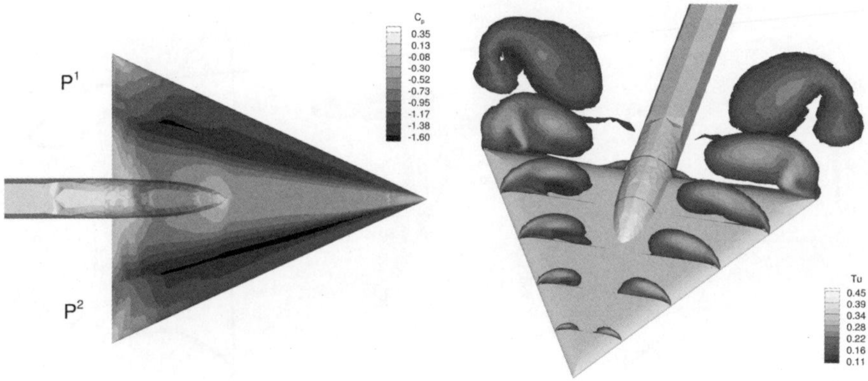

Fig. 20. VFE2: pressure coefficient and turbulence intensity contours of $\mathbb{P}^1$ and $\mathbb{P}^2$ solutions.

Acknowledgments

The authors acknowledge the financial support of the European Union, under the ADIGMA project. Furthermore, we express our gratitude to our co-workers, Andrea Crivellini and Nicoletta Franchina, for their contributions to the work here reported.

References

1. D. C. Wilcox, *Turbulence Modelling for CFD.* (DCW industries Inc., La Cañada, CA 91011, USA, 1993).
2. F. Bassi, A. Crivellini, S. Rebay, and M. Savini, Discontinuous Galerkin solution of the Reynolds-averaged Navier-Stokes and k-ω turbulence model equations, *Comput. & Fluids.* **34**, 507–540, (2005).
3. F. R. Menter, Two-equation eddy-viscosity turbulence models for engineering applications, *AIAA Journal.* **32**(8), 1598–1605, (1994).
4. A. Hellsten. On the solid-wall boundary condition of ω in the k-ω-type turbulence models. Technical Report B–50, Helsinky University of Technology, Laboratory of Aerodynamics, (1998).
5. F. Bassi, L. Botti, A. Crivellini, A. Ghidoni, and S. Rebay. D4.2.2–Investigation of Jacobian and Jacobian-free Newton-Krylov methods for implicit DG methods. Technical report, ADIGMA, (2009). http://www.dlr.de/as/adigma.
6. K. Wieghardt and W. Tillman. On the turbulent friction layer for rising pressure. Technical Memorandum 1314, NACA, (1951).
7. J. W. Slater. NPARC alliance CFD verification and validation Web site, (2003). http://www.grc.nasa.gov/WWW/wind/valid/archive.

8. F. Bassi, S. Rebay, G. Mariotti, S. Pedinotti, and M. Savini. A high-order accurate discontinuous finite element method for inviscid and viscous turbomachinery flows. In eds. R. Decuypere and G. Dibelius, *Proceedings of the 2nd European Conference on Turbomachinery Fluid Dynamics and Thermodynamics*, pp. 99–108, Antwerpen, Belgium (March 5–7, 1997). Technologisch Instituut.
9. F. Bassi and S. Rebay. A high order discontinuous Galerkin method for compressible turbulent flows. In eds. B. Cockburn, G. Karniadakis, and C.-W. Shu, *Discontinuous Galerkin Methods. Theory, Computation and Applications*, vol. 11, *Lecture Notes in Computational Science and Engineering*, pp. 77–88. Springer-Verlag, (2000).
10. F. Brezzi, M. Manzini, D. Marini, P. Pietra, and A. Russo, Discontinuous Galerkin approximations for elliptic problems, *Numer. Methods Partial Differential Equations*. **16**, 365–378, (2000).
11. D. N. Arnold, F. Brezzi, B. Cockburn, and D. Marini, Unified analysis of discontinuous Galerkin methods for elliptic problems, *SIAM J. Numer. Anal.* **39**(5), 1749–1779, (2002).
12. D. Hänel, R. Schwane, and G. Seider. On the accuracy of upwind schemes for the solution of the Navier–Stokes equations. AIAA Paper 87-1105 CP, AIAA, (1987).
13. L. Giraud, J. Langou, and M. Rozloznik. On the loss of orthogonality in the Gram-Schmidt orthogonalization process. Technical Report No. TR/PA/03/25, CERFACS, (2003).
14. G. S. Iannelli and A. J. Baker. A stiffly-stable implicit Runge–Kutta algorithm for CFD applications. AIAA Paper 88-0416, AIAA, (1988).
15. J. Lang and J. Verwer, ROS3P—An accurate third-order Rosenbrock solver designed for parabolic problems, *BIT*. **41**(4), 731–738, (2001).
16. N. Kroll, H. Bieler, H. Deconinck, V. Couaillier, H. van der Ven, and K. Sørensen, Eds., *ADIGMA - A European Initiative on the Development of Adaptive Higher-Order Variational Methods for Aerospace Applications*. vol. 113, *Notes on Numerical Fluid Mechanics and Multidisciplinary Design*, (Springer Berlin / Heidelberg, 2010). ISBN 978-3-642-03706-1.
17. V. Schmitt and F. Charpin. Pressure distributions on the ONERA-M6-wing at transonic Mach numbers. Advisory Report 138, AGARD, (1979).
18. Third AIAA Computational Fluid Dynamics Drag Prediction Workshop. http://aaac.larc.nasa.gov/tsab/cfdlarc/aiaa-dpw/Workshop3/ (June, 2006).
19. E. Lee-Rausch, N. Frink, D. Mavriplis, R. Rausch, and W. Milholen, Transonic drag prediction on a DLR-F6 transport configuration using unstructured grid solvers, *Computers & Fluids*. **38**, 511–532, (2009).
20. J. Chu and J. Luckring. Experimental surface pressure data obtained on 65° delta wing across Reynolds number and Mach number ranges. Technical memorandum 4645, NASA, (1996).
21. F. Bassi, A. Ghidoni, and S. Rebay, Optimal Runge-Kutta smoothers for the p-multigrid discontinuous Galerkin solution of the 1D Euler equations,

Journal of Computational Physics. **In Press, Corrected Proof**, –, (2010). ISSN 0021-9991. doi: DOI:10.1016/j.jcp.2010.04.030.
22. F. Bassi, A. Ghidoni, S. Rebay, and P. Tesini, High-order accurate p-multigrid discontinuous Galerkin solution of the Euler equations, *Int. J. Numer. Meth. Fluids*. **60**(8), 847–865, (2009).
23. F. Bassi, N. Franchina, A. Ghidoni, and S. Rebay, Spectral p-multigrid discontinuous Galerkin solution of the Navier-Stokes equations, *Int. J. Numer. Meth. Fluids*. (2010). accepted.
24. P. Tesini. *An h-multigrid approach for high-order discontinuous Galerkin methods*. PhD thesis, Università degli studi di Bergamo, Dipartimento di Ingegneria Industriale, Viale Marconi 5, 24044 Dalmine (BG), Italy, (2008).

CHAPTER 2

MASSIVELY PARALLEL SOLUTION TECHNIQUES FOR HIGHER-ORDER FINITE-ELEMENT DISCRETIZATIONS IN CFD

Laslo T. Diosady* and David L. Darmofal†

Massachusetts Institute of Technology (MIT),
Aerospace Computational Design Laboratory,
77 Massachusetts Ave. 33-207, Cambridge MA, 02139, USA
**diosady@mit.edu*
†darmofal@mit.edu

The purpose of this paper is to present techniques to solve higher-order finite element discretizations on massively parallel architectures. Implicit schemes are considered as a means of achieving mesh independent convergence rates for both time dependent problems and steady state solutions obtained through pseudo-transient continuation. Domain decomposition preconditioners are presented for the scalable parallel solution of the linear system arising at each iteration of a Newton-Krylov approach. Basic domain decomposition methods are presented along with theoretical results for simple model problems. Practical extensions of these algorithms for simulations of the Euler and Navier-Stokes equations are reviewed in reference to the theoretical results from the model problems. Extensions of some recently developed iterative substructuring algorithms are also proposed for the Euler and Navier-Stokes equations. Numerical examples using several domain decomposition algorithms are presented for a higher-order simulation of a convection-diffusion model problem.

1. Introduction

Today's most powerful supercomputers are able to reach a peak performance of more than one petaflop/s. However, peak performance has been reached by a continuing trend of parallelization with the most powerful machines now employing more than 100,000 processors. While several CFD codes have been used on large parallel systems with up to several thousand processors, Mavriplis notes: "The scalability of most [CFD] codes tops out around 512 cpus".[1] Developing CFD codes which are able to scale efficiently to tens or hundreds of thousands of processors remains a significant challenge.

A key use of massively parallel computers is to perform large-scale simulations in similar amount of time as typical industrial simulations on commodity hardware, through the use of parallelization. Thus, "optimal" algorithms are desired, for which the work scales linearly with the number of degrees of freedom. For iterative methods, for which the work associated with each iteration scales linearly with the number of degrees of freedom, optimality implies that the method converges at a rate independent of the size of the mesh. In the context of higher-order simulations, optimality also implies that the number of iterations is independent of the solution order. As the work associated with each iteration depends upon the number of degrees of freedom, the ability to perform large-scale simulations in reasonable time additionally requires that the work associated with each iteration may be performed in parallel across a large number of processors.

Two definitions of parallel scaling are common: "strong scaling" and "weak scaling". Strong scaling, discussed in reference to Amdahl's Law,[2] refers in general to parallel performance for fixed problem size, while weak scaling, discussed in reference to Gustafson's Law,[3] refers to parallel performance in terms of fixed problem size per processor. While the parallel performance of a particular CFD code depends upon an efficient implementation, the performance is limited by the scalability of the underlying algorithm. Thus, we focus primarily on the algorithmic aspects to ensure scalability. In the context of high-fidelity CFD simulations, we argue weak scaling is more important than strong scaling, as weak scaling relates closely to the ability of an algorithm to be optimal. Thus, unless otherwise stated we will use the term "scalable" to imply "weakly scalable". An iterative solution algorithm is said to be scalable if the rate of convergence is independent of the number of subdomains into which the mesh has been partitioned, for a fixed number of elements on each subdomain. Thus, for a fixed number of elements on each subdomain, a scalable algorithm may be viewed as being optimal on a macro scale. A scalable algorithm is truly optimal if the rate convergence is also independent of the number of elements on each subdomain.

For unsteady simulations, explicit methods have been touted as being highly parallelizable, as inter-processor communication is required only in updating ghosted data from neighbouring processors, while residual evaluations are trivially parallelized. While explicit methods are relatively simple to implement, the largest allowable time step is limited by the CFL condition, hence the number of iterations required for a particular simulation depends upon the mesh size. Thus, while explicit methods have the potential for very good strong scaling, these methods are not optimal. Implicit methods, on the other hand, do not have such a time step restriction. As a result, implicit methods have become the method of choice when the time step required for numerical stability is well below that required to resolve the unsteady features of the flow. Implicit schemes have also become widely used for the solution of steady state problems obtained through pseudo-transient continuation,[4] where time-stepping enables reliable convergence for nonlinear problems.[5–13] While most portions of an implicit code, such as residual and Jacobian evaluations, are trivially parallelized, implicit methods require at each iteration the solution of a globally coupled system of equations. Thus, implicit algorithms are optimal only if the globally coupled system may be solved in an optimal manner.

For aerodynamic problems, the most successful solution techniques have been nonlinear multigrid methods[5,14–18] and preconditioned Newton-Krylov methods.[6,10–13,19] Mavriplis showed that using a multigrid method as a preconditioner to a Newton-Krylov approach results in significantly faster convergence in terms of CPU time than a full nonlinear multigrid scheme.[20] Thus, in this work Newton-Krylov methods are considered, where the nonlinear system is solved using an approximate Newton method, while the linear system at each Newton iteration is solved using a preconditioned Krylov subspace method. In this context, multigrid methods may be viewed as one possible choice for the preconditioner. Thus, the development of an optimal solution method hinges on the ability to develop scalable preconditioners, to enable the efficient solution of large linear systems.

The desire to perform large scale simulations has led to an increased interest in domain decomposition methods for the solution of large algebraic systems arising from the discretization of PDE problems. The term domain decomposition in the engineering community has often been used simply to refer to the partitioning of data across a parallel machine. However, data parallelism alone is insufficient to ensure good parallel performance. In particular, the performance of a domain decomposition preconditioner for the solution of large linear systems is strongly coupled to the discretization and the underlying PDE problem. While high-fidelity simulations of aerodynamic flows involve solutions of the nonlinear compressible Euler and Navier-Stokes equations, performance of the algorithms developed for the systems resulting from the discretization of these equations are often analyzed in reference to simple scalar linear model equations for which the mathematical analysis is possible.

Early aerodynamic simulations involved potential flow calculations. Thus, the Poisson equation has often been used as a model. In particular, the elliptic nature of the Poisson equation may be seen as appropriate for the analysis of acoustic modes in low speed, incompressible flows. Convective modes, on the other hand are hyperbolic and thus a convection equation may be a more appropriate model for the analysis of these modes. A singularly perturbed convection-diffusion equation is often used as a model problem for high-speed compressible flows, where convective behaviour is dominant in most regions of the flow, while elliptic behaviour is dominant in the boundary layer region. Since much of the grid resolution is introduced in the boundary layer region, it is important to understand the elliptic behaviour present in these regions.

For elliptic PDEs, the Green's function extends throughout the entire domain decaying with increasing distance from the source. This implies that a residual at any point in the domain affects the solution everywhere else. In an unpreconditioned Krylov method, the application of the Jacobian matrix to a residual vector at each Krylov iteration exchanges information only to the extent of the numerical stencil. Thus, the number of iterations for an error to be felt across a domain of unit diameter is $O(\frac{1}{h})$, where h is the characteristic element size. In general, the convergence rate for symmetric problems in bounded by the condition number of the preconditioned system. An efficient preconditioner attempts to cluster the eigenvalues of the preconditioned system to ensure rapid convergence of the Krylov method. In particular, an efficient preconditioner for elliptic problems requires a means of controlling the lowest frequency error modes which extend throughout the domain.

While elliptic problems are characterized by Green's functions that extend throughout the entire domain, convection-dominated problems have a hyperbolic behaviour where the errors propagate along characteristics in the flow. Thus, for convection-dominated problems, the resulting discretization is strongly coupled along the characteristics with little dissipation of errors present especially across characteristics. Control of these errors is often accomplished by preconditioners that maintain strong coupling and often can be interpreted as increasing the propagation of errors out of the domain in the purely hyperbolic case.

As aerodynamic flows involve both elliptic and hyperbolic features, the most successful algorithms have combined effective solvers for elliptic and hyperbolic problems. For example multigrid methods have been used in combination with tri-diagonal line solvers.[15,16] The success of these algorithms may be attributed to the ability of line solvers to control error modes in strongly coupled directions (either along characteristics or in regions of high anisotropy), while low frequency errors are corrected through the multigrid process. An alternative approach which appears to be very successful for higher-order discretizations is a two-level method using an ILU(0) preconditioner with a minimum discarded fill ordering combined with a coarse grid correction.[19]

The development of efficient parallel preconditioners for aerodynamic flows builds upon successful algorithms in the serial context. While multigrid methods have been employed for large-scale parallel simulations,[15,18] care must be taken in forming the nested coarse grid problems to ensure good performance.[15] The domain decomposition preconditioners presented in this paper may be viewed as two-level preconditioners, where local solvers are employed on each subdomain, while specially constructed coarse spaces are used to ensure the control of low frequency (global) modes throughout the domain. In particular, successful multigrid and ILU preconditioners discussed in the serial context may be used as local solvers on each subdomain.

The purpose of this paper is twofold: first to provide the reader with an understanding of the performance of several successful solution algorithms on simple model problems; and second to discuss the extension of these algorithms to the solution of higher-order discretizations of convection-dominated flows of interest in the CFD community. In particular, we focus on describing the algorithms and give theoretical and numerical results where relevant. However, we refrain from providing proofs of the theoretical results, which may be found in the references provided. In Section 2, we present Schwarz methods in the context of the model problems, then review large-scale CFD applications of these algorithms. In Section 3, we present Schur complement techniques, while in Section 4 we discuss Neumann-Neumann methods. Finally, in Section 5, we present some numerical results discussing the algorithms presented.

2. Schwarz Methods

In this section, we present Schwarz methods, which are often referred to as overlapping methods. Schwarz methods can be traced back to 1870, when Schwarz described an iterative method for solving an elliptic PDE problem by alternatingly solving the problem in subdomains of the original domain using the solution from a previous iterate as the boundary condition. While this classical alternating Schwarz method was not used as numerical solution technique, it forms the basis for many successful domain decomposition algorithms. We present the basic ideas for the case of two subdomains then discuss the extension to the case of many subdomains. The presentation in this section closely follows that of Smith, Bjorstad and Gropp[21] and Toselli and Widlund[22] and we refer the reader to these books for a complete presentation.

2.1. *The case of two subdomains*

Consider the Poisson problem in a domain Ω:

$$-\Delta u = f \qquad \text{in } \Omega, \tag{1}$$

$$u = 0 \qquad \text{on } \partial\Omega. \tag{2}$$

We partition the domain Ω into two overlapping subdomains Ω_1' and Ω_2'. Given an iterate u^n, the Schwarz alternating method solves for u^{n+1} by solving successive Dirichlet problems in Ω_1' and Ω_2':

$$\begin{cases} -\Delta u^{n+1/2} = f & \text{in } \Omega_1', \\ u^{n+1/2} = 0 & \text{on } \partial\Omega_1' \cap \partial\Omega, \\ u^{n+1/2} = u^n & \text{on } \partial\Omega_1' \backslash \partial\Omega, \\ u^{n+1/2} = u^n & \text{in } \Omega_2' \backslash \Omega_1', \end{cases} \tag{3}$$

$$\begin{cases} -\Delta u^n = f & \text{in } \Omega_2', \\ u^{n+1} = 0 & \text{on } \partial\Omega_2' \cap \partial\Omega, \\ u^{n+1} = u^{n+1/2} & \text{on } \partial\Omega_2' \backslash \partial\Omega, \\ u^{n+1} = u^{n+1/2} & \text{in } \Omega_1' \backslash \Omega_2', \end{cases} \tag{4}$$

Consider a finite element discretization of (1)-(2). Given an appropriate bilinear form and basis, the corresponding discrete system of equations may be written as:

$$Au = f \tag{5}$$

where $u \in \mathbb{R}^n$ denotes the vector of discrete unknowns. We denote by u_1 and u_2 degrees of freedom corresponding to Ω_1' and Ω_2', respectively. Additionally, we denote by R_1 and R_2 the $\{0,1\}$ matrices, respectively, that extract degrees of freedom u_1 and u_2 from u (i.e. $u_i = R_i u, i \in \{1,2\}$). Using this notation, the discrete Schwarz alternating method may be written using the following steps:

$$u^{n+1/2} = u^n + R_1^T A_1^{-1} R_1 \left(f - Au^n\right), \quad (6)$$

$$u^{n+1} = u^{n+1/2} + R_2^T A_2^{-1} R_2 \left(f - Au^{n+1/2}\right). \quad (7)$$

Here $A_1 = R_1 A R_1^T$ and $A_2 = R_2 A R_2^T$ are simply the blocks extracted from A corresponding u_1 and u_2, respectively. Eliminating $u^{n+1/2}$ we see that the Schwarz alternating method is a Richardson iteration for the preconditioned system:

$$M^{-1} A u = M^{-1} f, \quad (8)$$

with the preconditioner given by:

$$M_{MSM}^{-1} = R_1^T A_1^{-1} R_1 + R_2^T A_2^{-1} R_2 \left(I - A R_1^T A_1^{-1} R_1\right). \quad (9)$$

This preconditioner is referred to as a multiplicative Schwarz method, thus we use the subscript $_{MSM}$. In the multiplicative Schwarz method the Dirichlet problem solved in Ω_2 depends upon the intermediate solution $u^{n+1/2}$ in Ω_1 and hence this algorithm is inherently sequential. As opposed to using the intermediate solution $u^{n+1/2}$ as the boundary condition in Ω_2, the previous iterate u^n may be used as boundary conditions for both Ω_1 and Ω_2, allowing the Dirichlet problems in Ω_1 and Ω_2 to be solved independently. This method, known as an additive Schwarz method, will in general not converge through a Richardson iteration, however may be used as an effective preconditioner for a Krylov method. We write the additive Schwarz preconditioner as:

$$M_{ASM}^{-1} = R_1^T A_1^{-1} R_1 + R_2^T A_2^{-1} R_2. \quad (10)$$

The adjectives additive and multiplicative refer to the propagation of the error, $u - u^n$, in the different Schwarz algorithms. Namely, the solution of the problem restricted to a subdomain may be viewed as a projection of the error to the finite element space orthogonal to the space defined by the degrees of freedom corresponding to that particular subspace. For additive methods, each subdomain problem is solved independently and thus the error is given by the sum of the projections corresponding to each subdomain. In multiplicative methods, the subdomain problems are solved sequentially, leading the error to be reduced as the product of two projections. In this paper we will present several preconditioners, involving both additive and multiplicative components, which are sometimes referred to as hybrid Schwarz methods. In general, we will use additive to refer to operations of these preconditioner which may be performed independently, while we use multiplicative to refer to sequential operations. We note that the convergence rate of the multiplicative Schwarz method relative the additive Schwarz method, is much like the performance of Gauss-Seidel versus Jacobi. Namely, the convergence rate of multiplicative Schwarz methods improve upon additive Schwarz methods by a constant factor.

2.2. *The case of many subdomains*

Both additive and multiplicative Schwarz methods are easily extended to the case of many subdomains. Consider a partition of the domain Ω into N nonoverlapping subdomains Ω_i, $i = 1, ..., N$. An overlapping partition of the domain is defined by extending each subdomain Ω_i by an amount δ to a region $\Omega_i^{'} \subset \Omega$. In practice, $\Omega_i^{'}$ may be defined by adding layers of elements from neighbouring subdomains to Ω_i. The additive Schwarz method involves the solution of N independent Dirichlet problems corresponding to each subdomain, which may be performed in parallel, by assigning a subdomain to each processor. Using the notation previously defined, we write the additive Schwarz preconditioner as:

$$M_{ASM}^{-1} = \sum_{i=1}^{N} R_i^T A_i^{-1} R_i, \tag{11}$$

As described in the case of two subdomains, the multiplicative Schwarz method is inherently sequential. In the case of many subdomains, parallelism is introduced using a colouring argument. Namely, each subdomain $\Omega_i^{'}$ is assigned to a "colour" corresponding to groups of subdomains which do not overlap. Subdomain problems corresponding to the same colour may be solved independently of one another. Thus, in the case of many subdomains, the multiplicative Schwarz method involves only a small number of sequential steps corresponding to each colour, as opposed to N steps corresponding to each subdomain. In order to achieve good performance, each processor should be assigned several subdomains, one corresponding to each colour. We note that each sequential step of the multiplicative Schwarz method involves a multiplication of the system matrix A in order to update the residual. However, usually only parts of the residual vector need to be updated at each iteration which may often be performed locally.

The basic forms of the additive and multiplicative Schwarz methods lack a global correction. Thus, for elliptic problems, these methods are not scalable. A coarse space capable of controlling low frequency modes can be introduced by considering a discretization of the original PDE on a coarse triangulation $\mathcal{T}_H$.[23] In general, the fine grid $\mathcal{T}_h$ does not need to be derived from a refinement of the coarse grid $\mathcal{T}_H$, only an interpolation operator from the fine space to the coarse space needs to be defined. We denote by R_0^T the interpolation operator from $\mathcal{T}_H$ to the finite element space defined on $\mathcal{T}_h$, where R_0 may be viewed as a restriction from the original finite element space to the coarse subspace. The coarse system matrix A_0 may be obtained either from discretizing the original PDE on $\mathcal{T}_H$ or through a restriction of the form $A_0 = R_0^T A R_0$. The additive Schwarz preconditioner with coarse grid correction is thus given by:

$$M_{ASM_0}^{-1} = R_0^T A_0^{-1} R_0 + \sum_{i=1}^{N} R_i^T A_i^{-1} R_i. \tag{12}$$

A simple variant of this preconditioner may be obtained by applying the coarse grid correction in a multiplicative manner.[21] Namely, this preconditioner involves two sequential steps: 1) the solution of the coarse grid problem followed by a corresponding update of the residual, 2) the solution of N independent subdomain problems. Similar variants of the multiplicative Schwarz method have also be developed.[24] The presence of the coarse space enables the additive Schwarz method to be scalable for elliptic problems. Namely, the condition number of the preconditioned system is given by $\kappa\left(M_{ASM_0}^{-1}A\right) \leq C\left(1+\frac{H}{\delta}\right)$, where H is the diameter of a subdomain Ω_i, while δ is the amount of overlap and C is a constant independent of H or h.[23,24] The condition number does not depend directly upon H but only upon the factor $\frac{H}{\delta}$. If the overlap is such that $\delta \geq cH$ for some constant c, the subdomains are said to have "generous" overlap. With generous overlap, the condition number of the preconditioned system becomes independent of $\frac{1}{H}$ and $\frac{H}{h}$ and the method is both scalable and optimal. On the other hand, we may consider the case where the overlap is defined by extending each nonoverlapping subdomain by a small number of element of the fine triangulation. In this case we have $\delta \geq ch$, and the condition number bound has the form $\kappa \leq C_1\left(1+\frac{H}{h}\right)$. Thus in the case of small overlap this type of preconditioner is scalable, but not optimal.

While originally presented for the solution of self-adjoint elliptic problems,[23] the analysis of Schwarz methods has been extended to linear convection-diffusion problems by Cai and collaborators.[24–28] For linear convection-diffusion problems, Schwarz methods with generous overlap and a coarse space have been shown to be both scalable and optimal, provided the diameter of the subdomains are sufficiently small.[24–26] Namely, if the Peclet number defined using the subdomain length scale, H, is sufficiently small, then the behaviour of the Schwarz method matches the symmetric, diffusion-dominated limit. In the convection-dominated limit, the errors are propagated along characteristics in the domain. Thus, the number of iterations required to converge is related to the number of subdomains through which a characteristic must cross before exiting a domain. Similar behaviour is observed for other domain decomposition methods for convection-dominated problems and this remains an open area of research.

In the case of unsteady convection-diffusion problems, solved using implicit time integration, analysis of additive and multiplicative Schwarz methods shows that a coarse space may not be necessary to guarantee scalability if the time step is sufficiently small relative the size of the subdomains.[27,28] This behaviour may be interpreted using physical intuition. Namely, for small time steps the evolution of the flow is mostly local, thus a coarse space is not required for the global control of error modes. From a linear algebra standpoint, the presence of the large temporal term leads to a diagonally dominant system, which tend to be easier to solve using iterative methods.

While initially analyzed for the solution of the systems of equations arising from linear continuous finite element discretizations, overlapping Schwarz methods have been extended to mixed finite element,[29] spectral element,[30] and discontinuous Galerkin discretizations.[31–35] Schwarz methods have also been applied to finite difference,[36] and finite volume discretizations.[11] For higher-order discretization, the overlapping regions may be defined by extending nonoverlapping domains by layers of nodes corresponding to the discrete unknowns.[30,37] However, for unstructured meshes, choosing an appropriate set of nodes may be non-trivial.[38] Thus, if only moderate polynomial orders are used, the overlapping regions are typically defined by adding layers of elements.

2.3. *Large scale CFD applications*

Overlapping additive Schwarz methods are the most widely used domain decomposition methods for CFD applications. Overlapping methods may be seen as particularly well suited to cell-centered finite-volume, or higher-order discontinuous Galerkin discretizations, where degrees of freedom are naturally associated with element interiors. Thus each elemental degree of freedom is "owned" by a single processor, while overlapping regions consist of elements owned by neighbouring processors. For these type of discretizations, we may also consider the special case of zero overlap, such that the

u_i's correspond to distinct degrees of freedom. In this case the additive Schwarz method reduces to a block Jacobi preconditioner with each block A_i corresponding to a single subdomain Ω_i. Similarly, the multiplicative Schwarz method reduces to a subdomain-wise block Gauss-Seidel preconditioner for A.

For node-based finite-volume, or continuous finite-difference discretizations, a nonoverlapping partitioning of the elements results in a "minimum-overlapping" partition of nodes. In a practical implementation, a nodal degree of freedom on the interface is assigned to a unique processor, which is updated by local solves corresponding to both sides of the interface. A variant, known as the restricted additive Schwarz method, updates only locally owned degrees of freedom, eliminating communication during the solution update.[39] Numerical results have shown that this method actually requires fewer iterations to converge than the basic additive Schwarz preconditioner for both scalar convection-diffusion,[39] and compressible Euler problems.[40]

The use of domain decomposition methods for large scale applications involves additional considerations in order to achieve good performance.[10] Large scale CFD applications may be both memory and CPU limited, making the exact solution of the local problems (corresponding to A_i^{-1}) using LU factorization intractable. Thus, the local solver may replaced with an iteration of an efficient serial preconditioner, such as an ILU factorization or a multigrid cycle. The performance of the Schwarz method will, in general, depend upon the strength of the local solver. For example, Venkatakrishnan showed significant improvement using block-ILU(0) as opposed to block-Jacobi for the local solvers for an additive Schwarz method with zero overlap.[6] ILU factorizations have been particularly popular as local solvers for additive Schwarz methods with and without a coarse correction.[6,10,11,40–43] Cai, Farhat and Sarkis also employed a preconditioned GMRES method to solve the local problem on each subdomain.[41,42] In particular, this allowed for different number of iterations to be used in each subdomain ensuring that each local problem was solved with sufficient accuracy.

The ability to achieve high performance for large scale simulations also requires an appropriate balance between local computation on each processor and relatively slow communication tasks.[10] As discussed previously, the case of generous overlap ensures that the preconditioner is optimal. However, if the overlap is generous, then the number of degrees of freedom in the overlap region of a subdomain is proportional to the volume of the subdomain. On the other hand, in the case of small overlap, where the overlap is defined by extending each subdomain by a few layers of elements, the number of degrees of freedom corresponding to the overlap region is proportional to the surface area of the subdomain. Thus if each subdomain is assigned to a single processor, the ratio of computation to communication may be much higher for the case of small overlap and thus potentially better performance may be achieved. Subdomain-wise block-Jacobi preconditioners have been used for discontinuous Galerkin discretization of the compressible Euler and Navier-Stokes equations on up to 512 processors.[43] Gropp et al. showed that adding a very small overlap results in a significant improvement in the number of iterations required to converge a finite volume discretization of inviscid compressible flows.[11] In particular, the lowest CPU times were achieved using an overlap regions of just two layers of elements.

For practical aerodynamic flows, the question remains whether a coarse space is necessary for a scalable preconditioner. For the solution of steady compressible Euler equations, Venkatakrishnan used a coarse space developed using an algebraic multigrid-type scheme.[6] In numerical simulations with up to 128 processors, Venkatakrishnan shows that the presence of the coarse grid gives some improvement in the performance of the preconditioner in terms of number of iterations, though this does not necessarily translate into faster solution time. Gropp et al. do not employ a coarse space, and show only modest increase in the number of linear iterations for strong scaling results from 32 to 128 processors.[11] In particular, Anderson, Gropp, and collaborators have performed large scale inviscid CFD simulations using over 3000 processors without employing a coarse space.[10,11,44] For these simulations, the use of a coarse space may be unnecessary due to the temporal terms present as a results of the pseudo-transient continuation used to arrive at steady state solutions.[4] For unsteady simulations for the compressible Navier-Stokes equations, Cai, Farhat, and Sarkis find only a small increase in the number of iterations for strong scaling results up to 512 subdomains without the presence of a coarse space.[40,42] Similarly, Persson showed good strong scaling performance up to 512 processors for the unsteady Navier-Stokes equations using a subdomain wise block-Jacobi preconditioner without a coarse space.[43] We note that this observation is consistent with the theoretical result for the time-dependent convection-diffusion problems, where a coarse space is not necessary if the time step is sufficiently small.

As the time step is allowed to increase, Persson showed that the performance of the preconditioner without a coarse space degrades significantly.[43] For steady state problems solved using a p-sequencing approach with little or no pseudo-temporal contributions, Diosady[45] showed very poor strong scaling using a similar preconditioner, particularly for viscous problems. In order to improve the parallel scaling of this preconditioner, Diosady presented a partitioning strategy weighted by the strength of the coupling between elements. A similar strategy was also employed by Persson.[43] However, the resulting partitions have larger surface area to volume ratios resulting in less computation per communication. While such a technique improves parallel performance on moderate number of processors, the use of a coarse space may be essential for obtaining a scalable method for steady state viscous flow problems on massively parallel systems.

3. Schur Complement Methods

In this section, we present Schur complement methods, also known as nonoverlapping or iterative substructuring methods. In general these methods reduce the globally coupled system of equations to a smaller system involving only the degrees of freedom associated with the interface between subdomains. We present the basic ideas for substructuring methods for a continuous finite element discretization in the case of two subdomains, and then discuss the extensions to the case of many subdomains. The presentation in this section closely follows that of Toselli and Widlund.[22] For a full presentation we refer to the books by Toselli and Widlund,[22] Quarteroni and Valli,[46] or Smith, Bjorstad and Gropp.[21]

3.1. *An interface problem*

Again, we consider the Poisson problem (1)-(2) in a domain Ω. We partition the domain Ω into two nonoverlapping subdomains Ω_1 and Ω_2, with $\Gamma = \partial\Omega_1 \cap \partial\Omega_2$ the interface between the two subdomains. We may rewrite (1)-(2) as an equivalent coupled problem:

$$-\Delta u_1 = f \qquad \text{in } \Omega_1, \tag{13}$$

$$u_1 = 0 \qquad \text{on } \partial\Omega_1 \cap \partial\Omega, \tag{14}$$

$$u_1 = u_2 \qquad \text{on } \Gamma, \tag{15}$$

$$\frac{\partial u_1}{\partial \boldsymbol{n}_1} = -\frac{\partial u_2}{\partial \boldsymbol{n}_2} \qquad \text{on } \Gamma, \tag{16}$$

$$-\Delta u_2 = f \qquad \text{in } \Omega_2, \tag{17}$$

$$u_2 = 0 \qquad \text{on } \partial\Omega_2 \cap \partial\Omega, \tag{18}$$

where $\boldsymbol{n}_i$ is the outward pointing normal vector from Ω_i. The solutions, u_i, $i = 1, 2$, of the coupled problem gives the restriction of the solution, u, to each subdomain Ω_i. The transmission conditions (15) and (16) ensure that $u_\Gamma := u_1 = u_2$ and $\lambda_\Gamma := \frac{\partial u_1}{\partial \boldsymbol{n}_1} = -\frac{\partial u_2}{\partial \boldsymbol{n}_2}$ on Γ. We note that if u_Γ is known then the u_i's may be obtained by solving independent problems in each subdomain with Dirichlet boundary condition on Γ:

$$-\Delta u_i = f \qquad \text{in } \Omega_i, \tag{19}$$

$$u_i = 0 \qquad \text{on } \partial\Omega_i \cap \partial\Omega, \tag{20}$$

$$u_i = u_\Gamma \qquad \text{on } \Gamma. \tag{21}$$

Alternatively, if λ_Γ is known then the u_i's may be obtained by solving independent problems with Neumann boundary conditions on Γ:

$$-\Delta u_i = f \qquad \text{in } \Omega_i, \tag{22}$$

$$u_i = 0 \qquad \text{on } \partial\Omega_i \cap \partial\Omega, \tag{23}$$

$$\frac{\partial u_i}{\partial \boldsymbol{n}_1} = \lambda_\Gamma \qquad \text{on } \Gamma, \tag{24}$$

Schur complement algorithms are based on a discrete equivalent of the coupled problem (13)-(18). Namely, the discrete problem may be reduced to a system corresponding only to discrete unknowns u_Γ or λ_Γ, on the interface Γ. Once u_Γ or λ_Γ are known the solution interior to each subdomain may be obtained by solving discrete equivalents of the Dirichlet problem (19)-(21) or Neumann problem (22)-(24). Methods which solve for the discrete unknowns corresponding to u_Γ are known as primal substructuring methods, while dual substructuring methods are based on solving the discrete equivalent of the flux λ_Γ.

We now derive a discrete equation for the interface state u_Γ. Once again we consider the discretization of (1)-(2), which results in the discrete system (5). We denote by $u^{(1)}$ and $u^{(2)}$ degrees of freedom associated with nodes on subdomains Ω_1 and Ω_2 respectively. Additionally we use subscript $_\Gamma$ to denote degrees on freedom associated with the interface Γ, while we use subscript $_I$ to denote degrees of freedom strictly interior to a particular subdomain. The discrete system of equations (5) may be written as:

$$\begin{bmatrix} A_{II}^{(1)} & 0 & A_{I\Gamma}^{(1)} \\ 0 & A_{II}^{(2)} & A_{I\Gamma}^{(2)} \\ A_{\Gamma I}^{(1)} & A_{\Gamma I}^{(2)} & A_{\Gamma\Gamma} \end{bmatrix} \begin{bmatrix} u_I^{(1)} \\ u_I^{(2)} \\ u_\Gamma \end{bmatrix} = \begin{bmatrix} f_I^{(1)} \\ f_I^{(2)} \\ f_\Gamma \end{bmatrix}, \tag{25}$$

where we note that we have explicitly enforced the discrete equivalent of the first transmission condition (15), namely $u_\Gamma := u_\Gamma^{(1)} = u_\Gamma^{(2)}$. Consider the following block factorization of the system matrix, A:

$$\begin{bmatrix} I_{II}^{(1)} & 0 & 0 \\ 0 & I_{II}^{(2)} & 0 \\ A_{\Gamma I}^{(1)} A_{II}^{(1)^{-1}} & A_{\Gamma I}^{(2)} A_{II}^{(2)^{-1}} & I_{\Gamma\Gamma} \end{bmatrix} \begin{bmatrix} A_{II}^{(1)} & 0 & 0 \\ 0 & A_{II}^{(2)} & 0 \\ 0 & 0 & S \end{bmatrix} \begin{bmatrix} I_{II}^{(1)} & 0 & A_{II}^{(1)^{-1}} A_{I\Gamma}^{(1)} \\ 0 & I_{II}^{(2)} & A_{II}^{(2)^{-1}} A_{I\Gamma}^{(2)} \\ 0 & 0 & I_{\Gamma\Gamma} \end{bmatrix}. \tag{26}$$

Where S is the Schur complement given by:

$$S = A_{\Gamma\Gamma} - \sum_{i=1}^{2} A_{\Gamma I}^{(i)} A_{II}^{(i)^{-1}} A_{I\Gamma}^{(i)}. \tag{27}$$

The corresponding inverse of A may be written as, A^{-1}:

$$\begin{bmatrix} I_{II}^{(1)} & 0 & -A_{II}^{(1)^{-1}} A_{I\Gamma}^{(1)} \\ 0 & I_{II}^{(2)} & -A_{II}^{(2)^{-1}} A_{I\Gamma}^{(2)} \\ 0 & 0 & I_{\Gamma\Gamma} \end{bmatrix} \begin{bmatrix} A_{II}^{(1)^{-1}} & 0 & 0 \\ 0 & A_{II}^{(2)^{-1}} & 0 \\ 0 & 0 & S^{-1} \end{bmatrix} \begin{bmatrix} I_{II}^{(1)} & 0 & 0 \\ 0 & I_{II}^{(2)} & 0 \\ -A_{\Gamma I}^{(1)} A_{II}^{(1)^{-1}} & -A_{\Gamma I}^{(2)} A_{II}^{(2)^{-1}} & I_{\Gamma\Gamma} \end{bmatrix}. \tag{28}$$

We note that the only globally coupled operation involved in computing the inverse given in (28) corresponds to solving a system with the Schur complement S. Namely, (25) may be solved using the following steps:

(1) Compute in parallel the Schur complement residual

$$g_{\Gamma} = f_{\Gamma} - \sum_{i=1}^{2} A_{\Gamma I}^{(i)} A_{II}^{(i)^{-1}} f_I^{(i)}. \tag{29}$$

(2) Solve the following global coupled Schur complement problem for u_{Γ}:

$$S u_{\Gamma} = g_{\Gamma}. \tag{30}$$

(3) Compute in parallel the subdomain interior degrees of freedom $u_I^{(i)}$:

$$u_I^{(i)} = A_{II}^{(i)^{-1}} \left(f_I^{(i)} + A_{I\Gamma}^{(i)} u_{\Gamma} \right), \qquad i = 1, 2 \tag{31}$$

We note that (31) is the discrete equivalent of the continuous Dirichlet problem (19)-(21). It remains to solve the Schur complement problem (30) for u_{Γ}. The Schur complement S may be too large to solve directly, thus a preconditioned Krylov method may be used to solve (30) iteratively. In the following section we discuss parallel preconditioners for the Schur complement problem (30). In particular, Schwarz methods discussed in Section 2 may also be used as preconditioners for the Schur complement, with the benefit of smaller Krylov vectors corresponding only to interface degrees of freedom.

3.2. *Classical substructuring methods*

In this section, we present classical substructuring methods, which are block Jacobi type preconditioners for (30) where the blocks are associated with subdomain faces, edges and vertices. The development of these type of preconditioners for symmetric elliptic problems is presented in a series of papers by Bramble, Pasciak, and Schatz.[47–50]

We consider groups of degrees of freedom on the interface Γ corresponding to the faces, edges, and vertices of subdomains. Namely, we denote by $\mathcal{F}_k$ the set of degrees of freedom interior to a subdomain face associated with exactly two subdomains Ω_i and Ω_j. Similarly, $\mathcal{E}_k$ denotes the set of degrees of freedom on a single edge between several subdomains, while $\mathcal{V}_k$ denotes the degrees of freedom associated with a single node at the cross-points between subdomains. We may consider rewriting the Schur complement matrix as:

$$S = \begin{bmatrix} S_{\mathcal{F}\mathcal{F}} & S_{\mathcal{F}\mathcal{E}} & S_{\mathcal{F}\mathcal{V}} \\ S_{\mathcal{E}\mathcal{F}} & S_{\mathcal{E}\mathcal{E}} & S_{\mathcal{E}\mathcal{V}} \\ S_{\mathcal{V}\mathcal{F}} & S_{\mathcal{V}\mathcal{E}} & S_{\mathcal{V}\mathcal{V}} \end{bmatrix}, \tag{32}$$

where $\mathcal{F}$, $\mathcal{E}$ and $\mathcal{V}$ correspond to the set of subdomain faces, edges and vertices. A simple block diagonal preconditioner for S may be given by dropping the off-diagonal blocks $S_{\mathcal{F}\mathcal{E}}$, $S_{\mathcal{F}\mathcal{V}}$, $S_{\mathcal{E}\mathcal{F}}$, $S_{\mathcal{E}\mathcal{V}}$, $S_{\mathcal{V}\mathcal{F}}$, and $S_{\mathcal{V}\mathcal{E}}$ corresponding to coupling between faces, edges, and vertices, as well as blocks in $S_{\mathcal{F}\mathcal{F}}$, $S_{\mathcal{E}\mathcal{E}}$, and $S_{\mathcal{V}\mathcal{V}}$ corresponding to the coupling between different faces, edges and vertices. We may write this block preconditioner as:

$$M^{-1} = \begin{bmatrix} \bar{S}_{\mathcal{F}\mathcal{F}}^{-1} & 0 & 0 \\ 0 & \bar{S}_{\mathcal{E}\mathcal{E}}^{-1} & 0 \\ 0 & 0 & \bar{S}_{\mathcal{V}\mathcal{V}}^{-1} \end{bmatrix}, \tag{33}$$

where $\bar{S}_{\mathcal{F}\mathcal{F}}$, $\bar{S}_{\mathcal{E}\mathcal{E}}$ and $\bar{S}_{\mathcal{V}\mathcal{V}}$ are the resulting block diagonal matrices. Several simplifications of this basic classical substructuring method exist that replace the blocks associated $\bar{S}_{\mathcal{E}\mathcal{E}}^{-1}$ and $\bar{S}_{\mathcal{V}\mathcal{V}}^{-1}$ with simple approximations, however we do not discuss these here. We note that the preconditioner (33) lacks a coarse space and hence is not scalable for elliptic problems. A coarse space may be added by considering the finite element discretization of the original problem on the coarse mesh whose elements are the subdomains. We may write this preconditioner as:

$$M^{-1} = \begin{bmatrix} \bar{S}_{\mathcal{F}\mathcal{F}}^{-1} & 0 & 0 \\ 0 & \bar{S}_{\mathcal{E}\mathcal{E}}^{-1} & 0 \\ 0 & 0 & \bar{S}_{\mathcal{V}\mathcal{V}}^{-1} \end{bmatrix} + \begin{bmatrix} \hat{R}_{\mathcal{F}}^T & \hat{R}_{\mathcal{E}}^T & I \end{bmatrix} A_H^{-1} \begin{bmatrix} \hat{R}_{\mathcal{F}} \\ \hat{R}_{\mathcal{E}} \\ I \end{bmatrix}, \tag{34}$$

where $\hat{R}_{\mathcal{F}}^T$ and $\hat{R}_{\mathcal{E}}^T$ are interpolation operators from the coarse finite element space to the faces and edges of the original finite element space. For the Poisson problem, the condition number of the preconditioned operator $M^{-1}S$ is bounded by $\kappa \leq C\left(1 + \log\left(\frac{H}{h}\right)\right)^2$. We note that this algorithm is scalable, but not optimal since the condition number (and hence convergence rate) depends upon $\frac{H}{h}$. However, the condition number depends only weakly upon $\frac{H}{h}$ and we say that the method is quasi-optimal. Many of the iterative substructuring algorithms presented will have similar condition number bounds. We do not discuss the proofs of these bounds, but refer the reader to the series of papers by Bramble, Pasciak, and Schatz[47–50] or Section 4.6 of Toselli and Widlund.[22]

While originally developed for scalar elliptic problems, algorithms in the spirit of classical substructuring methods have also been applied to systems of equations arising from CFD applications. Cai et al. discussed several classical substructuring variants along with overlapping Schwarz methods for a finite-difference discretization of convection-diffusion problems.[36,51] Gropp and Keyes developed a block triangular preconditioner for the streamfunction-vorticity formulation of two-dimensional flows.[52] Their preconditioner applied to the entire discrete system of equations (5) may be written as:

$$M^{-1} = \begin{bmatrix} A_{II} & A_{I\mathcal{E}} & A_{I\mathcal{V}} \\ 0 & \bar{A}_{\mathcal{E}\mathcal{E}} & A_{\mathcal{E}\mathcal{V}} \\ 0 & 0 & A_H \end{bmatrix}^{-1}. \tag{35}$$

3.3. *Approximate factorizations*

In the general case, the local solves corresponding to ${A_{II}^{(i)}}^{-1}$ in (29) and (31) may also be replaced with an approximate solver such as an ILU factorization or a multigrid cycle leading to an approximation of the Schur complement. Thus, steps corresponding to (29)-(31) may be replaced with approximate solvers to provide a preconditioner for the global problem (25). As with Schwarz methods the performance of Schur complement methods in general depend upon the choice of the approximate local solvers.

Barth et al developed a global preconditioner based on an approximate Schur complement for the solution of the conforming finite element discretization of the Euler equations.[9] Approximate Schur complements were formed by using an ILU preconditioned GMRES method for the solution of ${A_{II}^{(i)}}^{-1} A_{I\Gamma}^{(i)}$. Additional approximations were introduced to control the sparsity including element dropping and an approximate Schur complement formed by considering a small region of elements near the interface.[9] The approximate Schur complement problem was solved using a block preconditioned GMRES method, where the blocks correspond to groups of faces and edges. The blocks which correspond to groups of edges and faces across subdomains provide a global means of correcting low frequency modes, and hence no additional coarse space was required.[9] The use of GMRES for both the local and approximate Schur complement solves means that the preconditioner was non-stationary. Thus the global problem used the flexible variant of GMRES (fGMRES).[53] Barth presented weak scaling results on up to 64 processors, which showed slight performance degradation with increasing number of processors attributed to the growth of the maximum interface size in the partitioning of the domain.[9]

In the case of higher-order finite-difference or finite-volume discretizations it is often convenient to associate each degree of freedom uniquely to a particular processor. In this situation, u_Γ, corresponds to layers of nodes/elements in the interface region, which may be split into groups u_{Γ_i} associated with a particular processor. In particular the degrees of freedom u_{Γ_i} are chosen such that A corresponding to $u_I^{(i)}$ have non-zero columns corresponding only to $u_I^{(i)}$ and u_{Γ_i}.[54] Thus (25) may be rewritten as:

$$\begin{bmatrix} A_{II}^{(1)} & 0 & A_{I\Gamma_1}^{(1)} & 0 \\ 0 & A_{II}^{(2)} & 0 & A_{I\Gamma_2}^{(2)} \\ A_{\Gamma_1 I}^{(1)} & 0 & A_{\Gamma_1\Gamma_2} & A_{\Gamma_1\Gamma_2} \\ 0 & A_{\Gamma_2 I}^{(2)} & A_{\Gamma_1\Gamma_2} & A_{\Gamma_1\Gamma_2} \end{bmatrix} \begin{bmatrix} u_I^{(1)} \\ u_I^{(2)} \\ u_{\Gamma_1} \\ u_{\Gamma_2} \end{bmatrix} = \begin{bmatrix} f_I^{(1)} \\ f_I^{(2)} \\ f_{\Gamma_1} \\ f_{\Gamma_2} \end{bmatrix}. \tag{36}$$

The corresponding Schur complement problem is given by:

$$\begin{bmatrix} S_{\Gamma_1\Gamma_2} & A_{\Gamma_1\Gamma_2} \\ A_{\Gamma_1\Gamma_2} & S_{\Gamma_1\Gamma_2} \end{bmatrix} \begin{bmatrix} u_{\Gamma_1} \\ u_{\Gamma_2} \end{bmatrix} = \begin{bmatrix} g_{\Gamma_1} \\ g_{\Gamma_2} \end{bmatrix}, \tag{37}$$

where $S_{\Gamma_i\Gamma_i} = A_{\Gamma_i\Gamma_i} - A_{\Gamma_i I}^{(i)} A_{II}^{(i)^{-1}} A_{I\Gamma_i}^{(i)}$. A simple block Jacobi preconditioner may then be applied to solve (37). Unfortunately, the convergence rate of this method identical to that obtained when applying a subdomain-wise block Jacobi preconditioner to the full system (36). However, if approximate factorizations are used for the local solvers and Schur complement, then an algorithm involving an inner iteration on the approximate Schur complement problem may provide a preconditioner for the full system (36). Such an approach was used by Hicken and Zingg for the solution of a finite difference discretization of the Euler equations.[54] Their algorithm involved an ILU factorization as a local solver, and solved the block-Jacobi preconditioned approximate Schur complement problem using GMRES as a preconditioner to fGMRES. Numerical results showed good strong scaling performance on up to 48 processors.

4. Neumann-Neumann Methods

In this section we present Neumann-Neumann methods which are a class of preconditioners for the Schur complement problem (30).[55–57] While all of the methods discussed thus far have employed blocks of the fully assembled discrete system as preconditioners, Neumann-Neumann methods exploit the finite element residual assembly. Namely, the discrete system of equations (25) may be obtained by assembling contributions from each subdomain of the form:

$$A^{(i)} = \begin{bmatrix} A_{II}^{(i)} & A_{I\Gamma}^{(i)} \\ A_{\Gamma I}^{(i)} & A_{\Gamma\Gamma}^{(i)} \end{bmatrix}, \qquad f^{(i)} = \begin{bmatrix} f_I^{(i)} \\ f_\Gamma^{(i)} \end{bmatrix}, \qquad i = 1, 2 \tag{38}$$

where $A_{\Gamma\Gamma} = A_{\Gamma\Gamma}^{(1)} + A_{\Gamma\Gamma}^{(2)}$ and $f_\Gamma = f_\Gamma^{(1)} + f_\Gamma^{(2)}$. The local problems:

$$\begin{bmatrix} A_{II}^{(i)} & A_{I\Gamma}^{(i)} \\ A_{\Gamma I}^{(i)} & A_{\Gamma\Gamma}^{(i)} \end{bmatrix} \begin{bmatrix} u_I^{(i)} \\ u_\Gamma^{(i)} \end{bmatrix} = \begin{bmatrix} f_I^{(i)} \\ f_\Gamma^{(i)} + \lambda_\Gamma^{(i)} \end{bmatrix}, \qquad i = 1, 2 \tag{39}$$

correspond to a discrete equivalent of the Neumann problems (22)-(24).

The Schur complement, S, may be also be written as sum of subdomain-wise contributions $S = S^{(1)} + S^{(2)}$, where $S^{(i)} = A_{\Gamma\Gamma}^{(i)} - A_{\Gamma I}^{(i)} A_{II}^{(i)^{-1}} A_{I\Gamma}^{(i)}$. In the simplest form, Neumann-Neumann methods precondition $S = S^{(1)} + S^{(2)}$ with $M_{NN}^{-1} = S^{(1)^{-1}} + S^{(2)^{-1}}$. In practice, diagonal scaling matrices $D^{(i)}$ are used to average nodal values on Γ, such that the Neumann-Neumann preconditioner is given by:

$$M_{NN}^{-1} = \begin{bmatrix} D^{(1)} & D^{(2)} \end{bmatrix} \begin{bmatrix} S^{(1)} & 0 \\ 0 & S^{(2)} \end{bmatrix}^{-1} \begin{bmatrix} D^{(1)} \\ D^{(2)} \end{bmatrix}. \tag{40}$$

where the diagonal values of the scaling matrices are chosen such that at each node the $D^{(i)}$'s sum to 1. For problems with widely varying coefficients across subdomains, the choice of diagonal scaling matrices can significantly impact the performance of the preconditioner.[58]

In order to extend the Neumann-Neumann preconditioner to the case of many subdomains we introduce some additional notation which will be used throughout this section. Consider the partition of the domain Ω into N nonoverlapping subdomains Ω_i, $i = 1, ..., N$. We define $\Gamma_i = \partial\Omega_i \backslash \partial\Omega$, and $\Gamma = \cup_{i=1}^N \Gamma_i$. We define R_i as the $\{0, 1\}$ matrix such that $R_i u_\Gamma$ is the restriction from u_Γ to the degrees of freedom on Γ_i. We may write the global Schur complement system as in (30) with

$$S = \sum_{i=1}^{N} R_i^T S^{(i)} R_i. \tag{41}$$

The extension of the Neumann-Neumann method to the case of many subdomains may be written using the following compact notation:

$$M_{NN}^{-1} = \sum_{i=1}^{N} R_i^T D^{(i)} {S^{(i)}}^{-1} D^{(i)} R_i. \tag{42}$$

In the basic form given in (42), the Neumann-Neumann preconditioner lacks a coarse space and hence is not scalable.[57] Additionally, if a subdomain Ω_i is strictly interior to Ω (i.e $\partial\Omega_i \cap \partial\Omega = \emptyset$), then $S^{(i)}$ is singular, since Ω_i is a "floating" subdomain upon which Neumann boundary conditions are imposed on all of $\partial\Omega_i$. In this case, ${S^{(i)}}^{-1}$ may replaced with a suitable pseudo-inverse or approximate solver, however the performance of the preconditioner will depend upon the particular choice of pseudo-inverse.[57,59,60] The Balancing Domain Decomposition (BDD) method introduced by Mandel[60] addressed the lack of scalability and the issues associated with choosing a suitable pseudo-inverse for singular subdomains, by introducing a coarse space based on the null-spaces of the local Schur complements $S^{(i)}$. The coarse correction step which is applied in a multiplicative manner is known as balancing, and is the origin of the term Balancing Domain Decomposition. The corresponding condition number of the preconditioned system is given by $\kappa = C\left(1 + \log\left(\frac{H}{h}\right)\right)^2$ for the symmetric elliptic problems, where the constant C can be shown to be independent of the coefficients of the problem.[58–60]

The BDD method is closely related to a dual substructuring method known as the Finite Element Tearing and Interconnecting (FETI) method, originally introduced by Farhat and Roux.[61] As opposed to directly enforcing the transmission condition $u_\Gamma^{(1)} = u_\Gamma^{(2)} = u_\Gamma$ by subassembling the global system as in (25), FETI methods enforce the transmission condition through the use of Lagrange multipliers. The Schur complement problem corresponding to the interface degrees of freedom may be written in the following equivalent form:

$$\begin{bmatrix} S^{(1)} & 0 & {B^{(1)}}^T \\ 0 & S^{(2)} & {B^{(2)}}^T \\ B^{(1)} & B^{(2)} & 0 \end{bmatrix} \begin{bmatrix} u_\Gamma^{(1)} \\ u_\Gamma^{(2)} \\ \lambda_\Gamma \end{bmatrix} = \begin{bmatrix} g_\Gamma^{(1)} \\ g_\Gamma^{(2)} \\ 0 \end{bmatrix}, \tag{43}$$

where λ_Γ are Lagrange multipliers which are the discrete equivalent of the flux $\frac{\partial u}{\partial \boldsymbol{n}_1}$ on Γ. We note that $B^{(1)}$ and $B^{(2)}$ are matrices with values of $\{0, 1, -1\}$ which ensure the condition $u_\Gamma^{(1)} = u_\Gamma^{(2)}$ is enforced through the last block equation of (43). In Neumann-Neumann methods, (40), the fully assembled Schur complement system, (30), is preconditioned using the upper-diagonal block of (43) obtained by dropping the rows and columns corresponding to the Lagrange multiplier λ_Γ. In FETI methods, on the other hand, the system, (43), is reduced to a system corresponding to only the Lagrange multipliers λ_Γ, which is preconditioned using the local Schur complement matrices.

In this paper we do not present FETI methods in detail but note that the FETI and BDD method are closely related, and have similar eigenvalue spectra.[62,63] FETI methods are among the most widely used and well tested methods for structural mechanics problems. For example Bhardwaj et al. used FETI methods to solve structural mechanics problems on up to 1000 processors.[64] FETI methods have also been analyzed for the case where inexact solvers are used.[65]

4.1. *BDDC and FETI-DP*

The most advanced of the FETI and Neumann-Neumann class of methods are the dual-primal FETI (FETI-DP)[66,67] and the Balancing Domain Decomposition by Constraints (BDDC) method.[68,69] Like FETI and BDD, FETI-DP and BDDC methods are closely related and have essentially the same eigenvalue spectra.[70,71]

A key component of FETI-DP and BDDC methods involves enforcing the continuity of a small number of "primal" degrees of freedom across subdomains. Strictly enforcing the continuity of the primal degrees of freedom naturally introduces a coarse space ensuring that the FETI-DP and BDDC methods are scalable. Additionally, the constraint on the continuity of the local subdomain problems ensures that the local problems are not singular. On each subdomain the degrees of freedom $u_\Gamma^{(i)}$ are partitioned into primal and dual degrees of freedom $u_\Pi^{(i)}$ and $u_\Delta^{(i)}$, where the primal degrees of freedom correspond to nodal values at subdomain corners, or averages along subdomain edges or faces. As opposed to directly enforcing the transmission condition $u_\Gamma^{(1)} = u_\Gamma^{(2)} = u_\Gamma$ by subassembling the global system a partially subassembled system is obtained by enforcing the continuity of only the primal degrees of freedom $u_\Pi^{(1)} = u_\Pi^{(2)} = u_\Pi$. The corresponding subassembled problem may be written as:

$$\begin{bmatrix} S_{\Delta\Delta}^{(1)} & 0 & S_{\Delta\Pi}^{(1)} & B_{\Delta}^{(1)^T} \\ 0 & S_{\Delta\Delta}^{(2)} & S_{\Delta\Pi}^{(2)} & B_{\Delta}^{(2)^T} \\ S_{\Pi\Delta}^{(1)} & S_{\Pi\Delta}^{(2)} & S_{\Pi\Pi} & 0 \\ B_{\Delta}^{(1)} & B_{\Delta}^{(2)} & 0 & 0 \end{bmatrix} \begin{bmatrix} u_{\Delta}^{(1)} \\ u_{\Delta}^{(2)} \\ u_{\Pi} \\ \lambda_{\Delta} \end{bmatrix} = \begin{bmatrix} g_{\Delta}^{(1)} \\ g_{\Delta}^{(2)} \\ g_{\Pi} \\ 0 \end{bmatrix}, \tag{44}$$

where $S_{\Pi\Pi} = S_{\Pi\Pi}^{(1)} + S_{\Pi\Pi}^{(2)}$ and $f_{\Pi} = f_{\Pi}^{(1)} + f_{\Pi}^{(2}$, while $B_{\Delta}^{(i)}$ are chosen such that the last row enforces $u_{\Delta}^{(1)} = u_{\Delta}^{(2)}$.

In the FETI-DP methods, the partially assembled system, (44) is reduced to a system for the Lagrange multipliers λ_{Δ}, which is preconditioned by solving local constrained Neumann problems corresponding to $S_{\Delta\Delta}^{(i)}$. Once again, we do not describe the FETI-DP method in detail, but refer the reader to the references provided. In BDDC methods, the upper-diagonal block of partially assembled system, (44), is used to precondition the fully assembled Schur complement problem, (30), by averaging $u_{\Delta}^{(i)}$'s. We write the BDDC preconditioner as:

$$M_{BDDC}^{-1} = \begin{bmatrix} D_{\Delta}^{(1)} & D_{\Delta}^{(2)} & 0 \\ 0 & 0 & I_{\Pi} \end{bmatrix} \begin{bmatrix} S_{\Delta\Delta}^{(1)} & 0 & S_{\Delta\Pi}^{(1)} \\ 0 & S_{\Delta\Delta}^{(2)} & S_{\Delta\Pi}^{(2)} \\ S_{\Pi\Delta}^{(1)} & S_{\Pi\Delta}^{(2)} & S_{\Pi\Pi} \end{bmatrix}^{-1} \begin{bmatrix} D_{\Delta}^{(1)} & 0 \\ D_{\Delta}^{(2)} & 0 \\ 0 & I_{\Pi} \end{bmatrix}, \tag{45}$$

where $D_{\Delta}^{(i)}$ are diagonal scaling matrices corresponding to the dual degrees of freedom u_{Δ}. Prior to extending the BDDC method to the case of many subdomains, we introduce some additional notation. Let $R_{\Delta,i}$ be the $\{0,1\}$ matrix which extract degrees of freedom $u_{\Delta}^{(i)}$ from the globally assembled interface vector u_{Γ}, (i.e. $u_{\Delta}^{(i)} = R_{\Delta,i}u_{\Gamma}$). Similarly, we define R_{Π} to be the matrix such that $u_{\Pi} = R_{\Pi}u_{\Gamma}$, while $R_{\Pi,i}$ is defined such that $u_{\Pi}^{(i)} = R_{\Pi,i}u_{\Pi}$. The solution of the partially assembled system in the BDDC preconditioner may be written as the sum of independent constrained Neumann solves corresponding to $S_{\Delta\Delta}^{(i)}$ and a coarse solve involving only the primal degrees of freedom u_{Π}. Namely, we may write the BDDC preconditioner for the case of many subdomains as:

$$M_{BDDC}^{-1} = \Psi S_0^{-1} \Psi^{*^T} + \sum_{i=1}^{N} R_{\Delta}^{(i)^T} D_{\Delta}^{(i)} S_{\Delta\Delta}^{-1} D_{\Delta}^{(i)} R_{\Delta}^{(i)}, \tag{46}$$

where S_0, Ψ and Ψ^* are given by:

$$S_0 = \sum_{i=1}^{N} R_{\Pi,i}^T \left(S_{\Pi\Pi}^{(i)} - S_{\Pi\Delta}^{(i)} S_{\Delta\Delta}^{(i)^{-1}} S_{\Delta\Pi}^{(i)} \right) R_{\Pi,i}, \tag{47}$$

$$\Psi = R_{\Pi}^T + \sum_{i=1}^{N} R_{\Delta,i}^T D_{\Delta}^{(i)} S_{\Delta\Delta}^{(i)^{-1}} S_{\Delta\Pi}^{(i)} R_{\Pi,i}, \tag{48}$$

$$\Psi^* = R_{\Pi}^T + \sum_{i=1}^{N} R_{\Delta,i}^T D_{\Delta}^{(i)} S_{\Delta\Delta}^{(i)^{-T}} S_{\Pi\Delta}^{(i)^T} R_{\Pi,i}. \tag{49}$$

The BDDC and FETI-DP methods are amongst the most successful domain decomposition methods for second order elliptic problems and problems of structural mechanics. The analysis of these preconditioners have been extended to the case where inexact solvers are used for the local Dirichlet and constrained Neumann problems.[72–74] Additionally, several authors have presented multi-level versions of the BDDC method when the coarse problem corresponding to S_0 may be too large to solve exactly.[75–77] An adaptive method for adding primal degrees of freedom to ensure rapid convergence has also been presented.[78] Practical implementations of the FETI-DP method has been used to solve structural mechanics problem on up to 3000 processor.[79,80]

The extension of FETI and Neumann-Neumann methods, (and thus FETI-DP and BDDC) to convection-diffusion problems involves modifying the interface conditions for the local subdomain problems to ensure that these local problems are well posed in the convective limit. In particular, imposing Neumann conditions on the inflow portion of a subdomain may lead to a singular system. Achdou et al. replaced the Neumann-Neumann interface condition with a Robin-Robin interface condition,[81] ensuring that the local bilinear forms were coercive. A Fourier analysis, on a vertical strip partitioning of the domain, showed that in the convective limit, the resulting algorithm converges in a number of iterations equal to half the number of subdomains in the streamwise direction. The Robin-Robin interface conditions have been used along with a FETI method to solve linear convection-diffusion problems by Toselli.[82] Similarly, Tu and Li used the Robin-Robin interface condition to extend the BDDC method to convection-diffusion problems.[83] Tu and Li introduced additional primal degrees of freedom corresponding to "flux" constraints and showed that the resulting BDDC algorithm was scalable if the subdomain length scale, H, was sufficiently small relative the viscosity. Namely, in a manner analogous to additive Schwarz methods, the behaviour of BDDC preconditioner matches the symmetric, diffusion dominated limit if the subdomain Peclet number is sufficiently small.

Neumann-Neumann and FETI methods have in general not been used for large scale CFD simulations, however recent work is beginning to make these methods available to the systems of equations for compressible flows. Dolean and collaborators have extended the Robin-Robin interface condition to the isentropic Euler equations using a Smith factorization.[84,85] Yano and Darmofal used a generalization of the Robin-Robin interface condition to the Euler equations based on entropy symmetrization theory.[86,87] They solved a higher-order continuous finite element discretization for two-dimensional subsonic flow using a BDDC preconditioner with up to 128 subdomains.

The success of BDDC and FETI-DP preconditioner for structural mechanics problems, and the initial results of Yano and Darmofal motivates further research into attempting to apply these types of preconditioners to large scale CFD simulations. While originally developed for linear conforming finite element methods, Neumann-Neumann type preconditioners have been extended to mixed methods,[88,89] discontinuous Galerkin discretizations[90] and higher-order spectral element methods.[91,92] We note that Neumann-Neumann type preconditioners exploit the finite element construction of the discrete system of equations, where subdomain contributions provide a discrete analog of the continuous Neumann problems (22)-(24). For finite-difference or finite-volume discretizations which do not naturally have such a finite element construction the choice for the local discrete Neumann problems and the analogy to the continuous Neumann problem is unclear. These issues need to be addressed in the context of simple model problems prior to consider using Neumann-Neumann type methods for these types of discretizations.

5. Numerical Results

In this final section we present numerical results using different preconditioning methods discussed for the solution of a higher-order hybridizable discontinuous Galerkin (HDG) discretization. The HDG discretization was recently introduced for the solution of the Poisson problem[93] , then extended to convection-diffusion equations[94] and the compressible Euler and Navier-Stokes equations.[95] The HDG discretization is a mixed method where both the state variable and its gradient are approximated separately on each element. A unique value of the trace of the state variable is obtained by enforcing the continuity of the flux on element boundaries leading to a reduced system of equations where the only globally coupled degrees of freedom are associated with the trace values on element faces.

We solve the following convection-diffusion problem in the square domain $\Omega \in \mathbb{R}^2$ given by $\Omega = [0,1] \times [0,1]$:

$$\nabla \cdot (\boldsymbol{c}u) - \kappa \Delta u = f \qquad \text{in } \Omega \subset \mathbb{R}^2, \tag{50}$$

where $u(x,y)$ is the state, $\boldsymbol{c} = (1,0)$ is the convective velocity, and κ is the viscosity. Here f is a source function set such that the exact solution is given by:

$$u = e^{-y/\sqrt{\kappa \bar{x}}} \qquad \text{where } \bar{x} = x + 0.1. \tag{51}$$

The linear system resulting from the HDG discretization is solved using a right-preconditioned GMRES method. We examine the performance of three different parallel preconditioners: a minimum-overlap additive Schwarz preconditioner without a coarse space (ASM); a minimum-overlap additive Schwarz preconditioner with a coarse space added in a multiplicative manner (ASM_0); and a nonoverlapping BDDC preconditioner. As the globally coupled degrees of freedom of the HDG discretization correspond to the element edges the communication pattern in the application of the overlapping preconditioner is essentially the same as for a nonoverlapping method. A coarse space for the additive Schwarz preconditioner is defined using an algebraic multigrid approach where edges are agglomerated by using the graph partitioning algorithm ParMETIS,[96] resulting in an agglomeration of edges independent of the original partitioning of the domain. The number of agglomerated edges is chosen such that the resulting coarse space contains about twice the number of degrees of freedom as the corresponding coarse space for the BDDC preconditioner. For the BDDC preconditioner the coarse space is defined by choosing as primal degrees of freedom the average of the state along the interfaces between subdomains. The corresponding dual degrees of freedom have zero average on subdomain interfaces. A particular advantage of the BDDC preconditioner is the simple algebraic construction of the coarse space given the original partition of the domain. In the context of the HDG discretization, this is particularly important as multigrid type algorithms have not been studied for this type of discretization.

Numerical experiments are presented to show the performance of the three preconditioners over a large range of viscosity, κ, highlighting the difference between the diffusion- and convection-dominated limits. While CPU time is the most appropriate metric for the comparison of different algorithms, the CPU time is closely tied to a particular implementation of the algorithm. In order to avoid these implementation dependent comparisons, the performance of the preconditioners are presented in terms of the number of iterations required for the GMRES algorithm. The relative computational cost may be estimated by taking into consideration the cost of each Krylov iteration. In particular, a single Krylov iteration involves:

- ASM: one Jacobian multiplication and one local Dirichlet solve on each subdomain
- ASM_0: two Jacobian multiplications and one local Dirichlet solve on each subdomain and a global coarse solve
- BDDC: one Jacobian multiplication, two local Dirichlet solves and one local constrained Neumann solve on each subdomain and a global coarse solve.

For the numerical experiments presented, exact solvers are used for the local and global problems. In a practical setting, approximate solvers would be employed, and thus the relative cost of the preconditioners will, in general, depend upon the choice of approximate solver. In particular, for the BDDC preconditioner, if a triangular factorization is employed as the local solver, the two Dirichlet solves may be replaced by one forward- and one back-substitution resulting in a cost equivalent to only a single local solve. As a reference for comparing the different preconditioners, we may consider using a local solver which has a computational cost which is the same as the cost of applying the Jacobian matrix, while we assume that the cost of the global solve is insignificant in comparison to the local solves. Thus the relative cost of a Krylov iteration for the ASM, ASM_0 and BDDC preconditioners is approximately 2:3:3.

In the first numerical experiment, we solve the convection-diffusion problem, (50), on a structured mesh. The domain Ω is partitioned into N square subdomains in an $\sqrt{N} \times \sqrt{N}$ structured pattern. Locally, each subdomain consists of n elements obtained by dividing Ω into squares of equal size and splitting each square into two triangular elements. We examine the performance of the preconditioners varying N and n, for higher order solution with $p = 2$ and $p = 5$. Tables 1 and 2 show the number of GMRES iterations required to converged the l_2-norm of the residual by a factor of 10^4 for $\kappa = 1$ and $\kappa = 10^{-6}$, respectively.

Table 1. Number of GMRES iterations for $\kappa = 1$ on isotropic structured mesh.

		$p=2$			$p=5$		
N	n	ASM	ASM_0	BDDC	ASM	ASM_0	BDDC
4	128	23	21	5	28	25	5
16	128	45	31	8	55	38	10
64	128	86	41	9	103	51	11
256	128	168	47	10	201	59	11
1024	128	329	50	10	393	62	12
64	8	42	21	8	51	26	9
64	32	61	29	8	73	36	11
64	128	86	41	9	103	51	11
64	512	122	58	11	146	71	12
64	2048	172	82	11	205	100	13

Table 2. Number of GMRES iterations for $\kappa = 10^{-6}$ on isotropic structured mesh.

		$p=2$			$p=5$		
N	n	ASM	ASM_0	BDDC	ASM	ASM_0	BDDC
4	128	3	5	1	3	5	1
16	128	5	10	2	5	9	3
64	128	9	15	4	9	15	5
256	128	17	27	9	17	26	10
1024	128	33	48	18	33	47	19
64	8	9	12	4	9	11	4
64	32	9	14	4	9	14	5
64	128	9	15	4	9	15	5
64	512	9	15	5	9	15	6
64	2048	9	16	5	9	16	6

In the diffusion-dominated limit, ($\kappa = 1$), for a fixed number of elements per subdomain, $n = 128$, the performance of the ASM preconditioner degrades as the number of subdomains, N, increases. This behaviour is due to a lack of a coarse space able to control the lower frequency error modes. On the other hand, for the ASM_0 and BDDC preconditioners, the number of iterations appears to be bounded as the number of subdomains increases. For a fixed number of subdomains $N = 64$, the performance of the ASM_0 preconditioner degrades rapidly with increasing number of elements, due to the non-optimality of this preconditioner in the case of small overlap. On the other hand the number of iterations for the BDDC preconditioner which is quasi-optimal increases only slowly with increasing number of elements.

In the convection-dominated limit, ($\kappa = 10^{-6}$), all three preconditioners converge in a small number of iterations, proportional to the number of subdomains in the streamwise direction ($\sqrt{N}$). For this test case, the boundary layer region is not resolved, and hence diffusive effects are relevant only on the subdomains along the bottom wall. In particular, a coarse space is not justified as the ASM method converges in fewer iterations than the more expensive ASM_0 preconditioner. Additionally, we note that in the convection-dominated limit the number of iterations to converge appears essentially independent of the number of elements per subdomain for all three preconditioners.

While the results of Table 2 suggest that a coarse space may not be necessary for convection-dominated problems, the diffusive effects are masked by the lack of resolution in the boundary layer region. In practice, a significant portion of the mesh should be clustered near the bottom surface to ensure that the boundary layer region is fully resolved. In a second numerical experiment an anisotropic boundary layer mesh is employed, with uniform spacing in the x-direction and an exponential spacing in the y-direction. The aspect ratio of the elements at $y = 0$ is given by $AR = 1/\sqrt{Pe}$, where $Pe = |\boldsymbol{c}|/\kappa$ is the Peclet number. Table 3 shows the number of iterations required for the GMRES algorithm to converge by a factor of 10^4 for $\kappa = 10^{-6}$. As a significant portion of the mesh is in the boundary layer region, diffusive effects become more important. Compared to Table 2, the performance of the ASM preconditioner without a coarse space is seen to degrade relative to the ASM_0 and BDDC preconditioners.

In Table 4 we show the performance on both the isotropic and anisotropic meshes over a range of viscosities, for fixed N and n. On the isotropic meshes, the relative performance of the ASM preconditioner without coarse space improves rapidly as the viscosity is reduced. However, on these meshes the boundary layer region is under-resolved. On the other hand, for the anisotropic meshes on which the boundary layer is resolved a coarse space is important throughout the range of viscosities.

Table 3. Number of GMRES iterations for $\kappa = 10^{-6}$ on anisotropic structured mesh.

			$p=2$			$p=5$	
N	n	ASM	ASM_0	BDDC	ASM	ASM_0	BDDC
4	128	3	5	1	3	7	1
16	128	17	15	5	18	16	5
64	128	33	25	8	34	27	8
256	128	70	41	13	71	44	14
1024	128	170	75	35	172	80	36
64	8	18	15	8	19	16	8
64	32	24	18	8	24	20	8
64	128	33	25	8	34	27	8
64	512	50	35	8	51	38	8
64	2048	76	56	8	77	58	8

Table 4. Number of GMRES iterations on both isotropic and anisotropic meshes with $N = 256$, $n = 8$.

		$p = 2$			$p = 5$		
	κ	ASM	ASM_0	BDDC	ASM	ASM_0	BDDC
Isotropic Mesh	1	86	41	9	103	51	11
	10^{-1}	77	45	9	93	56	11
	10^{-2}	34	32	10	40	40	12
	10^{-4}	15	20	7	17	22	9
	10^{-4}	9	16	7	10	19	7
	10^{-5}	9	15	5	9	15	6
	10^{-6}	9	15	4	9	15	5
Anisotropic Mesh	1	86	41	9	103	51	11
	10^{-1}	90	42	11	111	54	13
	10^{-2}	71	39	11	81	47	14
	10^{-3}	53	31	9	57	34	11
	10^{-4}	43	29	8	44	31	9
	10^{-5}	37	27	9	37	29	9
	10^{-6}	33	25	8	34	27	8

Table 5. Number of GMRES iterations on unstructured anisotropic meshes with $\kappa = 10^{-3}$, $n \sim 370$.

	$p = 2$			$p = 5$		
N	ASM	ASM_0	BDDC	ASM	ASM_0	BDDC
4	64	56	12	84	76	15
16	162	111	35	196	146	44
64	418	214	79	491	270	94
256	> 1000	377	158	> 1000	461	187

In the final numerical experiment we show the performance of the three preconditioners on unstructured meshes. We solve the convection-diffusion problem with $\kappa = 10^{-3}$. A family of four anisotropic meshes with 1475, 5992, 23492, and 94313 elements were generated using the Bidimensional Anisotropic Mesh Generator (BAMG),[97] where the anisotropic metric was determined by the Hessian of the exact solution, (51). The meshes are partitioned using the ParMETIS package of Karypis,[96] into 4, 16, 64 and 256 subdomains, resulting in each subdomain having approximately 370 elements. Table 5 shows the resulting performance of the three preconditioners. Unfortunately, the performance of the preconditioners for the unstructured case is, in general, much poorer than the unstructured case. However, the importance of a coarse space is highlighted even in this test case.

In summary, the numerical results presented show that the ASM_0 and BDDC preconditioners equipped with coarse spaces perform much better than the ASM preconditioner without a coarse space even in the convection dominated limit. In particular, the performance of BDDC preconditioner is only weakly dependent upon the number of elements per subdomain, and thus is expected to perform better than the ASM_0 preconditioner as the size of the subdomains in increased. Finally, we note that for the numerical test cases presented the performance of both ASM_0 and BDDC preconditioners appear to be only weakly dependent upon p. In particular, for the convection-dominated, ($\kappa = 10^{-6}$), test cases the number of iterations for $p = 2$ and $p = 5$ are essentially the same. Thus, these types of preconditioners may be suited to higher-order CFD simulations.

References

1. D. J. Mavriplis, D. Darmofal, D. Keyes, and M. Turner. AIAA 2007-4084, (2007).
2. G. Amdahl. In *AFIPS Conference Proceedings*, vol. 30, pp. 483–485. AFIPS Press, Reston, Va, (1967).
3. J. L. Gustafson, *Communications of the ACM.* **31**, 532–533, (1988).
4. C. T. Kelley and D. E. Keyes, *SIAM J. Numer. Anal.* **35**(2), 508–523, (1998).
5. W. Anderson, R. Rausch, and D. Bonhaus. AIAA 1995-1740, (1995).
6. V. Venkatakrishnan. ICASE 95-28, (1995).
7. X.-C. Cai, W. D. Gropp, D. E. Keyes, and M. D. Tidriri. pp. 17–30. Proceedings of the International Workshop on Numerical Methods for the Navier-Stokes Equations, (1995).
8. D. J. Mavriplis. AIAA 1998-2966, (1998).
9. T. J. Barth, T. F. Chan, and W.-P. Tang, *Contemp. Math.* **218**, 23–41, (1998).
10. W. Gropp, D. K. Kaushik, B. F. Smith, and D. E. Keyes. In *HiPC '00: 7th Int. Conf. on HPC*, pp. 395–404. Springer-Verlag, (2000).
11. W. Gropp, D. Keyes, L. C. Mcinnes, and M. D. Tidriri, *Int. J. High Perform. Comput. Appl.* **14**(2), 102–136, (2000).
12. D. A. Knoll and D. E. Keyes, *J. Comput. Phys.* **193**(1), 357–397, (2004).
13. A. Nejat and C. Ollivier-Gooch. AIAA 2007-0719 (Jan., 2007).
14. D. J. Mavriplis, *J. Comput. Phys.* **145**, 141–165, (1998).
15. D. J. Mavriplis and S. Pirzadeh, *AIAA J. Aircraft.* **36**, 987–998, (1999).
16. K. J. Fidkowski, T. A. Oliver, J. Lu, and D. L. Darmofal, *J. Comput. Phys.* **207**(1), 92–113, (2005).
17. C. R. Nastase and D. J. Mavriplis, *J. Comput. Phys.* **213**(1), 330–357, (2006).
18. C. R. Nastase and D. J. Mavriplis. AIAA 2007-0512, (2007).
19. P.-O. Persson and J. Peraire, *SIAM J. Sci. Comput.* **30**(6), 2709–2722, (2008).
20. D. J. Mavriplis, *J. Comput. Phys.* **175**(1), 302–325, (2002).
21. B. Smith, P. Bjorstad, and W. Gropp, *Domain Decomposition: Parallel Multilevel Methods for Elliptic Partial Differential Equations.* (Cambridge University Press, New York, NY, 1996).

22. A. Toselli and O. Widlund, *Domain Decomposition Methods Algorithm and Theory.* (Springer-Verlag, 2005).
23. M. Dryja and O. Widlund. Tech report 339, Department of Computer Science, Courant Institute, (1987).
24. X.-C. Cai, *SIAM J. Sci. Comput.* **14**(1), 239–247, (1993).
25. X.-C. Cai. In *Third International Symposium on Domain Decomposition Methods for Partial Differential Equations*, pp. 232–244, Philidelphia, (1990).
26. X.-C. Cai. In *Domain-Based Parallelism and Problem Decomposition Methods in Computational Science and Engineering*, pp. 1–19. SIAM, (1995).
27. X.-C. Cai, *Numer. Math.* **60**, 41–61, (1991).
28. X.-C. Cai, *SIAM J. Sci. Comput.* **15**(3), 587–603, (1994).
29. S. C. Brenner, *Math. Comp.* **65**(215), 897–921, (1996).
30. M. A. Casarin, *SIAM J. Numer. Anal.* **34**(6), 2482–2502, (1997).
31. X. Feng and O. A. Karakashian, *SIAM J. Numer. Anal.* **39**(4), 1343–1365, (2002).
32. P. F. Antonietti and B. Ayuso. In *Domain Decomposition Methods in Science and Engineering*, vol. 60, pp. 185–192. Springer Berlin, (2008).
33. P. F. Antonietti and B. Ayuso, *Math. Model. Numer. Anal.* **41**(1), 21–54, (2007).
34. P. F. Antonietti and B. Ayuso. In *Communications in Computational Physics*, vol. 5, pp. 398–412, (2009).
35. C. Lasser and A. Toselli, *Math. Comp.* **72**, 1215–1238, (2003).
36. X.-C. Cai, W. D. Gropp, and D. E. Keyes, *J. Comput. Phys.* **157**, 1765–1774, (2000).
37. J. W. Lottes and P. F. Fischer, *J. Sci. Comput.* **24**(1), 45–78, (2005).
38. L. Olson, J. Hesthaven, and L. Wilcox. pp. 325–332. Domain Decomposition Methods in Science and Engineering XVI, (2007).
39. X.-C. Cai and M. Sarkis, *SIAM J. Sci. Comput.* **21**(2), 792–797, (1999).
40. X.-C. Cai, C. Farhat, and M. Sarkis, *Contemp. Math.* **218**, 479–485, (1998).
41. X.-C. Cai, C. Farhat, and M. Sarkis. ICASE 96-48, (1996).
42. X.-C. Cai, W. D. Gropp, D. E. Keyes, and M. D. Tidriri. In *Domain Decomposition Methods in Science and Engineering.* John Wiley & Sons, (1997).
43. P.-O. Persson. AIAA 2009-606, (2009).
44. W. K. Anderson, W. D. Gropp, D. K. Kaushik, D. E. Keyes, and B. F. Smith. In *Proceedings of SC99*, pp. 69–80. Portland, OR, (1999).
45. L. T. Diosady. A linear multigrid preconditioner for the solution of the Navier-Stokes equations using a discontinuous Galerkin discretization. Masters thesis, Mass. Inst. of Tech., CDO (May, 2007).
46. A. Quarteroni and A. Valli, *Domain Decomposition Methods for Partial Differential Equations.* (Oxford, New York, 1999).
47. J. H. Bramble, J. E. Pasciak, and A. H. Schatz, *Math. Comp.* **47**(175), 103–134 (July, 1986).
48. J. H. Bramble, J. E. Pasciak, and A. H. Schatz, *Math. Comp.* **49**(179), 1–16 (July, 1987).
49. J. H. Bramble, J. E. Pasciak, and A. H. Schatz, *Math. Comp.* **51**(184), 415–430 (October, 1988).

50. J. H. Bramble, J. E. Pasciak, and A. H. Schatz, *Math. Comp.* **53**(187), 1–24 (July, 1989).
51. X.-C. Cai, W. D. Gropp, and D. E. Keyes, *Numer. Math.* **61**, 153–169, (1992).
52. W. Gropp and D. Keyes, *Internat. J. Numer. Methods Fluids.* **14**, 147–165, (1992).
53. Y. Saad, *SIAM J. Sci. Comput.* **14**(2), 461–469, (1993).
54. J. E. Hicken and D. W. Zingg. AIAA 2007-4333, (2007).
55. J.-F. Bourgat, R. Glowinski, P. L. Tallec, and M. Vidrascu. In eds. T. Chan, R. Glowinski, J. Periaux, and O. Widlund, *Domain decomposition methods. Second international symposium on domain decomposition methods*, pp. 3–16. SIAM, (1988).
56. P. L. Tallec, Y. D. Roeck, and M. Vidrascu, *J. Comput. Appl. Math.* **34**, 93–117, (1991).
57. Y.-H. DeRoeck and P. LeTallec. In *Fourth International Symposium on Domain Decomposition Methods for Partial Differential Equations*, pp. 112–128, Philidelphia, PA, (1991). SIAM.
58. J. Mandel and M. Brezina, *Math. Comp.* **65**(216), 1387–1401, (1996).
59. M. Dryja, *Comm. Pure Appl. Math.* **48**, 121–155, (1995).
60. J. Mandel, *Comm. Numer. Methods Engrg.* **9**, 233–241, (1993).
61. C. Farhat and F.-X. Roux, *Internat. J. Numer. Methods Engrg.* **32**, 1205–1227, (1991).
62. C. Farhat, J. Mandel, and F.-X. Roux, *Comput. Methods Appl. Mech. Engrg.* **115**, 365–385, (1994).
63. Y. Fragakis and M. Papadrakakis, *Comput. Methods Appl. Mech. Engrg.* **192**, 3799–3830, (2003).
64. M. Bhardwaj, D. Day, C. Farhat, M. Lesoinne, K. Pierson, and D. Rixen, *Int. J. Numer. Meth. Engng.* **47**, 513–535, (2000).
65. A. Klawonn and O. B. Widlund, *SIAM J. Sci. Comput.* **22**(4), 1199–1219, (2000).
66. C. Farhat, M. Lesoinne, P. LeTallec, K. Pierson, and D. Rixen, *Internat. J. Numer. Methods Engrg.* **50**, 1523–1544, (2001).
67. J. Mandel and R. Tezaur, *Numer. Math.* **88**, 543–558, (2001).
68. C. R. Dohrmann, *SIAM J. Sci. Comput.* **25**(1), 246–258, (2003).
69. J. Mandel and C. R. Dohrmann, *Numer. Linear Algebra Appl.* **10**, 639–659, (2003).
70. J. Mandel, C. R. Dohrmann, and R. Tezaur, *Appl. Numer. Math.* **54**, 167–193, (2005).
71. J. Li and O. B. Widlund, *Internat. J. Numer. Methods Engrg.* **66**, 250–271, (2006).
72. C. R. Dohrmann, *Numer. Linear Algebra Appl.* **14**, 149–168, (2007).
73. J. Li and O. B. Widlund, *Comput. Methods Appl. Mech. Engrg.* **196**, 1415–1428, (2007).
74. A. Klawonn and O. Rheinbach, *Internat. J. Numer. Methods Engrg.* **69**, 284–307, (2007).
75. J. Mandel, B. Sousedik, and C. R. Dohrmann, *Lecture Notes in Computational Science and Engineering.* **60**, 287–294, (2008).

76. J. Mandel and B. Sousedik, *Computing.* **83**, 55–85, (2008).
77. X. Tu, *SIAM J. Sci. Comput.* **29**(4), 1759–1780, (2007).
78. J. Mandel and B. Sousedik, *Comput. Methods Appl. Mech. Engrg.* **196**, 1389–1399, (2007).
79. M. Bhardwaj, K. Pierson, G. Reese, T. Walsh, D. Day, K. Alvin, J. Peery, C. Farhat, and M. Lesoinne. In *Proceedings of the 2002 ACM/IEEE conference on supercomputing*, pp. 35–53, Baltimore, MD, (2002).
80. K. H. Pierson, G. M. Reese, M. K. Bhardwaj, T. F. Walsh, and D. M. Day. Sandia National Laboratories SAND2002-1371, (2002).
81. Y. Achdou, P. L. Tallec, F. Nataf, and M. Vidrascu, *Comput. Methods Appl. Mech. Engrg.* **184**, 145–170, (2000).
82. A. Toselli, *Comput. Methods Appl. Mech. Engrg.* **190**, 5759–5776, (2001).
83. X. Tu and J. Li, *Comm. Appl. Math. Comp. Sci.* **3**(1), 25–60, (2008).
84. V. Dolean, F. Nataf, and G. Rapin, *Comptes Rendus Mathematique.* **340**(9), 693 – 696, (2005).
85. V. Dolean and F. Nataf, *Math. Model. Numer. Anal.* **40**(4), 689–704, (2006).
86. M. Yano. Massively parallel solver for the high-order Galerkin least-squares method. Masters thesis, Mass. Inst. of Tech., CDO (May, 2009).
87. M. Yano and D. Darmofal, *Comput. Methods Appl. Mech. Engrg.* (2010).
88. X. Tu, *Electron. Trans. Numer. Anal.* **20**, 164–179, (2005).
89. X. Tu, *Electron. Trans. Numer. Anal.* **26**, 146–160, (2007).
90. M. Dryja, J. Galvis, and M. Sarkis, *J. Complexity.* **23**(4), 715–739, (2007).
91. A. Toselli and X. Vasseur, *IMA Journal on Numerical Analysis.* **24**, 123–156, (2004).
92. A. Klawonn, L. F. Pavarino, and O. Rheinbach, *Comput. Methods Appl. Mech. Engrg.* **198**, 511–523, (2008).
93. B. Cockburn, J. Gopalakrishnan, and R. Lazarov, *SIAM J. Numer. Anal.* **47** (2), 1319–1365, (2009).
94. N. Nguyen, J. Peraire, and B. Cockburn, *J. Comput. Phys.* **228**(9), 3232–3254, (2009).
95. J. Peraire, N. Nguyen, and B. Cockburn. AIAA 2010-363, (2010).
96. G. Karypis. Parmetis: Parallel graph partitioning and sparse matrix ordering library, (2006). http://glaros.dtc.umn.edu/gkhome/views/metis/parmetis.
97. F. Hecht. Bamg: Bidimensional anisotropic mesh generator, (1998). `http://www-rocq1.inria.fr/gamma/cdrom/www/bamg/eng.htm`.

CHAPTER 3

ERROR ESTIMATION AND *hp*-ADAPTIVE MESH REFINEMENT FOR DISCONTINUOUS GALERKIN METHODS

Tobias Leicht* and Ralf Hartmann†

German Aerospace Center (DLR),
Institute of Aerodynamics and Flow Technology,
Lilienthalplatz 7, 38108 Braunschweig, Germany
**tobias.leicht@dlr.de*
†ralf.hartmann@dlr.de

We present adjoint-based techniques to estimate the error of a numerical flow solution with respect to a given target quantity like an aerodynamic force coefficient. This estimate can be used to judge the overall accuracy of a computation, to enhance the computed value of the target quantity and to drive a solution-adaptive mesh refinement process. The error estimation procedure is extended to multiple target quantities.

The discontinuous ansatz spaces of the DG discretization allow for both element subdivision as well as a local increase of polynomial degrees for increasing the flow resolution. Targeting optimal rates of convergence, a smoothness estimation based on a truncated Legendre series expansion of the solution is employed to locally select the more promising strategy.

Numerical examples for inviscid, laminar viscous and turbulent viscous flows demonstrate the efficiency of the proposed algorithms.

1. Introduction

The past few years have seen considerable progress in the development of higher order discontinuous Galerkin (DG) methods for aerodynamic flows, see the references cited in this volume. Here, we are especially interested in the fact that DG methods offer a great flexibility in computing numerical solutions of selectable arbitrary design order. However, this design order only pays off if the problem at hand is smooth enough, as in that case the order of convergence is also increased. For most flow fields this is not the case, however. Thus, given that the computational cost per degree of freedom grows with the order of the method, a rather low order will often yield the best overall results. It is difficult to determine this order *a priori* and often the optimal order might vary throughout the computational domain. This gives rise to *hp*-adaptive methods which target the optimal mesh density and polynomial degree distribution to yield efficient results.

Aerodynamic force coefficients like the drag and lift as well as moment coefficients are important quantities in aerodynamic flow simulations. In addition to the exact approximation of these quantities it is of increasing importance, in particular in the field of uncertainty quantification, to estimate the error in the computed quantities. The finite element background of DG methods provides a substantial and rigorous error estimation framework. By employing a duality argument, error estimates can be derived for outputs such as aerodynamic force coefficients. The error estimate includes primal residuals multiplied by the solution to an adjoint problem related to the force coefficient. The error estimate can be decomposed into a sum of local adjoint-based indicators which can be employed to drive a goal-oriented adaptive mesh refinement algorithm specifically tailored to the accurate and efficient approximation of the aerodynamic force coefficient under consideration.

The approach of error estimation and goal-oriented mesh refinement for specific target quantities has been developed by Becker and Rannacher,[1,2] see also the work of Giles and Süli.[3] It has been transferred to compressible flows in the context of DG methods in Hartmann and Houston[4] for inviscid flows and extended in Lu[5] and Hartmann and Houston[6,7] to viscous laminar flows; we refer to Venditti and Darmofal[8] and Barth and Larson[9] for related work based on finite volume methods as well as to Pierce and Giles[10] who considered general discretizations. Subsequently, this approach has been combined with anisotropic hierarchic refinement for laminar compressible flows, see Refs. 11 and 12, and with a regeneration of output-adapted meshes using anisotropic mesh metrics by Oliver,[13] see also the related work of Venditti and Darmofal.[14] Furthermore, the adjoint-based error estimation and mesh refinement approach has been extended from single to multiple target quantities in Hartmann and Houston.[15,16] Whereas the above examples are based on body-aligned regular meshes, the adjoint-based mesh refinement has also been applied to embedded-boundary Cartesian meshes by Nemec *et al.*[17,18] and to the simplex cut-cell approach by Fidkowski and Darmofal.[19] It has been extended to 2d turbulent flows governed by the RANS equations and the Spalart-Allmaras turbulence model by Oliver.[13] Recently, results have been presented for 3d laminar[12] and turbulent flows.[20]

Ideally, the computational mesh and the polynomial degree should be adapted simultaneously, leading to so-called *hp*-adaptive methods. While *hp*-refinement has been developed for a range of problems[21–25] there are only few publications on *hp*-refinement in the context of compressible flows, e.g. the *hp*-adaptation for 2d inviscid flows by Wang and Mavriplis[26] and for 2d laminar flows in Hartmann and Houston.[27]

In this chapter we present an *hp*-adaptive discontinuous Galerkin method for inviscid, laminar and turbulent flows. To this end, we start by recalling the underlying equations and their discretization as well as the error estimation procedure for single as well as multiple target quantities. We then consider a local criterion used to choose between h-subdivision and p-enrichment which is then combined with the adjoint-based error indicators to yield an *hp*-adaptive algorithm. This can further be enhanced by considering anisotropic h-subdivision. Finally, several numerical examples for subsonic test cases and a transonic flow field demonstrate the potential of this approach.

2. Flow Problem and Its Discretization

We consider turbulent compressible flows governed by the Reynolds-averaged Navier-Stokes and the Wilcox k-ω turbulence model equations, RANS-$k\omega$ equations for short,

$$\nabla \cdot (\mathcal{F}^c(\mathbf{u}) - \mathcal{F}^v(\mathbf{u}, \nabla \mathbf{u})) - \mathcal{S}(\mathbf{u}, \nabla \mathbf{u}) = 0 \quad \text{in } \Omega \in \mathbb{R}^d, \tag{1}$$

as introduced in Chap. 1. For discretizing these equations on the domain Ω we assume that Ω can be subdivided into shape-regular meshes $\mathcal{T}_h = \{\kappa\}$ consisting of (possibly curved) quadrilateral or hexahedral elements κ. Furthermore, we assume that each $\kappa \in \mathcal{T}_h$ is an image of a fixed reference element $\hat{\kappa}$, that is, $\kappa = \sigma_\kappa(\hat{\kappa})$ for all $\kappa \in \mathcal{T}_h$, where $\hat{\kappa}$ is the open unit square in $\mathbb{R}^2$ and the open unit cube in $\mathbb{R}^3$ and σ_κ is a smooth bijective mapping. In order to allow boundary elements to be curved the mapping σ_κ is constructed based on employing a higher-order polynomial representation of the computational boundary. Furthermore, we also allow interior elements to be curved in order to avoid the intersection of curved boundary lines with interior elements which might occur, see Landmann *et al.*,[28] for meshes with highly stretched elements as typically used for turbulent flows.

On the reference element $\hat{\kappa}$ we define the space of complete polynomials $\mathcal{P}_p$ and the space of tensor-product polynomials $\mathcal{Q}_p$ of degree $p \geq 0$ as follows:

$$\mathcal{P}_p = \operatorname{span}\{\hat{\mathbf{x}}^\alpha : 0 \leq |\alpha| \leq p\},$$
$$\mathcal{Q}_p = \operatorname{span}\{\hat{\mathbf{x}}^\alpha : 0 \leq \alpha_i \leq p,\ 0 \leq i \leq d\}.$$

We now introduce the finite element space $\mathbf{V}_{h,p}$ consisting of discontinuous vector-valued piecewise polynomial functions of degree $p \geq 0$,

$$\mathbf{V}_{h,p} = \{\mathbf{v}_h \in [L_2(\Omega)]^n : \mathbf{v}_h|_\kappa \circ \sigma_\kappa \in [\mathcal{P}_p(\hat{\kappa})]^n \text{ or } [\mathcal{Q}_p(\hat{\kappa})]^n,\ \kappa \in \mathcal{T}_h\},$$

where h and p indicate the local elemental mesh spacing and polynomial degree, respectively, and are not necessarily uniform throughout the mesh, and n denotes the number of equations in Eq. (1), which depends on the space dimension and the specific turbulence model.

Then the discontinuous Galerkin discretization of Eq. (1) is given by: Find $\mathbf{u}_h \in \mathbf{V}_{h,p}$ such that

$$\mathcal{N}(\mathbf{u}_h, \mathbf{v}_h) = 0 \qquad \forall \mathbf{v}_h \in \mathbf{V}_{h,p}. \tag{2}$$

Here, the semilinear form $\mathcal{N}$ is as given in detail in Hartmann and Houston[29] augmented with an additional term discretizing the source term $\mathcal{S}$. We use the symmetric interior penalization (SIPG) scheme[29] and the second scheme (BR2) of Bassi and Rebay.[30,31] For turbulent flows the value of ω at walls is determined by Menter's boundary condition.[32]

3. Error Estimation and Local Error Indicators

In the following we consider the estimation of errors in target quantities like aerodynamic force and moment coefficients. We start with single target quantities before extending this approach to multiple quantities.

3.1. *Single target quantities*

Given a target quantity $J(\mathbf{u})$, a duality argument can be employed resulting in following error representation

$$J(\mathbf{u}) - J(\mathbf{u}_h) = -\mathcal{N}(\mathbf{u}_h, \mathbf{z}) \equiv \mathcal{R}(\mathbf{u}_h, \mathbf{z}) \approx \mathcal{R}(\mathbf{u}_h, \tilde{\mathbf{z}}_h), \tag{3}$$

see e.g. Becker and Rannacher[2] or Hartmann and Houston.[4] Here, the exact (and unknown) adjoint solution $\mathbf{z}$ is replaced by the solution $\tilde{\mathbf{z}}_h$ to following discrete adjoint problem: Find $\tilde{\mathbf{z}}_h \in \tilde{\mathbf{V}}_{h,p}$ such that

$$\mathcal{N}'[\mathbf{u}_h](\mathbf{w}_h, \tilde{\mathbf{z}}_h) = J'[\mathbf{u}_h](\mathbf{w}_h) \quad \forall \mathbf{w}_h \in \tilde{\mathbf{V}}_{h,p}. \tag{4}$$

A possible choice of the adjoint discrete function space is $\tilde{\mathbf{V}}_{h,p} = \mathbf{V}_{h,\tilde{p}} = \mathbf{V}_{h,p+1}$. The approximate error representation in Eq. (3) can then be localized[a]

$$J(\mathbf{u}) - J(\mathbf{u}_h) \approx \mathcal{R}(\mathbf{u}_h, \tilde{\mathbf{z}}_h) \equiv \sum_{\kappa \in \mathcal{T}_h} \tilde{\eta}_\kappa, \tag{5}$$

where $\tilde{\eta}_\kappa$ are the so-called adjoint-based indicators, also called dual-weighted residual (DWR) indicators,[2] which include the local residuals multiplied by the discrete adjoint solution. These indicators can be used to drive an adaptive mesh refinement algorithm tailored to the accurate and efficient approximation of the target quantity $J(\mathbf{u})$ under consideration. Finally, the approximate error representation Eq. (5) can be used to enhance the computed target quantity $J(\mathbf{u}_h)$. This yields

$$\tilde{J}(\mathbf{u}_h) = J(\mathbf{u}_h) + \mathcal{R}(\mathbf{u}_h, \tilde{\mathbf{z}}_h). \tag{6}$$

3.2. *Multiple target quantities*

The extension of the adjoint-based error estimation and mesh refinement approach to multiple target quantities has originally been considered for the inviscid Burgers' equation by Hartmann and Houston[15] and has been extended to two-dimensional viscous laminar compressible flows by Hartmann.[16] Using the technique introduced above, an estimation of the error in multiple quantities of interest, $J_i(\mathbf{u}), i = 1, \ldots, N$, would require the computation of the solutions $\tilde{\mathbf{z}}_{h,i} \in \tilde{\mathbf{V}}_{h,p}$ to N discrete adjoint problems,

$$\mathcal{N}'[\mathbf{u}_h](\mathbf{w}_h, \tilde{\mathbf{z}}_{h,i}) = J_i'[\mathbf{u}_h](\mathbf{w}_h) \quad \forall \mathbf{w}_h \in \tilde{\mathbf{V}}_{h,p}, \quad i = 1, \ldots, N,$$

and the evaluation of the error representation for each of the quantities,

$$J(\mathbf{u}) - J(\mathbf{u}_h) \approx \mathcal{R}(\mathbf{u}_h, \tilde{\mathbf{z}}_{h,i}), \quad i = 1, \ldots, N.$$

Instead, we compute the solution to following discrete error equation: Find $\tilde{\mathbf{e}}_h \in \tilde{\mathbf{V}}_{h,p}$ such that

$$\mathcal{N}'[\mathbf{u}_h](\tilde{\mathbf{e}}_h, \mathbf{w}_h) = \mathcal{R}(\mathbf{u}_h, \mathbf{w}_h) \quad \forall \mathbf{w}_h \in \tilde{\mathbf{V}}_{h,p}, \tag{7}$$

and evaluate following approximation to $J_i(\mathbf{u}) - J_i(\mathbf{u}_h)$,

$$J_i(\mathbf{u}) - J_i(\mathbf{u}_h) \approx J_i'[\mathbf{u}_h](\mathbf{e}) \approx J_i'[\mathbf{u}_h](\tilde{\mathbf{e}}_h), \quad i = 1, \ldots, N, \tag{8}$$

[a]Galerkin orthogonality is a global property for continuous finite element methods. For DG methods it is also a local property. Due to this fact it is not necessary to substract an additional lower order approximation to the adjoint solution as described in Ref. 2.

where $\mathbf{e} = \mathbf{u} - \mathbf{u}_h$. Furthermore, based on a suitable combination $J_c(\mathbf{u})$ of the original target quantities, we compute the solution to following discrete adjoint problem: Find $\tilde{\mathbf{z}}_{c,h} \in \tilde{\mathbf{V}}_{h,p}$ such that

$$\mathcal{N}'[\mathbf{u}_h](\mathbf{w}_h, \tilde{\mathbf{z}}_{c,h}) = J_c'[\mathbf{u}_h](\mathbf{w}_h) \quad \forall \mathbf{w}_h \in \tilde{\mathbf{V}}_{h,p}, \tag{9}$$

and evaluate the error estimate

$$J_c(\mathbf{u}) - J_c(\mathbf{u}_h) = \mathcal{R}(\mathbf{u}_h, \mathbf{z}_c) \approx \mathcal{R}(\mathbf{u}_h, \tilde{\mathbf{z}}_{c,h}) \equiv \sum_{\kappa \in \mathcal{T}_h} \tilde{\eta}_\kappa^c. \tag{10}$$

The combined target quantity $J_c(\mathbf{u})$ can be defined such that the error with respect to $J_c(\cdot)$ represents the sum of relative errors in the original target quantities

$$\sum_{i=1}^{N} |J_i(\mathbf{u}) - J_i(\mathbf{u}_h)| / |J_i(\mathbf{u}_h)| \tag{11}$$

or a weighted sum of absolute errors $\sum_{i=1}^{N} \alpha_i |J_i(\mathbf{u}) - J_i(\mathbf{u}_h)|$ with weighting factors $\alpha_i > 0$. The adjoint-based indicators, $\tilde{\eta}_\kappa^c$, obtained by localizing the estimate Eq. (10) can be used to drive an adaptive algorithm for the accurate and efficient approximation of all the target quantities, $J_i(\mathbf{u}), i = 1, \ldots, N$, under consideration.

Finally, we note that the error estimates Eq. (8) can be used to enhance the computed target quantities $J_i(\mathbf{u}_h)$, $i = 1, \ldots, N$, as follows

$$\tilde{J}_i(\mathbf{u}_h) = J_i(\mathbf{u}_h) + J_i'[\mathbf{u}_h](\tilde{\mathbf{e}}_h), \qquad i = 1, \ldots, N. \tag{12}$$

4. Adaptation Strategies

In an adaptive mesh refinement algorithm the local error indicators presented above can be exploited to select those elements which produce the largest error for refinement. This refinement can be done in one of two distinct ways:

(a) h-subdivision: The edges of selected elements are split in half, forming smaller child elements with a reduced local mesh size h.
(b) p-enrichment: The degree p of the local polynomial expansion on the selected elements is increased.

4.1. *Comparison of h- and p-type mesh refinement*

In both cases the local resolution is enhanced through the introduction of additional degrees of freedom. The interesting feature of the h-subdivision approach is its simplicity. Apart from the fact that a local refinement might introduce hanging nodes at the interface between differently refined areas of the mesh the solution process is unchanged and all local operators are the same as on the initial mesh. This yields an adaptation algorithm which requires very little change to the flow solver itself. On the other hand an enrichment of the polynomial degree is not very complicated in the DG context either. In fact, due to the discontinuous nature of the ansatz space without formal continuity requirements the local operators simply have to be applied corresponding to the selected polynomial degree, including the selection of numerical quadrature formulæ. For interface terms the quadrature formula should be selected according to the accuracy requirements of the higher degree neighboring element. Due to varying sizes of local data structures the implementation is slightly more involved than in the h-subdivision case, but the additional effort is small compared to the potential gain.

In order to decide which strategy is the more promising we recall a standard interpolation error estimate.[33] For a function $u \in H^k(\tilde{\Omega})$ the approximation error, i.e. the difference between the function and the best interpolation or projection $\Pi_h^p u$ onto the discrete space, is bounded by

$$\|u - \Pi_h^p u\|_{L_2(\tilde{\Omega})} \leq C \frac{h^{\min(p+1,k)}}{p^k} \|u\|_{H^k(\tilde{\Omega})}, \tag{13}$$

where C depends on the regularity of the (initial) mesh and the Sobolev index k, but not on h and p. Since this is only an interpolation error estimate, $\tilde{\Omega}$ might denote a subdomain or even a single element.

In the case of (global) h-subdivision only the mesh size h changes on the right hand side of Eq. (13). In the asymptotic range the error is reduced by a constant factor whenever the mesh size is reduced by a given factor.

For very smooth functions with large Sobolev index k increasing the polynomial degree enlarges not only the denominator but also adds additional powers of the small mesh size h to the numerator, which yields convergence rates that increase with each refinement step. This exponential convergence is the clear benefit of p-enrichment. However, for functions with limited smoothness and a small corresponding Sobolev index the positive effect in the numerator is lost after a certain refinement step once the polynomial degree p is as large as k. The denominator continues to grow and the refinement algorithm converges at the same algebraic rate of convergence k that the h-subdivision would achieve.

Clearly, p-enrichment is preferable in the case of smooth functions. For functions with limited smoothness, however, this technique does not perform better than h-subdivision with respect to the number of degrees of freedom required to obtain a given accuracy. In contrast to that, the work per degree of freedom as well as storage requirements for all but the simplest explicit solution schemes increase with the polynomial degree, whereas stability and robustness of the solvers usually decrease due to worse conditioning of the resulting algebraic problems. Due to these effects, h-subdivision is preferable in those cases.

4.2. *Combined hp-refinement*

In practice, only very few aerodynamic problems have highly smooth solutions. However, mostly the non-smooth behavior is only due to local phenomena at shocks or sharp (trailing) edges. Thus, it is appealing to consider a combination of the above techniques, in which h-subdivision is used to refine non-smooth regions of the flow due to its improved computational efficiency. During this process, the non-smooth behavior is localized to a decreasing part of the domain. In the rest of the flow field p-enrichment can be employed to reduce the error more efficiently. In the ideal case an hp-refinement technique can recover an exponential rate of convergence also for non-smooth solutions.

4.3. *hp-indicator in 1D*

In addition to the error indicators already available, an hp-adaptive strategy requires an hp-indicator used to locally choose between h- or p-type refinement. If the local Sobolev index was known this decision would be simple, according to the reasoning above. As the exact solution is unknown, however, this information is not available. hp-strategies suggested in literature are mainly from two categories. The first category employs several trial refinements or auxiliary problems using both h- and p-type refinements. Based on cost and merit functions the more competitive strategy is selected locally, see e.g. the work of Rachowicz *et. al*[24] and Kurtz and Demkowicz.[25] Methods from the second category analyze the evolution of the coefficients of a Legendre series expansion of the solution, see e.g. the work of Mavriplis[34] or Houston *et al.*[23] Based on this analysis the smoothness of the function is estimated, enabling a selection of p-enrichment for smooth functions and an h-subdivision otherwise. Similar results can be obtained by considering the moments of high vs. low order modes of the solution as in Wang and Mavriplis.[26] We will only consider the second strategy as it involves only marginal computational overhead. Please refer to Houston and Süli[22] for an overview of other possible techniques.

The basic principle of our hp-indicator is fairly simple and was first introduced by Mavriplis[35] in the context of a spectral element method. Following the arguments of Houston and Süli[22] we do not estimate the actual local Sobolev index. Instead, we only try to decide whether u is a real analytic function, i.e. whether $u \in C^{\infty}$ or not. For simplicity, we consider a one-dimensional example on the standard interval $\mathcal{I} = (-1, 1)$, other intervals can be mapped to this standard one by a linear function. If $u \in L_2(\mathcal{I})$ is analytic, we can express it in an infinite Legendre series expansion,

$$u = \sum_{i=0}^{\infty} a^{(i)} L^{(i)},$$

where $a^{(i)}$ denotes the coefficient of the Legendre polynomial $L^{(i)}$ of degree i. The L_2-norm is then given by

$$\|u\|_{L_2(\mathcal{I})} = \sqrt{\int_{\mathcal{I}} |u|^2 \, \mathrm{d}x} = \sqrt{\sum_{i=0}^{\infty} \left(a^{(i)}\right)^2 \frac{2}{2i+1}} = \sqrt{\sum_{i=0}^{\infty} \left(b^{(i)}\right)^2} \tag{14}$$

due to the orthogonality of the Legendre polynomials. As this infinite sum has to be bounded, the modified coefficients $b^{(i)}$ implicitly given by Eq. (14) have to decay exponentially fast after some index i_0, i.e.

$$|b^{(i)}| = |a^{(i)}| \sqrt{\frac{2}{2i+1}} \leq C \exp(-\sigma i) \qquad \forall i > i_0,$$

with a positive $\sigma > 0$. To create a practical algorithm we try to estimate σ based on available data. For that purpose we consider a truncated Legendre series up to the degree $i_p = p$ of the polynomial ansatz space. We then assume that the coefficients of the numerical solution are sufficiently close to the coefficients of the exact solution and that the last n coefficients are in the asymptotic range $i > i_0$. Performing a least squares fit of $\log(b^{(i)})$ vs. i we obtain an approximate decay coefficient $\tilde{\sigma}$. If this is sufficiently large, i.e. larger than some threshold parameter σ_0, the underlying function is assumed to be analytic, and a p-enrichment should be used to refine this element, otherwise h-subdivision is deemed appropriate. Due to the very limited number of available coefficients it is advisable to choose the number of coefficients as the total number of available coefficients, i.e. $n = p+1$. Coefficients that are close to zero should be filtered out, however. In practice, any coefficients with absolute value smaller than a given tolerance are simply omitted from the least squares fit in order to obtain an average decay rate which is appropriate for the remaining coefficients.

Although there is a solid motivation for this spectral analysis strategy the success of this approach is still questionable if it is applied to analyze solutions of low polynomial order with only three or even only two Legendre series coefficients for piecewise quadratic or linear ansatz functions, respectively. Nevertheless, numerical experiments indicate that this approach yields good results in practice.

4.4. *hp-indicator in multiple dimensions and for systems of equations*

For multi-dimensional domains the solution can be expanded in multi-dimensional hierarchical polynomials. On simplex meshes the Proriol-Koornwinder-Dubiner polynomials are an appropriate choice. Here, we consider only quadrilateral and hexahedral meshes, thus multi-dimensional Legendre polynomials are obtained by the product of the standard 1D polynomials in the individual coordinate directions on the reference element. A representative 1D spectrum can be computed through the accumulation of all coefficients of the Legendre polynomials of the corresponding multi-dimensional degree. In 3D, the representative coefficients $b^{(i)}_{3\mathrm{D,tp}}$ for a tensor-product ansatz space are given by

$$b^{(i)}_{3\mathrm{D,tp}} = \sum_{\max(j,k,l)=i} |a^{(j,k,l)}| \sqrt{\frac{2}{2j+1}\frac{2}{2k+1}\frac{2}{2l+1}},$$

where $a^{(j,k,l)}$ is the coefficient corresponding to the Legendre polynomial $L^{(j,k,l)}(\mathbf{x}) = L^{(j)}(x)L^{(k)}(y)L^{(l)}(z)$. For complete polynomial spaces the representative coefficients $b^{(i)}_{3\mathrm{D,cp}}$ are given by

$$b^{(i)}_{3\mathrm{D,cp}} = \sum_{j+k+l=i} |a^{(j,k,l)}| \sqrt{\frac{2}{2j+1}\frac{2}{2k+1}\frac{2}{2l+1}},$$

instead. This ensures that $b^{(p)}_{3\mathrm{D}}$ contains contributions from those ansatz functions which are included in the set of basis functions of degree p, but not in the set for $p-1$. Using these representative 1D coefficients the above algorithm can be used without further modifications.

For vector-valued problems the estimation can be performed for each component, the resulting decay coefficient is then chosen as the minimum individual value, favoring h-subdivision if this is more appropriate for at least one vector component of the solution field.[b]

In case an adjoint-based strategy is used for the error estimation process the solution to the adjoint problem can also be analyzed for its smoothness. h-subdivision should then only be chosen if both the adjoint solution and the flow solution are deemed not to be analytical functions. If an error estimation for several target quantities is computed, the solution to the discrete error equation (7) can be added to the solution of the flow problem which then represents a higher order estimate of the flow solution including an additional Legendre series coefficient and should thus be used instead of the original flow solution. This technique draws an additional gain from the computed auxiliary problems.

The choice of the parameter σ_0 gives some freedom to tune the resulting algorithm. Mavriplis[34] suggests $\sigma_0 = 1$. In our experience, higher values might yield even better results for clearly non-smooth cases like flow fields including shocks, whereas smaller values might be sufficient in other cases, but in general $\sigma_0 = 1$ seems to be an appropriate choice in most cases. Smaller values favor p-enrichment. As the estimates of the asymptotic decay rates are in general not very accurate this might not yield exponential convergence, but employing p-enrichment often yields better results in the initial, non-asymptotic phase of the refinement process, thus a specific tolerance might still be reached faster than with an algorithm that features exponential convergence asymptotically, but behaves inferior in the initial phase.

4.5. *Anisotropic h-subdivision*

Important flow features might be strongly anisotropic. In our element subdivison approach we do not strive to create a mesh of strongly stretched elements from an isotropic initial one, as one could do in remeshing algorithms. Nevertheless, we realize that on any given mesh the dominant part of the

[b]For the the mean flow conservative variables the solution might be regarded as a comparatively small perturbation around free stream values, thus it seems appropriate to include the coefficient of the constant mode in the least squares fit. For the turbulence model variables this is not the case, however. At least two coefficients are required to determine a decay rate, thus it would be impossible to use linear initial solutions if the constant mode was excluded from the smoothness analysis. For the sake of simplicity, we simply restrict the smoothness estimation to the mean flow conservative variables.

local error might be reduced by adding resolution in only one direction. Whereas it is possible to selectively increase the directional polynomial degree for tensor-product polynomial basis functions, see Georgoulis *et al.*,[36] this is not possible with complete polynomial basis functions. Thus, we restrict ourselves to the possibility of anisotropic h-subdivision.

Here, we will only consider a very simple heuristic local criterion to decide whether splitting just a subset of an element's edges and thus modifying the child elements' aspect ratios is preferable over splitting all edges. In the latter case the refinement is isotropic in the sense that child elements inherit the aspect ratio of the mother element. A more elaborate approach based on an anisotropic extension of the adjoint-based error estimate for the case of constant polynomial degrees was suggested by Richter[37] in the context of continuous finite elements and has been applied to DG methods in Ref. 12.

One of the most characteristic features of DG methods is the discontinuity of its discrete solutions across the faces between neighboring elements. In smooth parts of the solution these inter-element jumps tend to zero with successive mesh refinement as the solution is approximated with less error. Based on this observation it seems justified to assume that a large jump indicates a larger error as compared to a smaller jump, in particular a large jump over a face indicates that the mesh size perpendicular to this face is too coarse to sufficiently resolve the solution.

The average jump K_i of a function ϕ over the two opposite faces f_i^j, $j = 1, 2$, perpendicular to one coordinate direction i on the reference element can be evaluated as

$$K_i = \frac{\sum_j \left| \int_{f_i^j} [\phi] \, \mathrm{d}s \right|}{\sum_j \mathrm{meas}(f_i^j)}, \quad i = 1, 2, 3, \tag{15}$$

where $[\phi] = \phi^+ - \phi^-$ denotes the jump of a scalar function ϕ and $\int \cdot \, \mathrm{d}s$ indicates a curve or surface integral in two or three dimensions, respectively. Equation (15) provides three distinct values for each element. Let K_m denote the maximum value of K_i, $i = 1, 2, 3$. We want to refine along each direction in which the average jump is not considerably smaller than K_m, measured via a threshold factor $\theta > 1$, i.e. we refine along each direction l for which

$$\theta \, K_l > K_m, \quad l = 1, 2, 3.$$

Numerical experiments showed that $\theta = 5$ is a good choice for a range of test problems.

As the solution function is vector-valued in our case we replace the jump of a scalar function ϕ in Eq. (15) by an appropriate norm of the vector of jumps, for example the l_2-norm.

5. Adaptive Refinement Algorithm

In the following we describe the multi-target adjoint-based *hp*- and anisotropic refinement algorithm.

Algorithm 3.1. Adaptive algorithm for the accurate and efficient approximation of multiple target quantities $J_i(u)$, $i = 1, \ldots, N$:

(1) Construct an initial mesh $\mathcal{T}_h^{(0)}$, set $i = 0$.
(2) Compute $\mathbf{u}_h \in \mathbf{V}_{h,p}$, see Eq. (2), on the current mesh $\mathcal{T}_h^{(i)}$.
(3) Compute $\tilde{\mathbf{e}}_h \in \mathbf{V}_{h,\tilde{p}}$, see Eq. (7), on the same mesh $\mathcal{T}_h^{(i)}$ with $\tilde{p} > p$.
(4) Evaluate $J_i(\mathbf{u}) - J_i(\mathbf{u}_h) \approx J_i'[\mathbf{u}_h](\tilde{\mathbf{e}}_h) =: \psi_i$, $i = 1, \ldots, N$.
(5) If $|\psi_i| \leq \mathtt{TOL_i}$ for all $i = 1, \ldots, N$, then `STOP`.
(6) Compute $\tilde{\mathbf{z}}_{c,h} \in \mathbf{V}_{h,\tilde{p}}$, see Eq. (9), for the combined target quantity J_c on the same mesh $\mathcal{T}_h^{(i)}$ with $\tilde{p} > p$.
(7) Evaluate the approximate error representation $\sum_{\kappa \in \mathcal{T}_h^{(i)}} \tilde{\eta}_\kappa^c$, see Eq. (10).
(8) If $|\sum_{\kappa \in \mathcal{T}_h^{(i)}} \tilde{\eta}_\kappa^c| \leq \mathtt{TOL}$, then `STOP`.
(9) Select a fixed fraction of the total number of elements according to the largest values of $|\tilde{\eta}_\kappa^c|$.
(10) Decide upon *h*-subdivision or *p*-enrichment on the selected elements according to the *hp*-indicator, see Sec. 4. Perform the *p*-enrichment.
(11) On the elements selected for *h*-subdivision choose the specific anisotropic refinement case according to the anisotropic jump indicator Eq. (15). Perform the (anisotropic) *h*-subdivision which yields $\mathcal{T}_h^{(i+1)}$.
(12) Set $i = i + 1$ and `GOTO` (2).

$\mathtt{TOL_i}$ and `TOL` are given tolerances. Note, that for a single target algorithm, steps (3)-(5) are omitted, and in (6), (7) the solution $\tilde{\mathbf{z}}_h$ to Eq. (4) is computed and the error representation Eq. (5) evaluated.

6. Numerical Results

The performance of the adaptive algorithm and the underlying error estimates will now be demonstrated regarding a range of numerical examples for simple aerodynamic test cases. All computations have been performed using the DG flow solver PADGE,[38] which is based on a modified version of the deal.II finite element library.[39] Unless stated otherwise for specific results, the h-subdivision is performed in an isotropic way. In order to evaluate the approximation quality for different target quantities, reference values have been determined for all cases by means of an extrapolation procedure based on results from high order computations on fine meshes.

6.1. *Laminar subsonic flow around an airfoil*

As a first and simple test case we consider the laminar viscous flow at a free stream Mach number $M = 0.5$, a Reynolds number $\mathrm{Re} = 5000$ and an angle of attack $\alpha = 2°$ around the symmetric NACA0012 airfoil with a sharp trailing edge. We are interested in the accurate computation of the drag coefficient C_{D}.

Figure 1(a) shows the behavior of the adaptive algorithm for a single target quantity in combination with pure h-subdivision for piecewise linear (DG(1)) and piecewise quadratic (DG(2)) basis functions as well as in combination with the hp-adaptive strategy for two different values of the threshold parameter σ_0. In general, this test case features a particularly smooth flow field, thus the error in the drag coefficient is reduced rapidly as the mesh is refined. After the initial mesh, on which some random error cancellation occurs, the error for both types of h-refinement drop according to a straight line, indicating a constant order of convergence. Furthermore, the error for the higher order case is not only smaller but also drops faster, corresponding to an increased order of convergence. Due to the smoothness of the flow we expect higher order of convergence for higher order basis functions, thus this observation is in good agreement with expectations. The hp-adaptive algorithm for $\sigma_0 = 1$ shows an even more favorable behavior and yields a more efficient error reduction. The additional semi-log plot in Fig. 1(b) shows the error vs. the square root of degrees of freedom, i.e. vs. the equivalent one-dimensional number of degrees of freedom. There, the hp-adaptive algorithm yields approximately a straight line indicating exponential convergence, whereas the curves for constant orders of convergence show a reducing slope for increasing accuracy.

The second version of the hp-algorithm shown in Fig. 1 uses the theoretical minimum value $\sigma_0 = 0$ for the threshold parameter, which corresponds to an extreme favoring of p-enrichment. In fact, the number of elements is only increased by 6% over all refinement steps and after the fourth adapta-

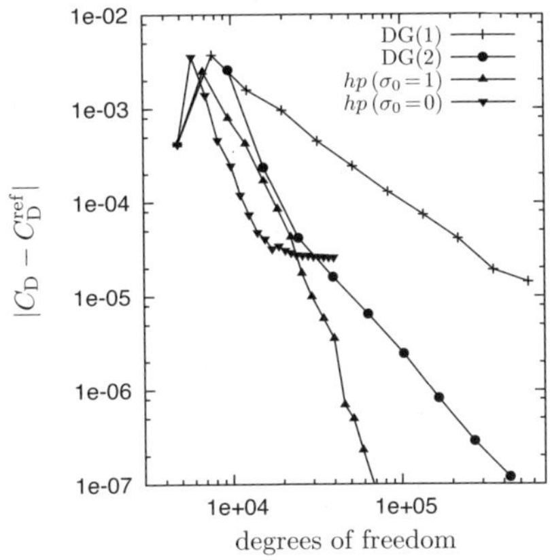

(a) computed values

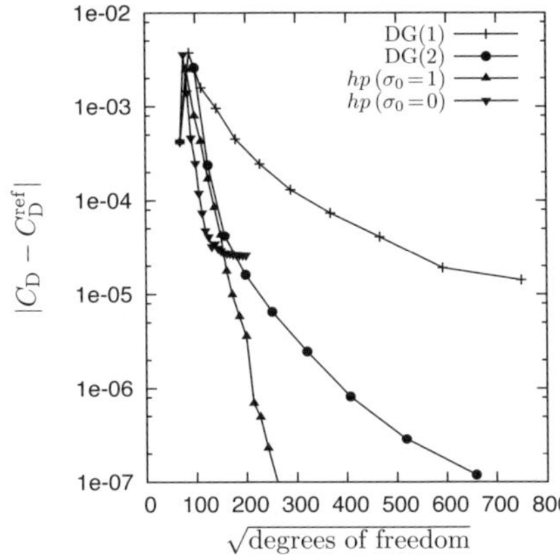

(b) computed values

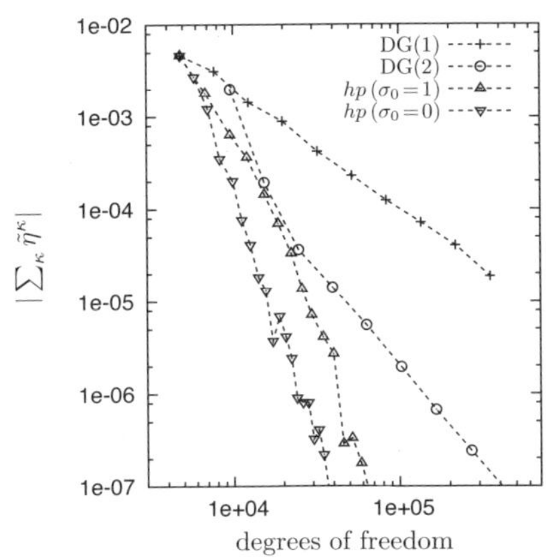

(c) error estimates

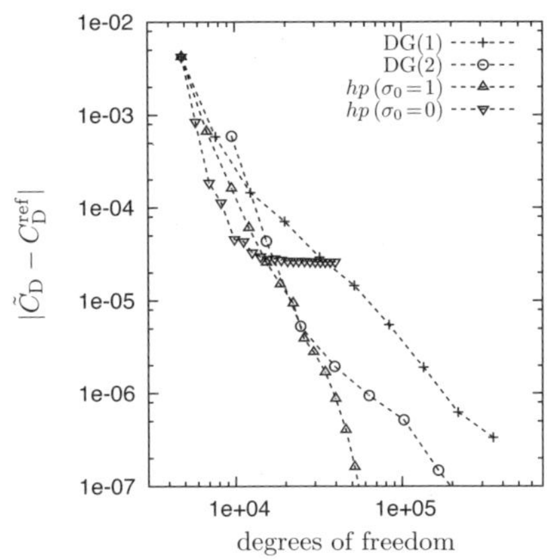

(d) enhanced values

Fig. 1. Laminar subsonic airfoil flow: Error and error estimates in the computed drag coefficient for a single target adaptive algorithm with constant as well as variable polynomial degree.

tion step, no h-subdivision is selected at all. Initially, the error is reduced even faster than in the $\sigma_0 = 1$ case. However, the reduction soon levels off and the error tends towards a constant value of approximately $2.6 \cdot 10^{-5}$. This is due to the fact that the computation uses a piecewise quadratic approximation of the airfoil boundary. Using h-subdivision this approximation is updated such that both endpoints of a boundary edge as well as one intermediate point coincide exactly with the analytical boundary description. Using only p-enrichment, the geometry approximation is not improved, thus yielding a different mesh convergent value for the drag coefficient. In fact, if the geometry is kept fixed for all adaptive strategies, a pure p-enrichment strategy converges fastest against the modified reference value. Increasing the order of the geometry approximation along with that of the ansatz space for the flow solution might avoid the geometry error. However, obtaining a good high-order representation of complex geometries is not a trivial task in practice.

This effect motivates the question whether geometrical effects should be included in the hp-selection strategy and in fact also in the error estimation procedure. Our experience, however, is that geometrical effects play a minor role as long as some h-subdivision is present, thus this aspect is not considered in our strategies.

Figure 1(c) gives a graphical illustration of the error estimates obtained for all three cases. For the $\sigma_0 = 0$ case this error estimate drops very rapidly, indicating that this algorithm actually converges very fast towards the reference value for the modified geometry. The other three curves are very close to the actual errors, indicating the high quality of the error estimates for this smooth flow.

Using the error estimates to enhance the computed values according to Eq. (6) yields the remaining errors depicted in Fig. 1(d). In all cases the enhanced values produce a smaller error in comparison with the originally computed ones, with the exception of the initial mesh. There, the error estimate does not show the error cancellation that happens by chance, thus the error estimates show a behavior which better corresponds to the theoretical one. Nevertheless, the enhanced value is actually slightly worse than the unmodified one. The overall improvement is especially prominent in the DG(1) case, less pronounced in the DG(2) case and even smaller in the hp-adaptive case. This is due to the fact that in all cases the polynomial degree used for the flow solution is increased by exactly one for the adjoint solution. For a linear function this enrichment is much larger compared to functions which are already of high order. Nevertheless, the estimates are still very efficient in driving an adaptive algorithm.

As a final aspect we note that the choice of an optimal or appropriate algorithm depends on the accuracy requirements. The advantage of the hp-adaptive algorithm compared to the simpler DG(2) algorithm is quite small initially and is only significant for very low levels of the error, i.e. for strict accuracy requirements. However, this is partly due to the very smooth behavior of this particular flow field. Furthermore, the relevant

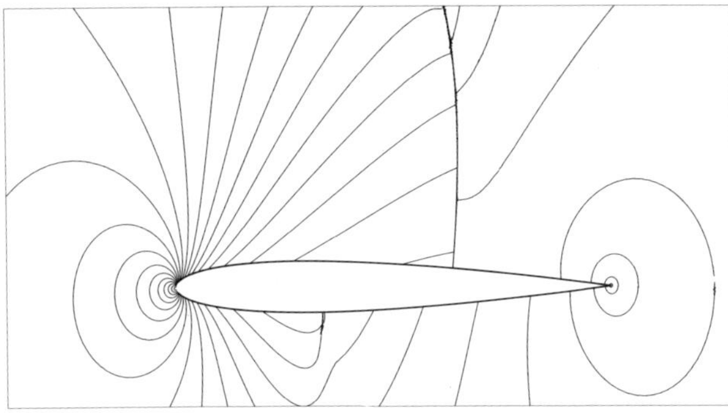

Fig. 2. Transonic airfoil flow: Pressure iso-lines around the NACA0012 airfoil.

accuracy level depends strongly on the case at hand, thus the difference might occur much earlier for more involved cases with complex geometries and flow features.

6.2. *Transonic flow around an airfoil*

As a second test case we consider the transonic flow around the NACA0012 airfoil at a Mach number $M = 0.8$ and an angle of attack $\alpha = 1.25°$. Fig. 2 illustrates the resulting flow field.

Due to the strong shock at the upper side as well as the weak one on the lower side of the airfoil the flow field is clearly not smooth. As discontinuities are more pronounced in the inviscid case due to the lack of the smoothing effect of viscosity we consider the more difficult case of the Euler equations. In order to stabilize the computation, an artificial viscosity term is added to the discrete equations. This shock capturing includes only the element terms of a diffusive operator, neglecting inter-element contributions. The strength of the viscosity is scaled with the residual of the Euler equations in strong form. This additional term represents an error for the numerical solution, but it is consistent as it vanishes for the exact solution which fulfills the Euler equations. Furthermore, this term converges to zero on refined meshes.

We are again interested in the drag coefficient, which in this case is strongly influenced by the position and strength of the shock at the upper side of the airfoil. Due to the fact that the solution is not differentiable at the shock we expect less than first order convergence for all DG methods on globally refined meshes, irrespective of the polynomial degree. This is verified in the left part of Fig. 3 which shows the convergence of the drag error

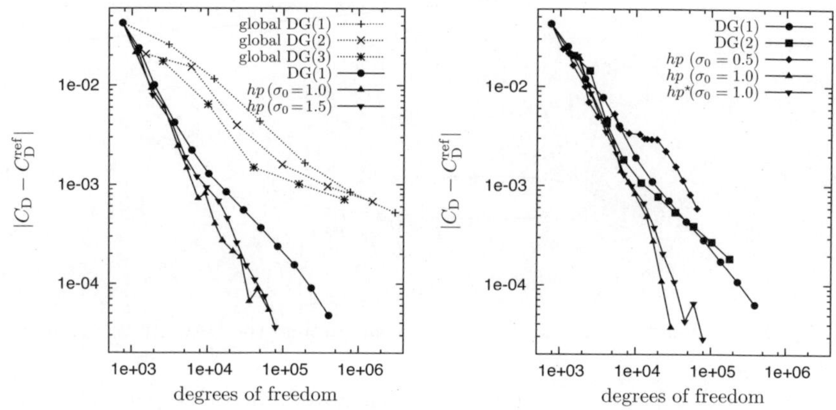

Fig. 3. Transonic airfoil flow: Error in the computed drag coefficient for global mesh refinement and a C_D-adaptive algorithm (left) as well as an adaptive algorithm for multiple target quantities (right), both with constant as well as variable polynomial degree.

for globally refined meshes employing piecewise linear, quadratic and cubic basis functions. The adaptive algorithm tailored at efficiently approximating C_D does a good job, as the drag coefficient converges significantly faster. Employing the *hp*-adaptive algorithm with $\sigma_0 = 1.5$ reduces the degrees of freedom required to obtain a given accuracy. After the final adaptation step the advantage has grown to a factor of five. Reducing the threshold to $\sigma_0 = 1.0$ creates an additional albeit small gain. As the convergence becomes less regular in this case it is probably preferable to use the slightly higher threshold value.

For this case we also consider error estimation for multiple target quantities. To this end, we consider the lift, drag and moment coefficient, thereby giving influence to basically the complete solution at the airfoil surface. As we are not interested in any particular accuracy combination we simply use the sum of relative errors, see Eq. (11), as a combined target quantity. Evaluating again the drag coefficient on the resulting mesh sequences yields the right part of Fig. 3. First, we note that the DG(1) *h*-adaptive version converges very similar to the version for a single target quantity, i.e. the quality of the error estimation and thus the created meshes does not notably deteriorate when considering multiple target quantities. The DG(2) *h*-adaptive case yields results which are similar to the DG(1) case, i.e. due to the limited smoothness of the solution it is not sufficient to simply increase the polynomial degree to yield a better adaptive algorithm, which is in line with the results for global refinement. Instead, it is necessary to consider a suitable combination of local h-subdivision at low polynomial degree near the shock and an increase in the polynomial degree away from the shock. The hp-adaptive algorithm with $\sigma_0 = 1.0$ does a very good job at that, reducing the required number of degrees of freedom by an order of magnitude. Lowering the threshold to $\sigma_0 = 0.5$ yields a poor convergence initially, which is significantly worse than the h-subdivision version for piecewise linear polynomials. In this case, the behavior improves drastically at some point, but in general the threshold should not be chosen too small, especially if considerable non-smoothness is expected.

As a last aspect, the right part of Fig. 3 compares the suggested hp-algorithm with a modified version (denoted by $hp^\star$), which does not utilize the solution to the discrete error equation in the decision process between h- and p-type refinement. The numerical results underline the suggestion to use the algorithm as proposed due to its improved behavior, although the differences are not very pronounced.

Finally, in Fig. 4 we compare the last adapted mesh for the multiple target quantity adaptive case with pure h-subdivision to the corresponding hp-adaptive version. Here, a darker color represents a higher polynomial degree. Please note that the dark regions in the left part are only due to a

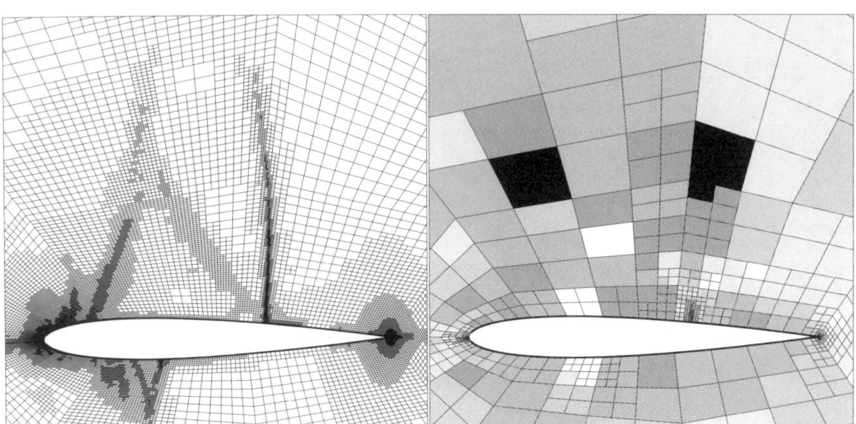

Fig. 4. Transonic airfoil flow: Final adapted meshes for h-subdivision (left) and hp-adaptation (right). In the hp case a darker color indicates a higher polynomial degree, ranging from one (white) to six (black).

high number of mesh lines. We observe, as expected, that h-subdivision is mainly present in the shock region as well as at the sharp trailing edge and in the curved leading edge region. Away from these features the resolution is mainly increased through a higher polynomial degree, with a maximum degree of six. Overall, the amount of h-subdivision, also near the shock but away from the surface of the airfoil, is surprisingly low. Although the number of degrees of freedom is reduced by a factor of slightly more than ten, the number of elements is almost forty times smaller in the hp case. These results support the statement that also in the transonic case accurate results can be obtained with a higher order discretization due to the stabilization introduced via the shock capturing term.

6.3. *Laminar flow around a delta wing*

As an example of a complex laminar flow field we consider the flow at a Mach number $M = 0.3$, a Reynolds number $\mathrm{Re} = 4000$ and an angle of attack $\alpha = 12.5°$ around a delta wing with sloped sharp leading edge and a blunt trailing edge, see Fig. 5. This test case has been considered in the European ADIGMA project[40] and in Leicht and Hartmann,[12] a similar case was earlier treated by Klaij *et al.*[41]

For the sake of brevity, we will only consider the error of different approximations of the lift coefficient C_L. Similar results have been obtained for the drag coefficient C_D. We start by computing the lift from the second order DG(1) flow solution on globally refined meshes starting from a very

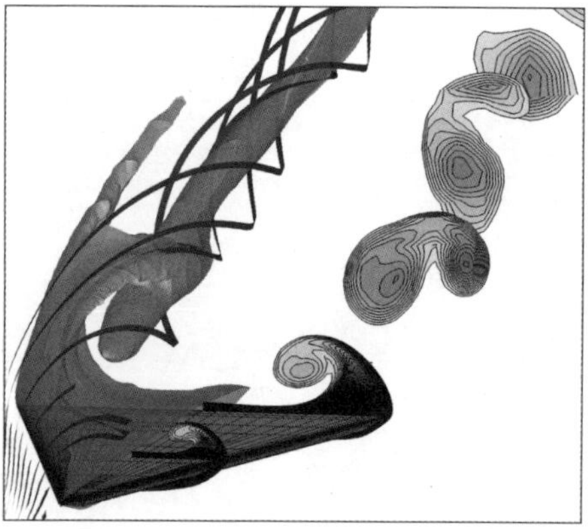

Fig. 5. Laminar flow around a delta wing: Solution plot showing streamlines and a Mach number iso-surface over the left half of the laminar delta wing as well as Mach number slices over the right half.

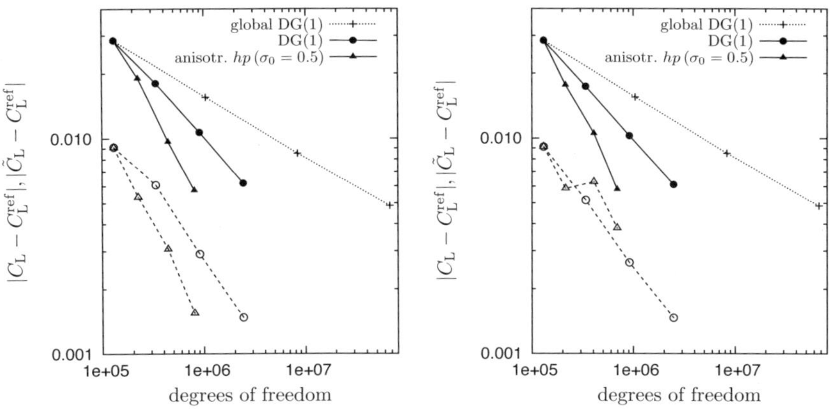

Fig. 6. Laminar flow around a delta wing: Error in the computed lift coefficient (filled symbols) and the enhanced lift coefficients (open symbols) for global mesh refinement and a C_L-adaptive algorithm (left) as well as an adaptive algorithm for multiple target quantities (right), both with constant as well as variable polynomial degree.

coarse initial mesh consisting of only 3264 elements for a half domain with symmetry boundary conditions. We then consider locally adaptive mesh refinement starting from the results on the initial coarse mesh.

The left part of Fig. 6 plots the error in the lift coefficient vs. the number of degrees of freedom for various refinement strategies. Compared to global mesh refinement, lift coefficients of a specific accuracy are obtained with considerably less degrees of freedom in the case of adjoint-based goal-oriented mesh refinement targeting an accurate prediction of the lift. In the case of *hp*-adaptive mesh refinement we expect and actually obtain quite a large fraction of h-subdivision due to two reasons. The sharp leading edges represent (geometric) singularities and thus create non-smooth local flow features. Furthermore, the vortices are smooth features if they are well resolved, but on coarse meshes the gradient is very large compared to the average value, thus they behave like non-smooth feature on a low resolution scale. Anticipating these effects, the *hp*-adaptive algorithm is based on a reduced threshold value of $\sigma_0 = 0.5$. Furthermore, it combined with anisotropic h-subdivision, yielding very efficient enrichments in both h and p. The resulting algorithm gives an additional gain over the DG(1) adaptive case, although computations with globally high order exhibit the same first order convergence as the DG(1) global refinement case.

Figure 6 also illustrates the errors of the enhanced lift coefficients obtained by adding the global error estimate to the computed lift coefficient. Already on the first adapted mesh the enhanced coefficient is almost as accurate as the values computed on the last adapted meshes. In addition to that, for this particular case the error estimation does not show any distinguished degradation in the case of *hp*-adaptive and anisotropically refined meshes. In fact, the overall quality of the error estimates even improves on finer meshes.

The right part of Fig. 6 repeats the same plots for adaptive algorithms based on multiple target quantities. As before, we choose the sum of relative errors in lift, drag and pitching moment. The side force, yaw and roll moment coefficients all vanish due to the symmetry of the flow. The obtained accuracy in the lift coefficients is very similar to that for the single target quantity case. In fact, the *hp*-adaptive algorithm yields even slightly better results due to exploiting the solution of the discrete error equation. Nevertheless, the error estimation itself is degraded in the *hp*-case, probably due to the reasons discussed for the laminar airfoil case.

6.4. *L1T2 three element high lift configuration*

Next we consider the turbulent flow at a free stream Mach number $M = 0.197$, a Reynolds number $\mathrm{Re} = 3.52 \cdot 10^6$ and an angle of attack $\alpha = 20.18°$ around the L1T2 three element airfoil, see Fig. 7. This case has been documented extensively in literature, in particular there is data of two wind tunnel experiments[42] which determined lift coefficients of $C_{\mathrm{L}}^1 = 4.110$ and $C_{\mathrm{L}}^2 = 4.075$, respectively. Considering that our computations were performed fully turbulent and did not involve any transition settings, our own reference value of $C_{\mathrm{L}}^{\mathrm{ref}} = 4.017$ seems to be quite close to those results.

An initial mesh has been obtained from a coarse level of a mesh designed for accurate computations with a finite volume code. The geometry is approximated with piecewise quartic polynomials based on a CAD representation of the geometry. This approximation is deemed accurate enough and thus kept constant under mesh refinement.

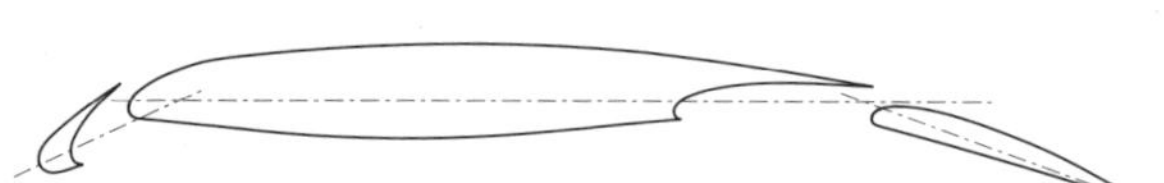

Fig. 7. Geometry of the L1T2 three element airfoil. The slat angle is 25°, the flap angle is 20°.

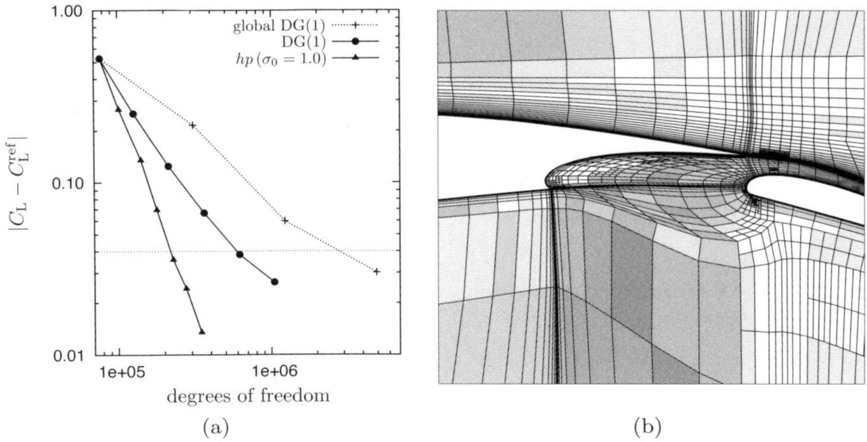

Fig. 8. L1T2 high lift configuration: (a) Error in the computed lift coefficient for global mesh refinement and a C_L-adaptive algorithm for DG methods with constant as well as variable polynomial degree; (b) Final *hp*-adapted mesh. A darker color indicates a higher polynomial degree, ranging from one (white) to seven (black).

We now compare the approximation of the lift coefficient on three different sequences of meshes shown in Fig. 8(a). The first one has been obtained by global uniform refinement of the initial mesh using piecewise linear ansatz functions, i.e. a formally second order DG method. The second method targets the accurate prediction of the lift coefficient. Although the initial mesh has been hand-tailored for this particular flow field, the locally adaptive second order DG(1) method can significantly reduce the degrees of freedom required to obtain a given accuracy. In this case an error of $4\cdot 10^{-2}$, indicated by the dotted line in Fig. 8(a), corresponds to a relative error of 1 %. In order to obtain this accuracy, the error of the initial mesh has to be reduced by more than an order of magnitude. Interpolating in the plot we note that the DG(1) *h*-adaptive algorithm reduces the degrees of freedom by a factor between four and five for this case. Using the *hp*-adaptive algorithm the total reduction is even beyond an order of magnitude, more than the increase in degrees of freedom from the initial to the final adapted mesh. It is hard to judge whether the algorithm shows exponential convergence. In any case, the efficiency is drastically improved.

Figure 8(b) shows a view of the final *hp*-adapted mesh in the main wing and flap region. The darker color indicating a higher polynomial degree

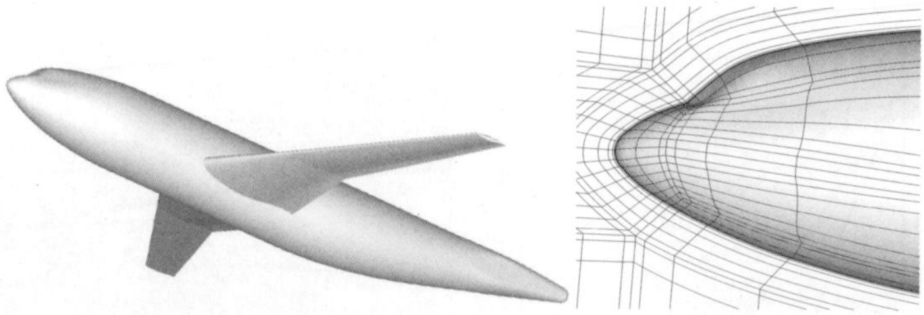
(a) (b)

Fig. 9. DLR-F6 wing-body configuration: (a) Geometry and (b) mesh with 50618 curved elements close to the nose.

can mostly be found in coarse elements away from the boundaries.[c]

6.5. *Subsonic turbulent flow around the DLR-F6 wing-body configuration*

In this final example we consider a turbulent flow at a Mach number $M = 0.5$, a Reynolds number $\mathrm{Re} = 5 10^6$ and an angle of attack $\alpha = -0.141°$ around the DLR-F6 wing-body configuration without fairing, see Fig. 9(a). This is a modification of a test case from the third Drag Prediction Workshop (DPW), where a fixed angle of attack has been assumed instead of a given target lift. Also, the Mach number has been reduced from originally $M = 0.75$ to $M = 0.5$ in order to obtain a subsonic flow. This test case has previously been considered in Refs. 20 and 27.

The original DPW mesh of about 3.2 million hexahedral elements has been agglomerated twice, yielding a coarse mesh of 50618 hexahedral elements. The additional points of the original mesh have been used to define curved elements, see Fig. 9(b), where the curved lines are represented by polynomials of degree 4.

On this mesh we first compute the DG(1) and DG(2) flow solutions on the coarse mesh and a once globally refined mesh. The resulting drag coefficients are given in Fig. 10(a). Due to the complexity of the problem, no rigorous convergence study is available for this case. Thus, the plot shows the computed values rather than the resulting errors due to the lack of a good reference value for the mesh-convergent drag. Nevertheless, we clearly see the advantage in terms of accuracy and degrees of freedom of using the

[c]The black areas near the wall are due to overlapping grid lines of strongly stretched elements.

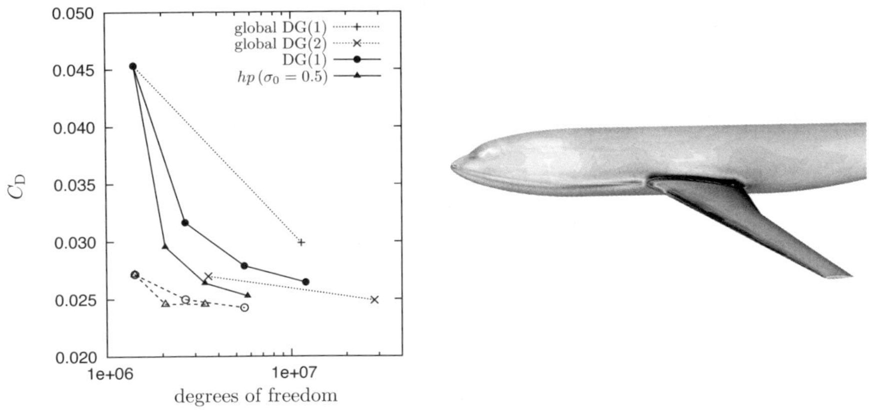

(a) (b)

Fig. 10. Turbulent flow around the DLR-F6 wing-body configuration: (a) Convergence of the drag coefficient (filled symbols) and the enhanced drag coefficient (open symbols) for global mesh refinement and a C_{D}-adaptive algorithm with constant as well as variable polynomial degree; (b) Density adjoint on a twice *hp*-adapted mesh.

discretization with the polynomial degree $p = 2$ over the discretization with the lower polynomial degree $p = 1$. Additionally, Fig. 10(a) shows the drag coefficient values for a DG(1)-adaptive algorithm as well as an *hp*-adaptive version with $\sigma_0 = 0.5$, both targeting the accurate prediction of the drag coefficient C_{D}. In both cases anisotropic *h*-subdivision is considered based on the jump indicator described in Sec. 15. Figure 10(a) clearly shows that a specific accuracy is reached with a significantly reduced number of degrees of freedom using the adjoint-based adaptive approaches compared to global mesh refinement. Furthermore, the *hp*-adaptive version is more efficient than the pure *h*-subdivision. In particular, after two *hp*-adaptive refinement steps the same accuracy is obtained as after three *h*-refinement steps. In the latter case, the number of degrees of freedom is larger by a factor of about 3. The enhanced drag coefficients seem to converge faster against a constant drag coefficient, indicating that the error estimation procedure still works well for this relatively complex flow.

Finally, Fig. 10(b) shows the density adjoint, i.e. the first component of the discrete adjoint solution on the second *hp*-adapted mesh. Large values of this variable indicate a large influence of the density on the drag.

7. Conclusion and Outlook

The adaptive algorithm presented in this chapter is able to produce nearly optimal meshes for the efficient approximation of given target quantities like aerodynamic coefficients. In principal, it is always possible to replace an automatic adaptation strategy with a good hand-crafted mesh. However, the extensive experience available for CFD meshes is mainly attributed to second-order finite volume schemes and does not necessarily cover higher order methods. Furthermore, variable order methods provide still another challenge to the manual creation of good meshes. Regarding these difficulties, the presented algorithm is a good alternative. Furthermore, it offers an additional gain through the availability of global error estimates in terms of the target quantities.

Whereas the presented examples were limited to quadrilateral and hexahedral meshes, the concepts for error estimation and *hp*-selection are directly applicable to general unstructured meshes. The anisotropic h-subdivision is an exception from that rule as the very concept is only applicable to tensor-product element types.

The current presentation concentrated on demonstrating the potential of the suggested algorithms to utilize the inherent flexibility of DG methods. In order to exploit that potential for practical applications, efficient solver algorithms for *hp*-adaptive meshes will have to be considered in future work. Depending on the solution strategy it will then be possible to compare the actual cost of different adaptive strategies in terms of CPU time and memory requirements. This will yield more accurate assessments of the relative benefits than those possible when only degrees of freedom are considered.

Acknowledgments

The authors would like to thank Paul Houston and Francesco Bassi for many inspiring and fruitful discussions. Furthermore, we would like to express our gratitude to our former co-workers, Joachim Held and Florian Prill, for their contributions to the `PADGE` code.[38]

References

1. R. Becker and R. Rannacher, *East–West J. Numer. Math.* **4**, 237–264, (1996).
2. R. Becker and R. Rannacher, *Acta Numerica.* **10**, 1–102, (2001).
3. M. Giles and E. Süli, *Acta Numerica.* **11**, 145–236, (2002).
4. R. Hartmann and P. Houston, *J. Comput. Phys.* **183**(2), 508–532, (2002).

5. J. Lu. *An a posteriori Error Control Framework for Adaptive Precision Optimization using Discontinuous Galerkin Finite Element Method.* PhD thesis, Massachusetts Institute of Technology, (2005).
6. R. Hartmann and P. Houston, *Int. J. Num. Anal. Model.* **3**(2), 141–162, (2006).
7. R. Hartmann, *Int. J. Numer. Meth. Fluids.* **51**(9–10), 1131–1156, (2006).
8. D. A. Venditti and D. L. Darmofal, *J. Comp. Phys.* **176**, 40–69, (2002).
9. T. Barth and M. Larson. In eds. R. Herbin and D. Kröner, *Finite Volumes for Complex Applications III: Problems and Perspectives.* Hermes Penton Science, London, (2002).
10. N. Pierce and M. Giles, *SIAM Review.* **42**(2), 247–264, (2000).
11. T. Leicht and R. Hartmann, *Int. J. Numer. Meth. Fluids.* **56**(11), 2111–2138 (April, 2008).
12. T. Leicht and R. Hartmann, Error estimation and anisotropic mesh refinement for 3d laminar aerodynamic flow simulations. *J. Comput. Phys.*, **229**(19), 7344–7360, (2010).
13. T. A. Oliver. *A High-Order, Adaptive, Discontinuous Galerkin Finite Element Method for the Reynolds-Averaged Navier-Stokes Equations.* PhD thesis, Massachusetts Institute of Technology, (2008).
14. D. A. Venditti and D. L. Darmofal, *J. Comp. Phys.* **187**, 22–46, (2003).
15. R. Hartmann and P. Houston. In eds. T. Y. Hou and E. Tadmor, *Hyperbolic problems: theory, numerics, applications*, pp. 579–588. Springer, (2003).
16. R. Hartmann, *SIAM J. Sci. Comput.* **31**(1), 708–731, (2008).
17. M. Nemec and M. J. Aftosmis. Error estimation and adaptive refinement for embedded-boundary Cartesian meshes. 45th AIAA Aerospace Sciences Meeting, (2007). AIAA 2007-4187.
18. M. Nemec, M. J. Aftosmis, and M. Wintzer. Adjoint-based adaptive mesh refinement for complex geometries. 46th AIAA Aerospace Sciences Meeting, (2008). AIAA Paper 2008-0725.
19. K. J. Fidkowski and D. L. Darmofal, *J. Comput. Phys.* **225**, 1653–1672, (2007).
20. R. Hartmann, J. Held, and T. Leicht, Adjoint-based error estimation and adaptive mesh refinement for the RANS and $\kappa - \omega$ turbulence model equations. *J. Comput. Phys.*, (2010). In press. DOI: 10.1016/j.jcp.2010.10.026
21. C. Schwab, *p- and hp- finite element methods – Theory and applications in solid and fluid mechanics.* (Oxford University Press, 1998).
22. P. Houston and E. Süli, *Comput. Methods Appl. Mech. Engrg.* **194**, 229–243, (2005).
23. P. Houston, B. Senior, and E. Süli, *Int. J. Numer. Meth. Fluids.* **40**, 153–169, (2002).
24. W. Rachowicz, D. Pardo, and L. Demkowicz, *Comput. Methods Appl. Mech. Engrg.* **195**, 4816–4842, (2006).
25. J. Kurtz and L. Demkowicz, *Comput. Methods Appl. Mech. Engrg.* **196**, 3534–3545, (2007).
26. L. Wang and D. J. Mavriplis, *J. Comput. Phys.* **228**(20), 7643–7661, (2009).

27. R. Hartmann and P. Houston. Error estimation and adaptive mesh refinement for aerodynamic flows. In ed. H. Deconinck, *VKI LS 2010-01: 36th CFD/ADIGMA course on hp-adaptive and hp-multigrid methods, Oct. 26-30, 2009.* Von Karman Institute for Fluid Dynamics, Rhode Saint Genèse, Belgium, (2010).
28. B. Landmann, M. Kessler, S. Wagner, and E. Krämer. A parallel discontinuous Galerkin code for the Navier-Stokes equations. 44th AIAA Aerospace Sciences Meeting and Exhibit, (2006). AIAA 2006-111.
29. R. Hartmann and P. Houston, *J. Comput. Phys.* **227**(22), 9670–9685, (2008).
30. F. Bassi, S. Rebay, G. Mariotti, S. Pedinotti, and M. Savini. In eds. R. Decuypere and G. Dibelius, *2nd European Conference on Turbomachinery Fluid Dynamics and Thermodynamics, Antwerpen, Belgium, March 5–7, 1997*, pp. 99–108. Technologisch Instituut, (1997).
31. F. Bassi, A. Crivellini, S. Rebay, and M. Savini, *Computers & Fluids.* **34**, 507–540, (2005).
32. F. R. Menter, *AIAA J.* **32**(8), 1598–1605, (1994).
33. P. Houston, C. Schwab, and E. Süli, *SIAM J. Numer. Anal.* **39**(6), 2133–2163, (2002).
34. C. Mavriplis, *Comput. Methods Appl. Mech. Engrg.* **116**, 77–86, (1994).
35. C. Mavriplis. A posteriori error estimators for adaptive spectral element methods. In ed. P. Wesseling, *Notes on numerical fluid mechanics*, vol. 29, pp. 333–342. Vieweg, (1990).
36. E. H. Georgoulis, E. Hall, and P. Houston, *Appl. Numer. Math.* **59**(9), 2179–2194, (2009).
37. T. Richter, *Int. J. Numer. Meth. Fluids.* **62**(1), 90–118, (2010).
38. R. Hartmann, J. Held, T. Leicht, and F. Prill, Discontinuous Galerkin methods for computational aerodynamics – 3D adaptive flow simulation with the DLR PADGE code. *Aerosp. Sci. Technol.*, **14**: 512–519, 2010.
39. W. Bangerth, R. Hartmann, and G. Kanschat, *ACM Transactions on Mathematical Software.* **33**(4), (2007).
40. N. Kroll. ADGIMA – A European project on the development of adaptive higher-order variational methods for aerospace applications. 47th AIAA Aerospace Sciences Meeting, (2009). AIAA 2009-176.
41. C. M. Klaij, J. J. W. van der Vegt, and H. van der Ven, *J. Comput. Phys.* **217**(2), 589–611, (2006).
42. I. R. M. Moir. AGARD Advisory Report 303, Advisory Group for Aerospace Research & Development, Neuilly-sur-Seine, (1994). Test case A2.

CHAPTER 4

A RUNGE-KUTTA BASED DISCONTINUOUS GALERKIN METHOD WITH TIME ACCURATE LOCAL TIME STEPPING

Gregor J. Gassner*, Florian Hindenlang†, Claus-Dieter Munz‡

Institute for Aerodynamics and Gasdynamics, Universität Stuttgart, Pfaffenwaldring 21, 70550 Stuttgart, Germany

**gassner@iag.uni-stuttgart.de*

†hindenlang@iag.uni-stuttgart.de

‡munz@iag.uni-stuttgart.de

An explicit one-step time discretization for discontinuous Galerkin schemes applied to advection-diffusion equations is proposed that is based on a predictor corrector approach. The predictor is local and takes only the time evolution of the data within the grid cell into account. For this continuous extension Runge-Kutta schemes are used. The advantage of the predictor corrector formulation is that the time evolution is done in one step and the data of the direct neighbors are needed only. Hence, the proposed discontinuous Galerkin scheme has the optimal locality within the whole time step. This is the basis to introduce a time consistent local time stepping in a way such that every grid cell may run with its own optimal time step as given by the local stability restriction. The time accuracy and the efficiency of the local time stepping is shown for linear and non-linear problems. Finally, the capability of the approach is demonstrated for a direct simulation of a three-dimensional jet within a natural gas injector, where the noise generated by the flow is investigated in addition.

1. Introduction

In a series of papers[1–5] Cockburn and Shu developed the Runge-Kutta discontinuous Galerkin (RKDG) framework. They used a high order accurate explicit Runge-Kutta scheme for time approximation. Due to the locality the RKDG scheme results in a very efficient method for massive parallel computations.[6] A characteristic of explicit time integration is that the maximum allowable time step is restricted to guarantee stability.

For a discontinuous Galerkin scheme the time step restriction has the form

$$\begin{aligned} \Delta t^a &\leq \frac{\alpha(p)}{\lambda^a_{max}} \left(\frac{\Delta x}{2\,p+1}\right) \text{ for advection,} \\ \Delta t^d &\leq \frac{\beta(p)}{\lambda^d_{max}} \left(\frac{\Delta x}{2\,p+1}\right)^2 \text{ for diffusion,} \\ \Delta t &\leq \left((\Delta t^a)^{-2} + (\Delta t^d)^{-2}\right)^{-\frac{1}{2}} \text{ for advection-diffusion,} \end{aligned} \tag{1}$$

where $\lambda^{a/d}_{max}$ denote the maximum eigenvalue of the advection Jacobians and diffusion matrices. The stability numbers α and β depend on the order of the underlying approximation space and on the explicit time discretization method. We note that the time step depends on the spatial resolution $\Delta x/(2\,p+1)$ analogously to the *CFL* condition in finite volume schemes. This condition relates the maximal possible time step (time resolution) to the spatial resolution, which is natural and physical meaningful for unsteady advection dominated problems. In the unsteady case, the physical meaningful time step determined for an unsteady problem is in the range of the explicit time step. For an approximation space with uniform spatial resolution and uniform distribution of maximal eigenvalues this yields an efficient method. However, practical problems of interest often have a strong inhomogeneous distribution of the maximal eigenvalues. Furthermore to save computational cost, the spatial resolution is non-uniformly distributed to take into account different solution behaviors during a simulation. Either constructing the approximation space in an initial phase of the simulation or in an adaptive fashion during the simulation, the magnitude of the local resolution can vary drastically in the computational domain. In relation to this, a major drawback of an explicit time discretization is that the minimum time step over the whole computational domain has to be used as a global time step to advance the solution in time. Thus advancing with the minimum time step may result in a drastic decrease in efficiency in such simulations.

An alternative to explicit time stepping is to use implicit time integration methods, such as *backward difference formulae* or *implicit Runge-Kutta* methods.[7] The advantage of these methods is that no theoretical stability limit for the maximal allowable time step exists. Thus, the time step is only restricted due to physical considerations and not due to stability limits. The drawback, however, is that large (non-linear) algebraic systems have to be solved with Newton iteration methods,[8] which causes a high computational cost per time step. An implicit time discretization therefore only pays off, if the maximal global explicit time step is 'small' compared to the physical meaningful time step. The definition of 'small' strongly depends on the problem to be solved.

If we focus on massive parallel large scale computations with $\mathcal{O}(1000)$ processors, it is clear that up to now the algorithms and methods to solve large non-linear algebraic systems are under-developed. An interesting alternative solution strategy for the large non-linear system is to use a pseudo time approach, see e.g. Ref. 9, where the system is solved in pseudo time with an explicit scheme. The advantage of this method is that the algorithm retains its explicit character which is good for the parallelizability, the drawback is that we have again some sort of (pseudo) time step restriction. An open topic is the use of (local) implicit convergence acceleration methods to increase the maximal allowed pseudo time step, which could be arbitrarily large as no accuracy in pseudo time is needed. Another approach that combines explicit and implicit techniques was recently introduced in Kanevsky *et al.*,[10] where the ODE for the time dependent DG degrees of freedom are solved with an *implicit-explicit Runge-Kutta* method.

All these methods share the commonality that the time step is chosen globally, i.e. only global time levels are considered. But, as in the case of a spatial varying solution where the spatial resolution of the scheme is adjusted, the typical solution features different time scales as well. Using a global time stepping method, either explicit, implicit or implicit-explicit, one has to resolve globally the local finest time scale of interest to get the desired accuracy, i.e. adapt the global time step to the accuracy requirements. To overcome this fundamental deficiency of global time integration, we propose in this paper an explicit time approximation with local time stepping. While local time stepping is a well known concept to accelerate the convergence to steady state, we are considering such an approach for unsteady problems only. Here, the local time stepping technique has to be time accurate which does not matter, if one is interested in steady state solutions. The time-accurate local time stepping is based on the ideas in Refs. 11–13. The novel idea is to combine this with a Runge-Kutta based time integration.

2. General Formulation

2.1. *The semi discrete form*

In the following we discuss the discontinuous Galerkin method. To keep matters simple, we restrict the discussion to a scalar conservation law of the form

$$u_t + \vec{\nabla} \cdot \vec{f} = 0, \tag{2}$$

with appropriate initial and boundary conditions in a domain $\Omega \times [0; T] \subseteq \mathbb{R}^d \times \mathbb{R}_0^+$. The flux function $\vec{f}\left(u, \vec{\nabla} u\right)$ is composed of two parts

$$\vec{f} = \vec{f}\left(u, \vec{\nabla} u\right) = \vec{f}^a\left(u\right) - \vec{f}^v\left(u, \vec{\nabla} u\right), \tag{3}$$

with

$$\vec{f}^v\left(u, \vec{\nabla} u\right) = \mu(u) \vec{\nabla} u. \tag{4}$$

The first step of our approximation is to subdivide the domain Ω in non-overlapping grid cells Q. For every grid cell, we use a local polynomial approximation of the form

$$u(\vec{x}, t)\big|_Q \approx u_Q(\vec{x}, t) = \sum_{j=1}^{N} \hat{u}_j^Q(t) \varphi_j^Q(\vec{x}) =: \underline{\hat{u}}^Q(t) \cdot \underline{\varphi}^Q(\vec{x}), \tag{5}$$

where $\{\varphi_j^Q(\vec{x})\}_{j=1,\dots,N}$ is a set of modal hierarchical normalized orthogonal basis functions, which we construct with a Gram-Schmidt orthogonalization algorithm for arbitrary (reference) grid cell types. Independent of the grid cell type, only complete order polynomial spaces are considered. The dimension of this space N and thus the number of time dependent degrees of freedom $\hat{u}_j^Q(t)$ depends on the polynomial degree p and the spatial dimension d

$$N = N(p, d) = \frac{(p + d)!}{p!d!} \tag{6}$$

The next step of our approximation is to define how the unknown degrees of freedom $\hat{u}_j^Q(t)$ are determined. The base of the considered discontinuous Galerkin method is a weak formulation. We insert the approximate solution (5) into the conservation law (2), multiply with a smooth test function $\phi = \phi(\vec{x})$ and integrate over Q. To keep notations short, we omit the index Q

for the approximate solution and the test function in case of uniqueness and introduce the following abbreviations for the volume and surface integrals

$$
\begin{aligned}
\langle a, b\rangle_Q &:= \int_Q a\, b\, d\vec{x} \qquad a, b \in L_2(Q), \\
\left\langle \vec{a}, \vec{b}\right\rangle_Q &:= \int_Q \vec{a}\cdot\vec{b}\, d\vec{x} \qquad \vec{a}, \vec{b} \in [L_2(Q)]^d, \\
\langle a, b\rangle_{\partial Q} &:= \oint_{\partial Q} a\, b\, ds.
\end{aligned} \tag{7}
$$

With these notations, we obtain in a first step the following formulation

$$\left\langle u_t + \vec{\nabla}\cdot\vec{f}, \phi\right\rangle_Q = 0. \tag{8}$$

We proceed with a first integration by parts

$$\langle u_t, \phi\rangle_Q + \left\langle \vec{f}\cdot\vec{n}, \phi\right\rangle_{\partial Q} - \left\langle \vec{f}^a - \mu\vec{\nabla}u, \vec{\nabla}\phi\right\rangle_Q = 0, \tag{9}$$

where $\vec{n}$ denotes the outward pointing normal vector. Using this variational formulation as a basis for the discontinuous Galerkin scheme results in non-optimal convergence behavior with respect to diffusion terms, as the scheme is not adjoint consistent, see e.g. Ref. 14. To overcome this problem, Ref. 15 introduced a mixed finite element approach, where they reformulated the second order problem into a first order system. However, the disadvantage of this approach is that auxiliary variables are introduced, resulting in an increase of the computational effort, especially for systems of equations. In Refs. 13,16 and 17 another variational formulation for diffusion problems is introduced, where the need for auxiliary variables is circumvented. We note that the volume integral still contains derivatives of the solution due to the second order nature of the diffusion flux. Thus, a second integration by parts for this term is possible, yielding the *ultra weak* DG formulation

$$
\begin{aligned}
\langle u_t, \phi\rangle_Q &+ \left\langle \vec{f}\cdot\vec{n}, \phi\right\rangle_{\partial Q} - \left\langle u, \mu\,\vec{n}\cdot\vec{\nabla}\phi\right\rangle_{\partial Q} \\
&- \left\langle \vec{f}^a, \vec{\nabla}\phi\right\rangle_Q + \left\langle u, \vec{\nabla}\cdot\left(\mu\vec{\nabla}\phi\right)\right\rangle_Q = 0.
\end{aligned} \tag{10}
$$

As the approximative solution is in general discontinuous across grid cell interfaces, the traces of the flux normal component $\vec{f}\cdot\vec{n}$ and the solution in the surface integrals $\langle.,.\rangle_{\partial Q}$ are not uniquely defined. To get a stable and accurate discretization, several choices for the numerical approximation are known. We refer to the book of Toro[18] for a comprehensive collection and description of Riemann problem based approximations of the advection fluxes $\vec{f}^a \cdot \vec{n}$. For the traces of the viscous components Gassner et al.[16] and Lörcher et al.[17] developed approximations based on diffusive generalized Riemann problems, which are used in this work. The extension to systems, such as the viscous terms of the compressible Navier-Stokes equations, is described in Ref. 13, yielding an interior penalty type approximation of optimal order of convergence and a physically motivated penalty constant.

If we insert the trace approximations into the ultra weak formulation, further reformulations of the discretization can be performed to avoid the costly computations of second order derivatives. We start with the discrete ultra weak form (10) and use *back* integration by parts twice to derive the *weak* DG formulation and the *strong* DG formulation

$$\begin{aligned} &\langle u_t, \phi\rangle_Q + \langle h, \phi\rangle_{\partial Q} - \left\langle w, \mu \vec{n}\cdot\vec{\nabla}\phi\right\rangle_{\partial Q} - \left\langle \vec{f}^a, \vec{\nabla}\phi\right\rangle_Q + \left\langle u, \vec{\nabla}\cdot\left(\mu\vec{\nabla}\phi\right)\right\rangle_Q \\ =&\langle u_t, \phi\rangle_Q + \langle h, \phi\rangle_{\partial Q} - \left\langle w - u^-, \mu \vec{n}\cdot\vec{\nabla}\phi\right\rangle_{\partial Q} - \left\langle \vec{f}^a - \mu\vec{\nabla}u, \vec{\nabla}\phi\right\rangle_Q \\ =&\left\langle u_t + \vec{\nabla}\cdot\vec{f}, \phi\right\rangle_Q + \left\langle h - \vec{f}^- \cdot\vec{n}, \phi\right\rangle_{\partial Q} - \left\langle w - u^-, \mu\vec{n}\cdot\vec{\nabla}\phi\right\rangle_{\partial Q}. \end{aligned} \tag{11}$$

We introduced

$$\begin{aligned} h &= h(u^-, u^+, \nabla u^-, \nabla u^+) \approx \vec{f}\cdot\vec{n}\big|_{\partial Q} \\ w &= w(u^+, u^-) \approx u\big|_{\partial Q}, \end{aligned} \tag{12}$$

to denote the numerical approximation of the traces. These traces typically depend on the values from inside the grid cell $(.)^-$ and on the values $(.)^+$ from the face sharing neighbor grid cells.

If we choose for each grid cell Q the test functions $\phi = \varphi_1^Q, ..., \varphi_N^Q$, we get N equations and consequently a solvable equation system for our N degrees of freedom $\hat{u}_1^Q, ..., \hat{u}_N^Q$. In this work, we use the strong DG formulation. Although the weak and strong DG formulation are mathematically equivalent, their properties can differ due to the implementation of the spatial integrations $\langle ., .\rangle_Q$ and $\langle ., .\rangle_{\partial Q}$. In contrast to the weak DG formulation, which is inherent exact conservative, the strong DG formulation is exact conservative if and only if the Gauss theorem

$$\left\langle \vec{\nabla}\cdot\vec{f}, 1\right\rangle_Q = \left\langle \vec{f}^- \cdot\vec{n}, 1\right\rangle_{\partial Q} \tag{13}$$

holds on a discrete level. As we aim to use the strong DG formulation throughout this work and as we insist on the exact conservation of our discretization, we propose to use a simple modification of the strong modal DG formulation to guarantee exact conservation.

We recall that we consider orthonormal basis and test functions in this work, which we construct by using a Gram-Schmidt orthogonalization algorithm starting from classic monomial bary-centric basis functions. This implies that the first basis function $\varphi_1^Q(\vec{x}) = 1/\sqrt{|Q|}$ is a constant, which means that the corresponding first degree of freedom $\hat{u}_1^Q(t)$ is related to the grid cell mean value of the DG approximation. Thus we can limit the modification of our DG formulation to the equation resulting from test function φ_1^Q. The proposed modified formulation reads as

$$\left\langle u_t, \varphi_1^Q \right\rangle_Q + \left\langle h, \varphi_1^Q \right\rangle_{\partial Q} = 0, \tag{14}$$

and

$$\left\langle u_t + \vec{\nabla} \cdot \vec{f}, \varphi_j^Q \right\rangle_Q + \left\langle h - \vec{f}^- \cdot \vec{n}, \varphi_j^Q \right\rangle_{\partial Q} - \left\langle w - u^-, \mu \vec{n} \cdot \vec{\nabla} \varphi_j^Q \right\rangle_{\partial Q} = 0, \tag{15}$$

for $j = 2, ..., N$. If we recall that φ_1^Q is constant, it is obvious that the term

$$-\left\langle w - u^-, \mu \vec{n} \cdot \vec{\nabla} \varphi_1^Q \right\rangle_{\partial Q} \tag{16}$$

cancels out. However, as mentioned above, it is not trivial that the term

$$\left\langle \vec{\nabla} \cdot \vec{f}, \varphi_1^Q \right\rangle_Q - \left\langle \vec{f}^- \cdot \vec{n}, \varphi_1^Q \right\rangle_{\partial Q} \tag{17}$$

cancels out and as such it is this forced modification that guarantees the exact conservation of our DG formulation, independent of the numerical integration. To maintain efficiency of our modal DG discretization we use a recently developed nodal based integration technique[19] for the approximation of the volume and surface integrals. Taking into account that the basis and test functions are orthonormal, the resulting semi discrete DG scheme reads as

$$\begin{aligned}
&(\hat{u}_1^Q)_t + \frac{1}{\sqrt{|Q|}} \langle h, 1 \rangle_{\partial Q} = 0, \\
&(\hat{u}_j^Q)_t + \left\langle \vec{\nabla} \cdot \vec{f}, \varphi_j^Q \right\rangle_Q + \left\langle h - \vec{f}^- \cdot \vec{n}, \varphi_j^Q \right\rangle_{\partial Q} - \left\langle w - u^-, \mu \vec{n} \cdot \vec{\nabla} \varphi_j^Q \right\rangle_{\partial Q} = 0, \\
&j = 2, ..., N,
\end{aligned} \tag{18}$$

which we rewrite in a more compact form as the following set of ODE's

$$\underline{\hat{u}}_t^Q = \mathcal{R}_V \left(\underline{\hat{u}}^Q, \underline{\varphi}^Q \right) + \mathcal{R}_S \left(\underline{\hat{u}}^Q, \underline{\hat{u}}^{Q^+}, \underline{\varphi}^Q \right), \tag{19}$$

where we collect all volume terms in $\mathcal{R}_V$ and all surface terms in $\mathcal{R}_S$. We indicate the dependence of the surface term on neighbor data by $\underline{\hat{u}}^{Q^+}$.

We note that the first equation of (18) is identical to the well known cell centered finite volume discretization. The second part of (18) is used to determine higher order contributions, which are used for the evaluation of the fluxes in the surface integral. This stands in stark contrast to high order finite volume schemes, where the needed higher order contributions for the evaluation of the surface integral are obtained by means of reconstruction and not by additional evolution equations.

2.2. *The fully discrete form*

The set of ODE's (19) can now be integrated, where the time interval $[0;T]$ is subdivided into time levels t_n, by using for instance the standard Runge-Kutta methods, resulting in the classic Runge-Kutta discontinuous Galerkin method.[1–5]

In this work, an approach introduced in Ref. 20 is presented. We start with a simple integration in time of the semi-discrete formulation (19) from time level t_n to time level t_{n+1}

$$\underline{\hat{u}}^Q_{n+1} - \underline{\hat{u}}^Q_n = \int_{t_n}^{t_{n+1}} \mathcal{R}_V\left(\underline{\hat{u}}^Q, \underline{\varphi}^Q\right) + \mathcal{R}_S\left(\underline{\hat{u}}^Q, \underline{\hat{u}}^{Q^+}, \underline{\varphi}^Q\right) dt. \qquad (20)$$

The most efficient way to approximate the time integrals is to use one dimensional Gauss quadrature. We get for instance for the volume integral the following approximation

$$\int_{t_n}^{t_{n+1}} \mathcal{R}_V\left(\underline{\hat{u}}^Q(t), \underline{\varphi}^Q\right) dt \approx \sum_{\ell=1}^{L_G} \mathcal{R}_V\left(\underline{\hat{u}}^Q(\tau_\ell), \underline{\varphi}^Q\right) \omega_\ell, \qquad (21)$$

where τ_ℓ and ω_ℓ are the Gauss positions and Gauss weights, respectively. The number of necessary Gauss points L_G and thus the number of necessary volume integral evaluations is the integer part of $\frac{\mathcal{O}_t}{2}$ for a given time order $\mathcal{O}_t$. The problem is, that the DG solution is only known at the 'old' time level t_n. However, the Gauss points are located in-between the time levels t_n and t_{n+1} and thus the solution needed for the evaluation of the volume and surface terms is not known. Lörcher *et al.*[11] proposed to use a space-time expansion in the barycenter, where the space-time derivatives and pure time derivatives are approximated with the so-called Cauchy-Kovalevskaya procedure, and used this auxiliary solution to evaluate the space-time integrals in the fully discrete DG formulation. Recently Dumbser *et al.*[21,22] proposed to use a locally implicit either continuous or discontinuous Galerkin time discretization to define an auxiliary solution. It is interesting to observe that all these different auxiliary solutions are approximations to the following *local Cauchy problem*: Find for every grid cell Q the function $v = v(\vec{x}, t)$ for $(\vec{x}, t) \in \mathbb{R}^d \times [0; \Delta t]$, which satisfies the initial value problem

$$\begin{aligned} v_t + \vec{\nabla} \cdot \vec{f}\left(v, \vec{\nabla} v\right) &= 0, \\ v(\vec{x}, t = 0) &= u^*(\vec{x}, t_n), \end{aligned} \tag{22}$$

where $u^*(\vec{x}, t_n)$ is the DG polynomial $u_Q(\vec{x}, t_n)$ of grid cell Q extended in $\mathbb{R}^d$. It is clear, that the fully discrete DG scheme (20) is afflicted with numerical errors. Hence, the exact solution v is not needed as an approximation v_Q with same accuracy (order) suffices to get a fully discrete scheme with the desired accuracy order in space and time. We therefore propose in this work to use an explicit local Runge-Kutta Galerkin discretization to construct an approximative solution to this local Cauchy problem. Accordingly to the semi discrete DG scheme described above we introduce an approximation with the same polynomial degree

$$v_Q(\vec{x}, t) = \sum_{j=1}^{N} \hat{v}_j^Q(t) \varphi_j^Q(\vec{x}) =: \underline{\hat{v}}^Q(t) \cdot \underline{\varphi}^Q(\vec{x}). \tag{23}$$

Inserting this into (22), multiplying by a test function and integrating over the grid cell Q yields the semi-discrete Galerkin formulation

$$\left\langle (v_Q)_t + \vec{\nabla} \cdot \vec{f}\left(v_Q, \vec{\nabla} v_Q\right), \phi \right\rangle_Q = 0, \tag{24}$$

and analogously the set of ODE's for the time dependent polynomial coefficients

$$\begin{aligned} \underline{\hat{v}}_t^Q &= \mathcal{R}_V\left(\underline{\hat{v}}^Q, \underline{\varphi}^Q\right), \\ \underline{\hat{v}}^Q(0) &= \underline{\hat{u}}^Q(t_n). \end{aligned} \tag{25}$$

We note that the auxiliary problem (22) does not involve DG data from neighbor grid cells. An integration by parts consequently does not change the Galerkin formulation (24), as the normal flux component in the surface integral is uniquely defined.

As stated above we aim to use a Runge-Kutta method to integrate (25) in time. However, to evaluate the space-time integrals in Eq. (20), a continuous approximation in time is needed. In Refs. 23 and 24 a special

Runge-Kutta based framework for the solution of such initial value problems was introduced, with the main feature that the approximation can be naturally extended to a time polynomial, hence the name *continuous extension Runge-Kutta (CERK) schemes*. The space-time-polynomial $\underline{\hat{v}}^Q(t)$ is computed according to

$$\begin{aligned} \underline{\hat{v}}^Q(t) &= \sum_{k=0}^{\mathcal{O}_t^*-1} c_k^t\, t^k, \\ c_k^t &= \frac{1}{(\Delta t)^k} \sum_{j=1}^{n_{\text{stages}}} b_{kj}^t \underline{\hat{k}}_j, \\ j &= 1, n_{\text{stages}} \\ &\left| \begin{array}{ll} \underline{\hat{v}}_{n,j}^Q &= \underline{\hat{u}}_n^Q + \Delta t \sum_{l=1}^{j-1} a_{jl} \underline{\hat{k}}_l, \\ \underline{\hat{k}}_j &= \mathcal{R}_V(\underline{\hat{v}}_{n,j}^Q, \underline{\varphi}^Q), \end{array} \right. \end{aligned} \tag{26}$$

where n_{stages} and a_{jl}, b_{kl}^t are the CERK coefficients depending on the time order $\mathcal{O}_t^*$. We have listed the coefficients for a second,third and forth order CERK scheme in Tables 1-4.

Higher order schemes can be found in Refs. 23 and 24. We observed, that for a desired time order $\mathcal{O}_t$ of the final scheme, we need one order less for the construction of the approximation of the local Cauchy problem

$$\mathcal{O}_t^* = \mathcal{O}_t - 1. \tag{27}$$

Table 1. Coefficients for $\mathcal{O}_t^* = 2$ with $n_{stages} = 2$ (Heun method).

j	$\mathbf{a_{j1}}$	k	$\mathbf{b^t_{k1}}$	$\mathbf{b^t_{k2}}$
		0	1	0
2	1	1	$-\frac{1}{2}$	$\frac{1}{2}$

Table 2. Coefficients for $\mathcal{O}_t^* = 3$ with $n_{stages} = 4$.

j	$\mathbf{a_{j1}}$	$\mathbf{a_{j2}}$	$\mathbf{a_{j3}}$	k	$\mathbf{b^t_{k1}}$	$\mathbf{b^t_{k2}}$	$\mathbf{b^t_{k3}}$	$\mathbf{b^t_{k4}}$
2	$\frac{12}{23}$			0	1	0	0	0
3	$-\frac{68}{375}$	$\frac{368}{375}$		1	$-\frac{65}{48}$	$\frac{529}{384}$	$\frac{125}{128}$	-1
4	$\frac{31}{144}$	$\frac{529}{1152}$	$\frac{125}{384}$	2	$\frac{41}{72}$	$-\frac{529}{576}$	$-\frac{125}{192}$	1

Table 3. Coefficients a_{jl} for $\mathcal{O}_t^* = 4$ with $n_{stages} = 6$.

j	$\mathbf{a_{j1}}$	$\mathbf{a_{j2}}$	$\mathbf{a_{j3}}$	$\mathbf{a_{j4}}$	$\mathbf{a_{j5}}$
2	$\frac{1}{6}$				
3	$\frac{44}{1369}$	$\frac{363}{1369}$			
4	$\frac{3388}{4913}$	$-\frac{8349}{4913}$	$\frac{8140}{4913}$		
5	$-\frac{36764}{408375}$	$\frac{767}{1125}$	$-\frac{32708}{136125}$	$\frac{210392}{408375}$	
6	$-\frac{1697}{18876}$	0	$\frac{50653}{116160}$	$\frac{299693}{1626240}$	$\frac{3375}{11648}$

Table 4. Coefficients b_{kj}^t for $\mathcal{O}_t^* = 4$ with $n_{stages} = 6$.

k	$\mathbf{b_{k1}^t}$	$\mathbf{b_{k2}^t}$	$\mathbf{b_{k3}^t}$	$\mathbf{b_{k4}^t}$	$\mathbf{b_{k5}^t}$	$\mathbf{b_{k6}^t}$
0	1	0	0	0	0	0
1	$-\frac{104217}{37466}$	0	$\frac{861101}{230560}$	$-\frac{63869}{293440}$	$-\frac{1522125}{762944}$	$\frac{165}{131}$
2	$\frac{1806901}{618189}$	0	$-\frac{2178079}{380424}$	$\frac{6244423}{5325936}$	$\frac{982125}{190736}$	$-\frac{461}{131}$
3	$-\frac{866577}{824252}$	0	$\frac{12308679}{5072320}$	$-\frac{7816583}{10144640}$	$-\frac{624375}{217984}$	$\frac{296}{131}$

This means that for a desired time order of $\mathcal{O}_t = 3$, we need a second order accurate CERK method to calculate the approximation v_Q. The evaluation of (20) with the approximation v_Q

$$\underline{\hat{u}}_{n+1}^Q - \underline{\hat{u}}_n^Q = \int_0^{\Delta t} \mathcal{R}_V\left(\underline{\hat{v}}^Q(t), \underline{\varphi}^Q\right) + \mathcal{R}_S\left(\underline{\hat{v}}^Q(t), \underline{\hat{v}}^{Q^+}(t), \underline{\varphi}^Q\right) dt, \qquad (28)$$

increases the time order $\mathcal{O}_t$ by 1. Summing up, we have shown how to use a Runge-Kutta method to construct a time continuous auxiliary solution and therewith a fully discrete DG scheme.

2.3. *The predictor-corrector formulation*

The computation of the auxiliary problem involves the evaluation of the volume integral term for every local Runge-Kutta stage. The evaluation of the fully discrete scheme involves an additional computation of the volume integral terms, evaluated at the time Gauss points

$$\int_{t_n}^{t_{n+1}} \mathcal{R}_V\left(\underline{\hat{u}}^Q(t), \underline{\varphi}^Q\right) dt \approx \int_0^{\Delta t} \mathcal{R}_V\left(\underline{\hat{v}}^Q(t), \underline{\varphi}^Q\right) dt \approx \sum_{\ell=1}^{L_G} \mathcal{R}_V\left(\underline{\hat{v}}^Q(\tau_\ell), \underline{\varphi}^Q\right) \omega_\ell. \tag{29}$$

At first sight it seems that the volume integral is calculated twice. If we recall the semi-discrete Galerkin formulation of the local Cauchy problem (24), we notice that the volume residual is related to the time derivative of the auxiliary solution

$$\begin{aligned} -\left\langle \vec{\nabla} \cdot \vec{f}\left(v_Q, \vec{\nabla} v_Q\right), \phi \right\rangle_Q &= \langle (v_Q)_t, \phi \rangle_Q, \\ \mathcal{R}_V\left(\underline{\hat{v}}^Q, \underline{\varphi}^Q\right) &= \underline{\hat{v}}_t^Q. \end{aligned} \tag{30}$$

Inserting this into (29) yields

$$\int_{t_n}^{t_{n+1}} \mathcal{R}_V\left(\underline{\hat{u}}^Q, \underline{\varphi}^Q\right) dt \approx \int_0^{\Delta t} \mathcal{R}_V\left(\underline{\hat{v}}^Q, \underline{\varphi}^Q\right) dt = \int_0^{\Delta t} \underline{\hat{v}}_t^Q \, dt = \underline{\hat{v}}^Q(\Delta t) - \underline{\hat{v}}^Q(0). \tag{31}$$

The strong variant of the fully discrete DG scheme (28) can now be simplified to

$$\underline{\hat{u}}_{n+1}^Q - \underline{\hat{u}}_n^Q = \underline{\hat{v}}^Q(\Delta t) - \underline{\hat{v}}^Q(0) + \int_0^{\Delta t} \mathcal{R}_S\left(\underline{\hat{v}}^Q(t), \underline{\hat{v}}^{Q^+}(t), \underline{\varphi}\right) dt. \tag{32}$$

Due to the construction of the auxiliary solution we have furthermore

$$\underline{\hat{v}}^Q(t=0) = \underline{\hat{u}}_n^Q. \tag{33}$$

Inserting this into the formulation (32) yields the *predictor-corrector formulation*

$$\underline{\hat{u}}_{n+1}^Q = \underline{\hat{v}}^Q(\Delta t) + \int_0^{\Delta t} \mathcal{R}_S\left(\underline{\hat{v}}^Q(t), \underline{\hat{v}}^{Q^+}(t), \underline{\varphi}\right) dt. \tag{34}$$

This formulation shows, that the DG solution at the new time level $\underline{\hat{u}}^{n+1}$ is determined by the value of the prediction at the new time level $\underline{\hat{v}}^Q(\Delta t)$ (note that the predictor does not take any neighbor data into account) corrected with the surface integral term, where information from the local and the neighbor grid cells is taken into account. We see that in this formulation, the volume integral of the local Runge-Kutta scheme is re-used.

3. Beyond the Global Time Integration Paradigm

Up to now, only global time levels are considered. But as discussed above, the fundamental problem of explicit time integration schemes is that for global stability, the minimum time step is required, rendering these type of schemes unpopular for the application of real life problems. In the next section the presented Runge-Kutta based time integration method is combined with a time accurate local time step algorithm, first introduced in Ref. 11. It is the general locality of the DG semi discrete formulation and the locality of our Runge-Kutta based fully discrete scheme, which allows this new time-marching technique: Each grid cell may evolve with its own *local time step* in a time-consistent manner.

3.1. *Time-accurate local time stepping*

We give up the assumption that all grid cells run with the same time step and skip the common time level. Let us denote the old local time level in grid cell Q_i by t_i^n. The degrees of freedom we store in a vector $\underline{\hat{u}}_i^n$ which then represents the solution at t_i^n in Q_i. According to the local stability restriction the approximation in Q_i may evolve in time with the local time step Δt_i^n. The next local time level in Q_i is then given by

$$t_i^{n+1} = t_i^n + \Delta t_i^n \tag{35}$$

and the space-time grid cell is denoted $Q_i^n = Q_i \times \left[t_i^n, t_i^{n+1}\right]$. The fully-discrete evolution equations for the DOF (28) or (34) have still the same form, since the predictor is completely local. The evolution of the DOF $\underline{\hat{u}}_i^n$ from level t_i^n to t_i^{n+1} is now done in two steps. The *predictor step* is directly calculated using the volume terms $\mathcal{R}_V$, since they depends only on local data. For the *corrector step*, the surface terms $\mathcal{R}_S$ have to be evaluated. They depend on data of the adjacent neighbors, i.e., on their predictive space-time approximation of the continuous extension Runge-Kutta Galerkin scheme. Thus the neighbor predictor must be available and therefore the succession of the time evolution of the elements has to be controlled.

First we take a careful look at the surface term $\mathcal{R}_S$ which has the following general form

$$\int_{t_i^n}^{t_i^{n+1}} \oint_{\partial Q_i} h(\vec{x}, t)\varphi_j(\vec{x})\, ds\, dt, \tag{36}$$

where $h(\vec{x},t)$ is a numerical flux depending on local and neighbor data. In the modal DG framework, the *spatial* integral can be either approximated with Gauss integration or with a nodal type integration (see Ref. 19 for details). Both approaches can be formulated in an abstract way

$$\int_{t_i^n}^{t_i^{n+1}} \sum_{k=1}^{M} \tilde{h}_k(t)\,\omega_k^j \; dt, \tag{37}$$

where $\tilde{h}_k(t)$ is the value of the function $h(\vec{x},t)$ at the specific k-th spatial Gauss (or interpolation) point $\vec{\xi}_k$. The weight ω_k^j contains the weights of the Gauss or either Nodal integration and the evaluation of the test function $\varphi_j(\vec{x})$ at $\vec{x}=\vec{\xi}_k$. We note that this weight does not depend on time and can be calculated once at the beginning of the calculation. Furthermore, it can be put out of the time integral and integration and summation can be changed

$$\sum_{k=1}^{M} \underbrace{\left(\int_{t_i^n}^{t_i^{n+1}} \tilde{h}_k(t)\,dt \right)}_{=:H_i(\vec{\xi}_k,[t_i^n,t_i^{n+1}])} \omega_k^j, \tag{38}$$

where $H_i(\vec{\xi}_j)$ is now the *time integrated flux* at evaluation point $\vec{\xi}_j$ of grid cell Q_i. This means that fluxes are evaluated from the predictor and first integrated in time and then in space. Especially, the time interval $[t_i^n, t_i^{n+1}]$ can be splitted in an arbitrary number of time intervals

$$H_i(\vec{\xi}_j,[t_i^n,t_i^{n+1}]) = \int_{t_i^n}^{t_i^{n+1}} \tilde{h}_j(t)\,dt = \int_{t_i^n}^{t_i^{\alpha 1}} \tilde{h}_j(t)\,dt + \int_{t_i^{\alpha 1}}^{t_i^{\alpha 2}} \tilde{h}_j(t)\,dt + \cdots + \int_{t_i^{\alpha m}}^{t_i^{n+1}} \tilde{h}_j(t)\,dt \tag{39}$$

before applying the space integration. This is crucial for the efficiency of the local time-stepping algorithm.

As an example we sketched the time evolution with four adjacent one-dimensional grid cells in Figures 1 – 4 starting from a common time level $t^0 = 0$.

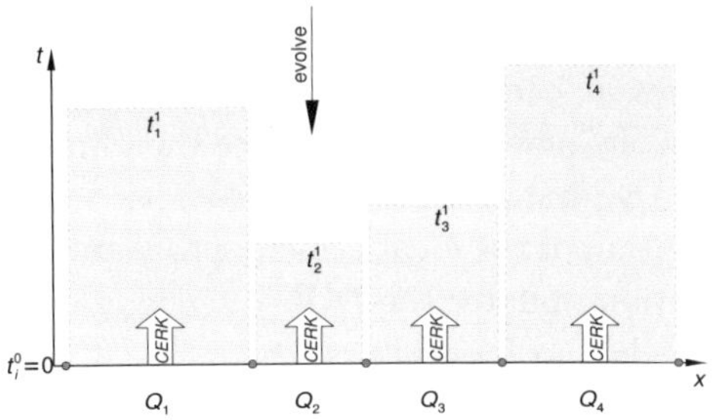

Fig. 1. Predictor for all elements.

After the determination of the local time steps, which are assumed to be different in our example due to the local stability restriction, the auxiliary local CERK Galerkin solutions are calculated in each grid cell using the methodology described in section 2.2. This results in a predictive approximate solution in all space-time cells $Q_i \times [t_i^0, t_i^1], i = 1 \ldots 4$ - see Figure 1. These space-time polynomials are stored. We note that after this step the degrees of freedom $\underline{\hat{u}}_i^0$ at the time level t_i^0 are not needed any longer and may be overwritten in the computer program. After the predictor is calculated for each element Q_i, the predictor solution at the new time level $\underline{\hat{v}}^{Q_i}(\Delta t)$ overwrites $\underline{\hat{u}}_i^0$, see Eq. (34). We call these values $\underline{\hat{u}}_i^*$.

Next, the surface flux contributions involving neighbor information have to be considered.

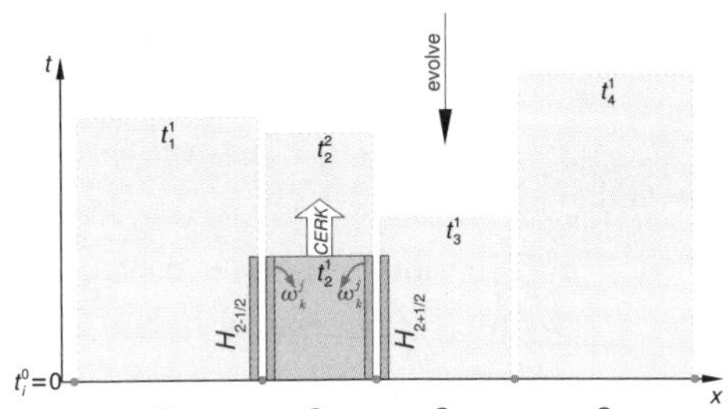

Fig. 2. Evolution of Q_2.

The local time-stepping algorithm relies on the following evolve condition. The update of the DOF can only be completed, if

$$t_i^{n+1} \leq \min\left\{t_j^{n+1}\right\}, \quad \forall j : Q_j \cap Q_i \neq \emptyset \qquad (40)$$

is satisfied. This condition guarantees that all the data for the interface fluxes are available. In our example Q_2 is the first to satisfy the evolve condition - see Figure 1.

The vertical bars in Figure 2 depict the flux time integral $H_{2\pm1/2}(\vec{\xi}, [t_2^0, t_2^1])$ for the right and left cell interface. The arguments for the numerical flux functions are obtained from the left and right space-time polynomials, i.e. the CERK Galerkin solutions. In order to keep this calculation exactly conservative as well as efficient, the contribution $H_{2\pm1/2}(\vec{\xi}, [t_2^0, t_2^1])$ is added simultaneously with the minus sign to the corresponding flux evaluation of the neighbors Q_1, Q_3, where it is stored in a container for the element side. The DOF of Q_2 are completed by applying the spatial integration at the right interface and the left interface, i.e. by multiplying the time integrated fluxes at each point with the integration weights, Eq. (38). The DOF of Q_2 at the new time level t_2^1 are then known and the procedure starts again as in the first time step: A new space-time polynomial is constructed via the solution of the local Cauchy problem (22) in $Q_2 \times [t_2^1, t_2^2]$ and the predictor solution updates the DOF $\hat{\underline{u}}_2^1$, now named $\hat{\underline{u}}_2^*$. We are then in the situation sketched in Figure 2.

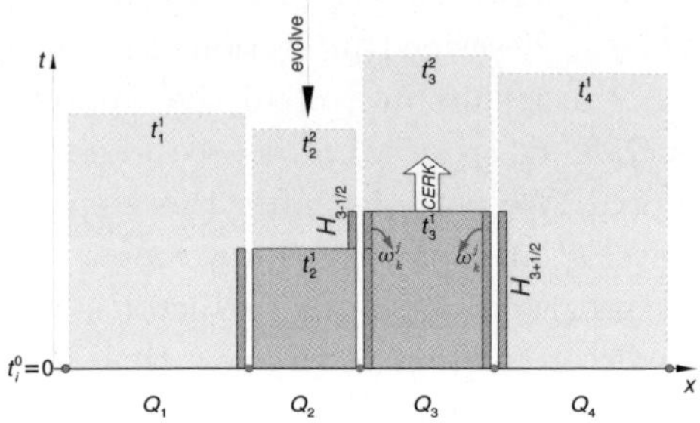

Fig. 3. Evolution of Q_3.

Next Q_3 satisfies the evolve condition - see Figure 2. As before, the predictor update was already done. But in this case, also a part of the flux contributions has already been computed during the previous evolution of Q_2. So only the missing flux contributions, $H_{3-1/2}(\vec{\xi}, [t_2^1, t_3^1])$ and $H_{3+1/2}(\vec{\xi}, [t_3^0, t_3^1])$, have to be added. In order to get the new DOF $\hat{\underline{u}}_1^1$, we finally apply the spatial integration. The time interval, for which the flux contribution at the interface shared by an element Q_i and an adjacent element Q_j has to be computed is generally

$$[t_{ij}^\star, t_i^{n+1}] = [max(t_i^n, t_j^n), t_i^{n+1}]. \tag{41}$$

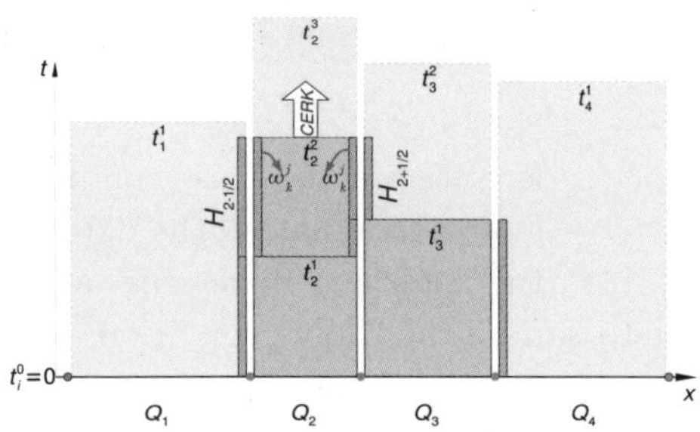

Fig. 4. Evolution of Q_2.

In this manner, the algorithm continues with the next element satisfying the evolve condition (40) - see Figure 4. So all elements are evolved in a suitable order by evaluating the surface terms of the right hand side of Eq. (34) effectively.

At each time, the interface fluxes are defined uniquely for both adjacent elements, making the scheme exactly conservative. The presented local time-stepping algorithm minimizes the total number of time steps for a computation with fixed end time. As outlined above, the spatial surface operator is only applied once per time step, in comparison to a classical (global) Runge-Kutta DG scheme, where the surface term is computed in each stage. The stability of the scheme is similar to the STE-DG scheme and was investigated in Ref. 11 for the global time stepping base scheme and the local time stepping variant and in Refs. 16 and17 for the diffusion equation. The stability numbers compare well with the stability numbers of global time stepping Runge-Kutta schemes with same time order.

4. Results

4.1. *Time accuracy*

A two-dimensional periodic linear advection problem is used for the time accuracy tests. A sinusoidal wave is transported periodically with the velocity $(0.5, 0.5)^T$ during the time interval $t = [0, 2]$, and a stretched mesh forces the local timesteps to be different, see Fig. 5. Although the mesh is not extremely stretched, the mean timestep is a factor of 1.7 times greater than the minimum timestep, showing the potential of the local time-stepping. The polynomial degree of the spatial approximation is kept constant to $p = 5$. In Table 5 the error to the exact solution and the experimental order of convergence is shown for different time orders. The problem is well resolved in space, thus the time accuracy can be clearly seen. The tests confirm that for a desired time accuracy of $\mathcal{O}_t$, it is sufficient that time accuracy of the predictor is one order less $\mathcal{O}_t^* = \mathcal{O}_t - 1$.

Table 5. Convergence of $p = 5$ in space at different time orders

		$\mathcal{O}_s = 6$, $\mathcal{O}_t^* = 5$		$\mathcal{O}_s = 6$, $\mathcal{O}_t^* = 2$		$\mathcal{O}_s = 6$, $\mathcal{O}_t^* = 3$	
N	n_{Cells}	L_2	EOC	L_2	EOC	L_2	EOC
4	16	3.33E-004		3.49E-004		3.33E-004	
8	64	9.16E-006	5.18	1.28E-005	4.76	9.18E-006	5.18
16	256	1.54E-007	5.90	1.32E-006	3.29	1.57E-007	5.87
32	1024	2.33E-009	6.05	1.67E-007	2.98	2.96E-009	5.73
64	4096	3.54E-011	6.04	2.10E-008	2.99	1.19E-010	4.63

4.2. *Accuracy for non-linear problems*

We consider in this subsection the two-dimensional non-linear Euler equations

$$U_t + \vec{\nabla} \cdot \vec{F}(U) = 0, \tag{42}$$

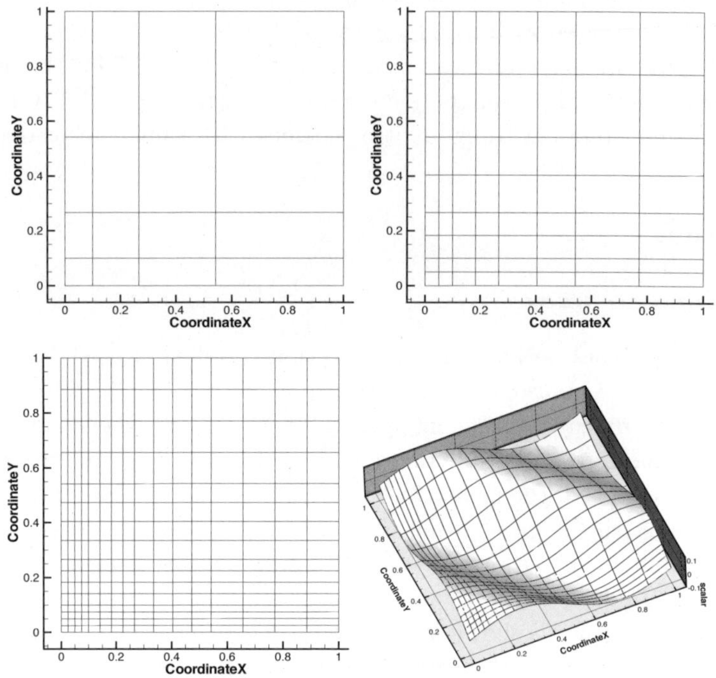

Fig. 5. Initial mesh and sequence of mesh refinements ($N = 4/8/16$), initialisation.

with the vector of the conservative variables $U = (\rho, \rho v_1, \rho v_2, \rho e)^T$ and the Euler fluxes $\vec{F} := (F_1, F_2)^T$:

$$F_l(U) = \begin{pmatrix} \rho\, v_l \\ \rho\, v_1 v_l + \delta_{1l}\, p \\ \rho\, v_2 v_l + \delta_{2l}\, p \\ \rho\, e v_l + p\, v_l \end{pmatrix}, \; l = 1, 2. \tag{43}$$

Here, we use the usual nomination of the physical quantities: ρ, $\vec{v} = (v_1, v_2)^T$, p, and e denote the density, the velocity vector, the pressure, and the specific total energy, respectively. Here the adiabatic exponent $\kappa = \frac{c_p}{c_v}$ with the specific heats c_p, c_v depend on the fluid, and are supposed to be constant for this test. The system is closed with the equation of state of a perfect gas:

$$p = \rho R T = (\kappa - 1)\rho(e - \frac{1}{2}\vec{v} \cdot \vec{v}), \qquad \text{and} \qquad e = \frac{1}{2}\vec{v} \cdot \vec{v} + c_v T. \tag{44}$$

Table 6. Convergence of $p = 5$ in space $\mathcal{O}_t^* = 2$ for the nonlinear problem.

		$\mathcal{O}_s = 6,\ \mathcal{O}_t^* = 5$		$\mathcal{O}_s = 6,\ \mathcal{O}_t^* = 2$	
N	n_{Cells}	$L_2(\rho)$	EOC	$L_2(\rho)$	EOC
7	49	1.19E-004	-	1.23E-004	-
14	196	2.26E-006	5.72	4.16E-006	4.88
28	784	3.39E-008	6.06	4.79E-007	3.12
56	3136	7.35E-010	5.79	6.02E-008	2.99

with the specific gas constant $R = c_p - c_v$. The considered test case is the two dimensional isentropic vortex convection problem of Hu and Shu[25]

$$
\begin{aligned}
\vec{r} &= (\, y - y_0 - v_{0,y}t \,,\, x - x_0 - v_{0,x}t \,)^T \,, \\
\delta v &= \frac{v_{max}}{2\pi} \exp\left(\frac{1 - \left(\frac{|\vec{r}|}{r_0}\right)^2}{2} \right), \\
\vec{v}(\vec{x},t) &= \vec{v}_0 + \delta v \cdot \vec{r}, \\
\frac{T}{T_0} &= 1 - \frac{\kappa - 1}{2} \left(\frac{\delta v}{c_o}\right)^2, \\
\rho(\vec{x},t) &= \rho_0 \left(\frac{T}{T_0}\right)^{\frac{1}{\kappa-1}}, \\
p(\vec{x},t) &= p_0 \left(\frac{T}{T_0}\right)^{\frac{\kappa}{\kappa-1}}.
\end{aligned}
\tag{45}
$$

For our test problem we chose the background flow $(\rho_0, \vec{v}_0^T, p_0) = (1., 1., 0., \frac{1}{\kappa})$, $\kappa = 1.4$, the initial center of the vortex $\vec{x}_0 = (0.5, 0.5)^T$, the amplitude of the vortex $v_{max} = 0.01$, the half width of the vortex $r_0 = 0.1$ and the end time of the simulation $t_{end} = 1.0$. The computational domain $\Omega := [0., 1.0]^2$ with periodic boundary conditions prescribed. The meshes and initial velocity contours are shown in figure 6 and the time accuracy is shown in Table 6 for $\mathcal{O}_t^* = 5$ and $\mathcal{O}_t^* = 2$.

4.3. *Application*

We investigate the aeroacoustics of a natural gas injector, using a fully unstructured mesh with 552,000 elements, a polynomial degree of $p = 3$ and $\mathcal{O}_t^* = 2$ for the predictor. The gas is injected at supersonic speed and the jet expands to ambient condition. At the injector outlet, fine shock structures have to be captured thus high spatial and temporal resolution is needed. Inside the jet, a medium resolution is used which is coarsened

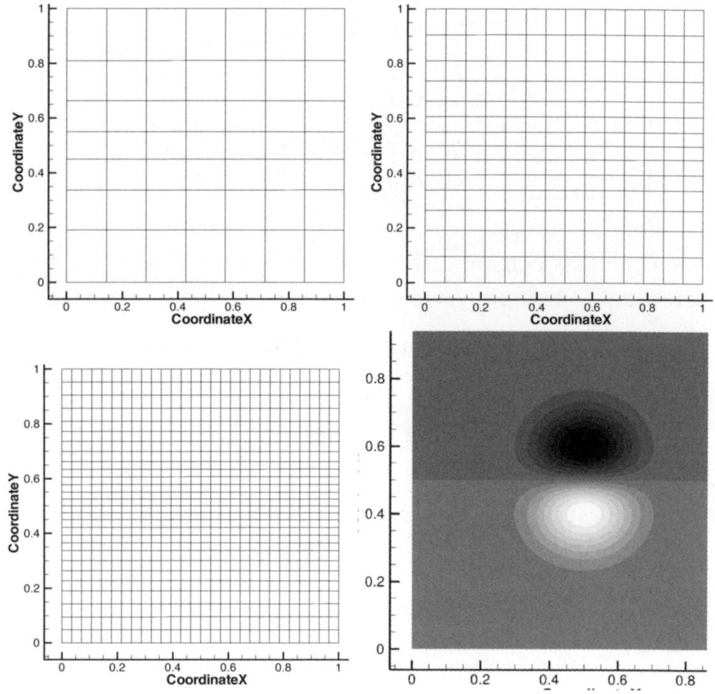

Fig. 6. Initial mesh and sequence of mesh refinements ($N = 7/14/28$), contours of v_x.

towards the outflow boundaries, see Fig 7. The three-dimensional Navier-Stokes equations are solved. For the shock capturing, artificial viscosity is applied locally to the troubled grid cells. The smallest timestep in this example is $\Delta t_{\min} = 1.12 \times 10^{-11}s$, which is due to the small grid size and high velocities as well as the artificial viscosity. The total number of timesteps for all grid cells performed for $\Delta T = 1\mu s$ is 1.8×10^8. Having $n_{\text{elems}} = 552,000$ elements, the total number of timesteps for a global time-stepping would be $n_{\text{elems}}\Delta T/\Delta t_{\min} = 4.9 \times 10^{10}$. Comparing this two numbers reveal, that the speed up due to the local time stepping compared to the global time stepping variant of the scheme is a factor of ~ 274. Figure 8 shows visualization of the instantaneous flow field, demonstrating the multi-scale character of this flow.

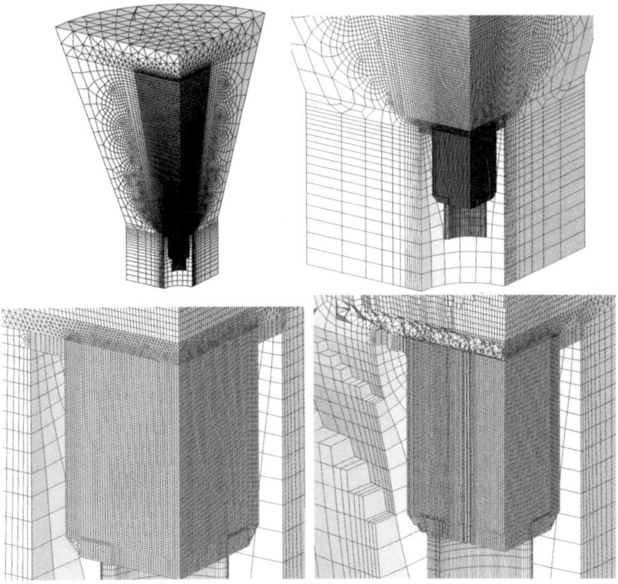

Fig. 7. Unstructured mesh for the natural gas injector.

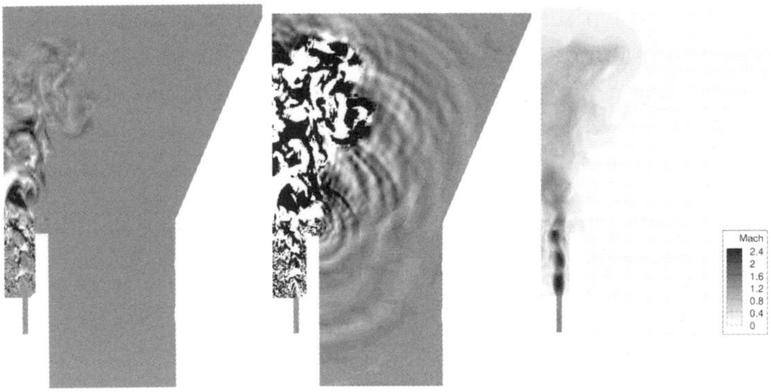

Fig. 8. diagonal cutting plane: Density gradient inside and outside of the jet and Mach number.

5. Conclusion

In this work we showed how to make use of the inherent locality of a Discontinuous Galerkin discretization to construct an explicit Runge-Kutta based predictor-corrector time integrator which allows time accurate local time stepping. The presented methodology yields an efficient solution method for unsteady 'advection dominated' problems and an interesting alternative to implicit or implicit/explicit time discretizations. The property 'advection dominated' depends heavily on the underlying problem. Focusing the discussion on the compressible Navier-Stokes equations, we have two effects which penalize an explicit time integration

- viscosity: if viscous physics dominate, the time step restriction behaves like $\Delta t \simeq \Delta x^2/(2\,p+1)^2$ causing a large number of time steps and thus high computational costs,
- compressibility: the maximum advection eigenvalue λ^a_{max} is composed of the flow speed magnitude and the speed of sound. If one is not interested in the propagation of the acoustic waves, the physical meaningful time step is determined by the flow speed. The ratio of flow speed and speed of sound is denoted by the Mach number Ma. The Mach number is used to characterize the compressibility of the flow, where low Mach numbers relate to low compressibility (a Mach number equal zero corresponds to the incompressible limit). For high Mach numbers the maximum advection eigenvalue is about the magnitude of the flow speed, yielding an explicit time step restriction in the range of the physical meaningful time step. For low Mach number flows, the eigenvalue is dominated by the speed of sound, yielding a time step which could be significantly lower than the physical time step. Generally, the efficiency of an explicit time discretization depends on the disparity of the problem inherent different time scales, namely the stiffness of the problem.

Thus, the presented time integration method and its application to the unsteady compressible Navier-Stokes equations is best suited for high Reynolds number and transonic and supersonic Mach number flows. The focus of our research is the application of this framework to large eddy simulation of such characterized problems.

Acknowledgments

The research presented in this paper was supported in parts by Deutsche Forschungsgemeinschaft (DFG), amongst others within the Schwerpunktprogramm 1276: MetStroem, the Bundesministerium für Bildung und Forschung (BMBF, Federal Ministry for Education and Research) in the HPC Software Initiative Projekt "STEDG: Hocheffiziente und skalierbare

Software für die Simulation turbulenter Strömungen in komplexen Geometrien" and the Cluster of Excellence in Simulation Technology (SimTech) at the University of Stuttgart.

References

1. B. Cockburn and C.-W. Shu, The Runge-Kutta local projection p^1-discontinuous Galerkin method for scalar conservation laws, *M^2AN.* **25**, 337–361, (1991).
2. B. Cockburn and C.-W. Shu, TVB Runge-Kutta local projection discontinuous Galerkin finite element method for conservation laws II: General framework, *Math. Comput.* **52**, 411–435, (1989).
3. B. Cockburn, S. Y. Lin, and C.-W. Shu, TVB Runge-Kutta local projection discontinuous Galerkin finite element method for conservation laws III: One dimensional systems, *J. Comput. Phys.* **84**, 90–113, (1989).
4. B. Cockburn, S. Hou, and C.-W. Shu, The Runge-Kutta local projection discontinuous Galerkin finite element method for conservation laws IV: The multidimensional case, *Math. Comput.* **54**, 545–581, (1990).
5. B. Cockburn and C.-W. Shu, The Runge-Kutta discontinuous Galerkin method for conservation laws V: Multidimensional systems, *J. Comput. Phys.* **141**, 199–224, (1998).
6. R. Biswas, K. Devine, and J. Flaherty, Parallel, adaptive finite element methods for conservation laws, *Appl. Numer. Math.* **14**, 255–283, (1994).
7. L. Wang and D. J. Mavriplis, Implicit solution of the unsteady Euler equations for high-order accurate discontinuous Galerkin discretizations, *J. Comput. Phys.* **225**, 1994–2015, (2007).
8. P.-O. Persson and J. Peraire. An efficient low memory implicit dg algorithm for time dependent problems. In *Proc. of the 44th AIAA Aerospace Sciences Meeting and Exhibit*, (2006).
9. J. J. W. van der Vegt and H. van der Ven, Space–time discontinuous Galerkin finite element method with dynamic grid motion for inviscid compressible flows: I. General formulation, *J. Comput. Phys.* **182**(2), 546–585, (2002).
10. A. Kanevsky, M. Carpenter, D. Gottlieb, and J. Hesthaven, Application of implicit-explicit high-order Runge-Kutta methods to discontinuous Galerkin schemes, *J. Comput. Phys.* **225**, 1753–1781, (2007).
11. F. Lörcher, G. Gassner, and C.-D. Munz, A discontinuous Galerkin scheme based on a space-time expansion. I. Inviscid compressible flow in one space dimension, *J. Sci. Comp.* **32**(2), 175–199, (2007).
12. M. Dumbser, M. Käser, and E. F. Toro, An arbitrary high order discontinuous Galerkin method for elastic waves on unstructured meshes V: Local time stepping and p-adaptivity, *Geophysical Journal International.* **171**, 695–717, (2007).
13. G. Gassner, F. Lörcher, and C.-D. Munz, A discontinuous Galerkin scheme based on a space-time expansion. II. Viscous flow equations in multi dimensions., *J. Sci. Comp.* **34**(3), 260–286, (2008).

14. D. N. Arnold, F. Brezzi, B. Cockburn, and L. D. Marini, Unified analysis of discontinuous Galerkin methods for elliptic problems, *SIAM J. Numer. Anal.* **39**(5), 1749–1779, (2002).
15. F. Bassi and S. Rebay, A high-order accurate discontinuous finite element method for the numerical solution of the compressible Navier-Stokes equations, *J. Comput. Phys.* **131**, 267–279, (1997).
16. G. Gassner, F. Lörcher, and C.-D. Munz, A contribution to the construction of diffusion fluxes for finite volume and discontinuous Galerkin schemes, *J. Comput. Phys.* **224**(2), 1049–1063, (2007).
17. F. Lörcher, G. Gassner, and C.-D. Munz, An explicit discontinuous Galerkin scheme with local time-stepping for general unsteady diffusion equations, *J. Comput. Phys.* **227**(11), 5649–5670, (2008).
18. E. Toro, *Riemann Solvers and Numerical Methods for Fluid Dynamics.* (Springer, June 1999).
19. G. Gassner, F. Lörcher, C.-D. Munz, and J. S. Hesthaven, Polymorphic nodal elements and their application in discontinuous Galerkin methods, *J. Comput. Phys.* **228**, 1573–1590, (2009).
20. G. Gassner, *Discontinuous Galerkin Methods for the Unsteady Compressible Navier-Stokes equations.* (Dr. Hut Verlag, 2009). http://elib.uni-stuttgart.de/opus/volltexte/2009/3948/.
21. M. Dumbser, D. S. Balsara, E. F. Toro, and C.-D. Munz, A unified framework for the construction of one-step finite-volume and discontinuous Galerkin schemes on unstructured meshes, *J. Comput. Phys.* **227**, 8209–8253, (2008).
22. M. Dumbser, C. Enaux, and E. F. Toro, Finite volume schemes of very high order of accuracy for stiff hyperbolic balance laws, *J. Comput. Phys.* **227**, 3971–4001, (2008).
23. B. Owren and M. Zennaro, Order barriers for continuous explicit runge-kutta methods, *Math. Comp.* **56**, 645–661, (1991).
24. B. Owren and M. Zennaro, Derivation of efficient continuous explicit runge-kutta methods, *SIAM J. Sci. Stat. Comput.* **13**, 1488–1501, (1992).
25. C. Hu and C.-W. Shu, Weighted essentially non-oscillatory schemes on triangular meshes, *J. Comput. Phys.* **1505**, 97–127, (1999).

CHAPTER 5

HIGH-ORDER DISCONTINUOUS GALERKIN METHODS FOR CFD

Jaime Peraire

Department of Aeronautics and Astronautics,
Massachusetts Institute of Technology,
Cambridge, MA 02139, USA
peraire@mit.edu

Per-Olof Persson

Department of Mathematics,
University of California, Berkeley,
Berkeley, CA 94720-3840, USA
persson@berkeley.edu

Discontinuous Galerkin (DG) methods have gained popularity in the CFD community because of their ability to discretize conservation laws with high-order accuracy on complex geometries. However, several issues need to be addressed before these schemes can replace traditional low-order methods, for example the high computational cost and the lack of robustness for nonlinear problems. We review a number of developments in our work on DG methods, including a mapping-based ALE formulation for deforming domains, the Compact Discontinuous Galerkin (CDG) method for low cost discretization of viscous terms, the stabilization of shocks and under-resolved features using artificial diffusion, and the ILU-multigrid preconditioner with automatic element ordering for Newton-Krylov solvers. We demonstrate the methods on a number of practical applications, including aeroacoustics, turbulent flows, and flapping flight.

1. Introduction

In recent years it has become clear that the current computational methods for scientific and engineering phenomena are inadequate for many challenging problems. Examples of these problems are wave propagation, turbulent fluid flow, as well as problems involving nonlinear interactions and multiple scales. This has resulted in a significant interest in so-called high-order accurate methods, which have the potential to produce more accurate and reliable solutions. A number of high-order numerical methods appropriate for flow simulation have been proposed, including finite difference methods,[1,2] high-order finite volume methods,[3,4] stabilized finite element methods,[5] Discontinuous Galerkin (DG) methods,[6–8] hybridized DG methods,[9–11] and spectral element/difference methods.[12,13] All of these methods have advantages in particular situations, but for various reasons most general purpose commercial-grade simulation tools still use traditional low-order methods.

Much of the current research is devoted to the discontinuous Galerkin method. This is partly because of its many attractive properties, such as a rigorous mathematical foundation, the ability to use arbitrary orders of discretization on general unstructured simplex meshes, and the natural stability properties for convective-diffusive operators. In this chapter, we describe our work on efficient DG methods for unsteady compressible flow applications, including deformable domains and turbulent flows.

2. Governing Equations

2.1. *The compressible Navier-Stokes equations*

The compressible Navier-Stokes equations are written as:

$$\frac{\partial \rho}{\partial t} + \frac{\partial}{\partial x_i}(\rho u_i) = 0, \tag{1}$$

$$\frac{\partial}{\partial t}(\rho u_i) + \frac{\partial}{\partial x_i}(\rho u_i u_j + p) = +\frac{\partial \tau_{ij}}{\partial x_j} \quad \text{for } i = 1, 2, 3, \tag{2}$$

$$\frac{\partial}{\partial t}(\rho E) + \frac{\partial}{\partial x_i}\left(u_j(\rho E + p)\right) = -\frac{\partial q_j}{\partial x_j} + \frac{\partial}{\partial x_j}(u_j \tau_{ij}), \tag{3}$$

where ρ is the fluid density, u_1, u_2, u_3 are the velocity components, and E is the total energy. The viscous stress tensor and heat flux are given by

$$\tau_{ij} = \mu \left(\frac{\partial u_i}{\partial x_j} + \frac{\partial u_j}{\partial x_i} - \frac{2}{3}\frac{\partial u_k}{\partial x_j}\delta_{ij} \right), \tag{4}$$

and

$$q_j = -\frac{\mu}{\mathrm{Pr}}\frac{\partial}{\partial x_j}\left(E + \frac{p}{\rho} - \frac{1}{2}u_k u_k \right). \tag{5}$$

Here, μ is the viscosity coefficient and $\mathrm{Pr} = 0.72$ is the Prandtl number which we assume to be constant. For an ideal gas, the pressure p has the form

$$p = (\gamma - 1)\rho \left(E - \frac{1}{2} u_k u_j \right), \tag{6}$$

where γ is the adiabatic gas constant.

2.2. *Turbulence modeling*

We consider two different approaches for the simulation of turbulent flows – Implicit Large Eddy Simulation (ILES) and the Reynolds Averaged Navier-Stokes (RANS) equations.

In LES modeling, the large scale flow features are resolved while the small scales are modeled. The rationale behind this is that the small scales are isotropic, carry less of the flow energy and therefore do not have as much influence on the mean flow, and can therefore be approximated or modeled. The effect of these subgrid scales (SGS) is approximated by an eddy viscosity which can be derived from a so-called SGS model or can be taken to be equal to the dissipation in the numerical scheme, which is the principle behind the ILES model.[14] Simulations based on ILES models often give very accurate predictions but are limited to low Reynolds number flows because of the high computational cost of resolving the large scale features of the flow.

For the RANS modeling, we add a turbulent dynamic (or eddy) viscosity μ_t to μ in the Navier-Stokes equations (1)-(3), and solve for the time-averaged values of the flow quantities ρ, ρu_i, and ρE. We use the Spalart-Allmaras One-Equation model for μ_t,[15] where a working variable $\tilde{\nu}$ is introduced to evaluate the turbulent dynamic viscosity. This new variable obeys the transport equation

$$\frac{D\tilde{\nu}}{Dt} = c_{b1}\tilde{S}\tilde{\nu} + \frac{1}{\sigma}\left[\nabla \cdot \left((\nu + \tilde{\nu}) \nabla \tilde{\nu} \right) + c_{b2} \left(\nabla \tilde{\nu} \right)^2 \right] - c_{w1} f_w \left[\frac{\tilde{\nu}}{d} \right]^2 . \tag{7}$$

For simplicity, the trip terms have been excluded here, meaning that we assume that the Reynolds numbers are large enough so that the flow over the entire body surface is turbulent. The turbulent dynamic viscosity is then calculated as

$$\mu_t = \rho \nu_t \quad , \quad \nu_t = \tilde{\nu} f_{v1} \quad , \quad f_{v1} = \frac{\chi^3}{\chi^3 + c_{v1}^3} \quad , \quad \chi = \frac{\tilde{\nu}}{\nu} . \tag{8}$$

The production term is expressed as

$$\tilde{S} = S + \frac{\tilde{\nu}}{\kappa^2 d^2} f_{v2} \quad , \quad S = \sqrt{2\Omega_{ij}\Omega_{ij}} \quad , \quad f_{v2} = 1 - \frac{\chi}{1 + \chi f_{v1}} . \tag{9}$$

Here $\Omega_{ij} = \frac{1}{2}(\partial u_i/\partial x_j - \partial u_j/\partial x_i)$ is the rotation tensor and d is the distance from the closest wall. The function f_w is given by

$$f_w = g\left[\frac{1+c_{w3}^6}{g^6+c_{w3}^6}\right]^{1/6} \quad , \quad g = r + c_{w2}(r^6 - r) \quad , \quad r = \frac{\tilde{\nu}}{\tilde{S}\kappa^2 d^2} \,. \tag{10}$$

The closure constants used here are $c_{b1} = 0.1355$, $c_{b2} = 0.622$, $c_{v1} = 7.1$, $\sigma = 2/3$, $c_{w1} = (c_{b1}/\kappa^2) + ((1+c_{b2})/\sigma)$, $c_{w2} = 0.3$, $c_{w3} = 2$, $\kappa = 0.41$.

2.3. *Mapping-based ALE formulation for deformable domains*

Here, we formulate the Navier-Stokes equations in an Arbitrary-Lagrangian Eulerian (ALE) framework, to allow for variable geometries.[16] We follow a similar approach to that presented in Ref. 2. That is, we construct a time dependent mapping between a fixed reference domain and the time varying physical domain. The original equations in the Eulerian domain are then transformed to the fixed reference configuration and the discretization is always carried out on the fixed domain. In order to ensure that constant solutions in the physical domain are preserved exactly, we introduce an additional scalar equation in which the Jacobian of the transformation is integrated numerically using the same spatial and time discretization schemes. This numerically integrated Jacobian is used to correct for integration errors in the conservation equations.

2.3.1. *The mapping*

Let the physical domain of interest be denoted by $v(t)$ and the fixed reference configuration be denoted by V (see Fig. 1). Also, let $\boldsymbol{N}$ and $\boldsymbol{n}$ be the outward unit normals in V and $v(t)$, respectively. We assume, for each time t, the existence of a smooth one-to-one time dependent mapping given by an isomorphism, $\mathcal{G}(\boldsymbol{X}, t)$, between V and $v(t)$. Thus a point $\boldsymbol{X}$ in V, is mapped to a point $\boldsymbol{x}(t)$ in $v(t)$, which is given by $\boldsymbol{x} = \mathcal{G}(\boldsymbol{X}, t)$. In addition, we assume that for all $\boldsymbol{X}$, $\boldsymbol{x} = \mathcal{G}(\boldsymbol{X}, t)$ is a smooth differentiable function of t. In order to transform the Navier-Stokes equations from the physical $(\boldsymbol{x}, t)$ domain to the reference $(\boldsymbol{X}, t)$ domain, we require some differential properties of the mapping. To this end, we introduce the mapping deformation gradient $\boldsymbol{G}$ and the mapping velocity $\boldsymbol{v}_G$ as

$$\boldsymbol{G} = \boldsymbol{\nabla}_X \mathcal{G} \quad , \quad \boldsymbol{v}_G = \left.\frac{\partial \mathcal{G}}{\partial t}\right|_X \,. \tag{11}$$

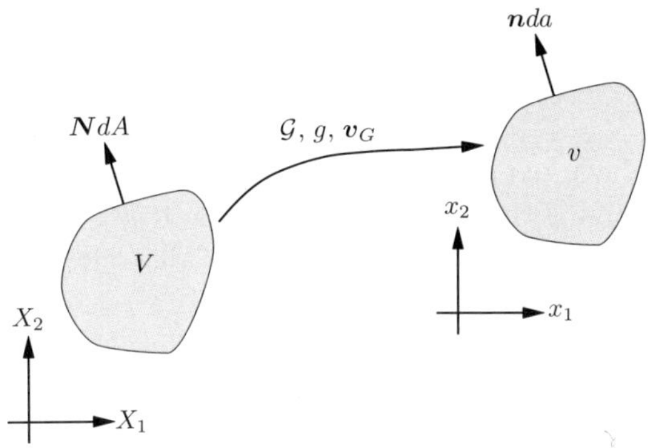

Fig. 1. Mapping between the physical and the reference domains.

In addition, we denote the Jacobian of the mapping by $g = \det(\boldsymbol{G})$. We note that corresponding infinitesimal vectors $d\boldsymbol{L}$ in V and $d\boldsymbol{l}$ in $v(t)$ are related by $d\boldsymbol{l} = \boldsymbol{G}d\boldsymbol{L}$. Also, the elemental volumes are related by $dv = gdV$. From this, we can derive an expression for the area change. Let $d\boldsymbol{A} = \boldsymbol{N}dA$ denote an area element which after deformation becomes $d\boldsymbol{a} = \boldsymbol{n}da$. We then have that, $dV = d\boldsymbol{L} \cdot d\boldsymbol{A}$ and $dv = d\boldsymbol{l} \cdot d\boldsymbol{a}$. Therefore, we must have that

$$\boldsymbol{n}\, da = g\boldsymbol{G}^{-T}\boldsymbol{N}dA \qquad \text{and} \qquad \boldsymbol{N}\, dA = g^{-1}\boldsymbol{G}^{T}\boldsymbol{n}\, da\ . \tag{12}$$

2.3.2. *Transformed equations*

As a starting point, we consider the compressible Navier-Stokes equations (1-3) in the physical domain $(\boldsymbol{x}, t)$, written as a system of conservation laws

$$\frac{\partial \boldsymbol{U}}{\partial t} + \boldsymbol{\nabla} \cdot \boldsymbol{F}(\boldsymbol{U}, \boldsymbol{\nabla U}) = 0\ , \tag{13}$$

where $\boldsymbol{U}$ is the vector of conserved variables and $\boldsymbol{F}$ is a generalized column flux vector which components are the physical flux vectors in each of the spatial coordinate directions. Here $\boldsymbol{F}$ incorporates both inviscid and viscous contributions. That is, $\boldsymbol{F} = \boldsymbol{F}^{\text{inv}}(\boldsymbol{U}) + \boldsymbol{F}^{\text{vis}}(\boldsymbol{U}, \boldsymbol{\nabla U})$ and $\boldsymbol{\nabla}$ represents the spatial gradient operator in the $\boldsymbol{x}$ variables.

In order to obtain the corresponding conservation law written in the reference configuration we re-write the above equation in an integral form as

$$\int_{v(t)} \frac{\partial \boldsymbol{U}}{\partial t}\, dv + \int_{\partial v} \boldsymbol{F} \cdot \boldsymbol{n}\, da = \boldsymbol{0}. \tag{14}$$

Note that the above expression follows directly from (13) by integrating over $v(t)$ and applying the divergence theorem. It is now possible to utilize the mapping and evaluate these integrals in the reference configuration. Consider first the second term,

$$\int_{\partial v} \boldsymbol{F} \cdot \boldsymbol{n}\, da = \int_{\partial V} \boldsymbol{F} \cdot (g\boldsymbol{G}^{-T}\boldsymbol{N})\, dA = \int_{\partial V} (g\boldsymbol{G}^{-1}\boldsymbol{F}) \cdot \boldsymbol{N}\, dA. \tag{15}$$

Similarly, the first integral is transformed by means of Reynolds transport theorem to give

$$\int_{v(t)} \frac{\partial \boldsymbol{U}}{\partial t}\, dv = \frac{d}{dt}\int_{v(t)} \boldsymbol{U}\, dv - \int_{\partial v} (\boldsymbol{U}\boldsymbol{v}_G) \cdot \boldsymbol{n}\, da \tag{16}$$

$$= \frac{d}{dt}\int_{V} g^{-1}\boldsymbol{U}\, dV - \int_{\partial V} (\boldsymbol{U}\boldsymbol{v}_G) \cdot (g\boldsymbol{G}^{-T}\boldsymbol{N})\, dA \tag{17}$$

$$= \int_{V} \left.\frac{\partial(g^{-1}\boldsymbol{U})}{\partial t}\right|_X dV - \int_{\partial V} (g\boldsymbol{U}\boldsymbol{G}^{-1}\boldsymbol{v}_G) \cdot \boldsymbol{N}\, dA\,. \tag{18}$$

Using the divergence theorem once again enables an equivalent local conservation law in the reference domain to be derived as,

$$\left.\frac{\partial \boldsymbol{U}_X}{\partial t}\right|_X + \boldsymbol{\nabla}_X \cdot \boldsymbol{F}_X(\boldsymbol{U}_X, \boldsymbol{\nabla}_X \boldsymbol{U}_X) = 0\,, \tag{19}$$

where the time derivative is at a constant $\boldsymbol{X}$ and the spatial derivatives are taken with respect to the $\boldsymbol{X}$ variables. The transformed vector of conserved quantities and corresponding fluxes in the reference space are

$$\boldsymbol{U}_X = g\boldsymbol{U}\,, \tag{20}$$

$$\boldsymbol{F}_X = g\boldsymbol{G}^{-1}\boldsymbol{F} - \boldsymbol{U}_X\boldsymbol{G}^{-1}\boldsymbol{v}_G\,, \tag{21}$$

or, more explicitly,

$$\boldsymbol{F}_X = \boldsymbol{F}_X^{\text{inv}} + \boldsymbol{F}_X^{\text{vis}}\,, \tag{22}$$

$$\boldsymbol{F}_X^{\text{inv}} = g\boldsymbol{G}^{-1}\boldsymbol{F}^{\text{inv}} - \boldsymbol{U}_X\boldsymbol{G}^{-1}\boldsymbol{v}_G\,, \tag{23}$$

$$\boldsymbol{F}_X^{\text{vis}} = g\boldsymbol{G}^{-1}\boldsymbol{F}^{\text{vis}}\,, \tag{24}$$

and by a simple application of the chain rule,

$$\boldsymbol{\nabla}\boldsymbol{U} = \boldsymbol{\nabla}_X(g^{-1}\boldsymbol{U}_X)\boldsymbol{G}^{-1} = (g^{-1}\boldsymbol{\nabla}_X\boldsymbol{U}_X + \boldsymbol{U}_X\boldsymbol{\nabla}_X(g^{-1}))\boldsymbol{G}^{-1}\,. \tag{25}$$

2.3.3. *Geometric conservation law*

It turns out that, for arbitrary mappings, a constant solution in the physical domain is not necessarily a solution of the discretized equations in the reference domain. Even though this error is typically very small for high order discretizations, the situation is quite severe for lower order approximations since the free-stream condition is not preserved identically. Satisfaction of the constant solution is often referred to as the Geometric Conservation Law (GCL) and is was originally discussed in Ref. 17. The source of the problem is the inexact integration of the Jacobian g of the transformation by the numerical scheme.

First, we note that using expressions (12) together with the divergence theorem, it is straightforward to prove the so-called Piola relationships, which hold for arbitrary vectors $\boldsymbol{W}$ and $\boldsymbol{w}$:

$$\boldsymbol{\nabla}_X \cdot \boldsymbol{W} = g \boldsymbol{\nabla} \cdot (g^{-1} \boldsymbol{G} \boldsymbol{W}) \quad , \quad \boldsymbol{\nabla} \cdot \boldsymbol{w} = g^{-1} \boldsymbol{\nabla}_X \cdot (g \boldsymbol{G}^{-1} \boldsymbol{w}) \; . \tag{26}$$

When the solution $\boldsymbol{U}$ is constant, say $\bar{\boldsymbol{U}}$, we have

$$\boldsymbol{\nabla}_X \cdot \boldsymbol{F}_X = g \boldsymbol{\nabla} \cdot (\boldsymbol{F} - \bar{\boldsymbol{U}} \boldsymbol{v}_G) = -g \bar{\boldsymbol{U}} \boldsymbol{\nabla} \cdot \boldsymbol{v}_G = -[\boldsymbol{\nabla}_X \cdot (g \boldsymbol{G}^{-1} \boldsymbol{v}_G)] \bar{\boldsymbol{U}} \; . \tag{27}$$

Therefore, for a constant solution $\bar{\boldsymbol{U}}$, equation (19) becomes

$$\left. \frac{\partial \boldsymbol{U}_X}{\partial t} \right|_X + \boldsymbol{\nabla}_X \cdot \boldsymbol{F}_X = g^{-1} \bar{\boldsymbol{U}}_x \left(\left. \frac{\partial g}{\partial t} \right|_X - \boldsymbol{\nabla}_X \cdot (g \boldsymbol{G}^{-1} \boldsymbol{v}_G) \right) \; . \tag{28}$$

We see that the right hand side is only zero if the equation for the time evolution of the transformation Jacobian g

$$\left. \frac{\partial g}{\partial t} \right|_X - \boldsymbol{\nabla}_X \cdot (g \boldsymbol{G}^{-1} \boldsymbol{v}_G) = 0 \; , \tag{29}$$

is integrated exactly by our numerical scheme. Since in general, this will not be the case, the constant solution $\bar{\boldsymbol{U}}_x$ in the physical space will not be preserved exactly.

An analogous problem was brought up in the formulation presented in Ref. 2. The solution proposed there was to add some corrections aimed at canceling the time integration errors. Here, we use a slightly different approach. The system of conservation laws (19) is replaced by

$$\left. \frac{\partial (\bar{g} g^{-1} \boldsymbol{U}_X)}{\partial t} \right|_X - \boldsymbol{\nabla}_X \cdot \boldsymbol{F}_X = \boldsymbol{0} \; , \tag{30}$$

where $\bar{g}$ is obtained by solving the following equation

$$\left.\frac{\partial \bar{g}}{\partial t}\right|_X - \boldsymbol{\nabla}_X \cdot (g\boldsymbol{G}^{-1}\boldsymbol{v}_G) = 0 \ . \tag{31}$$

We note that even though $\bar{g}$ is an approximation to g, when the above equation is solved numerically with the same numerical scheme employed for (30), its value will differ from that of g due to integration errors. It is straightforward to verify that (30) does indeed preserve a constant solution as desired. Finally, we point out that the fluxes in equation (31) do not depend on $\boldsymbol{U}$ and are only a function of the mapping. This has two implications. First, when the mapping is prescribed, equation (31) can be integrated independently to obtain $\bar{g}$ in time. Second, the fluxes in (31) do not require communication with the neighboring elements, thus simplifying its numerical solution.

3. Numerical Methods

3.1. *The compact Discontinuous Galerkin method*

In order to develop a Discontinuous Galerkin method, we consider a general system of conservation laws

$$\frac{\partial u}{\partial t} + \nabla \cdot F^{\text{inv}}(u) = \nabla \cdot F^{\text{vis}}(u, \nabla u) + S(u, \nabla u) \ , \tag{32}$$

in a domain Ω, with conserved state variables u, inviscid flux function F^{inv}, viscous flux function F^{vis} and source term S. We note that the governing equations described in the previous section can all be cast in this particular form. We eliminate the second order spatial derivatives of u by introducing additional variables $q = \nabla u$:

$$\frac{\partial u}{\partial t} + \nabla \cdot F^{\text{inv}}(u) = \nabla \cdot F^{\text{vis}}(u, q) + S(u, q) \ , \tag{33}$$

$$q - \nabla u = 0 \ . \tag{34}$$

Next, we consider a triangulation $\mathcal{T}_h$ of the spatial domain Ω and introduce the finite element spaces V_h and Σ_h as

$$V_h = \{v \in [L^2(\Omega)]^m \mid v|_K \in [\mathcal{P}_p(K)]^m, \ \ \forall K \in \mathcal{T}_h\} \ , \tag{35}$$

$$\Sigma_h = \{r \in [L^2(\Omega)]^{dm} \mid r|_K \in [\mathcal{P}_p(K)]^{dm}, \ \ \forall K \in \mathcal{T}_h\} \ , \tag{36}$$

where $\mathcal{P}_p(K)$ is the space of polynomial functions of degree at most $p \geq 0$ on triangle K, m is the dimension of u and d is the spatial dimension. We now consider DG formulations of the form: find $u_h \in V_h$ and $q_h \in \Sigma_h$ such that for all $K \in \mathcal{T}_h$, we have

$$\int_K q_h \cdot r \, dx = -\int_K u_h \nabla \cdot r \, dx + \int_{\partial K} \hat{u} r \cdot n \, ds, \qquad \forall r \in [\mathcal{P}_p(K)]^{dm}, \tag{37}$$

$$\int_K \frac{\partial u_h}{\partial t} v \, dx - \int_K F^{\text{inv}}(u_h) \cdot \nabla v \, dx + \int_{\partial K} \hat{F}^{\text{inv}} v \, ds = -\int_K F^{\text{vis}}(u_h, q_h) \cdot \nabla v \, dx + \int_{\partial K} \hat{F}^{\text{vis}} v \, ds + \int_K S(u_h, q_h) v \, dx, \qquad \forall v \in [\mathcal{P}_p(K)]^m. \tag{38}$$

Here, the numerical fluxes $\hat{F}^{\text{inv}}$, $\hat{F}^{\text{vis}}$ and $\hat{u}$ are approximations to $F^{\text{inv}} \cdot \boldsymbol{n}$, $F^{\text{vis}} \cdot \boldsymbol{n}$ and u, respectively, on the boundary ∂K of the element K and $\boldsymbol{n}$ is the unit normal to the boundary. As commonly done, the inviscid fluxes $\hat{F}^{\text{inv}}$ are approximated using an approximate Riemann solver. In most of our applications, we use the method due to Roe[18]. For the viscous fluxes $\hat{F}^{\text{vis}}$, we use the compact discontinuous Galerkin (CDG) scheme.[19]

The CDG scheme consists of a modification of the original Local Discontinuous Galerkin method.[20] This modification is aimed at making the scheme more compact and hence more attractive computationally, especially when dealing with implicit discretizations and implementations on parallel computers. In order to describe the CDG method, we first consider the LDG method for approximating $\hat{F}^{\text{vis}}$. In the LDG method, we choose $\hat{u}$ and $\hat{F}^{\text{vis}}$ according to

$$\hat{F}^{\text{vis}} = \{\{F^{\text{vis}}(u_h, q_h) \cdot \boldsymbol{n}\}\} + C_{11} [\![u_h \boldsymbol{n}]\!] + \boldsymbol{C}_{12} [\![F^{\text{vis}}(u_h, q_h) \cdot \boldsymbol{n}]\!] \tag{39}$$

$$\hat{u} = \{\{u_h\}\} - \boldsymbol{C}_{12} \cdot [\![u_h \boldsymbol{n}]\!] . \tag{40}$$

Here, $\{\{\ \}\}$ and $[\![\ ]\!]$ denote the average and jump operators across the interface.[19] The constant C_{11} can be chosen equal to zero, except at the boundary interfaces, leading to the so called minimum dissipation scheme.[21] On the other hand, $\boldsymbol{C}_{12}$ is chosen as $\boldsymbol{C}_{12} = \boldsymbol{n}^*$, where $\boldsymbol{n}^*$ is the unit normal to the interface taken with an arbitrary sign. The only constraint on the sign is that all the $\boldsymbol{C}_{12}$ vectors corresponding to the different faces of a given element do not all point either inwards or outwards. See Refs. 19,20 for additional details.

One of the major drawbacks of the LDG method is that it results in a scheme which is non-compact. This means that in the Jacobian of the residual some elements are not only connected to their neighbors but to the neighbors of its neighbors. This situation arises when structured or unstructured meshes are used as discussed in Ref. 19.

In the CDG method, equation (39) is replaced by

$$(\hat{F}^{\text{vis}})^e = \{\{F^{\text{vis}}(u_h, q_h^e) \cdot \boldsymbol{n}\}\} + C_{11}[\![u_h \boldsymbol{n}]\!] + \boldsymbol{C}_{12}[\![F^{\text{vis}}(u_h, q_h^e) \cdot \boldsymbol{n}]\!] \,. \quad (41)$$

The "edge" fluxes q_h^e are computed by solving the equation

$$\int_K q_h^e \cdot r\, dx = -\int_K u_h \nabla \cdot r\, dx + \int_{\partial K} \hat{u}^e r \cdot n\, ds, \quad \forall r \in [\mathcal{P}_p(K)]^{dm} \,, \quad (42)$$

where

$$\hat{u}_h^e = \begin{cases} \hat{u}_h & \text{on edge } e \text{, given by equation (40),} \\ u_h & \text{otherwise.} \end{cases} \quad (43)$$

The CDG scheme is a little more expensive than the original LDG method as it requires a different elemental value of q_h^e for each edge (or face) e of the element, but it results in a scheme with a sparser structure than the LDG method and element-wise compact connectivities. For more details we refer to Ref. 19.

3.2. *Stabilization by artificial diffusion*

Discontinuities and other under-resolved solution features pose considerable difficulties for most high-order methods. Several approaches inspired by the finite volume methodology have been proposed. The most straightforward approach consists of using some form of shock sensing to identify the elements lying in the shock region and reducing the order of the interpolating polynomial there.[22,23] Despite its simplicity, this approach may yield satisfactory answers, in particular when combined with adaptive mesh refinement procedures.

More sophisticated approaches exist for selecting the interpolating polynomial such as those based on weighted essentially non-oscillatory (WENO) concepts.[7,24,25] These methods allow for stable discretizations near discontinuities while still maintaining a high-order approximation. Although these methods have several attractive features, they have not yet been demonstrated in the practical unstructured mesh context using high-order approximation polynomials. Other interesting alternatives are based on applying filters to the solution,[26,27] the selective application of viscosity to the different spectral scales,[28] and methods based on reconstructing the solution from unlimited oscillatory solutions computed using a high-order method.[29,30] These methods hold the promise of yielding uniformly accurate solutions near the discontinuity in a pointwise sense. However, a number of issues still remain unresolved. In particular, the extension of these methods to multiple dimensions is an open question.

In Ref. 31, we proposed a new strategy inspired by the early artificial viscosity methods,[32] which has proved to be surprisingly effective in the context of high-order DG methods.[33–35] The rationale behind the method is to add viscosity to the original equations in order to spread discontinuities over a length scale that can be adequately resolved within the space of approximating functions. The goal is not to introduce discontinuities in the approximating space, but to resolve the sharp gradients existing in a viscous shock with continuous approximations. For low order approximations, such as piecewise constant and/or linear, this approach produces discontinuities which are spread over several cells (e.g. 2-4 cells). Therefore, it is considered to be inferior to the more established finite volume shock capturing approaches. This is because several cells are required to resolve a viscous shock profile with piecewise linear approximations.

However, for higher order polynomial approximations, the situation is different. The resolution of a piecewise polynomial of order p scales like $\delta \sim h/p$. This means that the amount of artificial viscosity required to resolve a shock profile is $\mathcal{O}(h/p)$. Keeping h fixed, the amount of artificial viscosity required scales like $1/p$ and the accuracy of the solution in the neighborhood of the shock becomes $\mathcal{O}(h/p)$. This compares favorably to the more standard approaches which are only $\mathcal{O}(h)$ accurate. In other words, for high order p, we can exploit sub-cell resolution and obtain shock profiles which are much thinner than the element size. Recall that in the more standard approaches, the order of the interpolating polynomial is reduced over the whole element and as a consequence, there is no hope for resolving the shock profile at a sub-cell level.

3.2.1. *Artificial viscosity models*

The most straight-forward artificial viscosity model is to add Laplacian diffusion to the system of conservation laws

$$\frac{\partial u}{\partial t} + \nabla \cdot F(u, \nabla u) = \nabla \cdot (\varepsilon \nabla u). \tag{44}$$

Here, the parameter ε controls the amount of viscosity. The shocks that may appear in the solution to this modified equation will spread over layers of thickness $\mathcal{O}(\varepsilon)$. Therefore, when attempting to approximate these solutions numerically, ε should be chosen as a function of the resolution available in the approximating space. Near the shocks, we take $\varepsilon = \mathcal{O}(h/p)$, where h is the element size and p is the order of the approximating polynomial. Away from discontinuities, where the unmodified solution is resolved, we want $\varepsilon = 0$.

Instead of the Laplacian term added to the right hand side of equation (44), one can also consider a physically inspired artificial viscosity term analogous to dissipative terms in the Navier-Stokes equations but with a viscosity coefficient which is determined by numerical considerations. Details about this physically inspired model can be found in Ref. 31. In our experience, the physical model and the Laplacian model of equation (44) perform similarly.

3.2.2. *Discontinuity sensor*

In order to avoid the use of artificial viscosity in resolved regions of the flow, we utilize a resolution sensor. We write the solution within each element in terms of a hierarchical family of orthogonal polynomials. In 1D, the solution is represented by an expansion in terms of orthonormal Legendre polynomials, whereas in multidimensions, an orthonormal Koornwinder basis[36] is employed within each triangle. For smooth solutions, the coefficients in the expansion are expected to decay very quickly. On the other hand, when the solution is not smooth, the strength of the discontinuity will dictate the rate of decay of the expansion coefficients. We express the solution of order p within each element in terms of an orthogonal basis as

$$u = \sum_{i=1}^{N(p)} u_i \psi_i \ , \tag{45}$$

where $N(p)$ is the total number of terms in the expansion and ψ_i are the basis functions. In addition, we also consider a truncated expansion of the same solution, only containing the terms up to order $p-1$, that is,

$$\hat{u} = \sum_{i=1}^{N(p-1)} u_i \psi_i \ . \tag{46}$$

Within each element Ω_e, the following resolution indicator is defined,

$$S_e = \frac{(u - \hat{u}, u - \hat{u})_e}{(u, u)_e} \ , \tag{47}$$

where $(\cdot, \cdot)_e$ is the standard inner product in $L_2(\Omega_e)$. In practice, we have found S_e to be a remarkably reliable indicator

Once the shock has been sensed, the amount of viscosity is taken to be constant over each element and determined by the following smooth function,

$$\varepsilon_e = \begin{cases} 0 & \text{if } s_e < s_0 - \kappa \ , \\ \frac{\varepsilon_0}{2}\left(1 + \sin\frac{\pi(s_e - s_0)}{2\kappa}\right) & \text{if } s_0 - \kappa \leq s_e \leq s_0 + \kappa \ , \\ \varepsilon_0 & \text{if } s_e > s_0 + \kappa \ . \end{cases} \tag{48}$$

Here, $s_e = \log_{10} S_e$ and the parameters $\varepsilon_0 \sim h/p$, $s_0 \sim 1/p^4$, and κ are chosen empirically sufficiently large so as to obtain a sharp but smooth shock profile.

3.3. *Parallel preconditioned Newton-Krylov solvers*

After discretization of (32) using the Compact Discontinuous Galerkin method and elimination of the variables associated with q_h within each element, we obtain system of coupled ordinary differential equations (ODEs) of the form:

$$\boldsymbol{M}\dot{\boldsymbol{u}} = \boldsymbol{r}(\boldsymbol{u}) \ , \tag{49}$$

where $\boldsymbol{u}(t)$ is a vector containing the degrees of freedom associated with u_h, which is represented using a nodal basis. The vector $\dot{\boldsymbol{u}}$ denotes the time derivative of $\boldsymbol{u}(t)$, $\boldsymbol{M}$ is the mass matrix, and $\boldsymbol{r}$ is the residual vector. We integrate (49) implicitly in time using either diagonal implicit Runge-Kutta methods (DIRK)[37] or the backward differentiation formulas (BDF).[38] Using Newton's method for solving the nonlinear systems of equations that arise, it is required to solve linear systems involving matrices of the form

$$\boldsymbol{A} \equiv \alpha_0 \boldsymbol{M} - \Delta t \frac{d\boldsymbol{r}}{d\boldsymbol{u}} \equiv \alpha_0 \boldsymbol{M} - \Delta t \boldsymbol{K} \ . \tag{50}$$

For simplicity of presentation, we assume that $\alpha_0 = 1$, which is the case for the first-order accurate backward Euler method. Other values of α_0, as required for higher-order methods, simply correspond to a scaling of the timestep Δt.

3.3.1. *Jacobian sparsity pattern*

The system matrix $\boldsymbol{A} = \boldsymbol{M} - \Delta t \boldsymbol{K}$ is sparse with a block-wise structure corresponding to the element connectivities. An example of a small triangular mesh with polynomial degrees $p = 2$ within each element is shown in Fig. 2. It is worth pointing out that the number of nonzero blocks in each

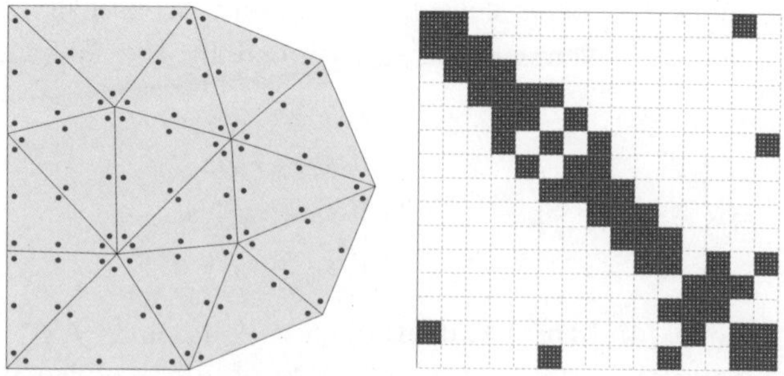

Fig. 2. An example mesh with elements of polynomial order $p = 2$, and the corresponding block matrix for a scalar problem.

row is equal to four in 2D and five in 3D, except for boundary elements. To be able to use machine optimized dense linear algebra routines, such as the BLAS/LAPACK libraries,[39] we represent $\boldsymbol{A}$ in an efficient dense block format, see Ref. 40 for details. We note that our CDG scheme actually produces sparser off-diagonal blocks than other known methods,[19] which we take advantage of in our implementation. However, in what follows, we assume for simplicity that all nonzero blocks are full dense matrices. The block with element indices $1 \leq i, j \leq n$ will be denoted by $\boldsymbol{A}_{ij}$, with n the total number of elements.

3.3.2. *Incomplete LU preconditioning*

It is clear that the performance of the iterative solvers will depend strongly on the timestep Δt. As $\Delta t \to 0$, the matrix $\boldsymbol{A}$ reduces to the mass matrix, which is block diagonal and inverted exactly by all preconditioners that we consider. However, as $\Delta t \to \infty$, the problem becomes harder and often not well-behaved. Physically, a small Δt means that the information propagation is local, while a large Δt means information is exchanged over large distances during the timestep. This effect, which is important when designing iterative methods, is even more important when we consider parallel algorithms since algorithms based on local information exchanges usually scale better than ones with global communication patterns.

When solving the system $\boldsymbol{Au} = \boldsymbol{b}$ using Krylov subspace iterative methods such as GMRES, it is essential to use a good preconditioner. This amounts to finding an approximation $\tilde{\boldsymbol{A}}$ to $\boldsymbol{A}$ which allows for a relatively inexpensive computation of $\tilde{\boldsymbol{A}}^{-1}\boldsymbol{p}$ for arbitrary vectors $\boldsymbol{p}$. One of the simplest choices that performs reasonably for many problems is the block-diagonal, or the block-Jacobi, preconditioner

$$\tilde{\boldsymbol{A}}^{\mathrm{J}}_{ij} = \begin{cases} \boldsymbol{A}_{ij} & \text{if } i = j \ , \\ \boldsymbol{0} & \text{if } i \neq j \ . \end{cases} \tag{51}$$

Clearly, $\tilde{\boldsymbol{A}}^J$ is cheap to invert compared to $\boldsymbol{A}$, since all the diagonal blocks are independent. However, unlike the point-Jacobi iteration, there is a significant preprocessing cost in the factorizations of the diagonal blocks $\tilde{\boldsymbol{A}}^J_{ij}$, which is comparable to the cost of more complex factorizations.[40]

A minor modification of the block-diagonal preconditioner is the block Gauss-Seidel preconditioner, which keeps the diagonal blocks plus all the upper triangular blocks:

$$\tilde{\boldsymbol{A}}^{\mathrm{GS}}_{ij} = \begin{cases} \boldsymbol{A}_{ij} & \text{if } i \leq j \ , \\ \boldsymbol{0} & \text{if } i > j \ . \end{cases} \tag{52}$$

The preprocessing cost is the same as before, and solving $\tilde{\boldsymbol{A}}^{\mathrm{GS}}\tilde{\boldsymbol{u}} = \boldsymbol{p}$ is only a constant factor more expensive than for $\tilde{\boldsymbol{A}}^J$. The Gauss-Seidel preconditioner can perform well for some simple problems, such as scalar convection problems, but in general it only gives a small factor of improvement over the block-diagonal preconditioner. Furthermore, the sequential nature of the triangular back-solve with $\tilde{\boldsymbol{A}}^{\mathrm{GS}}$ makes the Gauss-Seidel preconditioner hard to parallelize.

A more ambitious preconditioner with similar storage and computational cost is the block incomplete LU factorization $\tilde{\boldsymbol{A}}^{\mathrm{ILU}} = \tilde{\boldsymbol{L}}\tilde{\boldsymbol{U}}$ with zero fill-in. This block-ILU(0) algorithm corresponds to block-wise Gaussian elimination, where no new blocks are allowed into the matrix. This factorization can be computed with the following simple algorithm:

function $[\tilde{\boldsymbol{L}}, \tilde{\boldsymbol{U}}] \leftarrow \mathit{IncompleteLU}(\boldsymbol{A}, \mathit{mesh})$
$\tilde{\boldsymbol{U}} = \boldsymbol{A}, \tilde{\boldsymbol{L}} = \boldsymbol{I}$
for $j = 1, \ldots, n-1$
  for neighbors $i > j$ of j in *mesh*
    $\tilde{\boldsymbol{L}}_{ij} = \tilde{\boldsymbol{U}}_{ij}\tilde{\boldsymbol{U}}_{jj}^{-1}$
    $\tilde{\boldsymbol{U}}_{ii} = \tilde{\boldsymbol{U}}_{ii} - \tilde{\boldsymbol{L}}_{ik}\tilde{\boldsymbol{U}}_{ki}$

We also note here that the upper-triangular blocks of $\tilde{\boldsymbol{U}}$ are identical to those in $\boldsymbol{A}$, which reduces the storage requirements for the factorization. The back-solve using $\tilde{\boldsymbol{L}}$ and $\tilde{\boldsymbol{U}}$ has the same sequential nature as for

Gauss-Seidel, but it turns out that the performance of the block-ILU(0) preconditioner can be fundamentally better.

3.3.3. *Minimum discarded fill element ordering*

It is clear that the Gauss-Seidel and the incomplete LU factorizations will depend strongly on the ordering of the blocks, or the elements, in the mesh. This is because of the mesh ordering determines which element connections are kept and which are discarded when calculating $\tilde{A}$. In Refs. 40,41, we proposed a simple heuristic algorithm for finding appropriate element orderings. Our algorithm considers the fill that would be ignored if element j' was chosen as the pivot element at step j:

$$\Delta\tilde{\boldsymbol{U}}_{ik}^{(j,j')} = -\tilde{\boldsymbol{U}}_{ij'}\tilde{\boldsymbol{U}}_{j'j'}^{-1}\tilde{\boldsymbol{U}}_{j'k}, \qquad \text{for neighbors } i \geq j, k \geq j \text{ of element } j'. \tag{53}$$

The matrix $\Delta\tilde{\boldsymbol{U}}^{(j,j')}$ corresponds to fill that would be discarded by the ILU algorithm. In order to minimize these errors, we consider a set of candidate pivots $j' \geq j$ and pick the one that produces the smallest fill. As a measurement of the magnitude of the fill, or the corresponding weight, we take the Frobenius matrix norm of the fill matrix:

$$w^{(j,j')} = \|\Delta\tilde{\boldsymbol{U}}^{(j,j')}\|_{\mathrm{F}} \,. \tag{54}$$

As a further simplification, we note that

$$\|\Delta\tilde{\boldsymbol{U}}_{ik}^{(j,j')}\|_{\mathrm{F}} = \|-\tilde{\boldsymbol{U}}_{ij'}\tilde{\boldsymbol{U}}_{j'j'}^{-1}\tilde{\boldsymbol{U}}_{j'k}\|_{\mathrm{F}} \leq \|\tilde{\boldsymbol{U}}_{ij'}\|_{\mathrm{F}}\|\tilde{\boldsymbol{U}}_{j'j'}^{-1}\tilde{\boldsymbol{U}}_{j'k}\|_{\mathrm{F}} \,, \tag{55}$$

which means we can estimate the weight by simply multiplying the norms of the individual matrix blocks. By pre-multiplication of the block-diagonal, we can also avoid the matrix factor $\tilde{\boldsymbol{U}}_{j'j'}^{-1}$ above. See Ref. 41 for full pseudocode of the algorithm.

We note that for a pure upwinded scalar convective problem, the MDF ordering is optimal since at each step it picks an element that either does not affect any other elements (downwind) or does not depend on any other elements (upwind), resulting in a perfect factorization. But the algorithm works well for other problems too, including multivariate and viscous problems, since it tries to minimize the error between the exact and the approximate LU factorizations. It also takes into account the effect of the discretization (e.g. highly anisotropic elements) on the ordering. These aspects are harder to account for with methods that are based on physical observations, such as line-based orderings.[42–45]

3.3.4. *Coarse scale corrections and the ILU/Coarse scale preconditioner*

Multi-level methods, such as multigrid[46] solve the system $\boldsymbol{Au} = \boldsymbol{b}$ by introducing coarser level discretizations. This coarser discretization can be obtained either by using a coarser mesh (h-multigrid) or, for high-order methods, by reducing the polynomial degree (p-multigrid[43,47]). The residual is restricted to the coarse scale where an approximate error is computed, which is then applied as a correction to the fine scale solution. A few iterations of a smoother (such as Jacobi's method) are applied after and possibly before the correction to reduce the high-frequency errors.

For our DG discretizations, it is natural and practical to consider coarser scales obtained by reducing the polynomial order p. The problem size is highly reduced by decreasing the polynomial order to $p = 0$ or $p = 1$, even from moderately high orders such as $p = 4$. For very large problems it may be necessary to consider h-multigrid approaches to solve the coarse grid problem, but here we only use p-multigrid.

Furthermore, we have observed that we often get overall better performance by using a simple two-level scheme where the fine level corresponds to $p = 4$ and the coarse level is either $p = 1$ or $p = 0$ rather than a hierarchy of levels. Thus, our preconditioning algorithm solves the linear system $\boldsymbol{Au} = \boldsymbol{b}$ approximately using a single coarse scale correction,

0. $\boldsymbol{A}^{(0)} = \boldsymbol{P}^T\boldsymbol{AP}$	*Precompute coarse operator, block wise*
1. $\boldsymbol{b}^{(0)} = \boldsymbol{P}^T\boldsymbol{b}$	*Restrict residual element/component wise*
2. $\boldsymbol{A}^{(0)}\boldsymbol{u}^{(0)} = \boldsymbol{b}^{(0)}$	*Solve coarse scale problem*
3. $\boldsymbol{u} = \boldsymbol{P}\boldsymbol{u}^{(0)}$	*Prolongate solution element/component wise*
4. $\boldsymbol{u} = \boldsymbol{u} + \alpha\tilde{\boldsymbol{A}}^{-1}(\boldsymbol{b} - \boldsymbol{Au})$	*Apply smoother $\tilde{\boldsymbol{A}}$ with damping α*

Possible smoothers $\tilde{\boldsymbol{A}}$ include block Jacobi $\tilde{\boldsymbol{A}}^{\mathrm{J}}$ or Gauss-Seidel $\tilde{\boldsymbol{A}}^{\mathrm{GS}}$. It is also known that an ILU0 factorization can be used as a smoother for multigrid methods,[48,49] and it has been reported that it performs well for the Navier-Stokes equations, at least in the incompressible case using low-order discretizations.[50] Inspired by this, we use the block ILU0 factorization $\tilde{\boldsymbol{A}}^{\mathrm{ILU}}$ as a post-smoother for a two-level scheme. Our numerical experiments indicate that the block ILU0 preconditioner and the low-degree correction preconditioner complement each other. With our MDF element ordering algorithm, the ILU0 performs almost optimally for highly convective problems, while the coarse scale correction with either block diagonal, block GS, or block ILU0 post-smoothing, performs very well in the diffusive limit.

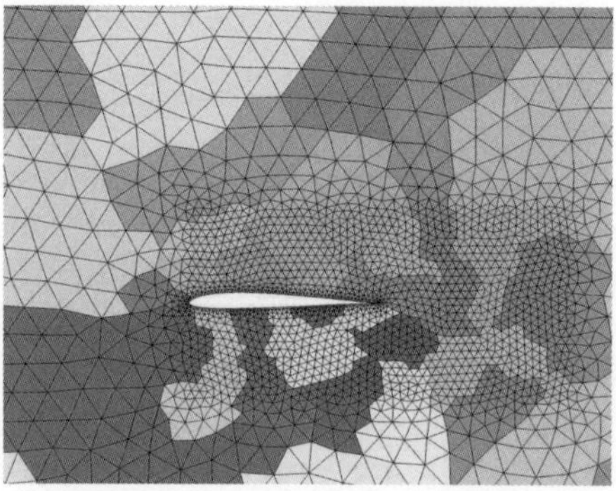

Fig. 3. Partitioning of the mesh elements.

The restriction/prolongation operator $\boldsymbol{P}$ is a block diagonal matrix with all the blocks being identical. The prolongation operator has the effect of transforming the solution from nodal basis to a hierarchical orthogonal basis function based on Koornwinder polynomials[36] and setting the coefficients corresponding to the higher modes equal to zero. The transpose of this operator is used for the restriction of the weighted residual and for the projection of the matrix blocks. For more details on these operators we refer to Ref. 51. We use a smoothing factor $\alpha = 1$ for the ILU0 smoother. We use a direct sparse solve for the linear system in step 2, and we note that the matrix $\boldsymbol{A}^{(0)}$ is usually magnitudes smaller than $\boldsymbol{A}$.

3.3.5. *Parallelization*

The domain is naturally partitioned by the elements to achieve load balancing and low communication volumes, see Fig. 3. Because of the element-wise compact stencil of the CDG scheme, only one additional layer of elements has to be maintained for each process in addition to the elements in the partition. The elements at the domain boundary are processed first, then the computed data is sent to neighboring processes using non-interrupting communications, and while the data is sent the interior elements are processed. This typically leads to algorithms where the communication costs are negligible for evaluations of the residual $\boldsymbol{r}(\boldsymbol{u})$, and therefore also for explicit time integration methods, as well as for the evaluation of the Jacobian matrix $\boldsymbol{K} = \partial \boldsymbol{r}/\partial \boldsymbol{u}$.

In the iterative implicit solvers, the matrix-vector products also scale well in terms of communication, by overlapping with the computation of the interior elements. However, a problem here is the fact that the matrix-vector product operations are highly memory intensive, and tend to give poor scaling on multicore processors. Another issue is the parallelization of the preconditioner, where in particular the Gauss-Seidel and the incomplete LU preconditioners have a highly serial structure that is hard to parallelize.

Our preferred approach for parallelization of the ILU factorization is to apply it according to the element orderings determined by the MDF algorithm, but ignoring any contributions between elements in different partitions. In standard domain decomposition terminology, this essentially amounts to a non-overlapping Schwartz algorithm with incomplete solutions.[52] It is clear that this approach will scale well in parallel, since each process will compute a partition-wise ILU factorization independent of all other processes.

To minimize the error introduced by separating the ILU factorizations, we use the ideas from the MDF algorithm to obtain information about the weights of the connectivities between the elements. By computing a weighted partitioning using the weight

$$C_{ij} = \|\boldsymbol{A}_{ii}^{-1}\boldsymbol{A}_{ij}\|_F \tag{56}$$

between element i and j, we obtain partitions that are less dependent on each other, and reduce the error from the decomposition. The drawback is that the total communication volume might be increased, but if desired, a trade-off between these two effects can be obtained by adding a constant C_0 to the weights. In practice, since the METIS software[53] used for the partitioning requires integer weights, we scale and round the C_{ij} values to integers between 1 and 100. It is clear that this method reduces to the block-Jacobi method as the number of partitions approaches the number of elements. However, in any practical setting, each partition will have at least 100's of elements, in which the difference between partition-wise block-ILU and block-Jacobi is quite significant.

4. Applications

In this section we present four representative applications of our high-order DG methods.

4.1. *Aeroacoustics and Kelvin-Helmholtz instability*

Our first example, which is adapted from Ref. 54, is an aeroacoustics problem with nonlinear interactions between a long-range wave and small-scale flow features. The flow has a Mach number of $M = 1/20$ in the double-periodic rectangular domain $-L \le x \le L = 1/M = 20$, $0 \le y \le 2L/5 = 8$. We use a Cartesian grid of 400-by-80 quadrilateral elements, with each element split into two triangles, giving a total of 64,000 elements. Within each elements we use polynomials of degree $p = 3$, giving a total of 640,000 DOFs per component, or 2.56 million DOFs for the compressible Navier-Stokes equations.

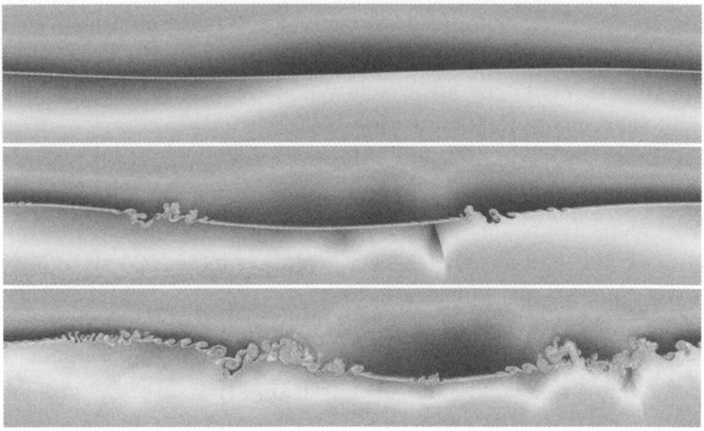

Fig. 4. The acoustic Kelvin-Helmholtz problem at three time instances, with color representing the density. The initially long-range acoustic wave forms a weak shock, which interacts with the density stratified flow to produce shear instabilities.

The initial conditions are based on a sawtooth-shaped density profile, which we smooth to allow an accurate high-order representation on the grid:

$$\Phi(y) = \frac{y}{10} - \frac{2}{5}(\mathrm{erf}(\delta(y - L/5)) - 1), \tag{57}$$

with grid resolution $\delta = N/p = 80/3$. We also define the long-range acoustic wave shape by

$$\Psi(x) = 1 + \cos\frac{\pi x}{L}\ . \tag{58}$$

The initial density, pressure, and velocities are then

$$\rho = 1 + 0.2M\Psi(x) + \Phi(y), \qquad u = \sqrt{\gamma}\Psi(x), \tag{59}$$

$$p = (1 + \gamma M\Psi(x))/M^2, \qquad v = 0, \tag{60}$$

with $\gamma = 1.4$. We solve the compressible Navier-Stokes equations, with a dynamic viscosity coefficient $\mu = 1/160$, corresponding to a Reynolds number $\text{Re} = 6,400$ based on the length of the domain. Since the grid is uniform and we wish to resolve the acoustic waves time-accurately, we use a standard explicit RK4 time integrator with timestep $\Delta t = 1.25 \cdot 10^{-4}$.

The resulting flow field is shown in Fig. 4, as color plots of the density at three time instances $t = 2.5$, $t = 7.5$, and $t = 12.5$. We note that the acoustic wave deforms into a weak shock, and that the density jump causes a Kelvin-Helmholtz type instability at the interface. Furthermore, although not clearly visible from these figures, there are highly complex interactions between the waves and the flow features, and capturing these accurately is one of our main motivations for using high-order methods.

4.2. *Implicit large Eddy simulations of flow past airfoil*

Next, we consider the transient flow around an SD7003 airfoil at an angle of attack of 4° and a Reynolds number of $60,000$. We study the formation of laminar separation bubbles and the related transition to turbulence by means of Implicit Large Eddy Simulations. Here we only show the partial results for a medium-sized grid with 52,800 tetrahedral elements and polynomial orders $p = 4$ within each element, giving a total of 1.848 million high-order nodes or 9.24 million degrees of freedom, for more details we refer to Ref. 55. The discretized equations are integrated in time using a two stage, A-stable, third-order accurate diagonal implicit Runge-Kutta (DIRK) method[37] with a non-dimensional timestep of $\Delta t = 0.01$.

Iso-surfaces of the q-criterion and the span-wise vorticity are shown in Fig. 4.2, and it is clear that complex three-dimensional structures are present. With the fifth-order accurate method in space, this relatively coarse mesh is able to accurately capture the average locations of separation, transition, and reattachment, as well as the average pressure and skin friction coefficient profiles along the foil, which can be seen in Fig. 4.2 together with comparison curves for the data by Galbraith & Visbal[56] and XFOIL.[57] The separation bubble is clearly visible in these profiles, with separation occurring on average at 21% of the chord, transition at 53% (as determined by the peak in boundary layer shape factor), and reattachment at 67% in the present simulations.

Table 1 gives a comparison with previously published results, as well as the mean lift and drag coefficients. TU-BS corresponds to the PIV experiments at the Technical University of Braunschweig Low-Noise Wind

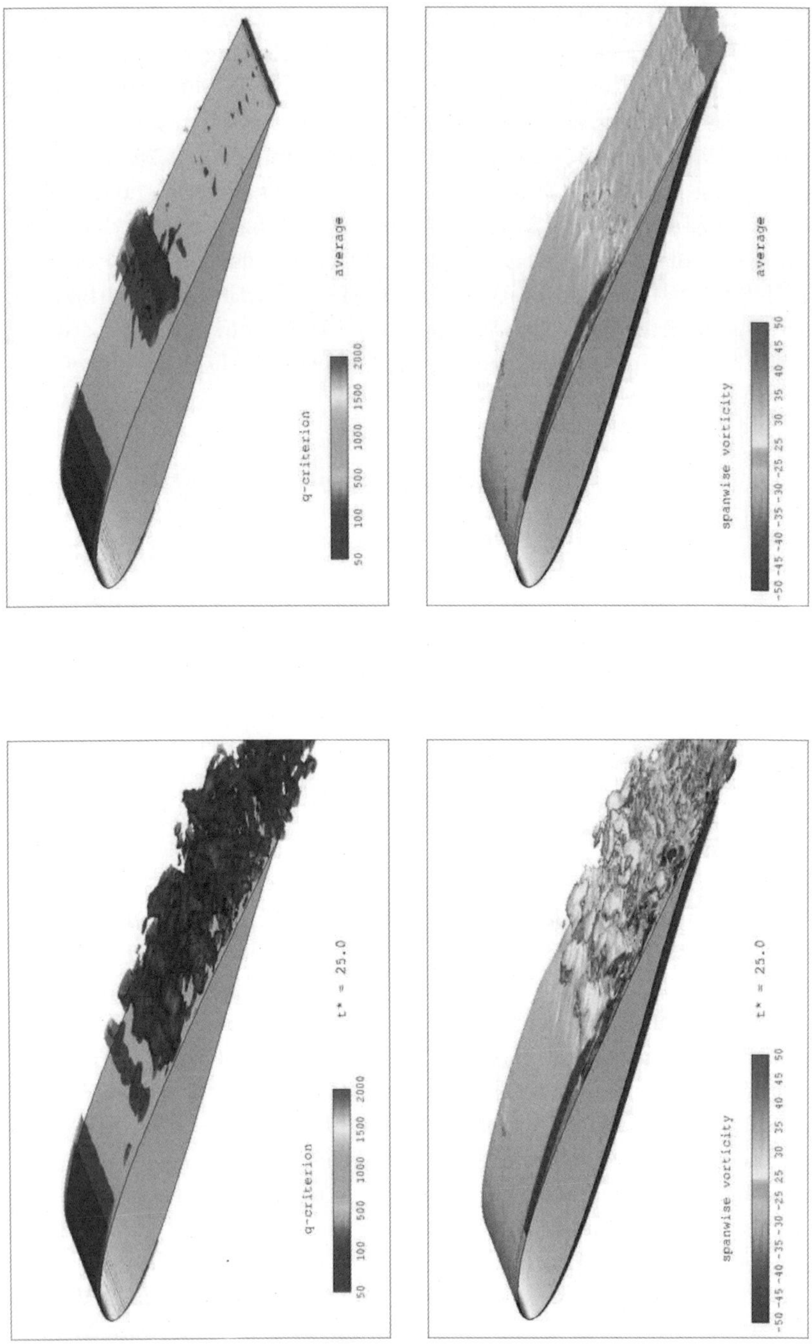

Fig. 5. Instantaneous (left) and average (right) iso-surfaces of q-criterion (top) and span-wise vorticity (middle) at Re = 60,000 with grid 2, $p = 4$ (from Ref. 55).

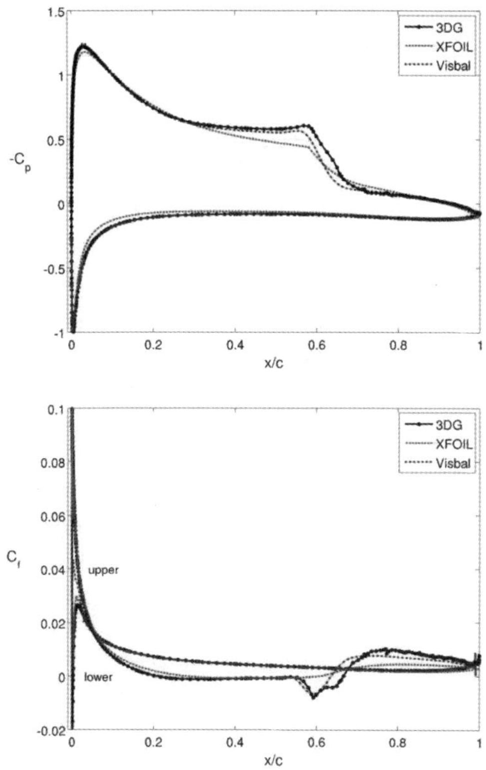

Fig. 6. Average pressure coefficient (top) and stream-wise skin friction coefficient (bottom) at Re = 60,000 on grid 2, $p = 4$. The dashed lines give XFOIL predictions at 3.37°, $N_{\text{crit}} = 7$. The dot-dashed lines show the ILES data of Galbraith & Visbal.[56]

Tunnel,[58] while HFWT is from the PIV experiments at the Air Force Research Lab Horizontal Free-SurfaceWater Tunnel.[59] The present results are well within variations between previously published works, which is notable because of the relatively coarse meshes used, showing that our DG method is particularly well-suited for simulation of these turbulent flows – including hard-to-capture Tollmien-Schlichting waves.[55]

4.3. *Transonic turbulent flows*

In the next example we demonstrate high-order DG methods for problems with shocks and the Spalart-Allmaras RANS turbulence model (7).

Table 1. Comparison of results at Reynolds 60,000 with grid 2, $p = 4$. The XFOIL data is for 3.37° angle of attack; TU-BS[58] and HFWT[59] correspond to PIV experiments; Visbal[56] is an ILES.

Source	Freestream Turbulence	Separation x_{sep}/c	Transition x_{tr}/c	Reattach-ment x_r/c	Bubble Length	$\overline{C_L}$	$\overline{C_D}$
TU-BS[58]	0.08 %	0.30	0.53	0.62	$0.32c$	-	-
HFWT[59]	$\sim 0.1\%$	0.18	0.47	0.58	$0.40c$	-	-
Visbal[56]	0	0.23	0.55	0.65	$0.42c$	-	-
XFOIL	$(N_{\text{crit}} = 7)$	0.28	0.58	0.61	$0.34c$	0.5624	0.0176
Present	0	0.21	0.53	0.67	$0.46c$	0.6122	0.0241

At the edge of the boundary layer, the profile of the eddy viscosity transitions to its free-stream value over a very narrow layer in which the curvature changes sign. Unless properly resolved, this may lead to non-smooth or even negative numerical values for the eddy viscosity variable. This may easily result in a sudden instability in the computations. It turns out that the thickness of this transition region is determined by the laminar viscosity and therefore it is extremely narrow and impractical to resolve in most cases. Our proposal is to address this issue by introducing a Laplacian artificial viscosity model to the diffusion term of the SA equation (7). The artificial viscosity model aims to stabilize the discretization of the continuous equation (7) in finite dimensional space, which then accommodates high-order approximations of RANS-SA equations on relatively coarse grids. We point out that the regions where the eddy viscosity profile is modified have a minor effect on the overall solution since they generally correspond to regions where the eddy viscosity is very small.

We stabilize our scheme using the artificial viscosity approach in section 3.2, and add two viscous models of the form:

$$F_{\text{stab}}(u, q) = \sum_{i=1}^{2} \frac{h}{p} \varepsilon(\psi(s^i(u))) F^m(u, q) \tag{61}$$

to the regular fluxes. Here, the *sensor variables* are the eddy viscosity $s^1(u) = \tilde{\nu}$ for the turbulence model and the density $s^2(u) = \rho$ for the shocks. As described earlier, the indicator $\psi(s) = \log_{10} E_H/E$ gives the ratio of high-frequency modes in the sensor s within each element, and the ε function gives a smooth transition from zero to one. For the viscosity models $F^1(u, q)$ and $F^2(u, q)$, we use simple Laplacian diffusion added to the turbulence model and to each of the Navier-Stokes equations, respectively.

Our first validation test is the turbulent flow past a flat plate[60] at $\text{Re}_x = 1.02 \cdot 10^7$. We use three grids with 10-by-16 elements (grid C), 19-by-31 elements (grid B), and 37-by-61 elements (grid A), and polynomial degrees

$p = 1, \dots, 4$ within each element. In addition we solve the problem using a grid with 72-by-120 elements and $p = 4$, to be used only for computing a reference solution for the convergence study.

The velocity profiles and the friction velocities $u_\tau = \sqrt{\nu \partial u / \partial y (y = 0)}$ are shown in Fig. 7, together with experimental data. Note how the high-order $p = 4$ method on the coarse grid C gives very good agreement with both the finer lower-order grids as well as with the experimental data. Furthermore, the third graph shows grid convergence in the computed drag forces for all grids at $p = 1, \dots, 4$. The slopes show good agreement with the expected dependency $\mathcal{O}(h^p)$ for differentiated quantities.

We also present results for a turbulent transonic flow past an RAE2822 airfoil at $M = 0.729$ and $\mathrm{Re} = 6.5 \cdot 10^6$. We use a single-block, two-dimensional C-grid with about 1,000 anisotropic triangular elements, and polynomial degrees $p = 4$. The grid is clustered around the leading and the trailing edges and around the airfoil surface to resolve the boundary layer on the airfoil, as well as around the shock. The first grid point off the wall is at a distance of 5×10^{-5} from the airfoil surface.

The resulting flow field is shown in Fig. 4.3. We note how the high-order stretched elements resolve the boundary layer even if the elements are large, and that the artificial viscosity approach resolves the shock with subgrid accuracy.

4.4. *Flapping elliptic wings*

Our final example is the transient laminar flow around a pair of flapping wings.[61] We consider a wing pair with an elliptical planform. The maximum normalized chord at the wing centerline is $c = 1$ and the wing tip-to-tip span is $b = 10$. The flapping motions occur symmetrically about a hinge located at the wing centerline. An HT13 airfoil is selected for the entire wing span, resulting in a maximum wing thickness of $t = 0.065$ at the wing centerline.

In order to obtain maximum geometrical flexibility, the equations are discretized on unstructured meshes of triangles and tetrahedra. We use the symmetry of the problem to only simulate one half of the domain, with a symmetry boundary condition at the cut plane. We generate all the surface meshes in parametrized form using the DistMesh triangular mesh generator.[62] The tetrahedral volume mesh is then generated by a Delaunay refinement based code.[63] The resulting mesh has about 43,000 nodes and 231,000 tetrahedral elements for the half-domain, which corresponds to 4.62 million high-order nodes with polynomial orders of degree $p = 3$. To

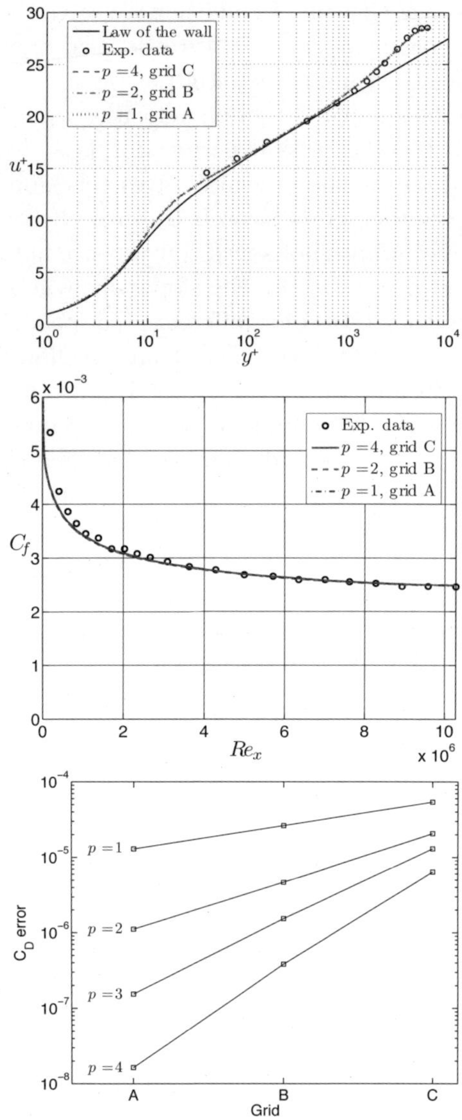

Fig. 7. The turbulent flow past a flat plate: (a) velocity profiles at $Re_x = 1.02 \times 10^7$, (b) skin friction coefficient as function of Re_x, and (c) errors in drag for the turbulent flow past a flat plate (from Ref. 35).

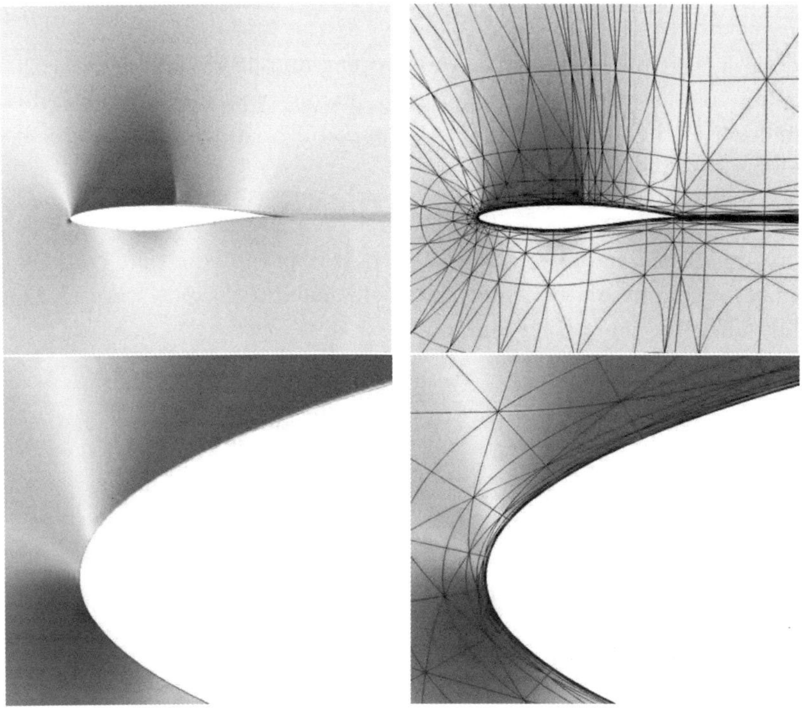

Fig. 8. Transonic flow ($M = 0.729, \mathrm{Re} = 6.5 \cdot 10^6$), with subcell resolution of shocks.

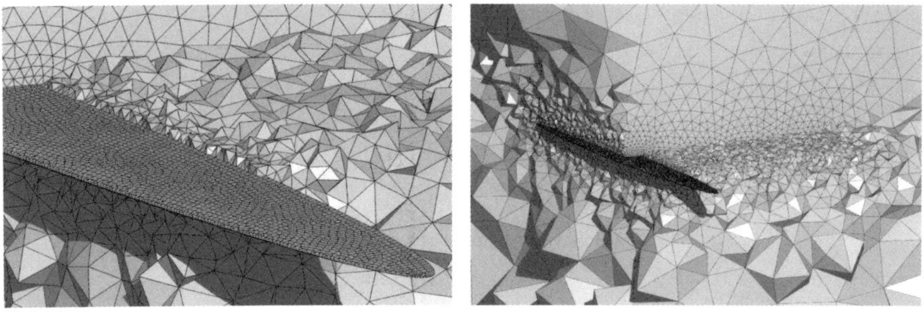

Fig. 9. A tetrahedral mesh for the domain around the elliptic wing pair.

account for the curved domain boundaries, we use the nonlinear elasticity approach that we proposed in Ref. 64. The mesh is shown in Fig. 9.

We prescribe the symmetric wing motion using a flapping angle at the wing centerline hinge given by

$$\phi(t) = A_\phi \cos \omega t \tag{62}$$

where t is the time, $A_\phi = 30°$ is the flapping amplitude and $w = 2\pi/20$ the flapping angular frequency. In addition, a wing twist angle is prescribed as a function of the span location. At the distance X from the centerline of the wing, the twist angle is

$$\theta(t, X) = \varepsilon\,(a(X) \cos \omega t + b(X) \cos \omega t)$$

where the *twist scaling factor* $\varepsilon \in [0, 1]$ is a parameter that controls the amount of spanwise twist, and the coefficients $a(X)$, $b(X)$ are chosen to locally align the wing with the flow:[61]

$$A(X) = -\frac{\sqrt{L^2 - X^2}}{4u_\infty L}\ , \qquad B(X) = \frac{X\phi_0\omega}{u_\infty}\ , \tag{63}$$

$$a(X) = \frac{-B}{A^2\omega^2 + 1}\ , \qquad b(X) = \frac{BA\omega}{A^2\omega^2 + 1}\ . \tag{64}$$

We note that this motion is not in any way an optimized flapping strategy, but it is adequate for the purposes of studying our computational models.

In order to develop an ALE formulation for this domain deformation, we need a smooth embedding of the flapping motion $\phi(t)$, $\theta(t, X)$ in the reference domain. That is, we need a smooth function $x = x(X, t)$ that maps the wing surface to the location given by $\phi(t)$, $\theta(t, X)$. We also prefer volume preserving mappings ($g = \det(G) = 1$) to simplify the ALE equations.

While there are many ways to find such a mapping, we use a shearing approach as follows. To begin with, the function $A(X)$ is not differentiable at $X = L$, and not even real-valued for $X > L$, so we need to modify the flapping motion to regularize this expression. We approximate

$$\sqrt{L^2 - X^2} \approx \frac{\arctan(r(L - X))}{\arctan(rL)} L \tag{65}$$

where $r = 1.2$ is a good choice. Note that this expression is also defined and smooth for $X > L$. The mapping function is then created using two combined shearing motions:

$$x(X, t) = \begin{bmatrix} X \cos \phi \\ Y \cos \theta \\ X \sin \phi + Y \sin \theta + Z \sec \phi \sec \theta \end{bmatrix}, \tag{66}$$

which gives a volume-preserving deformation gradient ($\det(G) = 1$):

$$G = \begin{bmatrix} \cos\phi & 0 & 0 \\ G_{21} & \cos\theta & 0 \\ G_{31} & G_{32} & \sec\phi\sec\theta \end{bmatrix}, \tag{67}$$

where all matrix entries as well as the grid velocity $\partial x/\partial t$ are found by analytical differentiation.

Our simulations are done at a Reynolds number of $\text{Re} = 3,000$ and a free-stream Mach number of $M = 0.1$. We use a third-order accurate diagonal implicit Runge-Kutta (DIRK) method[37] for time-integration, and polynomials of degree $p = 3$ within each tetrahedron for the space discretization. This gives a total of about 23 million degrees of freedom, and we integrate for three full flapping cycles using 600 implicit timesteps. We solve on a parallel computer with 32 nodes and a total of 256 computing processes, using the parallel Newton-Krylov methods described in section 3.3.

We show two representative test cases with different free-stream angle of attack α and twist scaling factor ε. Visualizations of these flow fields are shown in Fig. 10, where the Mach number is plotted as color on isosurfaces of the entropy. The flow plots show regions of flow separation and wake vortex structures. For the first case ($\alpha = 5°$, $\varepsilon = 0.5$), significant separa-

Angle of attack $\alpha = 5°$, twist multiplier $\varepsilon = 0.5$

Angle of attack $\alpha = 10°$, twist multiplier $\varepsilon = 1.0$

Fig. 10. The flow field around the flapping wing pair, visualized as Mach number color plots on isosurfaces of the entropy. The plots correspond to two representative cases of angle of attack and twist multiplier (top and bottom) and the times $t = 20$, $t = 25$, and $t = 30$ (left to right).

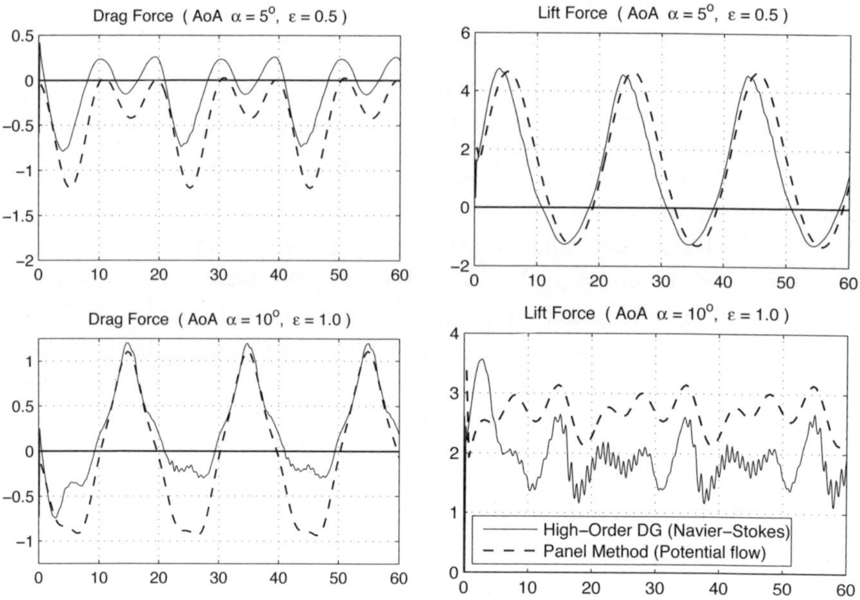

Fig. 11. The lift coefficients computed by the two simulation codes for the two cases considered ($\alpha = 5°$, $\varepsilon = 0.5$ and $\alpha = 10°$, $\varepsilon = 1.0$).

tion occurs over the entire wing during the mid-to-late downstroke. In the second case ($\alpha = 10°$, $\varepsilon = 1.0$), there is separation throughout the entire flapping cycle, in particular inboard of the wing.

The time evolution of the vertical and horizontal forces are shown in Fig. 11, along with a comparison with a simpler panel method code[65] based on a potential flow model. We note that the panel method code predicts the lift forces well in the first case, although it over-predicts the thrust production somewhat. In the second case, due to the large amount of separation during the downstroke, the force predictions do not agree well. Therefore, we can conclude that low-fidelity simulation tools can perform well for attached flows, but high-fidelity Navier-Stokes solvers are essential for predicting flows with large amounts of separation.

5. Acknowledgements

We would like to acknowledge our collaborators J. Bonet, M. Drela, E. Israeli, C. N. Nguyen, A. Uranga, and D. J. Willis for the many contributions to the work reported in this chapter.

References

1. S. K. Lele, Compact finite difference schemes with spectral-like resolution, *J. Comput. Phys.* **103**(1), 16–42, (1992).
2. M. R. Visbal and D. V. Gaitonde, On the use of higher-order finite-difference schemes on curvilinear and deforming meshes, *J. Comput. Phys.* **181**(1), 155–185, (2002).
3. T. J. Barth. Recent developments in high-order k-exact reconstruction on unstructured meshes, (1993). AIAA-93-0668.
4. A. Nejat and C. Ollivier-Gooch, A high-order accurate unstructured finite volume Newton-Krylov algorithm for inviscid compressible flows, *J. Comput. Phys.* **227**(4), 2582–2609, (2008).
5. T. Hughes, G. Scovazzi, and T. Tezduyar, Stabilized methods for compressible flows, *SIAM J. Sci. Comput.* **43**, 343–368, (2010).
6. W. H. Reed and T. R. Hill. Triangular mesh methods for the neutron transport equation. Technical Report Technical Report LA-UR-73-479, Los Alamos Scientific Laboratory, (1973).
7. B. Cockburn and C.-W. Shu, Runge-Kutta discontinuous Galerkin methods for convection-dominated problems, *J. Sci. Comput.* **16**(3), 173–261, (2001).
8. J. S. Hesthaven and T. Warburton, *Nodal discontinuous Galerkin methods*. vol. 54, *Texts in Applied Mathematics*, (Springer, New York, 2008). Algorithms, analysis, and applications.
9. B. Cockburn, J. Gopalakrishnan, and R. Lazarov, Unified hybridization of discontinuous Galerkin, mixed, and continuous Galerkin methods for second order elliptic problems, *SIAM J. Numer. Anal.* **47**(2), 1319–1365, (2009).
10. N. C. Nguyen, J. Peraire, and B. Cockburn, An implicit high-order hybridizable discontinuous galerkin method for the incompressible navier-stokes equations, *J. Comput. Phys.* (2010). To appear.
11. J. Peraire, N. C. Nguyen, and B. Cockburn. A hybridizable discontinuous galerkin method for the compressible euler and navier-stokes equations. In *48th AIAA Aerospace Sciences Conference, Orlando, Florida* (January, 2010). AIAA-2010-363.
12. Z. J. Wang, Spectral (finite) volume method for conservation laws on unstructured grids. Basic formulation, *J. Comput. Phys.* **178**(1), 210–251, (2002).
13. Y. Liu, M. Vinokur, and Z. J. Wang, Spectral difference method for unstructured grids. I. Basic formulation, *J. Comput. Phys.* **216**(2), 780–801, (2006).
14. J. P. Boris, *On Large Eddy Simulation using subgrid turbulence models*. (Springer-Verlag, New York, 1990). In J.L. Lumley, editor, Whither Turbulence? Turbulence at the Crossroads.
15. P. R. Spalart and S. R. Allmaras, A one-equation turbulence model for aerodynamic flows, *La Rech. Aérospatiale.* **1**, 5–21, (1994).
16. P.-O. Persson, J. Bonet, and J. Peraire, Discontinuous Galerkin solution of the Navier-Stokes equations on deformable domains, *Comput. Methods Appl. Mech. Engrg.* **198**(17-20), 1585–1595, (2009).

17. P. D. Thomas and C. K. Lombard, Geometric conservation law and its application to flow computations on moving grids, *AIAA J.* **17**(10), 1030–1037, (1979).
18. P. L. Roe, Approximate Riemann solvers, parameter vectors, and difference schemes, *J. Comput. Phys.* **43**(2), 357–372, (1981).
19. J. Peraire and P.-O. Persson, The compact discontinuous Galerkin (CDG) method for elliptic problems, *SIAM J. Sci. Comput.* **30**(4), 1806–1824, (2008).
20. B. Cockburn and C.-W. Shu, The local discontinuous Galerkin method for time-dependent convection-diffusion systems, *SIAM J. Numer. Anal.* **35**(6), 2440–2463, (1998).
21. B. Cockburn and B. Dong. An analysis of the minimal dissipation local discontinuous Galerkin method for convection–difussion problems. IMA Preprint Series # 2146, also presented at the 7th. World Congress on Computational Mechanics, Los Angeles, CA, June 16-22, 2006, (2006).
22. C. E. Baumann and J. T. Oden, A discontinuous *hp* finite element method for the Euler and Navier-Stokes equations, *Int. J. Numer. Methods Fluids.* **31**(1), 79–95, (1999). Tenth International Conference on Finite Elements in Fluids (Tucson, AZ, 1998).
23. A. Burbeau, P. Sagaut, and C.-H. Bruneau, A problem-independent limiter for high-order Runge-Kutta discontinuous Galerkin methods, *J. Comput. Phys.* **169**(1), 111–150, (2001).
24. C.-W. Shu and S. Osher, Efficient implementation of essentially nonoscillatory shock-capturing schemes, *J. Comput. Phys.* **77**(2), 439–471, (1988).
25. C.-W. Shu and S. Osher, Efficient implementation of essentially nonoscillatory shock-capturing schemes. II, *J. Comput. Phys.* **83**(1), 32–78, (1989).
26. I. Lomtev, C. B. Quillen, and G. E. Karniadakis, Spectral/*hp* methods for viscous compressible flows on unstructured 2D meshes, *J. Comput. Phys.* **144**(2), 325–357, (1998).
27. A. Kanevsky, M. H. Carpenter, and J. S. Hesthaven, Idempotent filtering in spectral and spectral element methods, *J. Comput. Phys.* **220**(1), 41–58, (2006).
28. E. Tadmor, Shock capturing by the spectral viscosity method, *Comput. Methods Appl. Mech. Engrg.* **80**(1-3), 197–208, (1990). Spectral and high order methods for partial differential equations (Como, 1989).
29. S. Hesthaven, J.S. Kaber and L. Lurati, Pade-legendre interpolants for gibbs reconstruction, *J. Sci. Comput.* (2005). (to appear).
30. G. May and A. Jameson. High-order accurate methods for high-speed flow. In *17th AIAA Computational Fluid Dynamics Conference, Toronto, Ontario* (June, 2005). AIAA-2005-5252.
31. P.-O. Persson and J. Peraire. Sub-cell shock capturing for discontinuous Galerkin methods. In *44th AIAA Aerospace Sciences Meeting and Exhibit, Reno, Nevada,* (2006). AIAA-2006-0112.
32. J. Von Neumann and R. D. Richtmyer, A method for the numerical calculation of hydrodynamic shocks, *J. Appl. Phys.* **21**, 232–237, (1950).

33. G. E. Barter. *Shock capturing with PDE-based artificial viscosity for an adaptive, higher-order discontinuous Galerkin finite element method.* PhD thesis, M.I.T. (June, 2008).
34. F. Bassi, A. Crivellini, S. Rebay, and M. Savini, Discontinuous Galerkin solution of the Reynolds-averaged Navier-Stokes and $k-\omega$ turbulence model equations, *Computers & Fluids.* **34**(4–5), 507–540, (2005).
35. N. C. Nguyen, P.-O. Persson, and J. Peraire. RANS solutions using high order discontinuous Galerkin methods. In *45th AIAA Aerospace Sciences Meeting and Exhibit, Reno, Nevada,* (2007). AIAA-2007-914.
36. T. H. Koornwinder. Askey-Wilson polynomials for root systems of type *BC*. In *Hypergeometric functions on domains of positivity, Jack polynomials, and applications (Tampa, FL, 1991)*, vol. 138, *Contemp. Math.*, pp. 189–204. Amer. Math. Soc., Providence, RI, (1992).
37. R. Alexander, Diagonally implicit Runge-Kutta methods for stiff o.d.e.'s, *SIAM J. Numer. Anal.* **14**(6), 1006–1021, (1977).
38. L. F. Shampine and C. W. Gear, A user's view of solving stiff ordinary differential equations, *SIAM Rev.* **21**(1), 1–17, (1979).
39. E. Anderson et al., *LAPACK Users' Guide.* (Society for Industrial and Applied Mathematics, Philadelphia, PA, 1999), third edition.
40. P.-O. Persson and J. Peraire, Newton-GMRES preconditioning for discontinuous Galerkin discretizations of the Navier-Stokes equations, *SIAM J. Sci. Comput.* **30**(6), 2709–2733, (2008).
41. P.-O. Persson. Scalable parallel Newton-Krylov solvers for discontinuous Galerkin discretizations. In *47th AIAA Aerospace Sciences Meeting and Exhibit, Orlando, Florida,* (2009). AIAA-2009-606.
42. C. R. Nastase and D. J. Mavriplis, High-order discontinuous Galerkin methods using an hp-multigrid approach, *J. Comput. Phys.* **213**(1), 330–357, (2006).
43. K. Fidkowski, T. Oliver, J. Lu, and D. Darmofal, p-multigrid solution of high-order discontinuous Galerkin discretizations of the compressible Navier-Stokes equations, *J. Comput. Phys.* **207**(1), 92–113, (2005).
44. G. Kanschat, Robust smoothers for high-order discontinuous galerkin discretizations of advection-diffusion problems, *J. Comput. Appl. Math.* **218** (1), 53–60, (2008).
45. L. T. Diosady and D. L. Darmofal, Preconditioning methods for discontinuous Galerkin solutions of the Navier-Stokes equations, *J. Comput. Phys.* **228** (11), 3917–3935, (2009).
46. W. Hackbusch, *Multigrid methods and applications.* vol. 4, *Springer Series in Computational Mathematics*, (Springer-Verlag, Berlin, 1985).
47. E. M. Rønquist and A. T. Patera, Spectral element multigrid. I. Formulation and numerical results, *J. Sci. Comput.* **2**(4), 389–406, (1987).
48. P. Wesseling. A robust and efficient multigrid method. In *Multigrid methods (Cologne, 1981)*, vol. 960, *Lecture Notes in Math.*, pp. 614–630. Springer, Berlin, (1982).
49. G. Wittum. On the robustness of ILU-smoothing. In *Robust multi-grid methods (Kiel, 1988)*, vol. 23, *Notes Numer. Fluid Mech.*, pp. 217–239. Vieweg, Braunschweig, (1989).

50. H. C. Elman, V. E. Howle, J. N. Shadid, and R. S. Tuminaro, A parallel block multi-level preconditioner for the 3D incompressible Navier-Stokes equations, *J. Comput. Phys.* **187**(2), 504–523, (2003).
51. W. L. Briggs, V. E. Henson, and S. F. McCormick, *A multigrid tutorial.* (Society for Industrial and Applied Mathematics (SIAM), Philadelphia, PA, 2000), second edition.
52. A. Toselli and O. Widlund, *Domain Decomposition Methods - Algorithms and Theory.* vol. 34, *Springer Series in Computational Mathematics*, (Springer, 2004).
53. G. Karypis and V. Kumar. METIS serial graph partitioning and fill-reducing matrix ordering. http://glaros.dtc.umn.edu/gkhome/metis/metis/overview.
54. C.-D. Munz, S. Roller, R. Klein, and K. J. Geratz, The extension of incompressible flow solvers to the weakly compressible regime, *Comput. & Fluids.* **32**(2), 173–196, (2003).
55. A. Uranga, P.-O. Persson, M. Drela, and J. Peraire, Implicit large eddy simulation of transition to turbulence at low Reynolds numbers using a discontinuous Galerkin method, *Int. J. Num. Meth. Eng.* (2010). To appear.
56. M. Galbraith and M. Visbal. Implicit Large Eddy Simulaion of low Reynolds number flow past the SD7003 airfoil. In *Proc. of the 46th AIAA Aerospace Sciences Meeting and Exhibit, Reno, Nevada, AIAA-2008-225*, (2008).
57. M. Drela. *XFOIL Users Guide, Version 6.94.* MIT Aeronautics and Astronautics Department, (2002).
58. R. Radespiel, J. Windte, and U. Scholz, Numerical and experimental flow analysis of moving airfoils with laminar separation bubbles, *AIAA Paper 2006-501.* (Jan. 2006.).
59. M. Ol, B. McAuliffe, E. Hanff, U. Scholz, and C. Kahler. Comparison of laminar separation bubbles measurements on a low Reynolds number airfoil in three facilities. In *Proc. of the 35th Fluid Dynamics Conference and Exhibit, Toronto, Ontario, Canada, AIAA-2005-5149*, (2005).
60. D. Coles and E. Hirst. Computation of turbulent boundary layers. In *AFOSR-IFP-Stanford Conference*, vol. II, CA, (1969). Stanford University.
61. P.-O. Persson, D. J. Willis, and J. Peraire. The numerical simulation of flapping wings at low reynolds numbers. In *48th AIAA Aerospace Sciences Meeting and Exhibit, Orlando, Florida*, (2010). AIAA-2010-724.
62. P.-O. Persson and G. Strang, A simple mesh generator in Matlab, *SIAM Rev.* **46**(2), 329–345, (2004).
63. K. Morgan and J. Peraire, Unstructured grid finite element methods for fluid mechanics, *Inst. of Physics Reviews.* **61**(6), 569–638, (1998).
64. P.-O. Persson and J. Peraire. Curved mesh generation and mesh refinement using Lagrangian solid mechanics. In *47th AIAA Aerospace Sciences Meeting and Exhibit, Orlando, Florida*, (2009). AIAA-2009-949.
65. D. J. Willis, J. Peraire, and J. K. White, A combined pFFT-multipole tree code, unsteady panel method with vortex particle wakes, *Internat. J. Numer. Methods Fluids.* **53**(8), 1399–1422, (2007).

CHAPTER 6

WEIGHTED NON-OSCILLATORY LIMITERS FOR RUNGE-KUTTA DISCONTINUOUS GALERKIN METHODS

Jianxian Qiu

Department of Mathematics, Nanjing University
Nanjing, Jiangsu, P.R. China, 210093
jxqiu@nju.edu.cn

The discontinuous Galerkin (DG) method is a spatial discretization procedure for hyperbolic conservation laws, which employs useful features from high resolution finite volume schemes, such as the exact or approximate Riemann solvers serving as numerical fluxes and limiters, which is termed as RKDG when TVD Runge-Kutta method is applied for time discretization. It has the advantage of flexibility in handling complicated geometry, *h*-*p* adaptivity, and efficiency of parallel implementation and has been used successfully in many applications. An important component of RKDG methods for solving conservation laws, with strong shocks in the solutions is a nonlinear limiter, which is applied to detect discontinuities and control spurious oscillations near such discontinuities. There are many limiters which exist in the literature, for example the minmod type limiters, the moment based limiters, the improved moment based limiters, the monotonicity-preserving (MP) limiters, modifications of MP limiters. In this chapter, we will review these limiters and describe a robust limiter, the weighted essentially non-oscillatory (WENO) type limiter, which was developed in the recently years.

Keywords: Runge–Kutta discontinuous Galerkin method, limiters, WENO finite volume scheme, high order accuracy

AMS(MOS) subject classification: 65M60, 65M99, 35L65

1. Introduction

The first discontinuous Galerkin (DG) method was introduced in 1973 by Reed and Hill,[1] in the framework of neutron transport (steady state linear hyperbolic equations). A major development of the DG method was carried out by Cockburn *et al.* in a series of papers,[2–6] in which they established a framework to easily solve *nonlinear* time dependent hyperbolic conservation laws:

$$\begin{cases} \partial_t u + \nabla \cdot f(u) = 0, \\ u(x,0) = u_0(x), \end{cases} \tag{1}$$

using explicit, nonlinearly stable high order Runge-Kutta time discretizations[7] and DG discretization in space with exact or approximate Riemann solvers as interface fluxes and total variation bounded (TVB) limiter[8] to achieve non-oscillatory properties for strong shocks. These schemes are termed Runge-Kutta discontinuous Galerkin (RKDG) methods.

The DG methods have the advantage of typical finite element methods in an easy handling of complicated geometry, arbitrary triangulations, and also the added advantage due to the discontinuous nature of the solution and the test function space, in an explicit time marching, local communications hence high efficiency in parallel implementation,[9] and easy *h-p* adaptivity. For these reasons, they have been widely used in applications, see for example the the survey paper,[10] and other papers in that Springer volume, which contains the conference proceedings of the First International Symposium on Discontinuous Galerkin Methods held at Newport, Rhode Island in 1999, the special issues for DG methods in Journal of Scientific Computing, V22-23 (1995), V40 (2009) and Computer Methods in Applied Mechanics and Engineering, V195, No. 25-28 (2006). The lecture notes[11] is a good reference for many details, as well as the extensive review paper.[12]

An important component of RKDG methods for solving conservation laws (1) with strong shocks in the solutions is a nonlinear limiter, which is applied to detect discontinuities and control spurious oscillations near such discontinuities. Many such limiters have been used in the literature on RKDG methods. For example, we mention the minmod type TVB limiter,[2–6] which is a slope limiter using a technique borrowed from the finite volume methodology; the moment based limiter[9] and an improved moment limiter,[13] which are specifically designed for discontinuous Galerkin methods and work on the moments of the numerical solution. These limiters tend to degrade accuracy when mistakenly used in smooth regions of the solution. There are also many limiters developed in the finite volume and finite difference literature, such as the various flux limiters,[14] the monotonicity-preserving (MP) limiters,[15] modifications of MP limiters,[16] which can be severed as limiters for DG methods. However, the limiters used to control spurious oscillations in the presence of strong shocks are less robust than the strategies of essentially non-oscillatory (ENO) and weighted ENO (WENO) finite volume and finite difference methods. In Refs. 17 and 18, Qiu *et al.* studied using WENO methodology as limiters for RKDG methods on structured and unstructured meshes. We adopted the following framework for limiter procedure:

(1) First we identify the "troubled cells", namely those cells which might need the limiting procedure;
(2) Then we replace the solution polynomials in those troubled cells by reconstructed polynomials which maintain the original cell averages (conservation), have the same orders of accuracy as before, but are less oscillatory.

This technique worked quite well in one and two dimensional test problems. In Refs. 19–21, this approach is further improved by using Hermite WENO (HWENO) rather than WENO methodology in the limiter so that a more compact stencil is used on both structured and unstructured meshes. The emphasis of the works[17–21] is on Step 2, where different WENO reconstruction strategies are considered. The work[22] is focused on Step 1, we systematically investigate and compare a few different limiter strategies as troubled-cell indicators, with an objective of obtaining the most efficient and reliable troubled-cell indicators to save computational cost.

The organization of this chapter is as follows. In section 2, we concentrate on Step 1 in the procedure above, and describe systematically a few discontinuity detecting methods as troubled-cell indicators. We use the usual WENO reconstructions based on cell averages of neighboring cells, such as in Refs. 23 and 24, to reconstruct the values of the solutions at certain Gaussian quadrature points in the troubled cells, and then rebuild the solution polynomials in those troubled cells from the original cell averages and the reconstructed values at the Gaussian quadrature points through a numerical integration for the moments. This turns out to be a robust way to retain the original high order accuracy of the discontinuous Galerkin method. We describe the details of this procedure in section 3. In section 4, we investigate the usage of the HWENO finite volume methodology as limiters for RKDG methods, following the idea in Ref. 17, with the goal of obtaining a robust and high order limiting procedure to simultaneously obtain uniform high order accuracy and sharp, non-oscillatory shock transition for RKDG methods. In section 5, we provide numerical examples to demonstrate the behavior of the DG methods with WENO type limiters with Runge-Kutta time discretizations. Concluding remarks are given in section 6.

2. Description of Troubled-Cell Indicators

In this section we review a few discontinuity detecting methods to identify troubled cells. We start with the description in the one-dimensional case and use the notations in Ref. 3, however we emphasize that the procedure described below does not depend on the specific basis chosen for the polynomials and works also in multi dimensions. We would like to solve the one-dimensional scalar conservation law:

$$\begin{cases} u_t + f(u)_x = 0 \\ u(x,0) = u_0(x). \end{cases} \tag{2}$$

The computational domain is divided into N cells with boundary points $0 = x_{\frac{1}{2}} < x_{\frac{3}{2}} < \cdots < x_{N+\frac{1}{2}} = L$. The points x_i are the centers of the cells $I_i = [xi - 1/2, x_{i+1/2}]$, and we denote the cell sizes by $\Delta x_i = x_{i+1/2} - x_{i-1/2}$ and the maximum cell size by $h = \max_i \Delta x_i$. The solution as well as the test function space is given by $V_h^k = \{p : p|_{I_i} \in P^k(I_i)\}$, where $P^k(I_i)$ is the space of polynomials of degree $\leq k$ on the cell I_i. We adopt a local orthogonal basis over I_i, $\{v_l^{(i)}(x),\ l = 0, 1, \cdots, k\}$, namely the scaled Legendre polynomials:

$$v_0^{(i)}(x) = 1, \qquad v_1^{(i)}(x) = \frac{x - x_i}{\Delta x_i/2}, \qquad v_2^{(i)}(x) = \left(\frac{x - x_i}{\Delta x_i/2}\right)^2 - \frac{1}{3}, \cdots$$

Then the numerical solution $u^h(x,t)$ in the space V_h^k can be written as:

$$u^h(x,t) = \sum_{l=0}^{k} u_i^{(l)}(t) v_l^{(i)}(x), \qquad \text{for } x \in I_i \tag{3}$$

and the degrees of freedom $u_i^{(l)}(t)$ are the moments defined by

$$u_i^{(l)}(t) = \frac{1}{a_l} \int_{I_i} u^h(x,t) v_l^{(i)}(x) dx, \qquad l = 0, 1, \cdots, k$$

where $a_l = \int_{I_i} (v_l^{(i)}(x))^2 dx$ are the normalization constants since the basis is not orthonormal. In order to determine the approximate solution, we evolve the degrees of freedom $u_i^{(l)}$:

$$\frac{d}{dt}u_i^{(l)} + \frac{1}{a_l}\Bigg(-\int_{I_i} f(u^h(x,t))\frac{d}{dx}v_l^{(i)}(x)dx + \hat{f}(u_{i+1/2}^-, u_{i+1/2}^+)v_l^{(i)}(x_{i+1/2})$$

$$-\hat{f}(u_{i-1/2}^-, u_{i-1/2}^+)v_l^{(i)}(x_{i-1/2})\Bigg) = 0, \qquad l = 0, 1, \cdots, k \tag{4}$$

where $u_{i+1/2}^{\pm} = u^h(x_{i+1/2}^{\pm}, t)$ are the left and right limits of the discontinuous solution u^h at the cell interface $x_{i+1/2}$, $\hat{f}(u^-, u^+)$ is a monotone flux (non-decreasing in the first argument and non-increasing in the second argument) for the scalar case and an exact or approximate Riemann solver for the system case. The integral term in (4) can be computed either exactly or by a suitable numerical quadrature accurate to at least $O(h^{k+l+2})$. The semidiscrete scheme (4), written as

$$u_t = L(u)$$

is then discretized in time by by a nonlinearly stable Runge-Kutta time discretization, e.g. the third order version:[7]

$$\begin{aligned}
u^{(1)} &= u^n + \Delta t L(u^n) \\
u^{(2)} &= \frac{3}{4}u^n + \frac{1}{4}u^{(1)} + \frac{1}{4}\Delta t L(u^{(1)}) \\
u^{n+1} &= \frac{1}{3}u^n + \frac{2}{3}u^{(2)} + \frac{2}{3}\Delta t L(u^{(2)}).
\end{aligned} \tag{5}$$

The method described above can compute solutions to (2) which are either smooth or have weak shocks and other discontinuities without further modification. If the discontinuities are strong, however, the scheme will generate significant oscillations and even nonlinear instability. To avoid such difficulties, a nonlinear limiter procedure is used after each Runge-Kutta inner stage (or after the complete Runge-Kutta time step) to control the numerical solution.

We will now review a few discontinuity detecting methods to identify troubled cells. Seven troubled cell indicators were described in Ref. 22, which are the minmod-based TVB limiter, moment limiter, modified moment limiter, the monotonicity-preserving limiter, a modification of the monotonicity-preserving limiter, a shock-detection technique and indicator based on Harten's sub-cell resolution idea.

(1) The minmod based TVB limiter.[3] Denote:

$$u_{i+1/2}^- = u_i^{(0)} + \tilde{u}_i, \qquad u_{i-1/2}^+ = u_i^{(0)} - \tilde{\tilde{u}}_i$$

From (3) we can see that

$$\tilde{u}_i = \sum_{l=1}^{k} u_i^{(l)} v_l^{(i)}(x_{i+1/2}), \qquad \tilde{\tilde{u}}_i = -\sum_{l=1}^{k} u_i^{(l)} v_l^{(i)}(x_{i-1/2}).$$

These are modified by either the standard minmod limiter[25]

$$\tilde{u}_i^{(mod)} = m(\tilde{u}_i, \Delta_+ u_i^{(0)}, \Delta_- u_i^{(0)}), \qquad \tilde{\tilde{u}}_i^{(mod)} = m(\tilde{\tilde{u}}_i, \Delta_+ u_i^{(0)}, \Delta_- u_i^{(0)}),$$

where the minmod function m is given by

$$m(a_1, a_2, \cdots, a_n) = \begin{cases} s \cdot \min_{1\le j\le n} |a_j| & \text{if } \operatorname{sign}(a_1) = \operatorname{sign}(a_2) = \cdots = \operatorname{sign}(a_n) = s, \\ 0 & \text{otherwise.} \end{cases} \tag{6}$$

or by the TVB modified minmod function[8]

$$\tilde{m}(a_1, a_2, \cdots, a_n) = \begin{cases} a_1 & \text{if } |a_1| \le Mh^2, \\ m(a_1, a_2, \cdots, a_n) & \text{otherwise.} \end{cases} \tag{7}$$

where $M > 0$ is a constant. The choice of M depends on the solution of the problem. For scalar problems it is possible to estimate M by the initial condition as in Ref. 3 (M is proportional to the second derivative of the initial condition at smooth extrema), however it is more difficult to estimate M for the system case. If M is chosen too small, more cells than necessary will be identified as troubled cells, thereby increasing the computational cost; however if M is chosen too large, spurious oscillations may appear.

(2) Moment limiter of Biswas, Devine and Flaherty.[9] We will denote this limiter as the BDF limiter.
The moment based limiter in Ref. 9 is given by

$$u_i^{(l),mod} = \frac{1}{2l-1} m\left((2l-1)u_i^{(l)}, u_{i+1}^{(l-1)} - u_i^{(l-1)}, u_i^{(l-1)} - u_{i-1}^{(l-1)}\right) \tag{8}$$

where m is again the minmod function (6). This limiter is applied adaptively. First, the highest-order moment $u^{(k)}$ is limited. Then the limiter is applied to successively lower-order moments when the next higher order moment on the interval has been changed by the limiting. For our purpose, when the BDF limiter (8) is enacted (returns other than the first argument) for the highest order moment, the cell is declared as a troubled cell and marked for further reconstruction.

(3) A modification of the moment limiter by Burbeau, Sagaut and Bruneau.[13] We will denote this modified moment limiter as the BSB limiter. For our purpose as a troubled-cell indicator, if both (8) and

$$\hat{u}_i^{(l),mod} = \frac{1}{2l-1} m((2l-1)u_i^{(l)}, u_{i+1/2}^{(l-1)+} - u_i^{(l-1)}, u_i^{(l-1)} - u_{i-1/2}^{(l-1)-}) \quad (9)$$

are enacted for the highest-order moment $u^{(k)}$, where

$$u_{i+1/2}^{(l-1)+} = u_{i+1}^{(l-1)} - (2l-1)u_{i+1}^{(l)}, \qquad u_{i-1/2}^{(l-1)-} = u_{i-1}^{(l-1)} + (2l-1)u_{i-1}^{(l)},$$

that is, if both $u_i^{(k),mod} \neq u_i^{(k)}$ and $\hat{u}_i^{(k),mod} \neq u_i^{(k)}$, then the cell is identified as a troubled cell, marked for further reconstruction.

(4) The monotonicity preserving (MP) limiter.[15] In Ref. 15, Suresh and Huynh designed a limiter to preserve accuracy near smooth extrema, which works well with Runge-Kutta time stepping for a class high-order monotonicity-preserving schemes. The interface values in these schemes are obtained by limiting a higher-order polynomial reconstruction. The key idea in that work is to distinguish between a smooth local extremum and a genuine $O(1)$ discontinuity.

For our purpose as a troubled-cell indicator, the MP limiter can be described as follows. First a median functions is defined as

$$\text{median}(x, y, z) = x + m(y - x, z - x) \quad (10)$$

where m is again the minmod function (6). If

$$u_{i+1/2}^- \neq \text{median}(u_{i+1/2}^-, u_{i+1/2}^{\min}, u_{i+1/2}^{\max}) \quad (11)$$

where

$$u_{i+1/2}^{\min} = \max[\min(u_i^{(0)}, u_{i+1}^{(0)}, u_{i+1/2}^{MD}), \min(u_i^{(0)}, u_{i+1/2}^{UL}, u_{i+1/2}^{LC})],$$

$$u_{i+1/2}^{\max} = \min[\max(u_i^{(0)}, u_{i+1}^{(0)}, u_{i+1/2}^{MD}), \max(u_i^{(0)}, u_{i+1/2}^{UL}, u_{i+1/2}^{LC})],$$

and

$$d_i = u_{i+1}^{(0)} - 2u_i^{(0)} + u_{i-1}^{(0)},$$

$$d_{i+1/2}^{M4X} = m(4d_i - d_{i+1}, 4d_{i+1} - d_i, d_i, d_{i+1}, d_{i-1}, d_{i+2}),$$

$$u_{i+1/2}^{MD} = \frac{1}{2}\left(u_i^{(0)} + u_{i+1}^{(0)} - d_{i+1/2}^{M4X}\right), \quad u_{i+1/2}^{UL} = u_i^{(0)} + \alpha(u_i^{(0)} - u_{i-1}^{(0)}),$$

$$u_{i+1/2}^{LC} = u_i^{(0)} + \frac{1}{2}\left(u_i^{(0)} - u_{i-1}^{(0)}\right) + \frac{\beta}{3} d_{i-1/2}^{M4X},$$

or if $u^+_{i-1/2}$ satisfies a similar (symmetric) condition, then the cell is identified as a troubled cell, marked for further reconstruction. We take the parameters $\alpha = 2$ and $\beta = 4$ in the numerical tests section 5, as suggested in Ref. 15.

(5) A modification of the MP limiter.[16] We will denote this modified MP limiter as the MMP limiter. In Ref. 16, Rider and Margolin presented a simple modification of the standard monotonicity-preserving limiter in Ref. 15. These modified MP limiters relax the relatively stringent condition of preserving monotonicity, while enforcing less-restrictive conditions. For our purpose as a troubled-cell indicator, the MMP limiter can be described as follows.

$$\phi = \min(1, \Delta\overline{u}^{\min}/\Delta_{\min}u). \tag{12}$$

where

$$\Delta\overline{u}^{\min} = u_i^{(0)} - \min(u_{i-1}^{(0)}, u_i^{(0)}, u_{i+1}^{(0)}),$$

$$\Delta_{\min}u = u_i^{(0)} - \min(u^+_{i-1/2}, u^-_{i+1/2}).$$

When $\phi \neq 1$, the limiter enacts, and the cell is identified as a troubled cell, marked for further reconstruction.

(6) A shock detection technique by Krivodonova, Xin, Remacle, Chevaugeon and Flaherty in Ref. 26. We will denote the troubled-cell indicator based on this technique as the KXRCF indicator.

The strategy in Ref. 26 is based on a strong superconvergence at the outflow boundary of each element in smooth regions for the discontinuous Galerkin method, to detect discontinuities and to lower the order of accuracy in the approximation there to avoid spurious oscillations near such discontinuities when solving hyperbolic systems of conservation laws.

For our purpose as a troubled-cell indicator, the KXRCF indicator can be described as follows. Partition the boundary of a cell I_i into two portions ∂I_i^- and ∂I_i^+, where the flow is into ($\vec{v}\cdot\vec{n} < 0$) and out of ($\vec{v}\cdot\vec{n} > 0$) I_i, respectively. The discontinuity detector in Ref. 26 is defined as

$$\mathcal{I}_i = \frac{\left| \int_{\partial I_i^-} (u^h|_{I_i} - u^h|_{I_{n_i}})\, ds \right|}{h^{(k+1)/2}\, |\partial I_i^-|\, \|u^h|_{I_i}\|}. \tag{13}$$

Here we choose h as the radius of the circumscribed circle in the element I_i, I_{n_i} is the neighbor of I_i on the side of ∂I_i^-, and the norm is based on the maximum norm taken at the integration quadrature points in two dimensions and based on an element average in one dimension. If $\mathcal{I}_i > 1$, the cell I_i is identified as a troubled cell, marked for further reconstruction.

(7) A troubled-cell indicator based on Harten's sub-cell resolution idea.[27] We will denote this indicator as the Harten indicator.
In Ref. 27, Harten introduced the notion of sub-cell resolution, which is based on the observation that, unlike point values, cell-averages of a discontinuous piecewise-smooth function contain information about the exact location of a discontinuity within the cell. For our purpose as a troubled-cell indicator, the Harten indicator can be described as follows. Let

$$F_i(z) = \frac{1}{\Delta x}\left\{\int_{x_{i-1/2}}^{z} u^h(x,t)|_{I_{i-1}}dx + \int_{z}^{x_{i+1/2}} u^h(x,t)|_{I_{i+1}}dx\right\} - u_i^{(0)},$$

where $u^h(x,t)|_{I_{i-1}}$ denotes the approximate solution u^h defined in the cell I_{i-1}, extended into cell I_i, and likewise $u^h(x,t)|_{I_{i+1}}$ is the approximate solution u^h defined in the cell I_{i+1}, extended into cell I_i. When

$$F_i(x_{i-1/2}) \cdot F_i(x_{i+1/2}) \leq 0,$$

I_i is suspected of having a discontinuity of $u^h(x,t)$ in its interior. However, this could also be a smooth extremum of the solution. To exclude the latter case, Harten[27] has a criterion comparing a minmod function of the first divided differences. We modify this criterion in the context of the RKDG method as follows. We compare the k-th moment $u_i^{(k)}$, which has the same magnitude of the k-th derivative of $u^h(x,t)$ modulo a constant, with that of the neighbors. Thus if

$$F_i(x_{i-1/2}) \cdot F_i(x_{i+1/2}) \leq 0, \quad \text{and} \quad |u_i^{(k)}| > \alpha|u_{i-1}^{(k)}|, \quad |u_i^{(k)}| > \alpha|u_{i+1}^{(k)}| \tag{14}$$

then the cell I_i is identified as a troubled cell, marked for further reconstruction. We suggest to take the constant $\alpha = 1.5$ in the numerical tests.

For the case of hyperbolic systems, to identify the troubled cells, we could use either a componentwise indicator or a characteristic one. The former works on each component of the solution and identifies a troubled cell when any component of the solution is flagging this cell as a troubled cell. The latter works in the local characteristic direction to do this identification. Their advantages and disadvantages are compared in Ref. 22 and the former is the choice there.

In Ref. 22, we have systematically studied and compared a few different troubled-cell indicators for the RKDG methods using WENO methodology as limiters. Extensive one- and two-dimensional simulations on the hyperbolic systems of Euler equations indicate that the minmod based TVB indicator (when the TVB constant M is suitably chosen), the KXRCF indicator by Krivodonova *et al.*,[26] and an indicator based on Harten's sub-cell resolution idea,[27] are better than other choices in all the test cases. Among these three there is no clear winner: any one of them would work better in some examples but not in all examples. All three of them should be suitable candidates for applications of the RKDG methods using WENO type reconstructions.

3. WENO Reconstruction as a Limiter for the RKDG Method

In this section, we will describe the procedure of WENO reconstruction as a limiter for the RKDG method in both the one dimensional and two dimensional cases. First, we use one of troubled cell indicators which are described in section 2 to identify the troubled cells. For the troubled cells, we would like to reconstruct the polynomial solution while retaining its cell average. In other words, we will reconstruct the degrees of freedom, or the moments for the troubled cell and retain only the cell average.

3.1. *WENO reconstruction in one dimensional case*

We have experimented with several different ways for this reconstruction and have settled in the following procedure. Let I_i be a troubled cell, we will reconstruct the degrees of freedom, or the moments, $u_i^{(l)}$ for the troubled cell I_i for $l = 1, \cdots, k$ and retain only the cell average $u_i^{(0)}$.

Step 1.1 Reconstruction of point values of u at the Gauss or Gauss-Lobatto quadrature points. For the $\mathbb{P}^k$ based DG (which is $(k+1)$-th order accurate), we need a Gauss or Gauss-Lobatto quadrature rule accurate to at least $O(h^{2k+2})$, and the order of accuracy for the WENO reconstruction must be at least $2k+1$. For this purpose, we would need to use the cell averages of the neighboring $2k+1$ cells $I_{i-k}, \ldots, I_{i+k}$ to reconstruct the point values of u at the Gauss or Gauss-Lobatto quadrature points. For examples, when $k = 1$ and $k = 2$, we use the following quadrature points:

(1) For the $\mathbb{P}^1$ case, we use the two-point Gauss quadrature points $x_{i-\sqrt{3}/6}$ and $x_{i+\sqrt{3}/6}$.
(2) For the $\mathbb{P}^2$ case, we use the four-point Gauss-Lobatto quadrature points $x_{i-1/2}$, $x_{i-\sqrt{5}/10}$, $x_{i+\sqrt{5}/10}$ and $x_{i+1/2}$.

The WENO reconstruction[28–30] is then performed:

- Step 1.1.1: We identify $k+1$ small stencils $S_j, j = 0, \cdots, k$, such that I_i belongs to each of them. Here we set $S_j = \cup_{l=0}^{k} I_{i+j-l}$. We denote by $\mathcal{T} = \cup_{j=0}^{k} S_j$ the larger stencil which contains all the cells from the $k+1$ smaller stencils.
  We have a k-th degree polynomial reconstruction denoted by $p_j(x)$, associated with each of the stencils $S_j, j = 0, \cdots, k$, such that the cell average of $p_j(x)$ in each of the cells in the stencil S_j agrees with the given cell average of u, i.e. $\frac{1}{\Delta x_{i+j-l}} \int_{I_{i+j-l}} p_j(x)dx = u_{i+j-l}^{(0)}, l = 0, \cdots, k$. We also have a higher order $(2k)$-th degree polynomial reconstruction denoted by Q(x), associated with the larger stencil $\mathcal{T}$, such that $\frac{1}{\Delta x_{i+l}} \int_{I_{i+l}} Q(x)dx = u_{i+l}^{(0)}, l = -k, \cdots, k$. The detail of the construction of $p_j(x)$ and $Q(x)$ can be found in Ref. 31.
- Step 1.1.2: We find the combination coefficients, also called linear weights, denoted by $\gamma_0, \cdots, \gamma_k$, which satisfy:

$$Q(x_G) = \sum_{j=0}^{k} \gamma_j p_j(x_G)$$

  where x_G is a Gauss quadrature point. Different quadrature points correspond to different linear weights. The value of the functions $Q(x)$ and $p_j(x), j = 0, \cdots, k$ at a Gaussian point x_G can be written as a linear combination of $u_i^{(0)}$ in the stencil. For example, when $k = 1$, with a uniform mesh, for $x_G = x_{i+\sqrt{3}/6}$, we have:

$$p_0(x_G) = -\frac{\sqrt{3}}{6} u_{i-1}^{(0)} + (1 + \frac{\sqrt{3}}{6}) u_i^{(0)},$$

$$p_1(x_G) = (1 - \frac{\sqrt{3}}{6}) u_i^{(0)} + \frac{\sqrt{3}}{6} u_{i+1}^{(0)},$$

$$Q(x_G) = -\frac{\sqrt{3}}{12} u_{i-1}^{(0)} + u_i^{(0)} + \frac{\sqrt{3}}{12} u_{i-1}^{(0)},$$

$$\gamma_0 = \frac{1}{2}, \qquad \gamma_1 = \frac{1}{2}.$$

The linear combination coefficients of the values of the functions $Q(x)$ and $p_j(x), j = 0, 1$ and the linear weights for the Gaussian point $x_{i-\sqrt{3}/6}$ are mirror symmetric with respect to those at $x_{i+\sqrt{3}/6}$, respectively.

For k=2, with a uniform mesh, for $x_G = x_{i+1/2}$, we have:

$$p_0(x_G) = \frac{1}{3}u_{i-2}^{(0)} - \frac{7}{6}u_{i-1}^{(0)} + \frac{11}{6}u_i^{(0)},$$

$$p_1(x_G) = -\frac{1}{6}u_{i-1}^{(0)} + \frac{5}{6}u_i^{(0)} + \frac{1}{3}u_{i+1}^{(0)},$$

$$p_2(x_G) = \frac{1}{3}u_i^{(0)} + \frac{5}{6}u_{i+1}^{(0)} - \frac{1}{6}u_{i+2}^{(0)},$$

$$Q(x_G) = \frac{1}{30}u_{i-2}^{(0)} - \frac{13}{60}u_{i-1}^{(0)} + \frac{47}{60}u_i^{(0)} + \frac{9}{20}u_{i+1}^{(0)} - \frac{1}{20}u_{i+2}^{(0)},$$

and

$$\gamma_0 = \frac{1}{10}, \qquad \gamma_1 = \frac{6}{10}, \qquad \gamma_2 = \frac{3}{10}.$$

For $x_G = x_{i+\sqrt{5}/10}$ we have:

$$p_0(x_G) = (-\frac{1}{60} + \frac{\sqrt{5}}{20})u_{i-2}^{(0)} + (\frac{1}{30} - \frac{\sqrt{5}}{5})u_{i-1}^{(0)} + (\frac{59}{60} + \frac{3\sqrt{5}}{20})u_i^{(0)},$$

$$p_1(x_G) = (-\frac{1}{60} - \frac{\sqrt{5}}{20})u_{i-1}^{(0)} + \frac{31}{30}u_i^{(0)} + (-\frac{1}{60} + \frac{\sqrt{5}}{20})u_{i+1}^{(0)},$$

$$p_2(x_G) = (\frac{59}{60} - \frac{3\sqrt{5}}{20})u_i^{(0)} + (\frac{1}{30} + \frac{\sqrt{5}}{5})u_{i+1}^{(0)} + (-\frac{1}{60} - \frac{\sqrt{5}}{20})u_{i+2}^{(0)},$$

$$Q(x_G) = \frac{1+6\sqrt{5}}{600}u_{i-2}^{(0)} - \frac{7+21\sqrt{5}}{300}u_{i-1}^{(0)} + \frac{313}{300}u_i^{(0)} + \frac{-7+21\sqrt{5}}{300}u_{i+1}^{(0)} + \frac{1-6\sqrt{5}}{600}u_{i+2}^{(0)},$$

and

$$\gamma_0 = \frac{91+9\sqrt{5}}{440}, \qquad \gamma_1 = \frac{129}{220}, \qquad \gamma_2 = \frac{91-9\sqrt{5}}{440}.$$

The linear combination coefficients of the values of the functions $Q(x)$ and $p_j(x), j = 0, 1, 2$ and the linear weights for the Gaussian points $x_{i-1/2}$ and $x_{i-\sqrt{5}/10}$ are mirror symmetric with respect to those at $x_{i+1/2}$ and $x_{i+\sqrt{5}/10}$, respectively.

- Step 1.1.3: We compute the smoothness indicator, denoted by β_j, for each stencil S_j, which measures how smooth the function $p_j(x)$ is in the target cell I_i. The smaller this smoothness indicator β_j, the smoother the function $p_j(x)$ is in the target cell. The smoothness indicators are the same for the reconstruction at all Gauss points in the same cell, thus significantly reducing the computational cost. As in Refs. 29 and 30, we are using the following smoothness indicator:

$$\beta_j = \sum_{l=1}^{k} \int_{I_i} \Delta x_i^{2l-1} \left(\frac{\partial^l}{\partial x^l} p_j(x) \right)^2 dx \,. \tag{15}$$

  In the actual numerical implementation the smoothness indicators β_j are written out explicitly as quadratic forms of the cell averages of u in the stencil, see Refs. 29–31 for details.
- Step 1.1.4. We compute the nonlinear weights based on the smoothness indicators:

$$\omega_j = \frac{\overline{\omega}_j}{\sum_l \overline{\omega}_l}, \qquad \overline{\omega}_j = \frac{\gamma_j}{\sum_l (\varepsilon + \beta_l)^2} \tag{16}$$

  where γ_j are the linear weights determined in Step 1.2 above, and ε is a small number to avoid the denominator to become zero. We are using $\varepsilon = 10^{-6}$ in all the computation. The final WENO approximation is then given by:

$$u_G \approx \sum_{j=0}^{k} \omega_j p_j(x_G) \,. \tag{17}$$

Step 1.2. We obtain the reconstructed moments based on the reconstructed point values $u(x_G)$ at the Gauss or Gauss-Lobatto quadrature points x_G and a numerical integration

$$u_i^{(l)} = \frac{\Delta x_i}{a_l} \sum_G w_G u(x_G) v_l^{(i)}(x_G), \qquad l = 1, \cdots, k.$$

Here w_G are the Gaussian quadrature weights for the Gaussian points x_G. The polynomial solution in this cell I_i is then obtained by (3) with these reconstructed moments $u_i^{(l)}$ for $l = 1, \cdots, k$ and the original cell average $u_i^{(0)}$.

Remark 2.1. For the $\mathbb{P}^2$ case, we can also reconstruct values of u at the three Gauss quadrature points by the fifth order WENO. But the linear weights at the middle Gaussian point x_j is negative. Although such negative weight case can be treated by the technique developed in Ref. 23, we have opted to use the four-point Gauss-Lobatto quadrature to guarantee positive linear weights.

Remark 2.2. It would seem to be more natural to reconstruct the moments $u_i^{(l)}$ for $l = 1, \cdots, k$ directly from the cell averages of neighboring cells. The procedure is similar to what is described above, with Step 1.1.2 replaced by

- Step 1.1.2': We find the combination coefficients, also called linear weights, denoted by $\gamma_0, \cdots, \gamma_k$, which satisfy:

$$\int_{I_i} Q(x) v_l^{(i)}(x) dx = \sum_{j=0}^{k} \gamma_j \int_{I_i} p_j(x) v_l^{(i)}(x) dx, \qquad l = 1, \cdots, k.$$

The final WENO approximation to the moments are then given by:

$$u_i^{(l)} \approx \frac{1}{a_l} \sum_{j=0}^{k} \omega_j \int_{I_i} p_j(x) v_l^{(i)}(x) dx, \qquad l = 1, \cdots, k.$$

and Step 1.2 is no longer needed. Indeed this approach works well for the $\mathbb{P}^1$ and $\mathbb{P}^2$ cases. Unfortunately, the linear weights for such reconstructions do not exist for the $\mathbb{P}^3$ case.

For the system cases, in order to achieve better qualities at the price of more complicated computations, the WENO reconstruction limiter is always used with a local characteristic field decomposition, see e.g. Ref. 31 for details.

3.2. *WENO reconstruction in two dimensional case*

In two spatial dimensional case, for the rectangular meshes, we choose to reconstruct values of the function u in troubled cells at the tensor product Gauss or Gauss-Lobatto points. We can use WENO reconstruction which was presented in Ref. 23.

For triangular meshes, given a triangulation consisting of cells $\triangle_j$, $\mathbb{P}^k(\triangle_j)$ denotes the set of polynomials of degree at most k defined on $\triangle_j$. Here k could actually change from cell to cell, but for simplicity we assume it is a constant over the whole triangulation. In the DG method, the solution as well as the test function space is given by $V_h^k = \{v(x,y) : v(x,y)|_{\triangle_j} \in \mathbb{P}^k(\triangle_j)\}$. We emphasize that the procedure described below does not depend on the specific basis chosen for the polynomials. We adopt a local orthogonal basis over a target cell, such as $\triangle_0$:

$\{v_l^{(0)}(x,y), l = 0, \ldots, K; K = (k+1)(k+2)/2 - 1\}$:

$$v_0^{(0)}(x,y) = 1,$$
$$v_1^{(0)}(x,y) = \frac{x - x_0}{\sqrt{|\triangle_0|}},$$
$$v_2^{(0)}(x,y) = a_{21}\frac{x - x_0}{\sqrt{|\triangle_0|}} + \frac{y - y_0}{\sqrt{|\triangle_0|}} + a_{22},$$
$$v_3^{(0)}(x,y) = \frac{(x - x_0)^2}{|\triangle_0|} + a_{31}\frac{x - x_0}{\sqrt{|\triangle_0|}} + a_{32}\frac{y - y_0}{\sqrt{|\triangle_0|}} + a_{33},$$
$$v_4^{(0)}(x,y) = a_{41}\frac{(x - x_0)^2}{|\triangle_0|} + \frac{(x - x_0)(y - y_0)}{|\triangle_0|} + a_{42}\frac{x - x_0}{\sqrt{|\triangle_0|}} + a_{43}\frac{y - y_0}{\sqrt{|\triangle_0|}} + a_{44},$$
$$v_5^{(0)}(x,y) = a_{51}\frac{(x - x_0)^2}{|\triangle_0|} + a_{52}\frac{(x - x_0)(y - y_0)}{|\triangle_0|} + \frac{(y - y_0)^2}{|\triangle_0|} + a_{53}\frac{x - x_0}{\sqrt{|\triangle_0|}} + a_{54}\frac{y - y_0}{\sqrt{\triangle_0}} + a_{55}, \ldots$$

where (x_0, y_0) and $|\triangle_0|$ are the barycenter and the area of the target cell $\triangle_0$, respectively. Then we would need to solve a linear system to obtain the values of $a_{\ell m}$ by the orthogonality property:

$$\int_{\triangle_0} v_i^{(0)}(x,y)\, v_j^{(0)}(x,y)\, dxdy = w_i \delta_{ij} \tag{18}$$

with $w_i = \int_{\triangle_0} \left(v_i^{(0)}(x,y)\right)^2 dxdy$.

The numerical solution $u^h(x,y,t)$ in the space V_h^k can be written as:

$$u^h(x,y,t) = \sum_{l=0}^{K} u_0^{(l)}(t)\, v_l^{(0)}(x,y), \qquad \text{for } (x,y) \in \triangle_0$$

and the degrees of freedom $u_0^{(l)}(t)$ are the moments defined by

$$u_0^{(l)} = u_0^{(l)}(t) = \frac{1}{w_l}\int_{\triangle_0} u^h(x,y,t)\, v_l^{(0)}(x,y) dxdy, \qquad l = 0, \cdots, K.$$

For the troubled cells, we reconstruct the polynomial solutions while retaining their cell averages. In other words, we reconstruct the degrees of freedom $u_0^{(l)}$, $l = 1, \ldots, K$ and retain only the cell average $u_0^{(0)}$.

For the $k = 1$ case, we summarize the procedure to reconstruct the first order moments $u_0^{(1)}$ and $u_0^{(2)}$ in the troubled cell $\triangle_0$ using the WENO

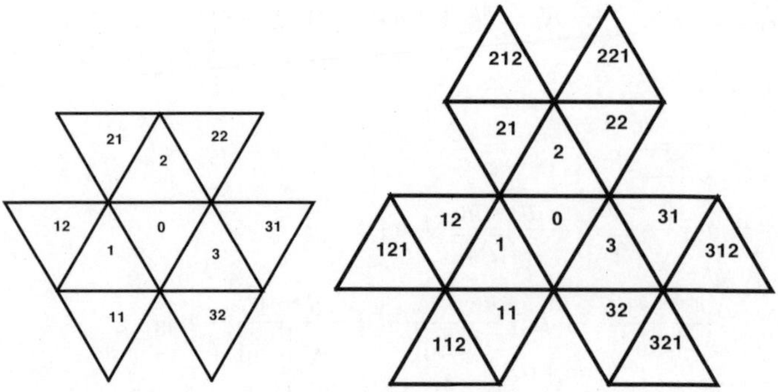

Fig. 1. The big stencil S for $k = 1$ (left). The big stencil T for $k = 2$ (right).

reconstruction procedure. For simplicity, we relabel the "troubled cell" and its neighboring cells as shown in Figure 1.

Step 2.1.1. We select the big stencil as $S = \{\triangle_0, \triangle_1, \triangle_2, \triangle_3, \triangle_{11}, \triangle_{12}, \triangle_{21}, \triangle_{22}, \triangle_{31}, \triangle_{32}\}$. Then we construct a quadratic polynomial $P(x, y)$ to obtain a third order approximation of u by requiring that it has the same cell average as u on the target cell $\triangle_0$, and matches the cell averages of u on the other triangles in the set $S \setminus \{\triangle_0\}$ in a least-square sense, see Ref 24.

Step 2.1.2. We divide S into nine smaller stencils:

$$S_1 = \{\triangle_0, \triangle_1, \triangle_2\}, \qquad S_2 = \{\triangle_0, \triangle_2, \triangle_3\}, \qquad S_3 = \{\triangle_0, \triangle_3, \triangle_1\},$$
$$S_4 = \{\triangle_0, \triangle_1, \triangle_{11}\}, \qquad S_5 = \{\triangle_0, \triangle_1, \triangle_{12}\}, \qquad S_6 = \{\triangle_0, \triangle_2, \triangle_{21}\},$$
$$S_7 = \{\triangle_0, \triangle_2, \triangle_{22}\}, \qquad S_8 = \{\triangle_0, \triangle_3, \triangle_{31}\}, \qquad S_9 = \{\triangle_0, \triangle_3, \triangle_{32}\}.$$

We then construct nine linear polynomials $q_i(x, y)$, $i = 1, \ldots, 9$, satisfying

$$\frac{1}{|\triangle_\ell|} \int_{\triangle_\ell} q_i(x, y) dx dy = \bar{u}_\ell, \qquad \text{for } \triangle_\ell \in S_i. \tag{19}$$

Step 2.1.3. We find the combination coefficients, also called linear weights, denoted by $\gamma_1^{(l)}, ..., \gamma_9^{(l)}$, $l = 1, 2$, satisfying

$$\int_{\triangle_0} P(x, y) v_l^{(0)}(x, y) dx dy = \sum_{i=1}^{9} \gamma_i^{(l)} \int_{\triangle_0} q_i(x, y) v_l^{(0)}(x, y) dx dy, \qquad l = 1, 2 \tag{20}$$

for the quadratic polynomial $P(x, y)$ defined before. The linear weights are achieved by asking for

$$\min \left(\sum_{i=1}^{9} (\gamma_i^{(l)})^2 \right), \qquad l = 1, 2. \tag{21}$$

By doing so, we can get the linear weights uniquely but can not guarantee their positivity. We use the method introduced in Refs. 23 and 24 to overcome this difficulty.

Step 2.1.4. We compute the smoothness indicators, denote by β_i, $i = 1, \ldots, 9$, for the smaller stencils S_i, $i = 1, \ldots, 9$, which measure how smooth the functions $q_i(x, y)$, $i = 1, \ldots, 9$ are in the target cell $\triangle_0$. The smaller these smoothness indicators, the smoother the functions are in the target cell. We use the same recipe for the smoothness indicators as in:[29]

$$\beta_i = \sum_{|\ell|=1}^{k} |\triangle_0|^{|\ell|-1} \int_{\triangle_0} \left(\frac{\partial^{|\ell|}}{\partial x^{\ell_1} \partial y^{\ell_2}} q_i(x, y) \right)^2 dxdy \tag{22}$$

where $\ell = (\ell_1, \ell_2)$.

Step 2.1.5. We compute the non-linear weights based on the smoothness indicators:

$$\omega_i = \frac{\bar{\omega}_i}{\sum_{\ell=1}^{9} \bar{\omega}_\ell}, \qquad \bar{\omega}_\ell = \frac{\gamma_\ell}{(\varepsilon + \beta_\ell)^2}. \tag{23}$$

Here ε is a small positive number to avoid the denominator to become zero. We take $\varepsilon = 10^{-6}$ in our computation.

The moments of the reconstructed polynomial are then given by:

$$u_0^{(l)} = \frac{1}{\int_{\triangle_0} (v_l^{(0)}(x, y))^2 \, dxdy} \sum_{i=1}^{9} \omega_i^{(l)} \int_{\triangle_0} q_i(x, y) \, v_l^{(0)}(x, y) dxdy, \qquad l = 1, 2. \tag{24}$$

For the $k = 2$ case, the procedure to reconstruct the first and second order moments $u_0^{(1)}$, $u_0^{(2)}$, $u_0^{(3)}$, $u_0^{(4)}$ and $u_0^{(5)}$ in the troubled cell $\triangle_0$ is analogous to that for the $k = 1$ case. The troubled cell and its neighboring cells are shown in Figure 1.

Step 2.2.1. We select the big stencil as $T = \{\triangle_0, \triangle_1, \triangle_2, \triangle_3, \triangle_{11}, \triangle_{12}, \triangle_{21}, \triangle_{22}, \triangle_{31}, \triangle_{32}, \triangle_{112}, \triangle_{121}, \triangle_{212}, \triangle_{221}, \triangle_{312}, \triangle_{321}\}$. Then we construct a fourth degree polynomial $Q(x, y)$ to obtain a fifth order approximation of u by requiring that it has the same cell average as u on the target cell $\triangle_0$ and matches the cell averages of u on the other triangles in the set $T \setminus \{\triangle_0\}$ in a least-square sense.

Step 2.2.2. We divide T into nine smaller stencils:

$$T_1 = \{\triangle_0, \triangle_1, \triangle_{11}, \triangle_{12}, \triangle_3, \triangle_{32}\}, \qquad T_2 = \{\triangle_0, \triangle_1, \triangle_{11}, \triangle_{12}, \triangle_2, \triangle_{21}\},$$
$$T_3 = \{\triangle_0, \triangle_2, \triangle_{21}, \triangle_{22}, \triangle_1, \triangle_{12}\}, \qquad T_4 = \{\triangle_0, \triangle_2, \triangle_{21}, \triangle_{22}, \triangle_3, \triangle_{31}\},$$
$$T_5 = \{\triangle_0, \triangle_3, \triangle_{31}, \triangle_{32}, \triangle_2, \triangle_{22}\}, \qquad T_6 = \{\triangle_0, \triangle_3, \triangle_{31}, \triangle_{32}, \triangle_1, \triangle_{11}\},$$
$$T_7 = \{\triangle_0, \triangle_1, \triangle_{11}, \triangle_{12}, \triangle_{112}, \triangle_{121}\},$$
$$T_8 = \{\triangle_0, \triangle_2, \triangle_{21}, \triangle_{22}, \triangle_{212}, \triangle_{221}\},$$
$$T_9 = \{\triangle_0, \triangle_3, \triangle_{31}, \triangle_{32}, \triangle_{312}, \triangle_{321}\}.$$

We can then construct quadratic polynomials $q_i(x, y)$, $i = 1, \ldots, 9$, which satisfy the following conditions

$$\frac{1}{|\triangle_\ell|} \int_{\triangle_\ell} q_i(x, y) dxdy = \bar{u}_\ell, \qquad \text{for } \triangle_\ell \in T_i. \tag{25}$$

The remaining steps 2.2.3, 2.2.4 and 2.2.5 are the same as those for the $k = 1$ case, respectively. Finally, the moments of the reconstructed polynomial are given by:

$$u_0^{(l)} = \frac{1}{\int_{\triangle_0} (v_l^{(0)}(x, y))^2 dxdy} \sum_{i=1}^{9} \omega_i^{(l)} \int_{\triangle_0} q_i(x, y) v_l^{(0)}(x, y) dxdy,$$
$$l = 1, 2, 3, 4, 5. \tag{26}$$

Remark 3.1. *If the troubled cell is near the boundary of computational domain, in order to guarantee enough stencils for reconstruction, we have to extend the stencils according to boundary condition.*

4. HWENO Reconstruction as a Limiter for the RKDG Method

4.1. *HWENO reconstruction in one dimensional case*

For the troubled cells, we would like to reconstruct the polynomial solution while retaining its cell average. In other words, we will reconstruct the degrees of freedom, or the moments, $u_i^{(l)}$ for the troubled cell I_i for $l = 1, \cdots, k$ and retain only the cell average $u_i^{(0)}$.

For the third order k=2 case, we summarize the procedure to reconstruct the first and second moments $u_i^{(1)}$ and $u_i^{(2)}$ for a troubled cell I_i using HWENO:

Step 3.1. Reconstruction of the first moment $u_i^{(1)}$ by HWENO.

(1) Given the small stencils $S_0 = \{I_{i-1}, I_i\}$, $S_1 = \{I_i, I_{i+1}\}$ and the bigger stencil $\mathcal{T} = \{S_0, S_1\}$, we construct Hermite quadratic reconstruction polynomials $p_0(x), p_1(x), p_2(x)$ and a fourth-degree reconstruction polynomial $q(x)$ such that:

$$\int_{I_{i+j}} p_0(x)dx = u_{i+j}^{(0)}a_0, \quad j=-1,0; \qquad \int_{I_{i-1}} p_0(x)v_1^{(i-1)}(x)dx = u_{i-1}^{(1)}a_1$$

$$\int_{I_{i+j}} p_1(x)dx = u_{i+j}^{(0)}a_0, \quad j=0,1; \qquad \int_{I_{i+1}} p_1(x)v_1^{(i+1)}(x)dx = u_{i+1}^{(1)}a_1$$

$$\int_{I_{i+j}} p_2(x)dx = u_{i+j}^{(0)}a_0, \quad j=-1,0,1$$

$$\int_{I_{i+j}} q(x)dx = u_{i+j}^{(0)}a_0, \quad j=-1,0,1; \qquad \int_{I_{i+j}} q(x)v_1^{(i+j)}(x)dx = u_{i+j}^{(1)}a_1,$$

$j = -1, 1$.

We now obtain:

$$\int_{I_i} p_0(x)v_1^{(i)}(x)dx = a_1\left(-2u_{i-1}^{(0)} + 2u_i^{(0)} - u_{i-1}^{(1)}\right)$$

$$\int_{I_i} p_1(x)v_1^{(i)}(x)dx = a_1\left(-2u_i^{(0)} + 2u_{i+1}^{(0)} - u_{i+1}^{(1)}\right)$$

$$\int_{I_i} p_2(x)v_1^{(i)}(x)dx = a_1\left(-u_{i-1}^{(0)} + u_{i+1}^{(0)}\right)/2$$

$$\int_{I_i} q(x)v_1^{(i)}(x)dx = a_l\left(\frac{15}{19}(u_{i-1}^{(0)} - u_{i+1}^{(0)}) - \frac{11}{38}(u_{i-1}^{(1)} + u_{i+1}^{(1)})\right).$$

(2) We find the combination coefficients, also called linear weights, denoted by γ_0, γ_1 and γ_2, satisfying:

$$\int_{I_i} q(x)v_1^{(i)}(x)dx = \sum_{j=0}^{2}\gamma_j \int_{I_i} p_j(x)v_1^{(i)}(x)dx$$

which leads to

$$\gamma_0 = \frac{11}{38}, \qquad \gamma_1 = \frac{11}{38}, \qquad \gamma_2 = \frac{8}{19}.$$

(3) We compute the smoothness indicator β_j by (15), and the nonlinear weights based on the smoothness indicators by (16). The first moment of the reconstructed polynomial is then given by:

$$u_i^{(1)} = \frac{1}{a_1}\sum_{j=0}^{2}\omega_j \int_{I_i} p_j(x)v_1^{(i)}(x)dx \tag{27}$$

Step 3.2. Reconstruction of the second moment $u_i^{(2)}$ by HWENO. When the first moment $u_i^{(1)}$ is needed we use the reconstructed one from Step 4.1.

(1) Given the small stencils $S_0 = \{I_{i-1}, I_i\}$, $S_1 = \{I_i, I_{i+1}\}$ and the bigger stencil $\mathcal{T} = \{S_0, S_1\}$, we construct Hermite cubic reconstruction polynomials $p_0(x), p_1(x), p_2(x)$ and a fifth-degree reconstruction polynomial $q(x)$ such that:

$$\int_{I_{i+j}} p_0(x)dx = u_{i+j}^{(0)}a_0, \quad \int_{I_{i+j}} p_0(x)v_1^{(i+j)}(x)dx = u_{i+j}^{(1)}a_1, \quad j = -1, 0$$

$$\int_{I_{i+j}} p_1(x)dx = u_{i+j}^{(0)}a_0, \quad \int_{I_{i+j}} p_1(x)v_1^{(i+j)}(x)dx = u_{i+j}^{(1)}a_1, \quad j = 0, 1$$

$$\int_{I_{i+j}} p_2(x)dx = u_{i+j}^{(0)}a_0, \ j = -1, 0, 1; \quad \int_{I_i} p_2(x)v_1^{(i)}dx = u_i^{(1)}a_1$$

$$\int_{I_{i+j}} q(x)dx = u_{i+j}^{(0)}a_0, \quad \int_{I_{i+j}} q(x)v_1^{(i+j)}(x)dx = u_{i+j}^{(1)}a_1, \quad j = -1, 0, 1,$$

which lead to

$$\int_{I_i} p_0(x)v_2^{(i)}(x)dx = a_2\left(\frac{15}{4}u_{i-1}^{(0)} - \frac{15}{4}u_i^{(0)} + \frac{11}{8}u_{i-1}^{(1)} + \frac{19}{8}u_i^{(1)}\right)$$

$$\int_{I_i} p_1(x)v_2^{(i)}(x)dx = a_2\left(-\frac{15}{4}u_i^{(0)} + \frac{15}{4}u_{i+1}^{(0)} - \frac{19}{8}u_i^{(1)} - \frac{11}{8}u_{i+1}^{(1)}\right)$$

$$\int_{I_i} p_2(x)v_2^{(i)}(x)dx = a_2\left(\frac{1}{2}u_{i-1}^{(0)} - u_i^{(0)} + \frac{1}{2}u_{i+1}^{(0)}\right)$$

$$\int_{I_i} q(x)v_2^{(i)}dx = a_2\left(\frac{73}{56}u_{i-1}^{(0)} - \frac{73}{28}u_i^{(0)} + \frac{73}{56}u_{i+1}^{(0)} + \frac{45}{112}u_{i-1}^{(1)} - \frac{45}{112}u_{i+1}^{(1)}\right).$$

(2) We find the linear weights denoted by γ_0, γ_1 and γ_2 satisfying

$$\int_{I_i} q(x)v_2^{(i)}dx(x) = \sum_{j=0}^{2}\gamma_j\int_{I_i} p_j(x)v_2^{(i)}(x)dx$$

which leads to

$$\gamma_0 = \frac{45}{154}, \qquad \gamma_1 = \frac{45}{154}, \qquad \gamma_2 = \frac{32}{77}.$$

(3) We compute the smoothness indicator β_j by (15). The nonlinear weights are then computed based on the smoothness indicators using (16). The second moment of the reconstructed polynomial is then given by:

$$u_i^{(2)} = \frac{1}{a_2} \sum_{j=0}^{2} \omega_j \int_{I_i} p_j(x) v_2^{(i)}(x) dx \tag{28}$$

4.2. HWENO reconstruction in two dimensional case

For the troubled cells, we reconstruct the polynomial solutions while retaining their cell averages. In other words, we reconstruct the degrees of freedom $u_0^{(l)}$, $l = 1, \ldots, K$ and retain only the cell average $u_0^{(0)}$.

For the $k = 1$ case, we summarize the procedure to reconstruct the first order moments $u_0^{(1)}$ and $u_0^{(2)}$ in the troubled cell $\triangle_0$ using the HWENO reconstruction procedure. The troubled cell and its neighboring cells are shown in Figure 1.

Step 4.1.1. We select the big stencil as $S = \{\triangle_0, \triangle_1, \triangle_2, \triangle_3\}$. Then we construct polynomial $P(x, y)$ to approximate u by requiring that it has the same cell average as $u^{(0)}$ on the target cell $\triangle_0$, and matches the cell averages of $u^{(0)}$, $u^{(1)}$ or $u^{(2)}$ on the other triangles in the set $S \setminus \{\triangle_0\}$ in a least square sense.

Step 4.1.2. We then construct six linear polynomials $q_i(x, y)$, $i = 1, \ldots, 6$, satisfying:

$$\frac{1}{|\triangle_\ell|} \int_{\triangle_\ell} q_i(x, y) dx dy = u_\ell^{(0)}, \tag{29}$$

$$\frac{1}{\int_{\triangle_{\ell_x}} (v_1^{(\ell_x)}(x, y))^2 dx dy} \int_{\triangle_{\ell_x}} q_i(x, y) v_1^{(\ell_x)}(x, y) dx dy = u_{\ell_x}^{(1)}, \tag{30}$$

$$\frac{1}{\int_{\triangle_{\ell_y}} (v_2^{(\ell_y)}(x, y))^2 dx dy} \int_{\triangle_{\ell_y}} q_i(x, y) v_2^{(\ell_y)}(x, y) dx dy = u_{\ell_y}^{(2)}. \tag{31}$$

For

$i = 1, \ell = 0, 1, 2; i = 2, \ell = 0, 2, 3; i = 3, \ell = 0, 3, 1; i = 4, \ell = 0, \ell_x = 1,$
$\ell_y = 1; i = 5, \ell = 0, \ell_x = 2, \ell_y = 2; i = 6, \ell = 0, \ell_x = 3, \ell_y = 3.$

Step 4.1.3. We find the combination coefficients, also called linear weights, denoted by $\gamma_1^{(l)}, ..., \gamma_6^{(l)}$, $l = 1, 2$, satisfying:

$$\int_{\triangle_0} P(x,y) v_l^{(0)}(x,y) dxdy = \sum_{i=1}^{6} \gamma_i^{(l)} \int_{\triangle_0} q_i(x,y) v_l^{(0)}(x,y) dxdy, \qquad l = 1, 2 \tag{32}$$

The linear weights are achieved by asking for

$$\min \left(\sum_{i=1}^{6} (\gamma_i^{(l)})^2 \right), \qquad l = 1, 2. \tag{33}$$

By doing so, we can get the linear weights uniquely but can not maintain them positively all the time, we can use the methods that produced in Ref. 23 and 24 to overcome this drawback.

Then we follow the step 2.1.4 and 2.1.5 to compute smoothness indicators and nonlinear weights, finally, the moments of the reconstructed polynomial are then given by:

$$u_0^{(l)} = \frac{1}{\int_{\triangle_0} (v_l^{(0)}(x,y))^2 \, dxdy} \sum_{i=1}^{6} \omega_i^{(l)} \int_{\triangle_0} q_i(x,y) \, v_l^{(0)}(x,y) dxdy, \qquad l = 1, 2. \tag{34}$$

For the $k = 2$ case, the procedure to reconstruct the first and second order moments $u_0^{(1)}$, $u_0^{(2)}$, $u_0^{(3)}$, $u_0^{(4)}$ and $u_0^{(5)}$ in the troubled cell $\triangle_0$ is analogous to that for the $k = 1$ case. The troubled cell and its neighboring cells are shown in Figure 1.

Step 4.2.1. We select the big stencil as $S = \{\triangle_0, \triangle_1, \triangle_2, \triangle_3, \triangle_{11}, \triangle_{12}, \triangle_{21}, \triangle_{22}, \triangle_{31}, \triangle_{32}\}$. Then we construct polynomial $Q(x,y)$ to approximate u by requiring that it has the same cell average as $u^{(0)}$ on the target cell $\triangle_0$ and matches the cell averages of $u^{(0)}$, $u^{(1)}$ or $u^{(2)}$ on the other triangles in the set $S \setminus \{\triangle_0\}$ in a least square sense.

Step 4.2.2. We can then construct quadratic polynomials $q_i(x,y)$, $i = 1, \ldots, 9$, which satisfy the following conditions:

$$\frac{1}{|\triangle_\ell|} \int_{\triangle_\ell} q_i(x,y) dxdy = u_\ell^{(0)}, \tag{35}$$

$$\frac{1}{\int_{\triangle_{\ell_x}} (v_1^{(\ell_x)}(x,y))^2 dxdy} \int_{\triangle_{\ell_x}} q_i(x,y) v_1^{(\ell_x)}(x,y) dxdy = u_{\ell_x}^{(1)}, \tag{36}$$

$$\frac{1}{\int_{\triangle_{\ell_y}} (v_2^{(\ell_y)}(x,y))^2 dxdy} \int_{\triangle_{\ell_y}} q_i(x,y) v_2^{(\ell_y)}(x,y) dxdy = u_{\ell_y}^{(2)}. \tag{37}$$

For

$i = 1, \ell = 0, 1, 11, 12, 3, 32;\ i = 2, \ell = 0, 1, 11, 12, 2, 21;$
$i = 3, \ell = 0, 2, 21, 22, 1, 12;\ i = 4, \ell = 0, 2, 21, 22, 3, 31;$
$i = 5, \ell = 0, 3, 31, 32, 2, 22;\ i = 6, \ell = 0, 3, 31, 32, 1, 11;$
$i = 7, \ell = 0, 1, 11, 12, \ell_x = 1, \ell_y = 1;\ i = 8, \ell = 0, 2, 21, 22, \ell_x = 2, \ell_y = 2;$
$i = 9, \ell = 0, 3, 31, 32, \ell_x = 3, \ell_y = 3.$

The remaining steps are the same as those for the $k = 1$ case. Finally, the moments of the reconstructed polynomial are given by:

$$u_0^{(l)} = \frac{1}{\int_{\triangle_0} (v_l^{(0)}(x,y))^2 dxdy} \sum_{i=1}^{9} \omega_i^{(l)} \int_{\triangle_0} q_i(x,y) v_l^{(0)}(x,y) dxdy, \quad l = 1, 2, 3, 4, 5. \tag{38}$$

5. Numerical Results

In this section we provide numerical results to demonstrate the performance of the WENO and HWENO reconstruction limiters for the RKDG methods on unstructured meshes described in section 3 and 4.

For accuracy test, we have tested many standard problems for accuracy, such as one and two dimensional linear advection, one and two dimensional nonlinear Burgers equation, and one and two dimensional nonlinear Euler equations. Both structure and unstructure meshes are used. To save space, we present only the results of the two dimensional nonlinear Euler equations on unstructure meshes with the TVB minmod limiter as troubled cell indicator as representative examples. We have used the TVB minmod limiter with a small $M = 0.01$ to identify troubled cells (this is close to a TVD limiter with $M = 0$), resulting in many good cells identified as troubled cells. In this way we can clearly see the effect of the WENO/HWENO reconstruction limiter on the accuracy of the RKDG method, namely the order of accuracy is maintained after the application of this limiter.

Example 1. We solve the following nonlinear system of Euler equations

$$\xi_t + f(\xi)_x + g(\xi)_y = 0 \tag{39}$$

with: $\xi = (\rho, \rho u, \rho v, E)^T$, $f(\xi) = (\rho u, \rho u^2 + p, \rho uv, u(E+p))^T$, $g(\xi) = (\rho v, \rho uv, \rho v^2 + p, v(E+p))^T$. Here ρ is the density, (u, v) is the velocity, E is the total energy, p is the pressure, which is related to the total energy

Table 1. 2D-Euler equations: initial data $\rho(x,y,0) = 1+0.2\sin(\pi(x+y))$, $u(x,y,0) = 0.7$, $v(x,y,0) = 0.3$, and $p(x,y,0) = 1$. Periodic boundary conditions in both directions. $t = 2.0$. L^1 and L^∞ errors. RKDG with WENO and HWENO limiters ($M = 0.01$) compared to RKDG without limiter. The mesh points on the boundary are uniformly distributed with cell length h.

		WENO limiter		HWENO limiter		without limiter	
	h	error	order	error	order	error	order
P^1	2/10	8.37E-2		1.11E-1		2.23E-2	
	2/20	3.50E-2	1.26	6.07E-2	0.88	5.42E-3	2.04
	2/40	1.25E-2	1.48	2.31E-2	1.39	1.29E-3	2.06
	2/80	3.83E-3	1.70	7.89E-3	1.55	3.27E-4	1.98
	2/160	1.16E-3	1.72	2.48E-3	1.67	8.48E-5	1.95
P^2	2/10	1.76E-2		1.33E-2		5.94E-3	
	2/20	3.47E-3	2.34	1.69E-3	2.98	1.14E-3	2.38
	2/40	4.94E-4	2.81	2.78E-4	2.60	1.94E-4	2.56
	2/80	6.60E-5	2.91	4.17E-5	2.74	2.87E-5	2.76
	2/160	7.09E-6	3.21	5.17E-6	3.00	3.62E-6	2.99

by $E = \frac{p}{\gamma-1} + \frac{1}{2}\rho(u^2+v^2)$ with $\gamma = 1.4$. The initial condition is set to be $\rho(x,y,0) = 1+0.2\sin(\pi(x+y))$, $u(x,y,0) = 0.7$, $v(x,y,0) = 0.3$, $p(x,y,0) = 1$, with a 2-periodic boundary condition. The exact solution is $\rho(x,y,t) = 1+0.2\sin(\pi(x+y-(u+v)t))$, $u = 0.7$, $v = 0.3$, $p = 1$. We compute the solution up to $t = 2$. The errors and numerical orders of accuracy for the RKDG method with WENO and HWENO limiters comparing with the original RKDG method without a limiter are shown in Table 1. We can see that the WENO and HWENO limiters keep the designed order of the original RKDG method, but have large numerical errors.

We now test the performance of the RKDG method with WENO and HWENO limiters for problems containing shocks.

Example 2. We solve the following one dimensional nonlinear system of Euler equations

$$u_t + f(u)_x = 0 \tag{40}$$

with

$$u = (\rho, \rho v, E)^T, \qquad f(u) = (\rho v, \rho v^2 + p, v(E+p))^T.$$

Here ρ is the density, v is the velocity, E is the total energy, p is the pressure, related to the total energy by $E = \frac{p}{\gamma-1} + \frac{1}{2}\rho v^2$ with $\gamma = 1.4$.

We use the following Riemann initial condition for the Lax problem:

$$(\rho, v, p) = (0.445, 0.698, 3.528)\,\text{for}\,x \le 0; \quad (\rho, v, p) = (0.5, 0, 0.571)\,\text{for}\,x > 0.$$

The computed density ρ is plotted at $t = 1.3$ against the exact solution. In this example we explore the effect of the TVB constant M in the minmod limiter to identify troubled cells. We observe that, with an increased M, we have fewer cells identified as troubled cells and subject to WENO or HWENO limiting, and the resolution of the contact discontinuity improves with an increased M. Thus we might want to choose a larger M within the range allowed by stability to minimize the number of troubled cells subject to WENO or HWENO limiting, both to save computational cost and to improve resolution at contact discontinuities. In Figure 2, we plot the densities by RKDG with WENO and HWENO limiters using $N = 200$ cells, and the time history of the "troubled cells", for the $M = 0.01$, $M = 1$ and $M = 50$ cases, and we only show the case for k=2 to save space. In the figures of solution, the solid line is the exact solution and squares are numerical solution (one point per cell). For time history of the "troubled cells", squares denote cells which are identified as "troubled cells" subject to WENO or HWENO limiting.

Example 3. We consider the interaction of blast waves of Euler equation (40) with the initial condition:

$$(\rho, v, p) = (1, 0, 1000) \text{ for } 0 \le x < 0.1;$$

$$(\rho, v, p) = (1, 0, 0.01) \text{ for } 0.1 \le x < 0.9;$$

$$(\rho, v, p) = (1, 0, 100) \text{ for } 0.9 \le x.$$

A reflecting boundary condition is applied to both ends. See Refs. 32 and 33. The computed density ρ is plotted at $t = 0.038$ against the reference "exact" solution, which is a converged solution computed by the fifth order finite difference WENO scheme[29] with 2000 grid points.

In Figure 3, we plot the densities by RKDG (k=2) with WENO and HWENO limiters using $N = 400$ cells, and the time history of the "troubled cells", for the $M = 0.01$, $M = 10$ and $M = 300$ cases. As before we explore the effect of the TVB constant M in the minmod limiter to identify troubled cells. We observe the same pattern as before, namely with an increased M we have fewer cells identified as troubled cells and subject to WENO or HWENO limiting, and the resolution of the numerical solution improves with an increased M up to a certain value (comparing for example the resolution of $M = 10$ with that of $M = 0.01$. Thus we might want to choose a larger M within the range allowed by stability to minimize the

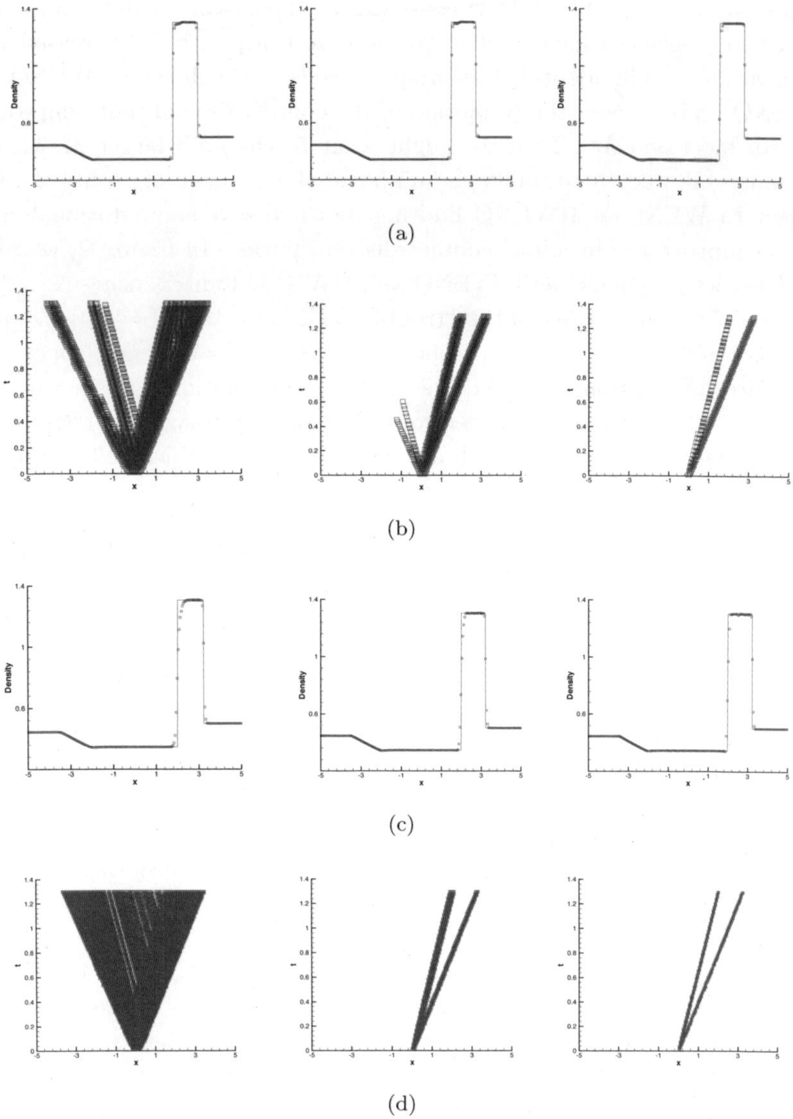

Fig. 2. Lax problem. Solution of density (a) and history of the "troubled cells" (b) by RKDG with WENO limiter. Solution of density (c) and history of the "troubled cells" (d) by RKDG with HWENO limiter. From left to right, the TVB constant $M = 0.01, 1$ and 50, respectively.

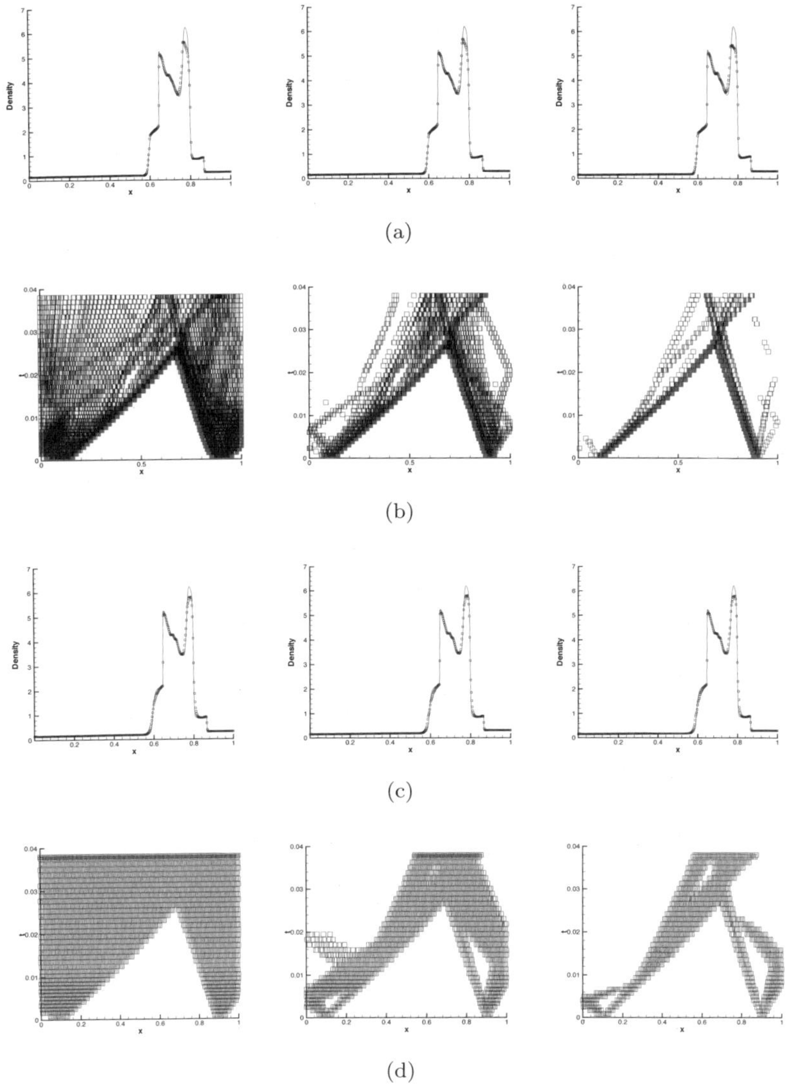

Fig. 3. Blast Wave problem. Solution of density (a) and history of the "troubled cells" (b) by RKDG with WENO limiter. Solution of density (c) and history of the "troubled cells" (d) by RKDG with HWENO limiter. From left to right, the TVB constant $M = 0.01, 10$ and 300, respectively.

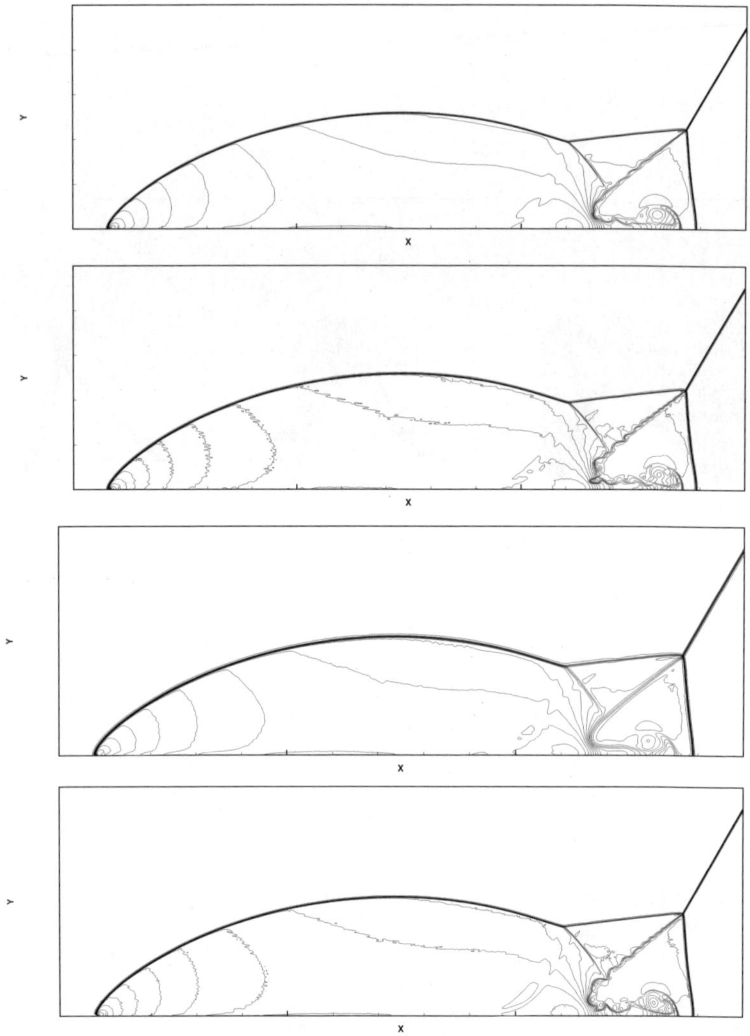

Fig. 4. Double Mach refection problem. 30 equally spaced density contours from 1.5 to 22.7. From top to bottom: The second order ($k = 1$) and the third order ($k = 2$) RKDG with the WENO limiter; The second order ($k = 1$) and the third order ($k = 2$) RKDG with the HWENO limiter

number of troubled cells subject to WENO or HWENO limiting, both to save computational cost and to improve resolution. However, if M is chosen too large, the improvement of resolution is not clear for this example. There is even some degradation of resolution for $M = 300$ comparing with that for $M = 10$.

Example 4. Double Mach reflection problem. This model problem is originally from Ref. 32. We solve the Euler equations (39) in a computational domain of a tube which contains a 30° wedge. The shock moves with a Mach number of 10, the undisturbed air ahead the shock has a density of 1.4 and a pressure of 1 and the left hand side of the shock has a density of 8, velocity of 8.25 and pressure of 116.5. The triangular meshes are generated by *EasyMesh*, with the mesh points on the boundary are uniformly distributed with cell length $h = 1/300$. The results are shown at $t = 0.2$. Two different orders of accuracy for the RKDG with both the WENO and the HWENO limiters, k=1 and k=2 (second and third order). We also tested TVB troubled cell indicator with three different values of the TVB constant, $M = 1$, $M = 50$ and $M = 100$, the resolution is slightly better for $M = 100$ than for $M = 1$ and $M = 50$, however this difference is not significant, we will only show results for the case of TVB constant $M = 100$ to save space. The simulation results are shown in Figure 4. All the figures are showing 30 equally spaced density contours from 1.5 to 22.7. Clearly, the resolution improves with an increasing k on the same mesh.

6. Concluding Remarks

In this chapter, we have described the limiters for the RKDG methods to solve conservation laws using finite volume high order WENO and HWENO reconstructions on structure and unstructured meshes. The idea is to first identify troubled cells subject to the WENO or HWENO limiting, using a troubled cell indicator, then reconstruct the polynomial solution inside the troubled cells by the WENO or HWENO reconstruction using the cell averages of neighboring cells or the cell averages and cell derivative averages of neighboring cells, while maintaining the original cell averages of the troubled cells. Numerical results are provided to show that the method is stable, accurate, and robust in maintaining accuracy. This limiter procedure can also be used for Local DG to solve convection-dominated problems.[33–35] Further work will be carried out an extension to three dimensional tetrahedral meshes using the WENO approaches in Refs. 36–38.

References

1. W. Reed and T. Hill. Triangular mesh methods for neutron transport equation. Technical report la-ur-73-479, Los Alamos Scientific Laboratory, Los Alamos, NM, (1973).
2. B. Cockburn and C.-W. Shu, The Runge-Kutta local projection P1-discontinuous Galerkin finite element method for scalar conservation laws, *Math. Model. Numer. Anal.*(M^2AN). **25**, 337–361, (1991).
3. B. Cockburn and C.-W. Shu, TVB Runge-Kutta local projection discontinuous Galerkin finite element method for conservation laws II: general framework, *Math. Comp.* **52**, 411–435, (1989).
4. B. Cockburn, S.-Y. Lin, and C.-W. Shu, TVB Runge-Kutta local projection discontinuous Galerkin finite element method for conservation laws III: one dimensional systems, *J. Comput. Phys.* **84**, 90–113, (1989).
5. B. Cockburn, S. Hou, and C.-W. Shu, The Runge-Kutta local projection discontinuous Galerkin finite element method for conservation laws IV: the multidimensional case, *Math. Comp.* **54**, 545–581, (1990).
6. B. Cockburn and C.-W. Shu, The Runge-Kutta discontinuous Galerkin method for conservation laws V: multidimensional systems, *J. Comput. Phys.* **141**, 199–224, (1998).
7. C.-W. Shu and S. Osher, Efficient implementation of essentially non-oscillatory shock-capturing schemes, *J. Comput. Phys.* **77**, 439–471, (1988).
8. C.-W. Shu, TVB uniformly high-order schemes for conservation laws, *Math. Comp.* **49**, 105–121, (1987).
9. R. Biswas, K. Devine, and J. Flaherty, Parallel, adaptive finite element methods for conservation laws, *Appl. Numer. Math.* **14**, 255–283, (1994).
10. B. Cockburn, G. Karniadakis, and C.-W. Shu, Eds. *Discontinuous Galerkin Methods: Theory, Computation and Applications*, number 11 in Lecture Notes in Computational Science and Engineering, Berlin, (2000). Springer.
11. B. Cockburn, *Discontinuous Galerkin methods for convection-dominated problems*, In eds. T. Barth and H. Deconinck, *High-Order Methods for Computational Physics*, vol. 9, *Lecture Notes in Computational Science and Engineering*, pp. 69–224. Springer, (1999).
12. B. Cockburn and C.-W. Shu, Runge-Kutta discontinuous Galerkin method for for convection-dominated problems, *J. Sci. Comput.* **16**, 173–261, (2001).
13. A. Burbeau, P. Sagaut, and C. Bruneau, A problem-independent limiter for high-order Runge-Kutta discontinuous Galerkin methods, *J. Comput. Phys.* **169**, 111–150, (2001).
14. P. K. Sweby, High resolution schemes using flux limiters for hyperbolic conservation laws, *SIAM J. Numer. Anal.* **21**, 995–1011, (1984).
15. A. Suresh and H. Huynh, Accurate monotonicity-preserving schemes with Runge-Kutta time stepping, *J. Comput. Phys.* **136**, 83–99, (1997).
16. W. Rider and L. Margolin, Simple modifications of monotonicity-preserving limiters, *J. Comput. Phys.* **174**, 473–488, (2001).
17. J. Qiu and C.-W. Shu, Runge-Kutta discontinuous Galerkin method using WENO limiters, *SIAM J. Sci. Comput.* **26**, 907–929, (2005).

18. C.-W. S. J. Zhu, J. Qiu and M. Dumbser, Runge-Kutta discontinuous Galerkin method using WENO limiters II: unstructured meshes,, *J. Comput. Phys.* **227**, 4330–4353, (2008).
19. J. Qiu and C.-W. Shu, Hermite WENO schemes and their application as limiters for Runge-Kutta discontinuous Galerkin method: one dimensional case, *J. Comput. Phys.* **193**, 115–135, (2004).
20. J. Qiu and C.-W. Shu, Hermite WENO schemes and their application as limiters for Runge-Kutta discontinuous Galerkin method II: two dimensional case, *Computer Fluids.* **34**, 642–663, (2005).
21. J. Q. J. Zhu, Hermite WENO schemes and their application as limiters for Runge-Kutta discontinuous Galerkin method III: Unstructured meshes, *J. Sci. Comput.* **39**, 293–321, (2009).
22. J. Qiu and C.-W. Shu, A comparison of troubled-cell indicators for Runge-Kutta discontinuous Galerkin mehtods using weighted essentially nonosillatory limiters, *SIAM J. Sci. Comput.* **27**, 995–1013, (2005).
23. C. H. J. Shi and C.-W. Shu, A technique of treating negative weights in WENO schemes, *J. Comput. Phys.* **175**, 108–127, (2002).
24. C. Hu and C.-W. Shu, Weighted essentially non-oscillatory schemes on triangular meshes, *J. Comput. Phys.* **150**, 97–127, (2007).
25. A. Harten, High resolution schemes for hyperbolic conservation laws, *J. Comput. Phys.* **49**, 357–393, (1983).
26. L. Krivodonova, J. Xin, J.-F. Remacle, N. Chevaugeon, and J. Flaherty, Shock detection and limiting with discontinuous Galerkin methods for hyperbolic conservation laws, *Appl. Numer. Math.* **48**, 323–338, (2004).
27. A. Harten, ENO schemes with subcell resolution, *J. Comput. Phys.* **83**, 148–184, (1989).
28. G. Jiang and C.-W. Shu, Weighted essentially non-oscillatory schemes, *J. Comput. Phys.* **115**, 200–212, (1994).
29. G. Jiang and C.-W. Shu, Efficient implementation of weighted ENO schemes, *J. Comput. Phys.* **126**, 202–228, (1996).
30. D. Balsara and C.-W. Shu, Monotonicity preserving weighted essentially non-oscillatory schemes with increasingly high order of accuracy, *J. Comput. Phys.* **160**, 405–452, (2000).
31. C. W. Shu, *Essentially non-oscillatory and weighted essentially non-oscillatory schemes for hyperbolic conservation laws*, In ed. A. Quarteroni, *High-Order Methods for Computational Physics*, vol. 1697, *Lecture Notes in Mathematics*, pp. 325–432. Springer, (1998).
32. P. Woodward and P. Colella, The numerical simulation of two-dimensional fluid flow with strong shocks, *J. Comput. Phys.* **54**, 115–173, (1984).
33. F. Bassi and S. Rebay, A high-order accurate discontinuous finite element method for the numerical solution of the compressible navier-stokes equations, *J. Comput. Phys.* **131**, 267–279, (1997).
34. B. Cockburn and C.-W. Shu, The local discontinuous Galerkin method for time-dependent convection diffusion systems., *SIAM J. Numer. Anal.* **35**, 2440–2463, (1998).

35. J. Q. J. Zhu, Local Runge-Kutta discontinous Glaerkin method using WENO type limiters for converction-diffusion equations, *J. Comput. Phys.* **to appear**.
36. M. Dumbser and M. Käser, Arbitrary high order non-oscillatory finite volume schemes on unstructured meshes for linear hyperbolic systems, *J. Comput. Phys.* **221**, 693–723, (2007).
37. V. T. M. Dumbser, M. Käser and E. Toro, Quadrature-free non-oscillatory finite volume schemes on unstructured meshes for nonlinear hyperbolic systems, *J. Comput. Phys.* **226**, 204–243, (2007).
38. Y.-T. Zhang and C.-W. Shu, Third order WENO scheme on three dimensional tetrahedral meshes, *Comm. Comput. Phys.* **5**, 836–848, (2009).

CHAPTER 7

A VENERABLE FAMILY OF DISCONTINUOUS GALERKIN SCHEMES FOR DIFFUSION REVISITED

Bram van Leer* and Marcus Lo†

Department of Aerospace Engineering,
University of Michigan, Ann Arbor, MI, USA
**bram@umich.edu*
†khlo@umich.edu

Rita Gitik

Ann Arbor, MI, USA
ritagtk7@gmail.com

Shohei Nomura

Toyota Motor Engineering & Manufacturing, Ann Arbor, MI, USA
shohei.nomura@tema.toyota.com

The oldest family of Discontinuous Galerkin schemes for diffusion includes two coupling or penalty terms, each with a free coefficient. The (σ, μ)-family has a rich structure that can be (but never was) thoroughly explored in the case of 1-D diffusion on a uniform grid, for polynomial degree $p = 1$. We use several guiding principles in the search for special schemes: maximizing low-frequency and high-frequency accuracy, achieving gradient-consistency, improving eigenvector structure and maximizing stability range. There are several surprises; in particular, the most promising scheme is not among the well-known ones but a fourth-order-accurate scheme found by the authors in 2005. It can be interpreted as a "recovery" scheme; this makes it immediately extendible to higher p and to multi-D diffusion on unstructured grids.

1. Introduction

The use of discontinuous basis functions in a Galerkin method, while a natural choice when discretizing advection operators,[1] is not obvious when dealing with diffusion operators. Precisely where the diffusion flux, proportional to the solution gradient, is to be computed, that is, at a cell interface, neither the discrete solution nor any of its derivatives are uniquely defined.

Over the past dozen years, a number of successful algorithms have emerged for computing a diffusion flux as part of a Discontinuous Galerkin (DG) discretization; these include Local DG (LDG),[2] the second method of Bassi and Rebay,[3] the method of Brezzi,[4] Recovery-based DG (RDG-2x, RDG-1x) by Van Leer *et al.*,[1,5] and Huynh's Poor-Man's Recovery.[6] Huynh's paper includes an extensive analysis and comparison of the new generation of DG diffusion methods regarding accuracy and stability.

In contrast, the first two decades of developing DG for diffusion were neither very productive nor very successful. Basically, four methods were launched, of which one was inconsistent and one was unstable. All four can be recognized as members of a two-parameter family, called the (σ, μ)-family in this chapter after,[7] which has remained largely unexplored. The reason for this disregard is very likely the hesitancy among finite-element analysts to leave the safe grounds of symmetric operators. Thus, the desire to construct mathematical proofs with ease may have been stifling the search for practical methods.

This chapter is entirely dedicated to the (σ, μ)-family. We shall consider the one-dimensional diffusion equation, adopt a piecewise-linear[a] basis ($p = 1$), and show the many possibilities included in this family, with regard to consistency, accuracy and stability. It turns out that two of the methods of the newer generation also belong to this venerable family. These 1-D methods for $p = 1$ can be interpreted using a reconstruction or *recovery* procedure, in a manner that immediately generalizes to higher-order bases and multiple dimensions; we consider these – RDG-1x and Bassi-Rebay-2 – the most valuable ones.

2. The (σ, μ)-family

Let us set the stage for the (σ, μ)-family. The equation to be approximated is the 1-D diffusion equation,

$$U_t = DU_{xx}, \tag{1}$$

where D is a constant; we shall consider semi-discretizations where the spatial operator is formulated on a uniform grid of cells I_j with width h, center $x_j = jh$ and faces $x_{j\pm\frac{1}{2}} = x_j \pm \frac{h}{2}$. The basis functions v_j^k, $k = 0, 1, .., p$, span

[a]The analysis for $p = 2$ is significantly more complicated because of the cubic eigenvalue equation; it is doubtful that such an analysis would be more productive than for $p = 1$.

the polynomial space of degree p in cell I_j and are zero elsewhere; to facilitate comparison with finite-volume methods let us assume they are Legendre polynomials in terms of the scaled local coordinate $\xi = (x - x_j)/(h/2)$, which maps I_j to the interval $(-1, 1)$. At each time level the approximate solution u_j in cell I_j is expressed in terms of these polynomials:

$$u_j(x) = \sum_{k=0}^{p} C_j^k v_j^k(x), \quad x \in I_j. \tag{2}$$

DG is a standard Galerkin method in the sense that the basis functions also serve as the test functions. The DG method is based on the following weak formulation, which includes integration by parts once:

$$\begin{aligned} \int_{I_j} v_j^k u_t dx &= D \int_{I_j} v_j^k u_{xx} dx \\ &= D \left(v_j^k u_x \Big|_{j+\frac{1}{2}} - v_j^k u_x \Big|_{j-\frac{1}{2}} \right) \\ &\quad - D \int_{I_j} (v_j^k)_x u_x dx, \quad k = 0, 1, ..., p. \end{aligned} \tag{3}$$

Satisfying these $p+1$ equations leads to finding expressions for the time derivatives of the $p+1$ coefficients C_j^k; these may for instance be fed to a time integrator such as a Runge-Kutta method.

The last member of Eq. (3) contains a cell-boundary term and a cell-interior integral. It is the former term that requires special attention, since its value is not uniquely defined. If we choose to evaluate the expression $v_j^k u_x$ at the interior of the cell faces, the update equations in adjacent cells will not be coupled, resulting in an inconsistent scheme. Some form of coupling must therefore be introduced. In the (σ, μ)-family of schemes, various terms containing jumps (denoted by square brackets, $[q] = q_+ - q_-$) and averages (denoted by angled brackets, $\langle q \rangle = (q_- + q_+)/2$) of u at the cell faces appear:

$$\begin{aligned} \int_{I_j} v_j^k u_t dx &= -D \left(\langle u_x \rangle [v_j^k] \Big|_{j+\frac{1}{2}} + \langle u_x \rangle [v_j^k] \Big|_{j-\frac{1}{2}} \right) && \text{(term 1)} \\ &\quad - D \int_{I_j} (v_j^k)_x u_x dx && \text{(term 2)} \\ &\quad + \sigma D \left(\langle (v_j^k)_x \rangle [u] \Big|_{j+\frac{1}{2}} + \langle (v_j^k)_x \rangle [u] \Big|_{j-\frac{1}{2}} \right) && \text{(term 3)} \\ &\quad - \frac{\mu D}{\Delta x} \left([v_j^k][u] \Big|_{j+\frac{1}{2}} + [v_j^k][u] \Big|_{j-\frac{1}{2}} \right). && \text{(term 4)} \end{aligned} \tag{4}$$

Term 1 on the right-hand side is needed to achieve consistency for $p > 0$. For example, for $k = 0$, with $[v_j^0]_{j\pm\frac{1}{2}} = \mp 1$ according to Eq. (6) below, it becomes

$$D\left(\langle u_x\rangle\Big|_{j+\frac{1}{2}} - \langle u_x\rangle\Big|_{j-\frac{1}{2}}\right). \tag{5}$$

The average of the solution-derivative values on both sides of an interface seems a good approximation to the diffusive flux, but it ignores the jump at the interface. This term and the interior integral (term 2) together yield an *inconsistent scheme.*

Term 3 is a penalty term that contains the solution jump and therefore couples the elements; with $\sigma = -1$, terms 1–3 form a symmetric operator. Among its eigenvalues, though, there are positive real ones, making the *symmetric scheme* unstable.

Term 4, Arnold's[8] interior penalty term of 1982, is necessary for both consistency and stability for $p = 0$ (one must choose $\mu = 1$), and may be used to manage stability for $p > 0$. In particular, for $\sigma = -1$ values of $\mu \geq 1$ yield stable schemes, all symmetric; with the minimum value $\mu = 1$ we speak of the *stabilized symmetric scheme.* On the other hand, flipping the sign of σ makes the μ-term unnecessary for stability, as discovered as late as 1997 by Baumann;[9] the *Baumann scheme* has $\sigma = +1,\ \mu = 0$.

The Inconsistent, Symmetric, Stabilized Symmetric and Baumann schemes were the meager yield of a two-decade search guided by finite-element theory. No one explored the entire (σ, μ)-family, because it was not recognized as a two-parameter family. Symmetry required σ to be fixed at the value -1; to explore any other value was a daring act, leading one onto dangerous terrain. In this regard the Baumann scheme with $\sigma = 1$, while itself not an attractive scheme, has had its merit as an eye-opener. In the following sections we shall explore the entire (σ, μ)-family, for $p = 1$.

3. Operator Form and Fourier Transform

The most important tool we have for investigating both stability and accuracy of discretizations is the Fourier analysis; we shall apply it to the (σ, μ)-family for $p = 1$.

First, we need to find the update operator corresponding to the weak form (4). In the case $p = 1$ there are two test functions:

$$v_j^0(x) = \begin{cases} 0, \ x < x_{j-\frac{1}{2}}, \\ 1, \ x \in I_j, \\ 0, \ x > x_{j+\frac{1}{2}}; \end{cases} \tag{6}$$

$$v_j^1(x) = \begin{cases} 0, & x < x_{j-\frac{1}{2}}, \\ x - x_j, & x \in I_j, \\ 0, & x > x_{j+\frac{1}{2}}. \end{cases} \tag{7}$$

The coefficients C_j^0 and C_j^1 of these basis functions can be identified as the average value and average gradient of the solution, so that

$$u(x) = \bar{u}_j + \frac{\overline{\Delta u}_j}{h}(x - x_j), \quad x \in I_j, \quad \forall j; \tag{8}$$

the quantity $\overline{\Delta u}_j$ is the "undivided gradient" in cell I_j. The quantities $\bar{u}_j$ and $\overline{\Delta u}_j$ can be retrieved from the solution as follows:

$$\bar{u}_j = \frac{1}{h}\int_{I_j} u(x)dx, \tag{9}$$

$$\overline{\Delta u}_j = \frac{12}{h^2}\int_{I_j} (x - x_j)u(x)dx. \tag{10}$$

Expanding scheme (4) for each of the test functions yields the following differential equations for updating $\bar{u}_j$ (this one is conservative) and $\overline{\Delta u}_j$:

$$\frac{\partial}{\partial t}\begin{pmatrix} \bar{u}_j \\ \overline{\Delta u}_j \end{pmatrix} = \frac{D}{h^2} M(T, \sigma, \mu) \begin{pmatrix} \bar{u}_j \\ \overline{\Delta u}_j \end{pmatrix}, \tag{11}$$

with

$$M(T, \sigma, \mu) = \begin{pmatrix} \mu(T - 2I + T^{-1}) & \frac{1-\mu}{2}(T - T^{-1}) \\ 6(\sigma + \mu)(T - T^{-1}) & 3(T - 2I + T^{-1}) - 3(\sigma + \mu)(T + 2I + T^{-1}) \end{pmatrix}; \tag{12}$$

here T represents forward translation by one cell,

$$Tq_j = q_{j+1}. \tag{13}$$

The eigenvalues of the matrix operator $M(T, \sigma, \mu)$ can be found through Fourier analysis. With

$$\begin{pmatrix} \bar{u}_j \\ \overline{\Delta u}_j \end{pmatrix} = e^{i\beta j}\begin{pmatrix} \bar{u}_0 \\ \overline{\Delta u}_0 \end{pmatrix} \tag{14}$$

the operator T reduces to scalar multiplication,

$$\hat{T} = e^{i\beta}; \tag{15}$$

the following Fourier symbols are frequently used in this and further sections:

$$\hat{T} - \hat{T}^{-1} = 2i\sin\beta, \tag{16}$$

$$\hat{T} - 2\hat{I} + \hat{T}^{-1} = -2(1-\cos\beta), \tag{17}$$

$$\hat{T} + 2\hat{I} + \hat{T}^{-1} = 2(1+\cos\beta). \tag{18}$$

For the matrix operator $M(T,\sigma,\mu)$ given by Eq. (12) the following Fourier transform is found:

$$\hat{M}(\beta,\sigma,\mu) = \begin{pmatrix} -2\mu(1-\cos\beta) & i(1-\mu)\sin\beta \\ 12i(\sigma+\mu)\sin\beta & -6(1+\sigma+\mu)+6(1-\sigma-\mu)\cos\beta \end{pmatrix}. \tag{19}$$

This matrix is an approximation of the exact differential operator; since the Fourier symbol of spatial differentiation is

$$\frac{\hat{\partial}}{\partial x} = \frac{i\beta}{\Delta x}, \tag{20}$$

the exact operator is

$$\hat{M}_{ex}(\beta) = -\beta^2 I. \tag{21}$$

4. The (σ,μ) Playing Field

Our exploration of the (σ,μ)-family will be guided by the map of Figure 1, showing the (σ,μ)-plane near the origin. Three lines of importance are drawn, their equations are:

(1) $\sigma+\mu=0$,
(2) $\sigma+\mu=\frac{5}{2}$,
(3) $\sigma-\mu=-2$.

We shall discuss these one by one.

4.1. *The line $\sigma+\mu=0$*

The line $\sigma+\mu=0$ divides the plane into stable and unstable domains. To understand this, consider the eigenvalues of $\hat{M}$; these satisfy the approximate characteristic equation

$$\lambda^2 + 12(\sigma+\mu)\lambda + 12(\sigma+\mu)\beta^2 + O(\beta^4) = 0. \tag{22}$$

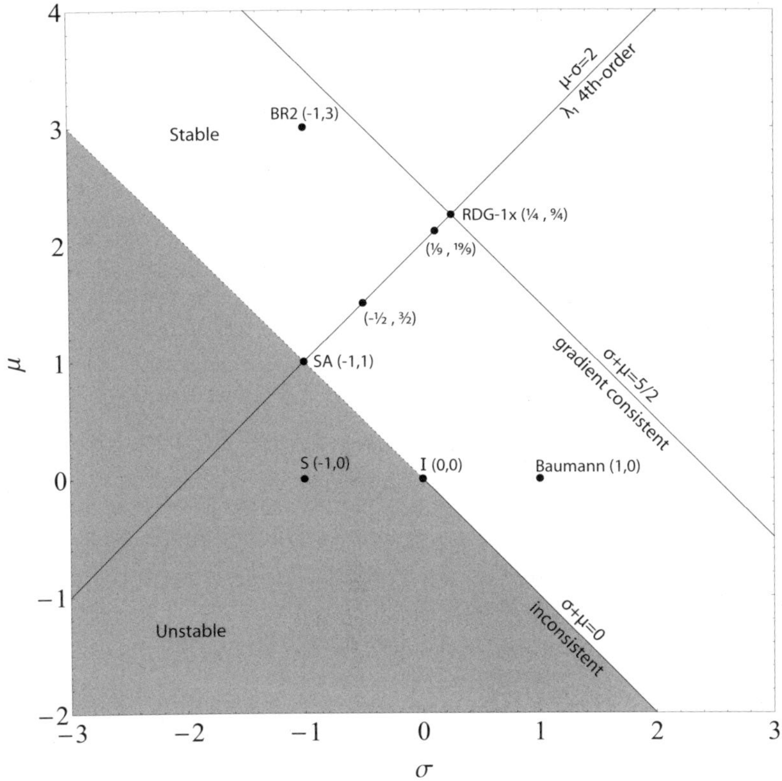

Fig. 1. The (σ, μ)-plane, showing the three important lines and several old and new schemes: Inconsistent (I), Symmetric (S), Stabilized Symmetric or Symmetric/Arnold (SA), Baumann, Recovery (RDG-1x), Sixth-Order $(\frac{1}{9}, \frac{19}{9})$, Efficient $(-\frac{1}{2}, \frac{3}{2})$, Bassi-Rebay-2 (BR2).

If $\sigma + \mu \neq 0$ it follows that

$$\lambda_1 = -\beta^2 + O(\beta^4), \tag{23}$$

$$\lambda_2 = -12(\sigma + \mu) + O(\beta^2). \tag{24}$$

The first eigenvalue is a low-frequency approximation of the exact spatial operator (21), the second eigenvalue may become accurate at high frequencies ($\beta > \pi$); for the sake of stability the latter must be nonpositive. Thus, stability is expected for

$$\sigma + \mu > 0; \tag{25}$$

an analysis of the full eigenvalues confirms this. In the stable open half-plane $\sigma + \mu > 0$ the first eigenvalue is generally second-order accurate, locally perhaps higher-order accurate.

The case $\sigma + \mu = 0$ needs to be investigated separately. In this case the diagonal elements of $\hat{M}$ become its eigenvalues, i.e.,

$$\lambda_1 = -2\mu(1 - \cos\beta), \tag{26}$$

$$\lambda_2 = -6(1 - \cos\beta). \tag{27}$$

From λ_1 it is seen that the scheme will be stable if $\mu \geq 0$, but it is consistent only if $\mu = 1$. In combination with $\sigma + \mu = 0$ this yields the stabilized symmetric scheme, shown here to be second-order accurate.

4.2. *The line* $\sigma + \mu = \frac{5}{2}$

If most of the schemes for $p = 1$ are merely second-order accurate, just as for $p = 0$, one may ask what use it is to add the linear basis function – it does not buy us anything. No wonder: the update equation for $\overline{\Delta u}_j$ appears to be *inconsistent* with the diffusion equation! One would expect the gradient update to approximate the third-order PDE

$$\frac{\partial}{\partial t}\left(\frac{\partial u}{\partial x}\right) = D\frac{\partial^3 u}{\partial x^3}; \tag{28}$$

therefore, if $\overline{\Delta u}$ is to remain an accurate approximation of the gradient of the solution, it must satisfy

$$\frac{1}{h}\frac{\partial \overline{\Delta u}_j}{\partial t} = D\left(\frac{\partial^3 u}{\partial x^3}\right)_j + O(h^2). \tag{29}$$

The actual update equation, as seen from Eqs. (11), (12), is

$$\begin{aligned}\frac{1}{h}\frac{\partial \overline{\Delta u}_j}{\partial t} &= D\left[6(\sigma+\mu)(T - T^{-1})\bar{u}_j + \{3(T - 2I + T^{-1})\right.\\ &\qquad \left. -3(\sigma+\mu)(T + 2I + T^{-1})\}\,\overline{\Delta u}_j\right]\\ &= D\left\{6(\sigma+\mu)([u]_{j+\frac{1}{2}} + [u]_{j-\frac{1}{2}}) + 3(T - 2I + T^{-1})\overline{\Delta u}_j\right\}. \end{aligned} \tag{30}$$

From the theory of recovery[1] we know that for $p = 1$ the interface jump of the solution scales with its third derivative:

$$[u] = -\frac{h^3}{15}\frac{\partial^3 u}{\partial x^3} + O(h^5); \tag{31}$$

inserting this into the update equation yields

$$\frac{1}{h}\frac{\partial\overline{\Delta u}_j}{\partial t} = D\left\{-\frac{4}{5}(\sigma+\mu)+3\right\}\left(\frac{\partial^3 u}{\partial x^3}\right)_j + O(h^2). \tag{32}$$

Upon comparing this with Eq. (29) we conclude that consistency requires $\sigma + \mu = \frac{5}{2}$.

A first inspection, disappointingly, shows that on the line $\sigma + \mu = \frac{5}{2}$ the schemes are still second-order accurate. Closer inspection, however, reveals a subtle difference between these schemes and those elsewhere in the plane: the amplitude of the second eigenvector is a mere[b] $O(\beta^6)$, to be compared to $O(\beta^4)$ or even $O(\beta^2)$ for the other schemes. In consequence, the gradient-consistent schemes update both $\bar{u}$ and $\overline{\Delta u}$ with second-order accuracy, right from the start of a simulation. In contrast, the other schemes are second-order accurate only in their description of the evolution of the first eigenvector, a particular combination of $\bar{u}$ and $\overline{\Delta u}$. But a fraction of the initial values, with magnitude $O(\beta^4)$ or $O(\beta^2)$, gets projected onto the second eigenvector and evolves inaccurately, creating errors of the order $O(\beta^4)$ or $O(\beta^2)$ in $\bar{u}$, and $O(\beta^3)$ or $O(\beta)$ in $\frac{1}{h}\overline{\Delta u}$.

In linear problems the second eigenvector is usually damped out rapidly owing to the large negative value of λ_2 for low frequencies, and not recreated. Once it has vanished the further evolution is due to the first eigenvector and is accurate for both $\bar{u}$ and $\overline{\Delta u}$.

The schemes satisfying the constraint $\sigma + \mu = \frac{5}{2}$ produce an accurately evolving gradient $\overline{\Delta u}/h$; this is useful if additional physics such as radiative transport requires a detailed and accurate knowledge of the subcell solution. But in this respect these schemes are still no better than a finite-volume scheme based on $p = 0$, with the gradient computed by finite-differencing.

A further truncation-error analysis[1] reveals that among the gradient-consistent schemes there is one for which λ_1 is fourth-order accurate, *viz.*, the scheme with

$$\sigma = \frac{1}{4},\ \mu = \frac{9}{4}. \tag{33}$$

It is a true fourth-order scheme, as the initial-value projection error (the part projected onto the second eigenvector) is only $O(\beta^6)$. This scheme happens to be interpretable as a Recovery scheme,[1] which makes it easily extendible to higher p, multiple dimensions and irregular grids, without necessitating further truncation-error or Fourier analysis.

[b] In one point of the line $\sigma + \mu = \frac{5}{2}$, *viz.*, $(-\frac{3}{7}, \frac{41}{14})$, the amplitude even drops to $O(\beta^8)$.

It is worth pointing out that by imposing gradient-consistency we have improved the eigenstructure of the schemes, without having to do the more complicated eigenvector analysis. Requiring consistency of solution-coefficient updates with higher-order PDE's could be a powerful tool in identifying superior DG schemes.

It may be anticipated that the fourth-order point $(\frac{1}{4}, \frac{9}{4})$ is not isolated but part of a locus of fourth-order schemes; this leads us to the third line.

4.3. *The line* $\sigma - \mu = -2$

The line $\sigma - \mu = -2$ is the locus of all schemes in the plane whose first eigenvalue is fourth-order accurate. It goes through the point representing the Recovery scheme $(\frac{1}{4}, \frac{9}{4})$ as well as through the Stabilized Symmetric scheme (-1,1); the latter one is the only point on the line where λ_1 is just second-order accurate. On the other hand, there is one point on the line, $(\frac{1}{9}, \frac{19}{9})$, where λ_1 is sixth-order accurate – but its initial projection error is still $O(\beta^4)$ in $\bar{u}$ and $O(\beta^3)$ in $\frac{1}{h}\overline{\Delta u}$. Almost all other points on the line yield these same projection errors, which leads us to expect they may not achieve a greater than third-order accuracy. In the Stabilized Symmetric scheme the amplitude of the second eigenvector is $O(\beta^2)$ in $\bar{u}$ and $O(\beta)$ in $\frac{1}{h}\overline{\Delta u}$; it will not perform better than the finite-volume scheme with $p = 0$. The Recovery scheme, with consistent gradient update, remains the best scheme on the line, with projection errors $O(\beta^6)$ in $\bar{u}$ and $O(\beta^5)$ in $\frac{1}{h}\overline{\Delta u}$; it also has good high-frequency accuracy.

Along this line, the high-frequency information provided by the second eigenvalue becomes poorer as the value of $\sigma+\mu$ decreases, especially when it drops below unity. In the high-frequency interval $[\pi, 2\pi]$ the exact damping factor (21) drops from -9.9 to -39.5, while λ_2 changes from -12 to $-12(\sigma+\mu)$; hence, values of $\sigma+\mu$ in the neighborhood of 3 should be favorable to high-frequency representation.

5. Stability Range

For large values of $\sigma + \mu$ the spectral radius of the operator $\hat{M}$ is set by the value of λ_2 at $\beta = 0$, i.e., $-12(\sigma + \mu)$; for small values of $\sigma + \mu$ it is set by the value of λ_2 at $\beta = \pi$. Analyzing $\hat{M}$ for $\beta = \pi$ yields that the largest negative eigenvalue always equals -12, so this sets the maximally stable time-step attainable in the (σ, μ)-family. The stability range suffers if $\sigma + \mu > 1$.

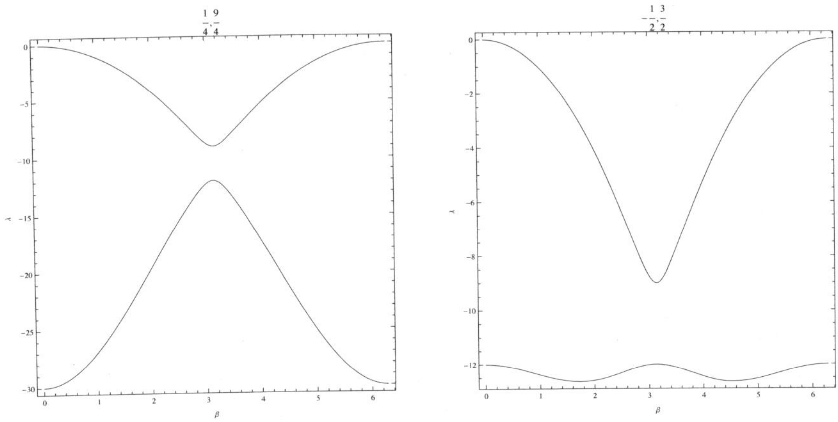

Fig. 2. Eigenvalues of two fourth-order (σ, μ)-schemes. Left: "Recovery" scheme $(\frac{1}{4}, \frac{9}{4})$; right: "Efficient" scheme $(-\frac{1}{2}, \frac{3}{2})$. The Recovery scheme's larger negative second eigenvalue causes it to have a smaller stability range, but better (subgrid) high-frequency accuracy (compare λ_2 to $-\beta^2$). The efficient scheme has its largest negative eigenvalue hovering around -12, allowing the maximum time-step available in the (σ, μ)-family; among the schemes of the fourth-order line having this stability range it is the one closest to the recovery scheme.

Figure 2 shows plots of the eigenvalues as functions of β for two different schemes on the fourth-order locus: the Recovery scheme $(\frac{1}{4}, \frac{9}{4})$ and the "Efficient" scheme with $\sigma + \mu = 1$, i.e., $(-\frac{1}{2}, \frac{3}{2})$, which will allow a 2.5× larger time step. Note that the Baumann scheme also has $\sigma + \mu = 1$, therefore also achieves the maximum stability range. Baumann's scheme, though, for $p = 1$ has an undamped mode ($\beta = \pi$) that makes it undesirable, as we shall see in Section 8.

6. Two Newer Schemes

Among the schemes of the new generation there are two that are members of the (σ, μ)-family. In the first place there is a fourth-order Recovery scheme (RDG-1x) featured in Ref. 5, based on the once-partially-integrated[c] weak form (3); for $p = 1$ it happens to be identical to the gradient-consistent scheme $(\frac{1}{4}, \frac{9}{4})$ found earlier in Ref. 1.

[c]Recovery schemes were originally based on the twice-partially-integrated weak form (34) for superior accuracy.[1] These RDG-2x schemes fall outside the (σ, μ)-family; see further Section 7.1.

Furthermore, there is the second scheme of Bassi and Rebay,[3] which for $p = 1$ is found to fall on (-1,3). This scheme is only second-order accurate and not gradient-consistent, but has a somewhat larger stability range than the fourth-order Recovery scheme. We shall discuss these schemes separately below.

6.1. *Recovery scheme* $(\frac{1}{4}, \frac{9}{4})$

In a Recovery scheme for $p = 1$ we recover from the linear distributions in two adjacent cells a single cubic function on the union of the cells, such that in each cell the linear distribution is an L_2-projection of (or least-squares fit to) the cubic; see Figure 3. The values of the recovered function and its derivative at the interface are used in computing the boundary term in the DG update equations; for RDG-1x this is the first term on the right-hand side of Eq. (3). This term is now so accurate that there is a mismatch with the interior integral (the second term), which is computed using a cellwise constant value for u_x. Since more accurate approximations of the solutions have already been obtained by recovery (one centered on the left interface and one on the right interface), we may as well reuse these. If we replace u in the interior integral by the algebraic average of the two recovered solutions, the scheme $(\frac{1}{4}, \frac{9}{4})$ results, originally derived by truncation-error and Fourier analysis.

As the Recovery Principle easily extends to higher p, a higher number of dimensions, and unstructured irregular grids, so does the 1-D scheme,

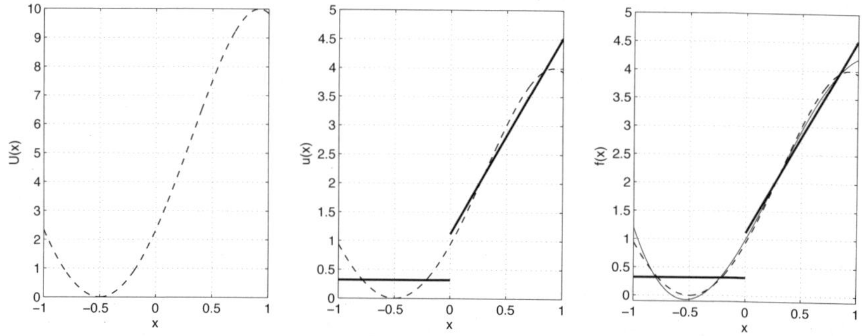

Fig. 3. Recovery in one dimension, $p = 1$. Shown are, from left to right, the original quartic initial values U (dashed), the piecewise linear discretization u (bold) together with U, and the cubic recovered distribution f (thin) together with u and U, on the adjacent intervals $(-1, 0)$ and $(0, 1)$. All three distributions yield the same value when their inner product is taken with either test function on either interval, making them indistinguishable in the weak sense.

without further appeal to truncation-error or Fourier analysis. Two ways of extending the scheme to two dimensions and triangular grids were discussed in Ref. 5, one smarter than the other; it later turned out that the best scheme became unstable for $p \geq 3$.

The reason for the instability can already be understood in the 1-D case. The recovered functions are accurate and well-behaved at the interface on which they are centered, but tend to swing more and more wildly at the far end of the abutting cells as p increases. It therefore is not a good idea to include those portions in the calculation of the interior integral. Lo[10] has indicated how to use the recovered information in a stable manner, at the cost of an extra recovery step. In his technique only the recovered interface values and gradient values are used to enhance the solution in a cell. Thus, combining the interface values with the linear interior solution affords a cubic approximation; including also the interface gradients affords a quintic approximation.

6.2. *Second Bassi-Rebay scheme (-1,3)*

The second scheme of Bassi and Rebay[3] is most easily understood in the interpretation of Huynh.[6] In this scheme it is first assumed that the value of the solution at the interface between two cells is the algebraic average of the left and right values; see Figure 4. Next, for $p = 1$, quadratic solution approximations are constructed in the adjacent cells, starting from the shared interface value and and incorporating data from the linear solution elements. There is some freedom in doing this; we choose to stay close to the recovery procedure and require that in each cell the linear distribution is the L_2-projection of (or least-squares fit to) the parabola.

The resulting piecewise quadratic solution is continuous, but its derivative jumps at the interface. The derivative value adopted to compute the diffusion flux is again the algebraic average of the left and right values. Computed this way, the interface values and gradient values are inserted into the twice-partially-integrated form (34) (see below) of the update equations; this produces the (-1,3)-scheme. The procedure can be easily generalized to higher p and multiple dimensions.

7. Schemes Outside the Family

The RDG-2x Recovery scheme,[1] Poor Man's Recovery scheme[6] and LDG scheme[2] all fall outside the (σ, μ)-family; we shall briefly discuss these here.

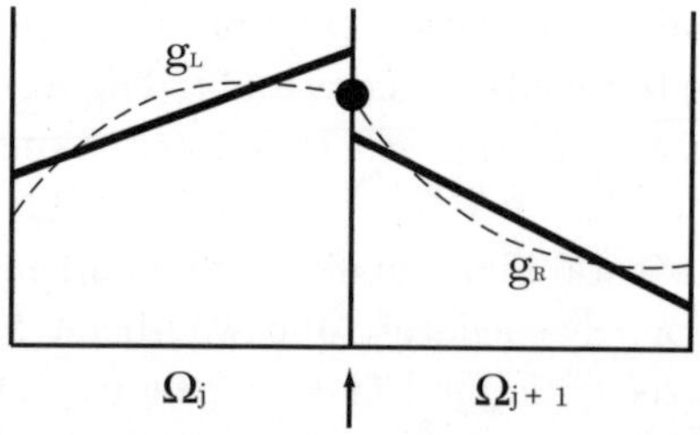

Fig. 4. Solution-reconstruction procedure in the Bassi-Rebay-2 (BR2) scheme; $p = 1$. The interface value (dot) is assumed to be the average of the left and right values. Starting from the interface point, parabolas are cast to the left (g_L) and right (g_R) that include information from the linear distributions. In the version shown, the parabolas are chosen such that the linear distributions are their L_2-projections, i.e., best fits in the least-squares sense. The average of the left and right derivatives at the interface is then used as the solution derivative in the update equation Eq. (34).

7.1. *RDG-2x*

The extra steps to improve the accuracy of the interior integral in RDG-1x are unnecessary, at least for linear diffusion, if the twice-partially-integrated form of the DG equations is used:

$$\int_{I_j} v_j^k u_t dx = D\left(\{v_j^k u_x - (v_j^k)_x u\}\Big|_{j+\frac{1}{2}} - \{v_j^k u_x - (v_j^k)_x u\}\Big|_{j-\frac{1}{2}}\right) + D\int_{I_j} (v_j^k)_{xx} u dx, \quad k = 0, 1, ..., p. \tag{34}$$

In this equation the interior integral is as accurate as can be; it can not be improved by including recovered information.[5]

the RDG-2x scheme results when the recovered cubic function described in Section 6.1 is inserted into the boundary term of Eq. (34). The scheme is fourth-order accurate like RDG-1x, but lies outside the (σ, μ)-family. As shown in Ref. 1, it includes $\sigma = -1$ and $\mu = \frac{9}{4}$, but requires an additional penalty term, proportional to $[v_x][u_x]$. Its stability, though, is still ruled by $\sigma + \mu$; with $\lambda_2 = -15$ for $\beta = 0$ the time-step range is close to maximal.

7.2. *Poor Man's Recovery scheme*

In Huynh's Poor Man's Recovery scheme the recovered solution for $p = 1$ is piecewise quadratic and continuous as in BR2, but the interface value is not *a priori* set; instead, it is chosen such that the derivative at the interface, too, becomes continuous. One may also say that the cubic used in

RDG-1x and RDG-2x is replaced by a continuously differentiable, piecewise quadratic function. The scheme has $\sigma = -1$, $\mu = 3$ just as BR2, but falls outside the family, requiring the same additional penalty term as RDG-2x. This scheme is potentially more accurate than BR2; for $p = 1$ it seems comparable and its stability range is the same. As a true Recovery scheme, Poor Man's Recovery easily generalizes to higher p and multiple dimensions.

7.3. *LDG*

The LDG scheme has two variants; in Huynh's interpretation, the interface value may be taken from the left cell and the derivative value from a quadratic reconstruction in the right cell, or the other way around. Neither variant has an obvious value of μ or σ, but the *average* of the results of the two variants can be identified with the choice $\sigma = -1$, $\mu = 6$. This average LDG scheme is similar to BR2 and Poor Man's in regard to order of accuracy, but its unnecessarily high value of μ is responsible for higher errors and smaller time-steps.

8. Numerical Test

We tested a number of schemes of the (σ, μ)-family with regard to their order of accuracy and stability range. The test problem was a 1-D Poisson problem on [0,1] with Dirichlet boundary conditions, solved by marching in time; the steady solution is

$$u(x, \infty) = 1 - x + \sin(2\pi x). \tag{35}$$

Time-stepping was done with the first-order "Forward Euler" method, since temporal accuracy was not required. All schemes were applied till the time derivative reached machine-zero, on a sequence of successively finer grids. The nondimensional time-step $D\Delta t/(\Delta x)^2$ or Von Neumann number (VNN) was run at the maximum value, found experimentally by increasing it with steps of 0.01. The results are summarized in the table shown as Figure 5.

We observe that the Symmetric/Arnold (= Stabilized Symmetric), Baumann and Efficient schemes boast the largest VNN values, as expected. The value is around 0.20, which may be compared to the value of 0.50 allowed by the classical second-order-accurate finite-volume method ($p = 0$).

As predicted, the VNN is inversely proportional to $\sigma + \mu$ for $\sigma + \mu > 1$; for instance, the Recovery scheme, with $\sigma + \mu = 2.5$, allows a VNN of

$0.20/2.5 = 0.08$. The CPU time needed for convergence on the finest grid would be expected to scale with the inverse of the VNN, but here other qualities of the schemes appear to enter and produce surprises, with the fourth-order recovery scheme being remarkably efficient in marching to a steady state, and the Baumann scheme showing an abysmal performance because of the undamped mode of wavelength $2h$.

Listed separately are the orders of accuracy found for $\bar{u}$ and $\overline{\Delta u}$; here, again, are some surprises. The nominally second-order schemes indeed show second-order accuracy in $\bar{u}$. The Symmetric/Arnold scheme clearly is the least accurate of all, with (predictably) only first-order accuracy for the gradient; Baumann and Bassi-Rebay-2 manage to achieve third-order accuracy for $\overline{\Delta u}$, which is not in contradiction with their $O(\beta^3)$ projection error. Among the schemes of the fourth-order line only the Recovery scheme (RDG-1x) shows fourth-order convergence for $\bar{u}$, and even fifth-order convergence for $\overline{\Delta u}$. The nominally fourth-order Efficient scheme and the nominally Sixth-Order scheme turn out to be no better than third-order accurate, apparently because of the $O(\beta^3)$ projection error in the gradient.

9. Conclusion

The entire (σ, μ)-family of DG diffusion schemes has been analyzed for the first time, without preconditions stemming from finite-element theory. The analysis is only for $p = 1$ and one dimension, but sheds light on the possibilities for higher p and multiple dimensions. In particular, the most promising member of the family, a fourth-order scheme with $\sigma = \frac{1}{4}$, $\mu = \frac{9}{4}$, initially derived by truncation-error and Fourier analysis, can be interpreted as a Recovery scheme (RDG-1x), which makes it immediately extendible to higher p, multiple dimensions and unstructured grids. The scheme has

Scheme	(σ, μ)	VNN	OAA of $\bar{u}$	OAA of $\overline{\Delta u}$	Time(s)
Symmetric Arnold	$-1, 1$	0.20	2	1	152
Baumann	$1, 0$	0.19	2	3	2372
RDG-1x	$\frac{1}{4}, \frac{9}{4}$	0.08	4	5	55
6th-order	$\frac{1}{9}, \frac{19}{9}$	0.09	3	3	149
Efficient	$-\frac{1}{2}, \frac{3}{2}$	0.19	3	3	107
BR2	$-1, 3$	0.10	2	3	294

Fig. 5. Stability range and order of error convergence in solving a Poisson-Dirichlet problem with various schemes of the (σ, μ)-family. VNN = maximum stable Von Neumann Number, OOA = order of accuracy, Time(s) = CPU time in seconds.

a stability limit that is only 40% of the maximum attainable within the family, but turns out to be by far the most efficient scheme for marching to a steady solution. Another modern scheme, by Bassi-Rebay, can also be identified as a member of the family, at least for $p = 1$; it does not appear to have special merits regarding accuracy or stability. A numerical comparison of the schemes based on a Poisson problem confirms the superiority of the fourth-order Recovery scheme.

References

1. B. van Leer and S. Nomura. Discontinuous Galerkin for diffusion. AIAA Paper 2005-5108, (2005).
2. B. Cockburn and C.-W. Shu, The Local Discontinuous Galerkin method for time-dependent convection-diffusion systems, *SIAM Journal on Numerical Analysis.* **35**, 2440–2463, (1998).
3. F. Bassi and S. Rebay, High-order accurate discontinuous finite element solution of the 2D Euler equations, *Journal of Computational Physics.* **138**, 251–285, (1997).
4. F. Brezzi, G. Manzini, D. Marini, P. Pietra, and A. Russo, Discontinuous Galerkin approximations for elliptic problems, *Numerical Methods for Partial Differential Equations.* **16**, 365–378, (2000).
5. M. Lo and B. van Leer. Analysis and implementation of the Recovery-based Discontinuous Galerkin method for diffusion. AIAA Paper 2009-3786, (2009).
6. H. T. Huynh. A reconstruction approach to high-order schemes including Discontinuous Galerkin for diffusion. AIAA Paper 2009-0403, (2009).
7. M. H. van Raalte. *Multigrid Analysis and Embedded Boundary Conditions for Discontinuous Galerkin Discretization.* PhD thesis, University of Amsterdam, (2004).
8. D. N. Arnold, An interior penalty finite element method with discontinuous elements, *SIAM Journal on Numerical Analysis.* **19**, 742–760, (1982).
9. C. E. Baumann. *An hp-adaptive discontinuous Finite Element Method for Computational Fluid Dynamics.* PhD thesis, University of Texas at Austin, (1997).
10. M. Lo. *Space-Time Discontinuous Galerkin for Diffusion and Advection based on Recovery.* PhD thesis, University of Michigan, (2010).

CHAPTER 8

$P_N P_M$ SCHEMES ON UNSTRUCTURED MESHES FOR TIME-DEPENDENT PARTIAL DIFFERENTIAL EQUATIONS

Michael Dumbser

Laboratory of Applied Mathematics, University of Trento, Via Mesiano 77, I-38100 Trento (TN), Italy
michael.dumbser@ing.unitn.it

We give a brief review of the $P_N P_M$ method on general unstructured triangular and tetrahedral meshes introduced in Ref. 1. The approach represents data in each cell by piecewise polynomials u_h of degree N and uses different piecewise polynomials w_h of degree $M \geq N$ to compute the fluxes and source terms. The polynomials w_h are obtained from u_h by a reconstruction operator that is based on a weak identity between u_h and w_h on a suitable neighborhood of each control volume. The approach generalizes classical finite volume schemes ($N = 0$) and high order discontinuous Galerkin finite element methods ($N = M$), which are both contained in the new approach as two special cases of a more general framework.

1. Introduction

1.1. *Historic background*

Conservation laws are among the most powerful physical and mathematical tools available nowadays to describe and model physical phenomena observed in the real world. According to the fundamental work of Emmy Noether[2] the principles of conservation are directly linked to symmetries and hence to the structure of space and time. While the mathematical formulation of the relevant conservation laws of solid and fluid mechanics is known for more than a century now, their accurate numerical solution poses still many difficulties even nowadays due to the strong nonlinearities present in the governing partial differential equations. It was Bernhard Riemann in his groundbreaking theoretical work[3] who discovered that nonlinear conservation laws can develop *discontinuous* solutions after some time even when starting from perfectly smooth initial data. It was only decades later that his theoretical results could be confirmed also experimentally. It was Sergeij Godunov in his pioneering work[4] who laid the basis of a continuing development of numerical techniques for the solution of nonlinear hyperbolic partial differential equations. The key idea of his work was to compute the flux across the boundary of two adjacent control volumes by solving a particular Cauchy problem of the governing PDE where the initial conditions are given by two piecewise constant states, a so-called Riemann problem. The resulting self-similar solution is then used to compute the flux at the element interface. It can be proven that the resulting Godunov scheme is the least dissipative monotone scheme at first order of accuracy. Other numerical flux functions based on so-called approximate Riemann solvers emerged later, in order to reduce the computational effort associated with the solution of the Riemann problem, see the pioneering works by Roe,[5] Harten–Lax–Leer,[6] Osher and Solomon,[7] Einfeldt *et al.*,[8] Toro *et al.*,[9] to name just a few. A comprehensive overview of Riemann solvers can be found in Ref. 10.

Unfortunately, it was proven by the same Godunov that there are no better than first order accurate linear schemes being monotone. The only possibility to circumvent the theorem is to develop schemes that are *nonlinear*. The first second order Godunov-type methods have been developed already more than 30 years ago in the works by Kolgan,[11] whose method was second order accurate in space, but only first order accurate in time and by the groundbreaking work of van Leer,[12] whose method was second order accurate in space and time. The key idea in those schemes consists in a reconstruction step that is Total Variation Diminishing (TVD). Since for the computation of turbulent flows even second order TVD schemes still contain too much dissipation, the quest towards higher order schemes continued and it was thanks to the landmark paper on essentially non-oscillatory (ENO) schemes by Harten *et al.*[13] that better than second order accurate schemes in space and time have been made available for the solution of nonlinear hyperbolic conservation laws. A much more efficient higher order scheme was introduced subsequently by Jiang and Shu[14] in their famous work on weighted ENO (WENO) methods. While the development of more accurate schemes for one-dimensional systems of conservation laws or for multi-dimensional systems using Cartesian meshes made quick progress, the development of high order finite volume schemes on general unstructured meshes was much slower. This was of course mainly due to the significantly higher computational effort, the necessity to rely on external mesh generation tools as well as the high algorithmic complexity of the underlying computer programs. Barth and Jespersen were the first to develop a second order accurate TVD finite volume scheme on unstructured triangular meshes[15] and Barth and Frederickson laid the grounds for higher order reconstruction schemes on unstructured meshes in Ref. 16. The first ENO methods on unstructured two-dimensional meshes emerged in Ref. 17 and Ref. 18 and the first unstructured WENO schemes have been developed in Ref. 19 and Ref. 20 and convection-diffusion problems on unstructured meshes with curved boundaries have been discussed in Ref. 21. However, all the aforementioned articles were strictly limited to the two-dimensional case. In three space dimensions, only very recently the first WENO schemes for nonlinear hyperbolic conservation laws on general unstructured tetrahedral meshes have been developed, see Refs. 22–24. A spectral finite volume method on unstructured meshes in two and three space dimensions has been presented in Ref. 25 and Ref. 26, respectively.

The most complicated part of a high order finite volume scheme on general unstructured meshes is the reconstruction operator. To avoid a cumbersome reconstruction step and still obtaining high order of accuracy in space, a different class of high order methods has been proposed and recently enjoys growing popularity in the scientific community: the discontinuous Galerkin (DG) finite element method. Originally applied to steady Neutron transport equations by Reed and Hill[27] it was later put on a solid mathematical basis by Lesaint and Raviart[28] and extended to general hyperbolic conservation laws by Cockburn and Shu in a prominent series of papers.[29–33] A cell entropy inequality was proven for arbitrary order DG schemes in Ref. 34, which is a very remarkable result that can not be obtained for finite volume schemes in this general form. While very successful for first order hyperbolic systems, the discretization of second order parabolic terms initially posed some difficulties. Bassi and Rebay were the first to solve the compressible Navier–Stokes equations with the DG method, see Ref. 35, followed by Baumann and Oden.[36] The introduction of the local DG scheme in Refs. 37 and 38 allowed to discretize even higher than second order derivative terms with the DG method, see Refs. 39 and 40. A unified analysis of several DG schemes for second order elliptic equations has been performed in Ref. 41.

An alternative DG scheme for the discretization of second order parabolic equations has been presented in Refs. 42 and 43 and is based on the solution of generalized Riemann problems of the underlying governing equations. The resulting DG method is similar to the penalty DG scheme, see Ref. 44 and 45, however, in the scheme of Gassner *et al.* for the first time the penalty constant is determined based on *physical* considerations by solving Riemann problems at the cell interface. In this sense, the method can be considered as an extension of the Godunov-type philosophy to diffusion equations. A similar idea has been used for high order finite volume schemes for nonlinear diffusion problems in Ref. 46.

The DG method has first been applied to hyperbolic PDE with non-conservative products in one and two space dimensions in Refs. 47 and 48, whereas the first three-dimensional DG scheme for PDE with non-conservative terms has been published in Ref. 49 in the general framework of $P_N P_M$ schemes discussed in detail later.

Most explicit DG schemes are based on TVD Runge-Kutta time integration schemes, as proposed by Shu and Osher in Refs. 50 and 51. However, also alternative explicit high order accurate time discretizations have been applied, for example Lax-Wendroff / ADER type one-step time discretizations, see Refs. 52–54 or also Adams-Bashforth-type time discretizations.[55] A very original fully implicit approach that uses a unified DG discretization in both space and time has been proposed in Refs. 56–59.

The idea of applying a reconstruction operator to the DG method in order to enhance accuracy was first introduced by Cockburn *et al.*[60] and further developed by Ryan *et al.*[61] However, they applied the reconstruction operator only at the final output time and therefore called their method a *postprocessing* technique for the DG finite element scheme. Obviously, this kind of accuracy enhancement becomes problematic on *coarse* meshes in space and time for general nonlinear time dependent problems, where temporal *and* spatial discretization errors accumulate during time stepping and thus information that is once lost due to any kind of discretization error can never be completely recovered. Therefore, Dumbser and Munz[62,63] were the first to propose the application of a reconstruction operator to the DG scheme at the beginning of *each* time step. The advantages of the proposed tensor product reconstruction on Cartesian grids were: First, the formal order of accuracy of a DG scheme using basis functions of degree N was increased to $M = 3N + 3$. Second, the resulting reconstructed DG scheme could be directly applied to the diffusion equation by simply using a central flux formulation, yielding a much larger stability limit than the classical local DG schemes.[37,38] This philosophy of using piecewise polynomials u_h of degree N to store and evolve data in each cell and to use piecewise polynomials w_h of degree $M \geq N$ to compute the fluxes and source terms was extended to general unstructured meshes in two and three space dimensions in Ref. 1 and the resulting approach was called the $P_N P_M$ method. In the special case of piecewise constant data representation $N = 0$ one recovers the classical finite volume method and when choosing $N = M$ one obtains the standard discontinuous Galerkin finite element scheme. For $N > 0$ and $M > N$ one obtains a new type of spatial discretization operator that can be seen as Hermite Finite Volume scheme or reconstructed discontinuous Galerkin method. Subsequent work included the extension of the $P_N P_M$ approach to systems with stiff source terms,[64] to non-conservative systems[48,49] as well as to systems with parabolic terms.[65,66]

Van Leer and co-workers[67,68] have recently proposed to apply a reconstruction operator to the DG method in order to discretize parabolic terms. However, in contrast to the $P_N P_M$ approach the reconstruction operator in the articles of van Leer and co-workers is only applied between two adjacent cells in order to compute a numerical diffusion flux at the element interface, whereas the $P_N P_M$ approach applies the reconstruction operator on a stencil composed of a suitable neighborhood of each element in order to *increase* locally the degree of the polynomial approximation inside each element.

Nonlinear versions of reconstruction operators are also applied to DG schemes in order to serve as limiters, as in the well-known HWENO approach introduced by Qiu and Shu[69,70] and as also recently furthered by Balsara *et al.*[71] An extension of HWENO limiters to unstructured meshes can be found in Ref. 72.

1.2. *Governing PDE*

In this chapter we consider general nonlinear time-dependent partial differential equations (PDE) in multiple space dimensions of the following general form

$$\frac{\partial Q}{\partial t} + \nabla \cdot F(Q, \nabla Q) + B(Q) \cdot \nabla Q = S(Q), \tag{1}$$

where Q is the vector of state, F is a nonlinear flux tensor, that may also depend on the gradient of Q in order to take into account viscous effects, $B(Q) \cdot \nabla Q$ is a non-conservative term and S denotes the vector of nonlinear algebraic source terms, which may also be *stiff*. Furthermore, we will denote by $J = \partial F / \partial Q$ the Jacobian of the flux F with respect to Q, $D_n = \partial(F \cdot \vec{n}) / \partial \nabla Q \cdot \vec{n}$ the Jacobian of the flux with respect to ∇Q in direction of the unit normal vector $\vec{n}$ and finally we will use the notation $A = J + B$. We will assume that the matrix $A \cdot \vec{n}$ is hyperbolic for all unit normal vectors $\vec{n}$, i.e. that all of its eigenvalues are real and there exists a full set of linearly independent eigenvectors.

In what follows, we present each single step of the high order one-step $P_N P_M$ method for the solution of Eq. (1). The general $P_N P_M$ reconstruction operator on general unstructured meshes in two and three space dimensions is described in section 2 and the high order one-step time discretization that allows also for a high order accurate discretization of stiff source terms according to[73] is described in detail in section 3. The fully discrete one-step $P_N P_M$ approach is given in section 4 and in section 5 some applications to the compressible Navier–Stokes equations, the viscous and resistive MHD equations as well as to the fully three-dimensional Baer–Nunziato model of compressible multi-phase flows are presented. The chapter is rounded-off by some concluding remarks given in section 6.

2. The Unified $P_N P_M$ Reconstruction Operator on General Unstructured Meshes

The spatial discretization of Eq. (1) used in this work is based on the $P_N P_M$ reconstruction operator first introduced on unstructured meshes in Ref. 1. In this section, we only give a short overview over the $P_N P_M$ reconstruction operator and for details we refer the reader to the publication of[1] and references therein. The computational domain Ω is discretized by conforming elements T_i that are chosen to be triangles in 2D and tetrahedrons in 3D, although also other, more general, element shapes would be possible. Each element is indexed by a single mono-index i ranging from 1 to the total number of elements N_E. The union of all elements is called the triangulation (2D) or the tetrahedrization (3D) of the domain, respectively,

$$\mathcal{T}_\Omega = \bigcup_{i=1}^{N_E} T_i. \tag{2}$$

At the beginning of a time-step, the numerical solution of (1) for the state vector Q, denoted by u_h, is represented by piecewise polynomials of degree N from the space V_h, spanned by the basis functions $\Phi_l = \Phi_l(\vec{x})$, i.e. at $t = t^n$ we have for each element

$$u_h(\vec{x}, t^n) = \sum_l \Phi_l(\vec{x}) \hat{u}_l^n. \tag{3}$$

From the polynomials u_h, we then reconstruct piecewise polynomials w_h of degree $M \geq N$ from the space W_h, spanned by the basis functions

$\Psi_l = \Psi_l(\vec{x})$:

$$w_h(\vec{x}, t^n) = \sum_l \Psi_l(\vec{x}) \hat{w}_l^n. \tag{4}$$

According to Ref. 1, the Ψ_l are chosen to be orthogonal and are identical with the Φ_l up to polynomial degree N. We note that the actual choice for the basis functions is not important, but only the choice of the approximation spaces V_h and W_h, i.e. the choice of the piecewise polynomial degrees N and M. However, the choice of an orthogonal basis used here leads to simple reconstruction equations and to diagonal element mass matrices, which makes the practical computation easier. To obtain the reconstruction polynomial w_h on element T_i, we now choose a reconstruction stencil

$$\mathcal{S}_i = \bigcup_{k=1}^{n_e} T_{j(k)} \tag{5}$$

that contains a total number of n_e elements. Here $1 \leq k \leq n_e$ is a local index, counting the elements in the stencil, and $j = j(k)$ is the mapping from the local index k to the global indexation of the elements in $\mathcal{T}_\Omega$. For ease of notation, we write in the following only j, meaning $j = j(k)$.

In the present paper we need the following three operators:

$$\langle f, g \rangle_{T_i} = \int_{t^n}^{t^{n+1}} \int_{T_i} (f(\vec{x}, t) \cdot g(\vec{x}, t))\, dV\, dt, \tag{6}$$

$$[f, g]_{T_i}^t = \int_{T_i} (f(\vec{x}, t) \cdot g(\vec{x}, t))\, dV, \tag{7}$$

$$\{f, g\}_{\partial T_i} = \int_{t^n}^{t^{n+1}} \int_{\partial T_i} (f(\vec{x}, t) \cdot g(\vec{x}, t))\, dS\, dt, \tag{8}$$

The first operator defines a space-time scalar product of two functions f and g over the space-time element $T_i \times \left[t^n; t^{n+1}\right]$, the second operator defines the standard spatial scalar product of f and g over the spatial element T_i, and the last operator defines a product of f and g over the space-time boundary element $\partial T_i \times \left[t^n; t^{n+1}\right]$. The notation $\langle f, g \rangle$ and $[f, g]^t$, i.e. without the index T_i, is used to define scalar products on the space-time *reference* element $T_E \times [0; 1]$ and on the spatial reference element

T_E at time t, respectively. The spatial reference element T_E is defined as the unit simplex with vertices $(0,0)$, $(1,0)$, $(0,1)$ in two space dimensions and vertices $(0,0,0)$, $(1,0,0)$, $(0,1,0)$ and $(0,0,1)$ in three space dimensions, respectively.

The reconstruction is now obtained via L_2-projection of the (unknown) piecewise polynomials w_h from the space W_h into the space V_h on each stencil $\mathcal{S}_i$, i.e. we require a *weak identity* between u_h and w_h in each stencil element as follows:

$$[\Phi_k, w_h]_{T_j}^{t^n} = [\Phi_k, u_h]_{T_j}^{t^n}, \qquad \forall T_j \in \mathcal{S}_i. \tag{9}$$

During the reconstruction step, the polynomials w_h are continuously extended over the whole stencil $\mathcal{S}_i$. After reconstruction, the piecewise polynomials w_h are again restricted onto each element T_i. The number of elements in the stencils are chosen in such a way that the number of equations in (9) is *larger* that the number of degrees of freedom in the space W_h. Eq. (9) constitutes thus an overdetermined linear algebraic equation system for the coefficients of w_h and is solved using a constrained least squares technique based, see Refs. 1 and 22. The linear constraint is that Eq. (9) is at least *exactly* satisfied for $T_j = T_i$, i.e. inside the element T_i under consideration:

$$[\Phi_k, w_h]_{T_i}^{t^n} = [\Phi_k, u_h]_{T_i}^{t^n}. \tag{10}$$

The constraint (10) is incorporated in the least squares problem using a standard Lagrangian multiplier technique, see Ref. 22 for details. The integral on the left hand side in (9) is computed using classical multidimensional Gaussian quadrature of appropriate order, see Ref. 74. The integral on the right hand side can be computed analytically and involves the standard element mass-matrix.

The resulting $P_N P_M$ least squares reconstruction operator can be interpreted as a generalization of the k-exact reconstruction proposed for pure finite volume schemes in the pioneering work of Ref. 16 and further discussed in Ref. 75.

3. An Approach for High-Order One-Step Time-Discretization of PDE Systems with Stiff Source Terms

Our high order one-step time discretization is based on a *local weak* formulation of the governing PDE (1) which is used to solve a *local* Cauchy problem in the small, with the reconstruction polynomial w_h as initial condition. Since this local solution is only used as a predictor, similar to the time-evolution to the half time level in the MUSCL method of Ref. 12, no coupling to the neighbor elements is needed. Note that this is a major difference with respect to the global space-time DG schemes of Refs. 56 and 57. To that purpose we start from the strong formulation of PDE (1) and transform the PDE into the reference coordinate system $(\vec{\xi}, \tau)$ of the space-time reference element $T_E \times [0;1]$ with $\vec{\xi} = (\xi, \eta, \zeta)$ and ∇_ξ being the nabla operator in the $\xi - \eta - \zeta$ reference system and $t = t^n + \tau \Delta t$:

$$\frac{\partial}{\partial \tau} Q + \nabla_\xi \cdot F^* \left(Q, \nabla Q\right) = S^* - B^*(Q) \cdot \nabla Q, \tag{11}$$

with

$$F^* := \Delta t \, F(Q, \nabla Q) \mathcal{J}^T, \quad S^* := \Delta t S(Q), \quad B^* := \Delta t B(Q), \tag{12}$$

and

$$\nabla Q = \mathcal{J}^T \, \nabla_\xi Q, \quad \mathcal{J} = \frac{\partial \vec{\xi}}{\partial \vec{x}}. \tag{13}$$

We now multiply Eq. (11) by a space-time test function $\theta_k = \theta_k(\vec{\xi}, \tau)$ from the space of piecewise space-time polynomials of degree M and integrate over the space-time reference control volume $T_E \times [0;1]$ to obtain the following weak formulation:

$$\left\langle \theta_k, \frac{\partial}{\partial \tau} q_h \right\rangle + \left\langle \theta_k, \nabla_\xi \cdot F^* \left(q_h, \nabla q_h\right) \right\rangle = \left\langle \theta_k, P^* \left(q_h, \nabla q_h\right) \right\rangle, \tag{14}$$

where we have used the abbreviation

$$P^* = P^*(Q, \nabla Q) = S^*(Q) - B^*(Q) \cdot \nabla Q. \tag{15}$$

Integration by parts of the first term in time allows us to introduce the initial condition $w_h(\vec{x}, t^n)$ in a weak form and leads to

$$[\theta_k, q_h]^1 - [\theta_k w_h]^0 - \left\langle \frac{\partial}{\partial \tau} \theta_k, q_h \right\rangle + \left\langle \theta_k, \nabla_\xi \cdot F^* \left(q_h, \nabla q_h\right) \right\rangle = \left\langle \theta_k, P^*(q_h, \nabla q_h) \right\rangle. \tag{16}$$

For the numerical solution of Eq. (16) and its gradient as well as for the flux tensor and the source term we use the ansatz

$$q_h = q_h(\vec{\xi}, \tau) = \sum_l \theta_l(\vec{\xi}, \tau) \hat{q}_l := \theta_l \hat{q}_l, \tag{17}$$

$$\nabla_\xi q_h = \nabla_\xi q_h(\vec{\xi}, \tau) = \sum_l \theta_l(\vec{\xi}, \tau) \hat{q}'_l := \theta_l \hat{q}'_l, \tag{18}$$

$$\mathcal{F}_h^* = \mathcal{F}_h^*(\vec{\xi},\tau) = \sum_l \theta_l(\vec{\xi},\tau)\hat{\mathcal{F}}_l := \theta_l \hat{\mathcal{F}}_l, \tag{19}$$

$$\mathcal{P}_h^* = \mathcal{P}_h^*(\vec{\xi},\tau) = \sum_l \theta_l(\vec{\xi},\tau)\hat{P}_l := \theta_l \hat{P}_l, \tag{20}$$

using the same space-time basis functions θ_l as used for the test functions. To facilitate notation, from now on we use the Einstein summation convention throughout the paper, which implies summation over indices appearing twice. Using a weak identity between the ansatz (18) and the gradient of q_h it is easy to show that the degrees of freedom $\hat{q}'_l$ of the gradient can be computed from the degrees of freedom $\hat{q}_l$ of the state by a simple matrix-vector multiplication as

$$\hat{q}'_k = \langle \theta_k, \theta_m \rangle^{-1} \langle \theta_m, \nabla_\xi \theta_l \rangle \hat{q}_l. \tag{21}$$

We use the nodal space-time basis and test functions θ_k proposed in Ref. 1, since this has shown to be computationally more efficient than a modal basis, which requires a more expensive L_2-projection. For an efficient implementation on Cartesian meshes, see Ref. 76.

In the nodal space-time framework we therefore compute the degrees of freedom of the interpolants for the flux and the source term simply as

$$\hat{\mathcal{F}}_l = F^*\left(\hat{q}_l, \hat{q}'_l\right), \qquad \hat{\mathcal{P}}_l = P^*(\hat{q}_l, \hat{q}'_l). \tag{22}$$

To solve the weak form (16) we insert (17)-(20) into (16) and then use the simple and robust fixed-point iteration scheme originally proposed in Ref. 1:

$$\left([\theta_k, \theta_l]^1 - \left\langle \frac{\partial}{\partial \tau}\theta_k, \theta_l \right\rangle \right) \hat{q}_l^{i+1} = [\theta_k, \psi_m]^0 \hat{w}_m^n + \langle \theta_k, \theta_l \rangle \; \hat{P}_l^i - \langle \theta_k, \nabla_\xi \theta_l \rangle \cdot \hat{\mathcal{F}}_l^i, \tag{23}$$

or in a more convenient matrix shorthand notation

$$K_1 \hat{q}_l^{i+1} = F_0 \hat{w}_m^n + M\hat{P}_l^i - K_\xi \cdot \hat{\mathcal{F}}_l^i, \tag{24}$$

with

$$K_1 = F_1 - K_\tau = \left([\theta_k, \theta_l]^1 - \left\langle \frac{\partial}{\partial \tau}\theta_k, \theta_l \right\rangle \right), \tag{25}$$

$$F_0 = [\theta_k, \psi_m]^0, \tag{26}$$

$$M = \langle \theta_k, \theta_l \rangle, \tag{27}$$

$$K_\xi = \langle \theta_k, \nabla_\xi \theta_l \rangle. \tag{28}$$

If the source term $S(Q)$ is stiff, it has to be taken implicitly in (23). Further details of this algorithm in the case of stiff source terms are given in Refs. 64 and 73.

4. The Fully Discrete $P_N P_M$ Method

Applying the operator $\langle \Phi_k, \cdot \rangle_{T_i}$ to PDE (1) one obtains

$$\left\langle \Phi_k, \frac{\partial}{\partial t} Q \right\rangle_{T_i} + \langle \Phi_k, \nabla \cdot F(Q, \nabla Q) + B(Q) \cdot \nabla Q \rangle_{T_i} = \langle \Phi_k, S(Q) \rangle_{T_i} . \quad (29)$$

For the first term in Eq. (29) we approximate Q with u_h from the space V_h and perform integration by parts in time. Note that Φ_k does not depend on time. For all the other terms in Eq. (29) the vector Q is approximated by the solution q_h of the local space-time Galerkin predictor of Section 3. Since q_h usually exhibits jumps at the element interfaces we integrate the second term by splitting it in its smooth part on the domain $T_i \backslash \partial T_i$ and the jump term on the boundary ∂T_i. For the jump term, we use the strategy of path-conservative schemes introduced in Refs. 77–79. The latter references are based on the theory of Dal Maso, Le Floch and Murat (DLM) which defines weak solutions in the presence of non-conservative products, see Ref. 80, which reduces to the classical Rankine-Hugoniot relations in the case $A = J$, i.e. when B vanishes. For a thorough discussion of problems inherent in path-conservative schemes for the case $B \neq 0$ see Ref. 81. According to the DLM theory, the following generalized Rankine-Hugoniot relations hold across an isolated discontinuity propagating with speed σ:

$$\int_{s=0}^{1} \left(A(\mathbf{\Psi}(Q^-, Q^+, s)) \cdot \vec{n} - \sigma \mathbf{I} \right) \frac{\partial \mathbf{\Psi}}{\partial s} ds = 0, \quad (30)$$

where $\mathbf{I}$ is the unit matrix and $\mathbf{\Psi} = \mathbf{\Psi}(Q^-, Q^+, s)$ is a *path* that links the left state Q^- with the right state Q^+ at the discontinuity by a Lipschitz continuous function in phase-space. We have $0 \leq s \leq 1$ and $\mathbf{\Psi}(Q^-, Q^+, 0) = Q^-$ and $\mathbf{\Psi}(Q^-, Q^+, 1) = Q^+$. Here, we will use for the sake of simplicity the segment path

$$\mathbf{\Psi}(Q^-, Q^+, s) = Q^- + s \left(Q^+ - Q^- \right) . \quad (31)$$

If Eq. (1) is a conservation law, i.e. in the case $B = 0$, the generalized Rankine-Hugoniot relations (30) reduce to the classical ones, *independent* of the choice of the path $\mathbf{\Psi}$. In Refs. 48 and 49 it has been shown that path-conservative $P_N P_M$ schemes automatically reduce to $P_N P_M$ schemes for classical conservation laws presented in Ref. 1.

Using the framework of path-conservative schemes, we hence obtain the following family of fully discrete one-step $P_N P_M$ schemes for PDE (1):

$$\left[\Phi_k, u_h^{n+1}\right]_{T_i}^{t^{n+1}} - \left[\Phi_k, u_h^n\right]_{T_i}^{t^n} + \langle \Phi_k, \nabla F(q_h, \nabla q_h) + B(q_h) \cdot \nabla q_h \rangle_{T_i \backslash \partial T_i} + \left\{ \Phi_k, \mathcal{D}^-_{i+\frac{1}{2}}(q_h^-, \nabla q_h^-, q_h^+, \nabla q_h^+) \cdot \vec{n} \right\}_{\partial T_i} = \langle \Phi_k, S(q_h) \rangle_{T_i}, \tag{32}$$

where q_h^- and ∇q_h^- denote the boundary extrapolated data and gradient from within element T_i and q_h^+ and ∇q_h^+ denote the boundary extrapolated data and gradient from the neighbor, respectively. $\mathcal{D}^-_{i+\frac{1}{2}}$ is a simple Rusanov-type jump term, including the convective and the viscous terms as well as the non-conservative product, see Refs. 48 and 65:

$$\mathcal{D}^-_{i+\frac{1}{2}} \cdot \vec{n} = \frac{1}{2}\left(F(q_h^+, \nabla q_h^+) - F(q_h^-, \nabla q_h^-)\right) \cdot \vec{n} + \frac{1}{2}\left(\tilde{B} \cdot \vec{n} - s\mathbf{I}\right)\left(q_h^+ - q_h^-\right), \tag{33}$$

with

$$s = \left(|\lambda_A^{\max}| + 2\eta|\lambda_D^{\max}|\right), \tag{34}$$

and

$$\tilde{B} \cdot \vec{n} = \int_{s=0}^{1} B\left(\mathbf{\Psi}(q_h^-, q_h^+, s)\right) \cdot \vec{n} ds. \tag{35}$$

Here, $|\lambda_A^{\max}|$ is the maximum absolute value of the eigenvalues of the left and right matrices $A(q_h^-, \nabla q_h^-) \cdot \vec{n}$ and $A(q_h^+, \nabla q_h^+) \cdot \vec{n}$ and $|\lambda_D^{\max}|$ is the maximum absolute value of the eigenvalues of the two matrices $D_n(q_h^-, \nabla q_h^-)$ and $D_n(q_h^+, \nabla q_h^+)$. Following the ideas developed in Ref. 42 η can be computed from the solution of the generalized diffusive Riemann problem as

$$\eta = \frac{2N+1}{h\sqrt{\frac{1}{2}\pi}}, \tag{36}$$

where the characteristic length h is taken to be the twice the distance between the barycenter of the element and the barycenter of the edge/face for which the flux is to be computed.

In our practical implementation, the matrix $\tilde{B} \cdot \vec{n}$ is evaluated in a purely numerical way using a G-point Gauss-Legendre quadrature formula of appropriate degree, as proposed in Refs. 49,82, and 83, i.e. we approximate

$\tilde{B} \cdot \vec{n}$ as

$$\tilde{B} \cdot \vec{n} = \sum_{j=1}^{G} \omega_j B\left(\boldsymbol{\Psi}(q_h^-, q_h^+, s_j)\right) \cdot \vec{n}, \tag{37}$$

where s_j are the points of the Gaussian quadrature rule in the unit interval $I = [0;1]$ and ω_j are the associated weights. In Refs. 49, 82 and 83 it has been shown experimentally that the use of three Gaussian quadrature points is sufficient, i.e. we use $G = 3$ with

$$s_1 = \frac{1}{2}, \quad s_{2,3} = \frac{1}{2} \pm \frac{\sqrt{15}}{10}, \qquad \omega_1 = \frac{8}{18}, \quad \omega_{2,3} = \frac{5}{18}. \tag{38}$$

4.1. *Algorithm summary*

The fully-discrete $P_N P_M$ method described in detail previously can be put into the following abstract form:

- **Reconstruction step.** Recovery of piecewise polynomials w_h of degree $M \geq N$ from the original piecewise polynomials u_h of degree N at time level t_n, where $\mathcal{R}_h$ denotes the reconstruction operator,

$$w_h^n = \mathcal{R}_h\left(u_h^n\right). \tag{39}$$

  R_h reduces to the identity operator for pure DG schemes, where $N = M$ and hence $w_h^n = u_h^n$.
- **Predictor step or data evolution step.** Here, we compute a solution *in the small* of all the element-local Cauchy problems with initial condition given by the reconstruction polynomials w_h^n at time t_n:

$$q_h = \mathcal{E}_h\left(w_h^n\right), \tag{40}$$

  where $\mathcal{E}_h$ denotes the element-local space-time DG scheme given by Eq. (24).
- **Fully-discrete one-step time evolution.** We write Eq. (32) in the following short-hand notation:

$$u_h^{n+1} = u_h^n + \mathcal{P}_{\mathcal{N}}^{\mathcal{M}}\left(q_h, \nabla q_h\right). \tag{41}$$

Since the predictor solution q_h is a function of space *and* time, the time-integration contained in the operator $\mathcal{P}_{\mathcal{N}}^{\mathcal{M}}\left(q_h, \nabla q_h\right)$ of Eq. (41) can be carried out directly in one single step.

5. Applications

5.1. *Compressible Navier–Stokes equations*

5.1.1. *Governing PDE*

The three-dimensional compressible Navier–Stokes equations for a Newtonian fluid with heat conduction can be cast in a special form of (1) with $B(Q) = 0$ by defining the vector of conserved variables as

$$Q = (\rho, \rho\vec{v}, \rho E)\,, \tag{42}$$

and the flux tensor as

$$F(Q, \nabla Q) = \begin{pmatrix} \rho\vec{v}^T \\ \rho\vec{v}^T\vec{v} + \sigma(Q, \nabla Q) \\ \vec{v}^T(\mathbf{I}\rho E + \sigma(Q, \nabla Q)) - \kappa\nabla T \end{pmatrix}, \tag{43}$$

where the stress tensor σ is given under Stokes' hypothesis by

$$\sigma = \left(p + \frac{2}{3}\mu\nabla\cdot\vec{v}\right)\mathbf{I} - \mu\left(\nabla\vec{v} + \nabla\vec{v}^T\right). \tag{44}$$

To close the system, we use the equation of state of an ideal gas

$$\frac{p}{\rho} = RT, \qquad p = (\gamma - 1)\left(\rho E - \frac{1}{2}\rho\vec{v}^{\,2}\right), \tag{45}$$

and Sutherland's law for the viscosity

$$\mu(T) = \mu_0 \left(\frac{T}{T_0}\right)^{\beta} \frac{T_0 + s}{T + s}. \tag{46}$$

The heat conduction coefficient κ is linked to the viscosity by the Prandtl number Pr

$$\kappa = \frac{\mu\gamma c_v}{Pr}, \qquad \text{with} \qquad c_v = \frac{1}{\gamma - 1}R. \tag{47}$$

Here, γ denotes the ratio of the specific heats at constant pressure c_p and at constant volume c_v and R is the gas constant.

5.1.2. *Numerical convergence study*

To study the accuracy of our numerical method, we present the results of a test case carried out in Ref. 65. It consists of a test case with an artificial exact solution Q_e of (1) that is obtained by balancing the left hand side of (1) with a source term $S_e(\vec{x}, t)$ on the right hand side, i.e. we have

$$\frac{\partial Q_e}{\partial t} + \nabla \cdot F(Q_e, \nabla Q_e) = S_e(\vec{x}, t). \tag{48}$$

The exact solution of the problem is defined in terms of primitive variables $U = (\rho, \vec{v}, p)^T$ as

$$U_e = \left(\rho_b + \rho_0 \cos(\vec{k} \cdot \vec{x} - \omega t), \vec{v}_0 \sin(\vec{k} \cdot \vec{x} - \omega t), p_b + p_0 \sin(\vec{k} \cdot \vec{x} - \omega t)\right)^T . \tag{49}$$

From (49) we can compute Q_e and ∇Q_e and insert them into (48) in order to compute S_e, which is only a function of position $\vec{x}$ and time t. To test the accuracy of the schemes for rather viscous flows at low Reynolds numbers we set the constants in Sutherland's law (46) to $\mu_0 = 10^{-1}$, $s = 1$, $T_0 = 1$ and $\beta = 1.5$. The Prandtl number is fixed to the constant value $Pr = 0.7$, the ratio of specific heats is chosen as $\gamma = 1.4$ and the heat capacity at constant volume is chosen as $c_v = 1$. We solve (1) on the periodic computational domain $\Omega = [0; 10] \times [0; 10]$ until time $t_e = 0.5$. For the exact solution in primitive variables U_e given by (49) we use the constants $\rho_b = 1$, $p_b = \frac{1}{\gamma}$, $\rho_0 = 0.5$, $\vec{v}_0 = \frac{1}{4}(1,1)^T$, $p_0 = 0.1$, $\vec{k} = \frac{2\pi}{10}\,(1,1)^T$ and $\omega = 2\pi$. The source term S_e can then be easily computed using a computer algebra package. The results for all third to sixth order P_NP_M schemes are presented in Table 1, where the pure finite volume schemes $N = 0$ can be found on the left of the Table and the pure DG methods $N = M$ are on the diagonal. We note that the pure finite volume schemes reach an observable order of accuracy that is closer to $M + \frac{1}{2}$ rather than the optimal accuracy $M + 1$. This also seems to be true for some of the odd order DG methods. For the intermediate P_NP_M schemes $N > 0$, $M > N$, however, we always observe that the optimal order of accuracy $M + 1$ is reached, for odd as well as for even order schemes. The results of the presented convergence study seem to justify the choice of a rather simple viscous Rusanov-type flux (33) at least for the new intermediate class of P_NP_M schemes with $N > 0$ and $M > N$, rather than the use of the more sophisticated lifting operators as proposed in[35] or the local DG schemes.[37] The computations have been carried out on one core of an Intel Dual Core machine with 4 GB of RAM and 2.5 GHz clock speed. From the CPU times reported in Table 1 we can deduce that the new intermediate P_NP_M schemes are definitely more efficient than classical finite volume schemes of the same order and that in all cases they are also computationally cheaper than pure discontinuous Galerkin finite element schemes ($N = M$). The time step has been chosen in all our computations as

$$\Delta t = \frac{\text{CFL}}{2N+1} \cdot \frac{h}{|\lambda_c^{\max}| + 2|\lambda_v^{\max}| \frac{2N+1}{h}}, \tag{50}$$

which is consistent with the choice of η for the viscous part of the Rusanov flux, see Refs. 42 and 84. For a von-Neumann stability analysis of the general $P_N P_M$ schemes see Ref. 1 and for a stability analysis of the viscous flux see Refs. 42 and 84. For the compressible Navier–Stokes equations the maximum convective eigenvalue is $|\lambda_A^{\max}| = |\vec{v}| + c$ with the sound speed $c^2 = \gamma RT$ and the maximum viscous eigenvalue is $|\lambda_D^{\max}| = \max(\frac{4}{3}\frac{\mu}{\rho}, \frac{\gamma\mu}{Pr\rho})$.

5.1.3. *Laminar high Reynolds number boundary layer flow*

Here, we solve the compressible Navier–Stokes equations at low Mach number for the classical flow problem of a laminar but high Reynolds number flow past a flat plate, see Ref. 65 for more details about this test problem. The boundary layer equations for the flat plate with zero angle of attach read

$$f''' + ff'' = 0, \qquad h'' + fh' = 0, \tag{51}$$

and can be solved with any standard ODE solver. In this paper we use the time-DG ODE solver proposed in Ref. 65, since the method is only a special case of the local space-time Galerkin predictor approach given by Eq. (24). The setup is as follows: the computational domain is $\Omega = [-0.5; 2] \times [0; 0.05]$ and is discretized with 1430 triangular elements. At $y = 0$ we impose a solid, adiabatic wall boundary condition in the interval from $x = 0$ to $x = 2$. The chosen Reynolds number of $Re = 10^6$ is high and thus leads to a very thin boundary layer, which makes the use of heavily stretched meshes necessary. At $x = 1$ the aspect ratio of the triangles at the wall is $1 : 320$. We use $Pr = 1$, $\gamma = 1.4$ and a linear viscosity law with $\beta = 2$, $s = 0$ and $\mu_0 = 3 \cdot 10^{-7}$. The Mach number is chosen as $M_\infty = 0.3$ by setting $\rho_\infty = 1$, $u_\infty = 0.3$, $v_\infty = 0$ and $p_\infty = 1/\gamma$. The initial condition is given by the freestream data. We use a P_3P_5 scheme and let the method run towards a steady state, which is reached after about 500,000 timesteps. In Fig. 1 a zoom into the unstructured triangular mesh is shown, together with the contour levels of the horizontal velocity component u. In Fig. 2 a comparison with the Blasius solution is made for the velocity profile at $x = 0.7$ and for the skin friction coefficient from $x = 0$ to $x = 1$. An excellent agreement between the numerical solution and the Blasius reference solution can be noted, despite the heavily stretched mesh. This confirms the accuracy of the proposed reconstruction algorithm, which performs reconstruction in the *reference space* rather than in physical space. For more details on this topic see Ref. 23. The reference solution has been computed solving (51)

Table 1. Numerical convergence study of $P_N P_M$ schemes from third to sixth order of accuracy in space and time applied to the 2D compressible Navier–Stokes equations. Error norms refer to variable u and the CPU times for each method (printed in bold letters) are shown for the computation on the finest mesh.

N_G	L^2	$\mathcal{O}_{L^2}$	L^2	$\mathcal{O}_{L^2}$	L^2	$\mathcal{O}_{L^2}$	L^2	$\mathcal{O}_{L^2}$	L^2	$\mathcal{O}_{L^2}$	L^2	$\mathcal{O}_{L^2}$
$\mathcal{O}3$	P_0P_2		P_1P_2		P_2P_2							
24 /16	5.11E-03		2.12E-03		1.35E-03							
32 /24	2.31E-03	2.8	6.19E-04	3.0	3.24E-04	3.5						
64 /32	3.35E-04	2.8	2.65E-04	3.0	1.35E-04	3.0						
128/64	5.70E-05	2.6	3.31E-05	3.0	2.24E-05	2.6						
CPU	**3011s**		**1355s**		**3621s**							
$\mathcal{O}4$	P_0P_3		P_1P_3		P_2P_3		P_3P_3					
24 /16	1.10E-03		1.26E-03		3.04E-04		1.67E-04					
32 /24	3.61E-04	3.9	2.59E-04	3.9	5.93E-05	4.0	3.20E-05	4.1				
64 /32	2.77E-05	3.7	8.76E-05	3.8	1.89E-05	4.0	1.04E-05	3.9				
128/64	2.49E-06	3.5	5.24E-06	4.1	1.09E-06	4.1	6.62E-07	4.0				
CPU	**5279s**		**2303s**		**6224s**		**12910s**					
$\mathcal{O}5$	P_0P_4		P_1P_4		P_2P_4		P_3P_4		P_4P_4			
24 / 8	6.13E-04		5.74E-03		2.14E-03		8.21E-04		5.17E-04			
32 /16	1.58E-04	4.7	1.93E-04	4.9	7.88E-05	4.8	2.74E-05	4.9	1.34E-05	5.3		
64 /24	5.25E-06	4.9	2.67E-05	4.9	1.19E-05	4.7	3.76E-06	4.9	1.38E-06	5.6		
128/32	2.14E-07	4.6	7.07E-06	4.6	2.84E-06	5.0	8.90E-07	5.0	2.88E-07	5.5		
CPU	**12532s**		**293s**		**751s**		**1842s**		**2965s**			
$\mathcal{O}6$	P_0P_5		P_1P_5		P_2P_5		P_3P_5		P_4P_5		P_5P_5	
24 / 4	1.45E-04		1.07E-02		1.97E-02		1.07E-02		4.26E-03		3.20E-03	
32 / 8	2.89E-05	5.6	3.05E-04	5.1	7.55E-04	4.7	3.05E-04	5.1	1.10E-04	5.3	8.19E-05	5.3
64 /16	5.12E-07	5.8	6.43E-06	5.6	1.76E-05	5.4	6.43E-06	5.6	1.58E-06	6.1	9.03E-07	6.5
128/24	1.21E-08	5.4	5.79E-07	5.9	1.68E-06	5.8	5.79E-07	5.9	1.26E-07	6.2	6.31E-08	6.6
CPU	**16267s**		**215s**		**558s**		**1057s**		**1719s**		**2498s**	

with a P_9 time-DG method using a time step of $\Delta\eta = 10^{-3}$ and a classical shooting algorithm to fit the boundary conditions $f(0) = 0$, $f'(0) = 0$, $f'(\infty) = 1$, $h(0) = 1$ and $h(\infty) = 0$.

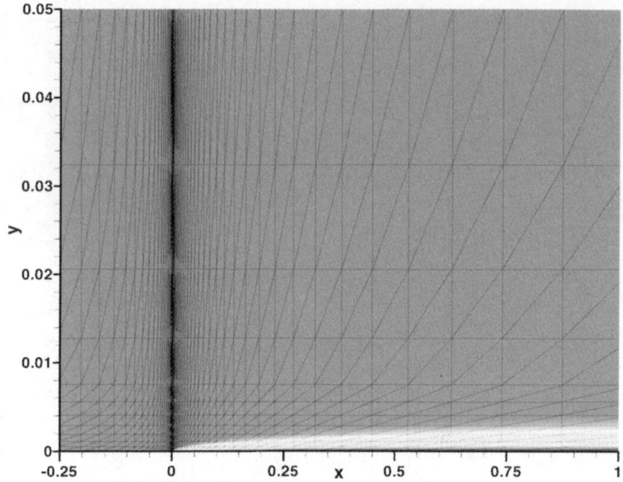

Fig. 1. High Reynolds number computation of a laminar boundary layer over a flat plate ($\alpha = 0°$, $Re = 10^6$, $M_\infty = 0.3$): Unstructured triangular mesh with color contours of the horizontal velocity component u.

5.1.4. *Compressible mixing layer in 2D*

The previous test cases were either stationary or had a very simple analytical solution. In this section we consider a rather complex test problem that was proposed by Colonius *et al.*[85] in two space dimension and that was successively extended to three space dimensions by Babucke *et al.* in.[86] It concerns the time-dependent flow of a compressible mixing layer. The upper horizontal velocity is $u_\infty = 0.5$ and the lower one is $u_{-\infty} = 0.25$. The velocity ratio λ is defined as $\lambda = u_\infty/u_{-\infty}$. The free stream density and pressure are $\rho_\infty = \rho_{-\infty} = 1$ and $p_\infty = p_{-\infty} = 1/\gamma$, respectively, with $\gamma = 1.4$. The vorticity thickness at the inflow, with respect to which all lengths are made dimensionless, is

$$\delta(x_0) = \frac{u_\infty - u_{-\infty}}{\max\left(\left.\frac{\partial u}{\partial y}\right|_{x=x_0} \right)} := 1, \tag{52}$$

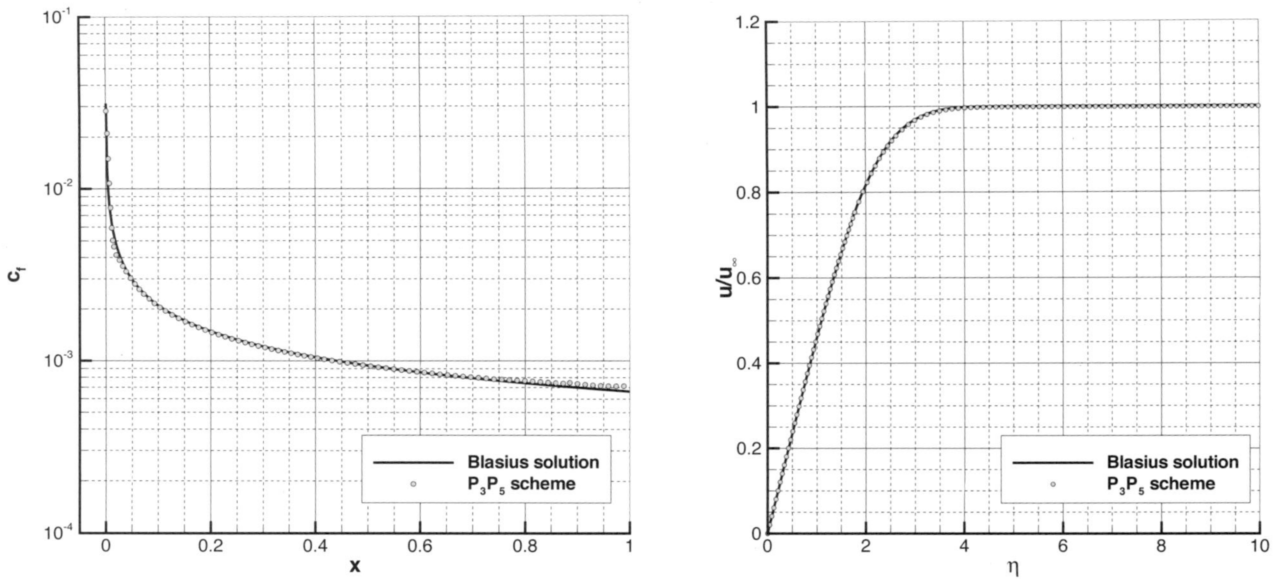

Fig. 2. High Reynolds number computation of a laminar boundary layer over a flat plate ($\alpha = 0°$, $Re = 10^6$, $M_\infty = 0.3$): Distribution of the skin friction coefficient c_f (left) and velocity profile at $x = 0.7$ (right) compared with the Blasius solution.

and the Reynolds number based on this vorticity thickness is

$$Re_\delta = \frac{\rho_\infty u_\infty \delta(x_0)}{\mu_\infty} = 500. \qquad (53)$$

Again, a linear viscosity law with $\beta = 2$, $s = 0$ and $\mu_0 = \mu_\infty$ is chosen and the Prandtl number is $Pr = 1$. With this choice for the viscosity law, the initial condition of the problem is given by the solution of the boundary layer equations (51), see Ref. 65 for more details. The flow is perturbed at the inflow using perturbations that come from a linear stability analysis, see Refs. 85 and 86 for more details, where either the inviscid Rayleigh equations or the full viscous Orr–Sommerfeld equations have been solved. Both waveforms of the perturbations — inviscid and viscous — have been reported in Ref. 86 and the inviscid ones have been used in our simulation. The fundamental angular frequency of the mixing layer is $\omega_0 = 0.0501 \cdot 2\pi = 0.3147876$ and the perturbations are a linear superposition of the fundamental frequency ω_0 and the first three subharmonics $\omega_0/2$, $\omega_0/4$ and $\omega_0/8$. According to Colonius *et al.*[85] the phases of the three subharmonics are shifted by -0.028, 0.141 and 0.391 with respect to the fundamental perturbation so that the distance between the pairings is minimized.

In this test case we compare three representative sixth order accurate $P_N P_M$ schemes with each other, namely the pure finite volume scheme P_0P_5, the pure DG method P_5P_5 and the new intermediate class of schemes on the example of a P_3P_5 method, i.e. a scheme that uses piecewise cubic data representation and piecewise quintic polynomials for the flux computations. The computational domain is $\Omega = [0; 800] \times [-320; 320]$, discretized with 395158 triangular elements in the finite volume case (P_0P_5) and with 50306 triangular elements in the case of the other schemes. The characteristic mesh spacing at $y = 0$ is $h = 0.15$ for the finite volume scheme (as in Refs. 85 and 86) and $h = 0.6$ for the other methods. Computations are performed up to a final time of $t_e = 1596.8 = 80T_f$ which corresponds to 80 forcing periods T_f of the fundamental frequency.

The wallclock times necessary to perform the computation on 128 CPUs of the HLRB2 supercomputer of the *Leibniz Rechenzentrum* in München, Germany were: P_0P_5 14.75 h, P_5P_5 8 h and P_3P_5 5 h. The vorticity contours obtained with all three schemes at time $t = 68T_f$ are shown in Fig. 4 and are compared with the reference solution of Colonius *et al.*[85] An excellent agreement of all three of our solutions with the one of Colonius *et al.* can be noted. For a comparison with the reference solution of Babucke *et al.* see.[65] The time histories of the horizontal velocity component u in four of

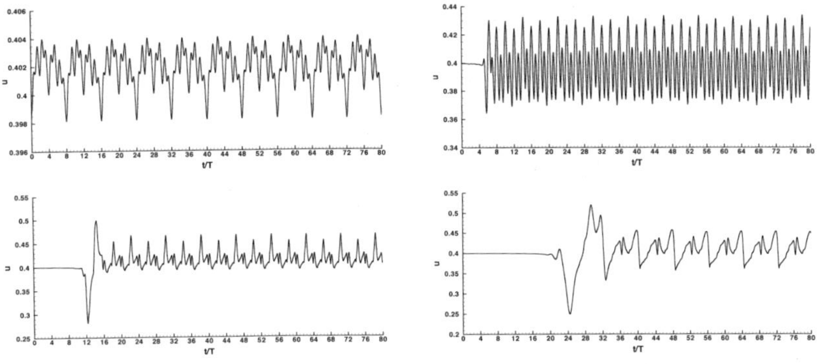

Fig. 3. Temporal signals of the horizontal velocity u on the x-axis at positions $x = 0$, $x = 45$, $x = 100$ and $x = 200$.

the observation points specified by Ref. 85 are depicted in Fig. 3 and agree qualitatively well with the ones shown in Ref. 85.

5.2. *The Baer–Nunziato model of compressible multi-phase flow*

5.2.1. *Governing PDE*

A Baer–Nunziato type model for compressible two-phase flow with interphase drag and pressure relaxation is given by the following system of equations, see:[87–89]

$$\left.\begin{aligned}
&\frac{\partial}{\partial t}\left(\phi_1\rho_1\right) + \nabla\cdot\left(\phi_1\rho_1\vec{v}_1\right) = 0,\\
&\frac{\partial}{\partial t}\left(\phi_1\rho_1\vec{v}_1\right) + \nabla\cdot\left(\phi_1\rho_1\vec{v}_1^{T}\vec{v}_1\right) + \nabla\phi_1 p_1 = p_I\nabla\phi_1 - \lambda\left(\vec{v}_1 - \vec{v}_2\right),\\
&\frac{\partial}{\partial t}\left(\phi_1\rho_1 E_1\right) + \nabla\cdot\left(\left(\phi_1\rho_1 E_1 + \phi_1 p_1\right)\vec{v}_1\right) = -p_I\partial_t\phi_1 - \lambda\,\vec{v}_I\cdot\left(\vec{v}_1 - \vec{v}_2\right),\\
&\frac{\partial}{\partial t}\left(\phi_2\rho_2\right) + \nabla\cdot\left(\phi_2\rho_2\vec{v}_2\right) = 0,\\
&\frac{\partial}{\partial t}\left(\phi_2\rho_2\vec{v}_2\right) + \nabla\cdot\left(\phi_2\rho_2\vec{v}_2^{T}\vec{v}_2\right) + \nabla\phi_2 p_2 = p_I\nabla\phi_2 - \lambda\left(\vec{v}_2 - \vec{v}_1\right),\\
&\frac{\partial}{\partial t}\left(\phi_2\rho_2 E_2\right) + \nabla\cdot\left(\left(\phi_2\rho_2 E_2 + \phi_2 p_2\right)\vec{v}_2\right) = p_I\partial_t\phi_1 - \lambda\,\vec{v}_I\cdot\left(\vec{v}_2 - \vec{v}_1\right),\\
&\frac{\partial}{\partial t}\phi_1 + \vec{v}_I\nabla\phi_1 = \mu(p_1 - p_2).
\end{aligned}\right\} \tag{54}$$

This system has first been solved by $P_N P_M$ schemes in Ref. 49. We use the so-called stiffened equation of state (EOS) for each phase:

$$e_k = \frac{p_k + \gamma_k\pi_k}{\rho_k(\gamma_k - 1)}. \tag{55}$$

Here, ϕ_k denotes the volume fraction of phase k, ρ_k is the density, $\vec{v}_k$ is the velocity vector, $E_k = e_k + \frac{1}{2}\vec{v}_k^2$ and e_k are the phase specific total and

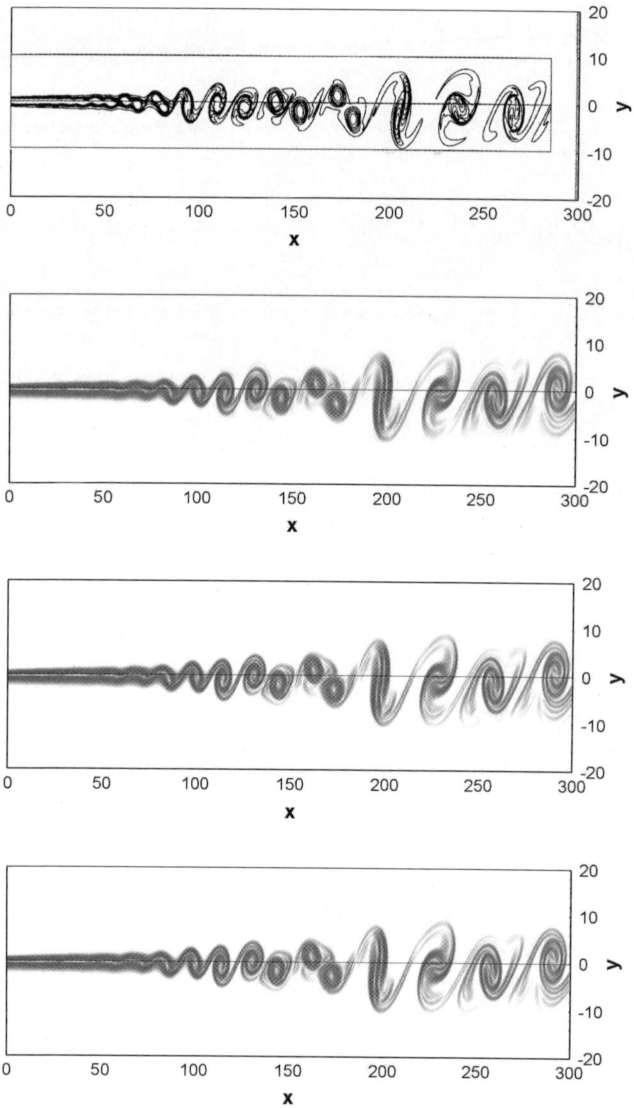

Fig. 4. Spanwise vorticity contours at time $t/T_f = 68$. Row 1: Reference solution of Colonius *et al.*[85] Row 2: P_0P_5 scheme. Row 3: P_3P_5 scheme. Row 4: P_5P_5 scheme.

internal energies, respectively. The parameters λ and μ take into account friction and pressure relaxation, respectively, between the two phases. For the interface velocity and pressure $\vec{v}_I$ and p_I we choose $\vec{v}_I = \vec{v}_1$ and $p_I = p_2$ respectively, according to Ref. 87, although other choices are possible, see e.g. the paper by Saurel and Abgrall.[88] The vector of state Q is

$$Q = (\phi_1\rho_1,\, \phi_1\rho_1\vec{v}_1,\, \phi_1\rho_1E_1,\, \phi_2\rho_2,\, \phi_2\rho_2\vec{v}_2,\, \phi_2\rho_2E_2,\, \phi_1)\,. \tag{56}$$

System (54) is already written in the general form of Eq. (1).

5.2.2. *A spherical explosion problem*

The initial condition of this test problem is

$$(\rho_1, \vec{v}_1, p_1, \rho_2, \vec{v}_2, p_2, \phi_1) = \begin{cases} (800, 0, 500, 1.5, 0, 2, 0.4) & \text{if } r \leq 0.5, \\ (1000, 0, 600, 1.0, 0, 1, 0.3) & \text{if } r > 0.5, \end{cases} \tag{57}$$

with the following parameters: $\gamma_1 = 3$, $\pi_1 = 100$, $\gamma_2 = 1.4$, $\pi_2 = 0$, $\lambda = \mu = 0$. The 3D computational domain is composed of a half-sphere with radius $R = 0.8$ in the half-space $x > 0$. A characteristic mesh spacing of $h = 1/130$ is used, which leads to a mesh containing 9,446,328 tetrahedrons, a segment of which is depicted in Fig. 5. We use a third order P_0P_2 WENO scheme, see Refs. 22 and 23 for details. The computation has been performed on 510 CPUs and took about 12h wall-clock time. A reference solution has been computed solving a reduced 1D system with geometric reaction source terms, see Ref. 49 for details. A comparison between the 3D computation and the reference solution is shown in Fig. 6. We observe

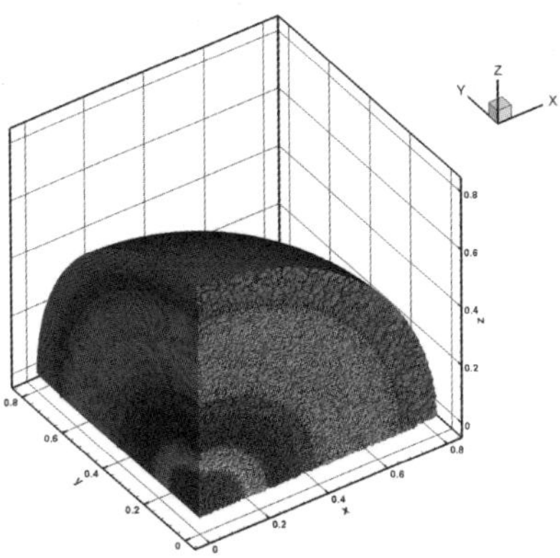

Fig. 5. Unstructured tetrahedral mesh for the 3D explosion problem. Only the segment with $x > 0$, $y > 0$ and $z > 0$ is shown. The contour colors represent ρ_g at $t = 0.18$.

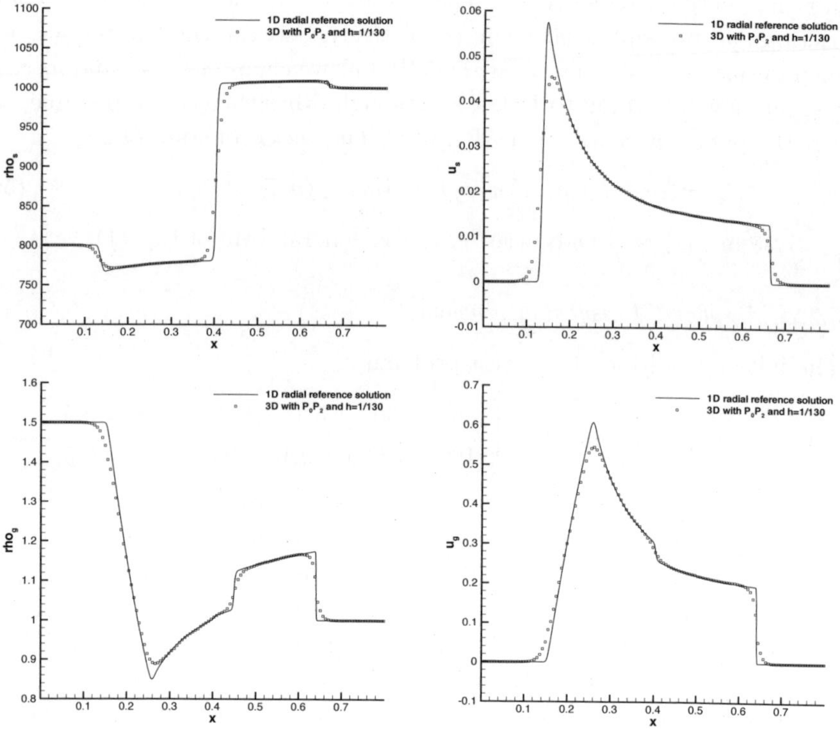

Fig. 6. Results for the 3D explosion problem. A cut along the x-axis is shown together with the 1D radial reference solution.

a very good agreement between the numerical and the reference solution and all flow features are reasonably well resolved.

6. Concluding Remarks

The $P_N P_M$ method is a new family of Godunov-type schemes which include classical finite volume as well as high order DG finite element schemes in a unified framework. The method is applicable to very general time-dependent partial differential equations of the form (1) that may contain at the same time viscous terms, non-conservative products as well as stiff source terms. In this chapter we have shown applications to the compressible Navier–Stokes equations as well as to the fully three-dimensional Baer–Nunziato model of compressible multi-phase flows.

For further applications, in particular to MHD equations, to geophysical flows or to the equations of nonlinear elasticity, the reader is referred to the literature.[1,48,49,64,66]

Future extensions will concern the introduction of general equations of state, as well as a jump term $D^{-}_{i+\frac{1}{2}}$ that is based on a *complete* Riemann solver, as recently proposed in Ref. 90. Further research will also be devoted to limiters for the general class of $P_N P_M$ schemes on unstructured meshes as well as to the introduction of a time-accurate local time-stepping scheme as used in Refs. 43, 84, 91 and 92.

References

1. M. Dumbser, D. S. Balsara, E. F. Toro, and C. D. Munz, A unified framework for the construction of one–step finite–volume and discontinuous Galerkin schemes, *Journal of Computational Physics.* **227**, 8209–8253, (2008).
2. E. Noether, Invariante Variationsprobleme, *Nachrichten von der Gesellschaft der Wissenschaften zu Göttingen, mathematisch-physikalische Klasse.* pp. 235–257, (1918).
3. B. Riemann, Über die Fortpflanzung ebener Luftwellen von endlicher Schwingungsweite, *Abhandlungen der Königlichen Gesellschaft der Wissenschaften zu Göttingen.* **8**, 43–65, (1860).
4. S. K. Godunov, Finite difference methods for the computation of discontinuous solutions of the equations of fluid dynamics, *Mathematics of the USSR: Sbornik.* **47**, 271–306, (1959).
5. P. L. Roe, Approximate Riemann solvers, parameter vectors, and difference schemes, *Journal of Computational Physics.* **43**, 357–372, (1981).
6. A. Harten, P. D. Lax, and B. van Leer, On upstream differencing and godunov-type schemes for hyperbolic conservation laws, *SIAM Review.* **25** (1), 35–61, (1983).
7. S. Osher and F. Solomon, Upwind difference schemes for hyperbolic conservation laws, *Math. Comput.* **38**, 339–374, (1982).
8. B. Einfeldt, C. D. Munz, P. L. Roe, and B. Sjögreen, On godunov-type methods near low densities, *Journal of Computational Physics.* **92**, 273–295, (1991).
9. E. F. Toro, M. Spruce, and W. Speares, Restoration of the contact surface in the Harten-Lax-van Leer Riemann solver, *Journal of Shock Waves.* **4**, 25–34, (1994).
10. E. F. Toro, *Riemann Solvers and Numerical Methods for Fluid Dynamics.* (Springer, 2009), third edition.
11. V. P. Kolgan, Application of the minimum-derivative principle in the construction of finite-difference schemes for numerical analysis of discontinuous solutions in gas dynamics, *Transactions of the Central Aerohydrodynamics Institute.* **3**(6), 68–77, (1972). in Russian.

12. B. van Leer, Towards the ultimate conservative difference scheme V: A second order sequel to Godunov's method, *Journal of Computational Physics.* **32**, 101–136, (1979).
13. A. Harten, B. Engquist, S. Osher, and S. Chakravarthy, Uniformly high order accurate essentially non-oscillatory schemes III, *Journal of Computational Physics.* **71**, 231–303, (1987).
14. G. S. Jiang and C. W. Shu, Efficient implementation of weighted ENO schemes, *Journal of Computational Physics.* **126**, 202–228, (1996).
15. T. Barth and D. Jespersen, The design and application of upwind schemes on unstructured meshes, *AIAA Paper 89-0366.* pp. 1–12, (1989).
16. T. J. Barth and P. O. Frederickson, Higher order solution of the Euler equations on unstructured grids using quadratic reconstruction, *AIAA paper no. 90-0013* (28th Aerospace Sciences Meeting January. 1990).
17. R. Abgrall, On essentially non-oscillatory schemes on unstructured meshes: analysis and implementation, *Journal of Computational Physics.* **144**, 45–58, (1994).
18. T. Sonar, On the construction of essentially non-oscillatory finite volume approximations to hyperbolic conservation laws on general triangulations: polynomial recovery, accuracy and stencil selection, *Computer Methods in Applied Mechanics and Engineering.* **140**, 157–181, (1997).
19. C. Hu and C. W. Shu, Weighted essentially non-oscillatory schemes on triangular meshes, *Journal of Computational Physics.* **150**, 97–127, (1999).
20. O. Friedrich, Weighted essentially non-oscillatory schemes for the interpolation of mean values on unstructured grids, *Journal of Computational Physics.* **144**, 194–212, (1998).
21. C. Ollivier-Gooch and M. Van Altena, A high-order-accurate unstructured mesh finite-volume scheme for the advection-diffusion equation, *Journal of Computational Physics.* **181**, 729–752, (2002).
22. M. Dumbser and M. Käser, Arbitrary high order non-oscillatory finite volume schemes on unstructured meshes for linear hyperbolic systems, *Journal of Computational Physics.* **221**, 693–723, (2007).
23. M. Dumbser, M. Käser, V. A. Titarev, and E. F. Toro, Quadrature-free non-oscillatory finite volume schemes on unstructured meshes for nonlinear hyperbolic systems, *Journal of Computational Physics.* **226**, 204–243, (2007).
24. Y. T. Zhang and C. W. Shu, Third order WENO scheme on three dimensional tetrahedral meshes, *Communications in Computational Physics.* **5**, 836–848, (2009).
25. Z. Wang, Spectral finite volume method for conservation laws on unstructured grids - basic formulation, *Journal of Computational Physics.* **178**, 210–251, (2002).
26. Y. Liu, M. Vinokur, and Z. Wang, Spectral finite volume method for conservation laws on unstructured grids V: Extension to three-dimensional systems, *Journal of Computational Physics.* **212**, 454–472, (2006).
27. W. Reed and T. Hill. Triangular mesh methods for neutron transport equation. Technical Report LA-UR-73-479, Los Alamos Scientific Laboratory, (1973).

28. P. Lesaint and P. Raviart. On a finite element method for solving the neutron transport equation. In ed. C. de Boor, *Mathematical Aspects of Finite Elements in Partial Differential Equations*, pp. 89–145. Academic Press, New York, (1974).
29. B. Cockburn and C. W. Shu, The Runge-Kutta local projection P1-Discontinuous Galerkin finite element method for scalar conservation laws, *Mathematical Modelling and Numerical Analysis.* **25**, 337–361, (1991).
30. B. Cockburn and C. W. Shu, TVB Runge-Kutta local projection discontinuous Galerkin finite element method for conservation laws II: general framework, *Mathematics of Computation.* **52**, 411–435, (1989).
31. B. Cockburn, S. Y. Lin, and C. W. Shu, TVB Runge-Kutta local projection discontinuous Galerkin finite element method for conservation laws III: one dimensional systems, *Journal of Computational Physics.* **84**, 90–113, (1989).
32. B. Cockburn, S. Hou, and C. W. Shu, The Runge-Kutta local projection discontinuous Galerkin finite element method for conservation laws IV: the multidimensional case, *Mathematics of Computation.* **54**, 545–581, (1990).
33. B. Cockburn and C. W. Shu, The Runge-Kutta discontinuous Galerkin method for conservation laws V: multidimensional systems, *Journal of Computational Physics.* **141**, 199–224, (1998).
34. G. Jiang and C. W. Shu, On a cell entropy inequality for discontinuous Galerkin methods, *Mathematics of Computation.* **62**, 531–538, (1994).
35. F. Bassi and S. Rebay, A high-order accurate discontinuous finite element method for the numerical solution of the compressible Navier-Stokes equations, *Journal of Computational Physics.* **131**, 267–279, (1997).
36. C. E. Baumann and T. J. Oden, A discontinuous hp finite element method for the Euler and the Navier–Stokes equations, *International Journal for Numerical Methods in Fluids.* **31**, 79–95, (1999).
37. B. Cockburn and C. W. Shu, The local discontinuous Galerkin method for time-dependent convection diffusion systems, *SIAM Journal on Numerical Analysis.* **35**, 2440–2463, (1998).
38. B. Cockburn and C. W. Shu, Runge-Kutta discontinuous Galerkin methods for convection-dominated problems, *Journal of Scientific Computing.* **16**, 173–261, (2001).
39. J. Yan and C. W. Shu, A local discontinuous Galerkin method for KdV-type equations, *SIAM Journal on Numerical Analysis.* **40**, 769–791, (2002).
40. D. Levy, C. W. Shu, and J. Yan, Local discontinuous Galerkin methods for nonlinear dispersive equations, *Journal of Computational Physics.* **196**, 751–772, (2004).
41. D. Arnold, F. Brezzi, B. Cockburn, and L. Marini, Unified analysis of discontinuous galerkin methods for elliptic problems, *SIAM Journal on Numerical Analysis.* **39**, 1749–1779, (2002).
42. G. Gassner, F. Lörcher, and C. D. Munz, A contribution to the construction of diffusion fluxes for finite volume and discontinuous Galerkin schemes, *Journal of Computational Physics.* **224**, 1049–1063, (2007).
43. F. Lörcher, G. Gassner, and C.-D. Munz, An explicit discontinuous Galerkin scheme with local time-stepping for general unsteady diffusion equations, *J. Comput. Phys.* **227**(11), 5649–5670, (2008).

44. R. Hartmann and P. Houston, Symmetric interior penalty DG methods for the compressible navier–stokes equations I: Method formulation, *Int. J. Num. Anal. Model.* **3**, 1–20, (2006).
45. R. Hartmann and P. Houston, An optimal order interior penalty discontinuous galerkin discretization of the compressible navierstokes equations, *Journal of Computational Physics.* **227**, 9670–9685, (2008).
46. E. F. Toro and A. Hidalgo, ADER finite volume schemes for nonlinear reaction-diffusion equations , *Applied Numerical Mathematics.* **59**, 73–100, (2009).
47. S. Rhebergen, O. Bokhove, and J. van der Vegt, Discontinuous Galerkin finite element methods for hyperbolic nonconservative partial differential equations, *Journal of Computational Physics.* **227**, 1887–1922, (2008).
48. M. Dumbser, M. Castro, C. Parés, and E. F. Toro, ADER schemes on unstructured meshes for nonconservative hyperbolic systems: Applications to geophysical flows, *Computers and Fluids.* **38**, 1731–1748, (2009).
49. M. Dumbser, A. Hidalgo, M. Castro, C. Parés, and E. F. Toro, FORCE schemes on unstructured meshes II: Non–conservative hyperbolic systems, *Computer Methods in Applied Mechanics and Engineering.* **199**, 625–647, (2010).
50. C. W. Shu and S. Osher, Efficient implementation of essentially non-oscillatory shock capturing schemes, *Journal of Computational Physics.* **77**, 439–471, (1988).
51. C. W. Shu and S. Osher, Efficient implementation of essentially non-oscillatory shock capturing schemes II, *Journal of Computational Physics.* **83**, 32–78, (1989).
52. J. Qiu, M. Dumbser, and C. W. Shu, The discontinuous Galerkin method with Lax-Wendroff type time discretizations, *Computer Methods in Applied Mechanics and Engineering.* **194**, 4528–4543, (2005).
53. M. Dumbser and C. D. Munz, Building blocks for arbitrary high order discontinuous Galerkin schemes, *Journal of Scientific Computing.* **27**, 215–230, (2006).
54. A. Taube, M. Dumbser, D. S. Balsara, and C. D. Munz, Arbitrary high order discontinuous Galerkin schemes for the magnetohydrodynamic equations, *Journal of Scientific Computing.* **30**, 441–464, (2007).
55. T. Warburton and G. Karniadakis, A Discontinuous Galerkin Method for the Viscous MHD Equations, *Journal of Computational Physics.* **152**, 608–641, (1999).
56. J. J. W. van der Vegt and H. van der Ven, Space–time discontinuous Galerkin finite element method with dynamic grid motion for inviscid compressible flows I. general formulation, *Journal of Computational Physics.* **182**, 546–585, (2002).
57. H. van der Ven and J. J. W. van der Vegt, Space–time discontinuous Galerkin finite element method with dynamic grid motion for inviscid compressible flows II. efficient flux quadrature, *Comput. Methods Appl. Mech. Engrg.* **191**, 4747–4780, (2002).

58. J. Sudirham, J. van der Vegt, and R. van Damme, Space–time discontinuous Galerkin method for advection–diffusion problems on time–dependent domains, *Applied Numerical Mathematics.* **56**, 1491–1518, (2006).
59. C. Klaij, J. V. der Vegt, and H. V. der Ven, Space–time discontinuous Galerkin method for the compressible Navier–Stokes equations, *Journal of Computational Physics.* **217**, 589–611, (2006).
60. B. Cockburn, M. Luskin, C. W. Shu, and E. Suli, Enhanced accuracy by post-processing for finite element methods for hyperbolic equations, *Mathematics of Computation.* **72**, 577–606, (2003).
61. J. Ryan, C. W. Shu, and H. Atkins, Extension of a post-processing technique for the discontinuous Galerkin method for hyperbolic equations with applications to an aeroacoustic problem, *SIAM Journal on Scientific Computing.* **26**, 821–843, (2005).
62. M. Dumbser, *Arbitrary High Order Schemes for the Solution of Hyperbolic Conservation Laws in Complex Domains.* (Shaker Verlag, Aachen, 2005).
63. M. Dumbser and C. D. Munz. Arbitrary high order Discontinuous Galerkin schemes. In eds. S. Cordier, T. Goudon, M. Gutnic, and E. Sonnendrucker, *Numerical Methods for Hyperbolic and Kinetic Problems*, IRMA Series in Mathematics and Theoretical Physics, pp. 295–333. EMS Publishing House, (2005).
64. M. Dumbser and O. Zanotti, Very high order PNPM schemes on unstructured meshes for the resistive relativistic mhd equations, *Journal of Computational Physics.* **228**, 6991–7006, (2009).
65. M. Dumbser, Arbitrary high order PNPM schemes on unstructured meshes for the compressible Navier–Stokes equations, *Computers & Fluids.* **39**, 60–76, (2010).
66. M. Dumbser and D. S. Balsara, Unstructured high-order one-step PNPM schemes for the viscous and resistive MHD equations, *Computer Modelling in Engineering & Sciences.* **54**, 301–333, (2009).
67. B. van Leer and S. Nomura. Discontinuous Galerkin for diffusion. In *Proceedings of 17th AIAA Computational Fluid Dynamics Conference (June 6–9 2005), AIAA-2005-5108*, (2005).
68. M. van Raalte and B. van Leer, Bilinear forms for the recovery-based discontinuous Galerkin method for diffusion, *Communications in Computational Physics.* **5**, 683–693, (2009).
69. J. Qiu and C. W. Shu, Hermite WENO schemes and their application as limiters for Runge-Kutta discontinuous Galerkin method: one-dimensional case, *Journal of Computational Physics.* **193**, 115–135, (2003).
70. J. Qiu and C. W. Shu, Hermite WENO schemes and their application as limiters for Runge-Kutta discontinuous Galerkin method II: two dimensional case, *Computers and Fluids.* **34**, 642–663, (2005).
71. D. S. Balsara, C. Altmann, C. D. Munz, and M. Dumbser, A sub-cell based indicator for troubled zones in RKDG schemes and a novel class of hybrid RKDG+HWENO schemes, *Journal of Computational Physics.* **226**, 586–620, (2007).

72. H. Luo, J. Baum, and R. Löhner, A Hermite WENO-based limiter for discontinuous Galerkin method on unstructured grids, *Journal of Computational Physics*. **225**, 686–713, (2007).
73. M. Dumbser, C. Enaux, and E. F. Toro, Finite volume schemes of very high order of accuracy for stiff hyperbolic balance laws, *Journal of Computational Physics*. **227**, 3971–4001, (2008).
74. A. Stroud, *Approximate Calculation of Multiple Integrals*. (Prentice-Hall Inc., Englewood Cliffs, New Jersey, 1971).
75. N. Petrovskaya, Discontinuous Weighted Least-Squares Approximation on Irregular Grids, *CMES - Computer Modeling in Engineering & Sciences*. **32**, 69–84, (2008).
76. D. S. Balsara, T. Rumpf, M. Dumbser, and C. D. Munz, Efficient, high accuracy ADER-WENO schemes for hydrodynamics and divergence-free magnetohydrodynamics, *Journal of Computational Physics*. **228**, 2480–2516, (2009).
77. I. Toumi, A weak formulation of roe's approximate riemann solver, *Journal of Computational Physics*. **102**, 360–373, (1992).
78. C. Parés, Numerical methods for nonconservative hyperbolic systems: a theoretical framework, *SIAM Journal on Numerical Analysis*. **44**, 300–321, (2006).
79. M. Castro, J. Gallardo, and C. Parés, High-order finite volume schemes based on reconstruction of states for solving hyperbolic systems with nonconservative products. applications to shallow-water systems, *Mathematics of Computation*. **75**, 1103–1134, (2006).
80. G. D. Maso, P. LeFloch, and F. Murat, Definition and weak stability of nonconservative products, *J. Math. Pures Appl.* **74**, 483–548, (1995).
81. M. Castro, P. LeFloch, M. Muñoz-Ruiz, and C. Parés, Why many theories of shock waves are necessary: Convergence error in formally path-consistent schemes, *Journal of Computational Physics*. **227**, 8107–8129, (2008).
82. A. Canestrelli, A. Siviglia, M. Dumbser, and E. F. Toro, A well-balanced high order centered scheme for nonconservative systems: Application to shallow water flows with fix and mobile bed, *Advances in Water Resources*. **32**, 834–844, (2009).
83. A. Canestrelli, M. Dumbser, A. Siviglia, and E. F. Toro, Well-balanced high-order centered schemes on unstructured meshes for shallow water equations with fixed and mobile bed, *Advances in Water Resources*. **33**, 291–303, (2010).
84. G. Gassner, F. Lörcher, and C. D. Munz, A discontinuous Galerkin scheme based on a space-time expansion II. viscous flow equations in multi dimensions., *Journal of Scientific Computing*. **34**, 260–286, (2008).
85. T. Colonius, S. K. Lele, and P. Moin, Sound generation in a mixing layer, *Journal of Fluid Mechanics*. **330**, 375–409, (1997).
86. A. Babucke, M. Kloker, and U. Rist, DNS of a plane mixing layer for the investigation of sound generation mechanisms, *Computers and Fluids*. **37**, 360–368, (2008).

87. M. Baer and J. Nunziato, A two-phase mixture theory for the deflagration-to-detonation transition (DDT) in reactive granular materials, *J. Multiphase Flow.* **12**, 861–889, (1986).
88. R. Saurel and R. Abgrall, A multiphase godunov method for compressible multifluid and multiphase flows, *Journal of Computational Physics.* **150**, 425–467, (1999).
89. A. Murrone and H. Guillard, A five equation reduced model for compressible two phase flow problems, *Journal of Computational Physics.* **202**, 664–698, (2005).
90. M. Dumbser and E. F. Toro, A simple extension of the Osher Riemann solver to non-conservative hyperbolic systems, *Journal of Scientific Computing.* submitted to.
91. M. Dumbser, M. Käser, and E. F. Toro, An arbitrary high order discontinuous Galerkin method for elastic waves on unstructured meshes V: Local time stepping and p-adaptivity, *Geophysical Journal International.* **171**, 695–717, (2007).
92. A. Taube, M. Dumbser, C. D. Munz, and R. Schneider, A high order discontinuous Galerkin method with local time stepping for the Maxwell equations, *Int. J. of Numerical Modelling: Electronic Networks, Devices and Fields.* **22**, 77–103, (2009).

CHAPTER 9

HIGH-ORDER FINITE-VOLUME DISCRETIZATION OF THE EULER EQUATIONS ON UNSTRUCTURED MESHES

Ollivier-Gooch* and Michalak†

Carl Ollivier-Gooch and Chris Michalak,
Advanced Numerical Simulation Laboratory,
The University of British Columbia,
6250 Applied Science Lane, Vancouver, BC V6T 1Z4, Canada
** cfog@mech.ubc.ca*
† chris@quantalgo.com

High-order accurate methods are intended to produce more accurate solutions to complex problems for given computing resources. This chapter describes solutions to two key problems in high-order finite-volume methods for inviscid flow simulation: monotonicity and efficient steady-state convergence. We show how to apply TVD limiters to preserve monotonicity with high-order finite-volume methods. The limiter must be inactive away from discontinuities to maintain accuracy there: we address this through a new limiter function and a strategy for selective limiting. We also show that high-order finite-volume schemes (and indeed low-order schemes as well) converge more rapidly when the full high-order Jacobian is available. The cost of computing the Jacobian is equivalent to 10–30 residual evaluations, but the more rapid convergence of the GMRES inner iterations when the explicit Jacobian is used more than offsets the cost of computing the Jacobian. On the same mesh, a fourth-order scheme with an explicit Jacobian matrix uses about 20% more CPU time and four times as much memory as a second-order matrix-free scheme with a first-order preconditioner. For a given level of accuracy, however, the second-order scheme will require at least one more level of mesh refinement, making the memory comparison quite close.

1. Introduction

The history of scientific computing in general and of computational fluid dynamics in particular is the story of a relentless pursuit of highly-accurate solutions to increasingly complex problems with efficient use of computing resources. This desire for highly-accurate solutions is the main general motivation for the development of high-order accurate methods (by which we mean specifically in this chapter third- and fourth-order accurate methods). Our work focuses on high-order finite-volume methods. In our opinion, high-order methodology in general is not presently mature enough to conclude which family of schemes is technically superior, or even to make general statements about the trade-offs between them. On a practical note, however, we feel that high-order finite-volume methods provide a more natural upgrade path from second order to high order for users with an existing finite-volume code, because their existing code can be upgraded instead of being completely re-written.

The first work on high-order finite-volume methods in a production CFD code was, to the authors' knowledge, the high-order structured mesh discretization included in INS3D,[1] which used third-order upwinding for the convective terms (though with single point flux quadrature, which is strictly only third-order accurate in one dimension), and a second-order centered discretization of the viscous terms. This work, as well as more recent studies,[2,3] have shown that high-order methods are superior to second-order methods — both in accuracy on a given mesh and in CPU time to compute a solution of given accuracy — in computing aerodynamic lift and drag coefficients on structured meshes.

Since the pioneering work of Barth,[4] a number of researchers have studied high-order finite-volume methods for computational aerodynamics using unstructured meshes.[5–8] Much of this work has been based, as ours is, on the use of k-exact reconstruction to attain high-order accuracy with a limiter employed to enforce monotonicity. Other researchers[9–11] favor the essentially non-oscillatory (ENO) approach to reconstruction. Regardless of reconstruction approach, the other challenges in creating a genuinely high-order solver remain the same: accurate flux integration, boundary treatment, and so on.

This chapter will focus primarily on two important aspects of high-order finite volume methods: monotonicity and rapid convergence. After setting the stage in Sec. 2 with a brief summary of least-squares reconstruction, we describe in Sec. 3 the proper design and application of a limiter for a high-order scheme to enforce monotonicity without impacting accuracy in smooth parts of the flow; in our opinion, this is an area where finite-volume schemes have a distinct advantage over other families of high-order schemes. Section 4 describes our usage of the full high-order Jacobian to improve convergence to steady state. This is clearly a time-memory trade-off; our experience is that the trade-off is a good one, when a fourth-order explicit Jacobian scheme is compared with a standard second-order matrix-free method at comparable solution accuracy. Section 5 presents some illustrative results. Finally, in Sec. 6, we discuss the overall current status and future directions for high-order finite-volume methods.

2. Reconstruction

The mathematical basis for higher-order accurate least-squares reconstruction is well understood and has been thoroughly explained in numerous places in the literature.[4,12,13] Here we will provide only a brief summary to provide a basis for discussion of advanced topics that depend on the details of reconstruction.

2.1. *Mathematical basis*

The solution is represented within each control volume by the Taylor series expansion

$$\begin{aligned} U_i^R(x-x_i, y-y_i) = U|_i &+ \left.\frac{\partial U}{\partial x}\right|_i (x-x_i) + \left.\frac{\partial U}{\partial y}\right|_i (y-y_i) \\ &+ \left.\frac{\partial^2 U}{\partial x^2}\right|_i \frac{(x-x_i)^2}{2} + \left.\frac{\partial^2 U}{\partial x\,\partial y}\right|_i (x-x_i)(y-y_i) \\ &+ \left.\frac{\partial^2 U}{\partial y^2}\right|_i \frac{(y-y_i)^2}{2} + \cdots \end{aligned} \tag{1}$$

where U_i is the value of the reconstructed solution and $\frac{\partial^{k+l} U_i}{\partial x^k \partial y^l}$ are its derivatives at the reference point (x_i, y_i) of control volume i. The coefficients of the polynomial are computed so that the mean value of the solution in the control volume is conserved and the reconstruction approximates nearby control volume averages. This leads to a constrained least-squares problem for the coefficients in the expansion:

$$\begin{bmatrix} 1 & \overline{x}_i & \overline{y}_i & \overline{x^2}_i & \overline{xy}_i & \overline{y^2}_i & \cdots \\ \hline w_{i1} & w_{i1}\widehat{x}_{i1} & w_{i1}\widehat{y}_{i1} & w_{i1}\widehat{x^2}_{i1} & w_{i1}\widehat{xy}_{i1} & w_{i1}\widehat{y^2}_{i1} & \cdots \\ w_{i2} & w_{i2}\widehat{x}_{i2} & w_{i2}\widehat{y}_{i2} & w_{i2}\widehat{x^2}_{i2} & w_{i2}\widehat{xy}_{i2} & w_{i2}\widehat{y^2}_{i2} & \cdots \\ w_{i3} & w_{i3}\widehat{x}_{i3} & w_{i3}\widehat{y}_{i3} & w_{i3}\widehat{x^2}_{i3} & w_{i3}\widehat{xy}_{i3} & w_{i3}\widehat{y^2}_{i3} & \cdots \\ \vdots & \vdots & \vdots & \vdots & \vdots & \vdots & \ddots \\ w_{iN} & w_{iN}\widehat{x}_{iN} & w_{iN}\widehat{y}_{iN} & w_{iN}\widehat{x^2}_{iN} & w_{iN}\widehat{xy}_{iN} & w_{iN}\widehat{y^2}_{iN} & \cdots \end{bmatrix} \begin{pmatrix} U \\ \frac{\partial U}{\partial x} \\ \frac{\partial U}{\partial y} \\ \frac{1}{2}\frac{\partial^2 U}{\partial x^2} \\ \frac{\partial^2 U}{\partial x\,\partial y} \\ \frac{1}{2}\frac{\partial^2 U}{\partial y^2} \\ \vdots \end{pmatrix}_i = \begin{pmatrix} \overline{U}_i \\ \hline w_{i1}\overline{U}_1 \\ w_{i2}\overline{U}_2 \\ w_{i3}\overline{U}_3 \\ \vdots \\ w_{iN}\overline{U}_N \end{pmatrix} \tag{2}$$

where the weights w_i can be used to emphasize geometrically nearby data:

$$w_{ij} = \frac{1}{|\vec{x}_j - \vec{x}_i|^n}; \tag{3}$$

typically $n \in [0, 2]$. Equation 2 contains two sets of geometric terms: the moments of a control volume about its own reference point

$$\overline{x^n y^m}_i \equiv \frac{1}{A_i} \int_{V_i} (x - x_i)^n (y - y_i)^m dA$$

and the moments of a control volume j about the reference point of control volume i, calculated by using the parallel axis theorem:

$$\begin{aligned}\widehat{x^n y^m}_{ij} &\equiv \frac{1}{A_j} \int_{V_j} ((x - x_j) + (x_j - x_i))^n \cdot ((y - y_j) + (y_j - y_i))^m \, dA \\ &= \sum_{l=0}^{m} \sum_{k=0}^{n} \frac{m!}{l!\,(m-l)!} \frac{n!}{k!\,(n-k)!} (x_j - x_i)^k \cdot (y_j - y_i)^l \cdot \overline{x^{n-k} y^{m-l}}_j .\end{aligned}$$

The first row in Eq. 2 is the mean constraint, which can be eliminated analytically to leave an unconstrained least-squares problem for the derivatives.

Since the matrix contains only geometric terms, is identical for each solution variable in a given control volume and does not change between iterations, substantial savings in computational time can be achieved by pre-computing and storing the pseudoinverse of the reconstruction matrix for each control volume. Given a singular value decomposition of a reconstruction matrix A, the pseudoinverse $A^\dagger$ can easily be obtained:

$$\begin{aligned} A &= U \Sigma V^T \\ A^\dagger &= V \Sigma^\dagger U^T \end{aligned} \tag{4}$$

where the diagonal entries of Σ are the singular values of A, the columns of U and V are the left and right singular vectors, and $\Sigma^\dagger$ is a diagonal matrix containing the reciprocal values of Σ.

Once this pre-computation has been performed, the reconstruction coefficients in each control volume can be obtained by performing a matrix-vector multiplication; the number of columns in this matrix equals the number of control volumes in the reconstruction stencil, while the number of rows equals the number of required reconstruction coefficients. From these coefficients the reconstructed values of the solution can easily be computed at the flux quadrature points.

2.2. *Conditioning of the least-squares system*

Numerically, the conditioning of the least-squares system can still be problematical. Specifically, as written in Eq. 2, the least-squares problem for isotropic stencils has a condition number of order h^{p-1}, where p is the polynomial degree of the reconstruction polynomial, because of differences in column scaling.[a] To eliminate this problem, we scale each column of the matrix by dividing by the largest magnitude of any of its entries $s_j = \max_i |A_{ij}|$:

$$\tilde{A}_{ij} = \frac{A_{ij}}{s_j};$$

a similar effect could be achieved by scaling each column by h^{r-1}, where r is the degree of the derivative term multiplying that column. The result of the scaled least-squares problem is now the desired derivatives multiplied by the scaling factors, and we remove this effect by scaling the rows of the resultant pseudo-inverse matrix.

$$A_{ij}^{\dagger} = \frac{\tilde{A}_{ij}^{\dagger}}{s_i}.$$

Figure 1 shows by numerical experiment that $\tilde{A}$ is indeed much better conditioned than A. A series of meshes were generated to be geometrically similar, in the sense of having proportional length scale at the same point. The condition number of the reconstruction problem in each control volume was computed as a side effect of the singular value decomposition. For the unscaled case, the condition number is highest, as expected, for small cells. With column scaling, condition number is nearly uniform in the interior of the mesh, with higher condition number for the asymmetric stencils near the boundaries.

3. Limiting for Discontinuous Solutions

Arguably the most significant numerical challenge for high-order methods in inviscid computational aerodynamics — regardless of the numerical approach — is maintaining monotonicity near shocks and other discontinuities. The underlying mathematical problem, of course, is that all common numerical methods are based on some local expansion of the solution using

[a]Using normal equations to solve the least squares system makes this worse, as the condition number of the normal equations is the square of the condition number of the rectangular system.

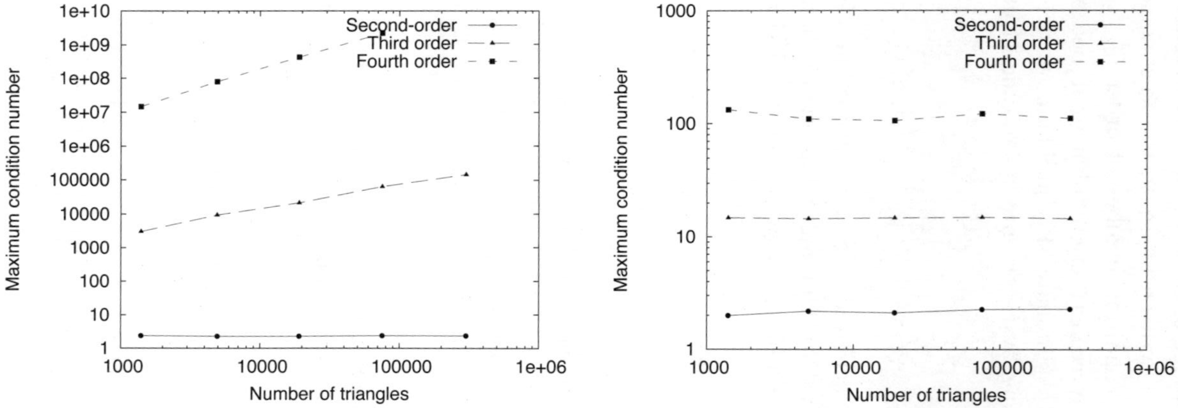

(a) Without column scaling.

(b) With column scaling.

Fig. 1. Impact of mesh size on condition number of least-squares reconstruction problem.

smooth functions; with the exception of spectral methods, these local expansions take the form of Taylor series, which intrinsically cannot converge to a discontinuous solution. High-order finite-volume methods fall into two broad categories with respect to discontinuity handling: essentially non-oscillatory (ENO) and total-variation diminishing (TVD) schemes. These two families of schemes take very different mathematical approaches to preventing overshoots.

The essentially non-oscillatory schemes,[9,14–16] including weighted (WENO) variants,[10,11,17,18] avoid producing overshoots by using only smooth data in the reconstruction. In principle, (W)ENO schemes can maintain high-order accuracy for all control volumes whose solution values are not intermediate values within a shock wave, including control volumes containing smooth extrema. These schemes produce excellent results for unsteady problems, with sharp resolution of moving shock waves and contact discontinuities. As a rule, however, these schemes do not converge rapidly, if at all, to steady state, and (W)ENO reconstruction can be quite expensive per iteration because of the need to compute multiple reconstructions for each control volume.

Total-variation diminishing schemes, introduced originally by Sweby,[19] make no attempt to avoid overshoots in the original reconstruction and instead post-process the reconstruction to eliminate overshoots using some sort of limiter; examples of and comparison between limiters can by found in numerous places in the literature [Refs. 20–24, among others]. In the form in which TVD schemes are usually applied for unstructured meshes, the reconstruction in each control volume is constrained so that the solution values at all flux integration points lie between the smallest and largest control volume average solution values for that control volume and its face neighbors. In practice, limiting is generally required for all control volumes whose stencil spans a discontinuity and at smooth extrema, resulting in loss of full-order accuracy over more of the mesh than a (W)ENO scheme. If a differentiable limiter is used, TVD schemes can converge quite rapidly to steady state, and less work is required per iteration than for (W)ENO schemes.

This section describes the extension of standard TVD limiting schemes to high-order unstructured mesh calculations. We begin with a review of the canonical Barth–Jespersen limiter for second-order TVD schemes before exploring how to compute and apply a limiting function to avoid unnecessary loss of accuracy for a high-order scheme.

3.1. *Applying limiter to a high-order reconstruction*

Barth[22] introduced the first limiter for unstructured grids. The scheme consists of finding a limiter value Φ_i for each primitive flow variable in each control volume that will limit the gradient in the piecewise-linear reconstruction of the solution. For second-order, if the reference location $\vec{x}_i$ is taken to be the control volume centroid, the point-wise value $U|_{\vec{x}_i}$ is equal to the control volume average $\overline{U}_i$. This leads to a limited reconstruction of the form

$$U_i^R(\vec{x}-\vec{x}_i,\Phi_i)=\bar{U}_i+\Phi_i\bigtriangledown U_i\cdot(\vec{x}-\vec{x}_i),\ \Phi\in[0,1]$$

The goal is to find the largest Φ_i which prevents the formation of local extrema at the flux integration Gauss points. The following procedure is used by Barth and Jespersen:

(1) Find the largest negative $(\delta U_i^{min} = \min_j\left(\overline{U}_j-\overline{U}_i\right))$ and positive $(\delta U_i^{max} = \max_j\left(\overline{U}_j-\overline{U}_i\right))$ difference between the solution in the immediate neighbors j and the current control volume i.
(2) Compute the unconstrained reconstructed value at each Gauss point used in flux integration $(U_{ik}=U_i^R\left(\vec{x}_k-\vec{x}_i\right))$.
(3) Compute a maximum allowable value of Φ_{ik} for each Gauss point k.

$$\Phi_{ik}=\begin{cases}\min\left(1,\frac{\delta U_i^{max}}{U_{ik}-\overline{U}_i}\right), & \text{if } U_{ik}-\overline{U}_i>0\\ \min\left(1,\frac{\delta U_i^{min}}{U_{ik}-\overline{U}_i}\right), & \text{if } U_{ik}-\overline{U}_i<0\\ 1, & \text{if } U_{ik}-\overline{U}_i=0\end{cases}$$

(4) Select $\Phi_i=\min(\Phi_{ik})$.
(5) Compute the limited reconstruction $U_i^R(\vec{x}-\vec{x}_i,\Phi_i)$ at Gauss points and use this value in flux integration.

Clearly, steps 1, 3, and 4 of this scheme introduce non-differentiability in the computation of the reconstructed function. Consequently, the flux is also non-differentiable, resulting in a scheme with known serious problems with steady-state convergence. In practice, the non-differentiability of step 3 causes the greatest degradation in convergence performance. We will return to this point in Sec. 3.2.

In extending the limiting procedure to third- and fourth-order accurate schemes, we first combine the reconstruction series expansion of Eq. 1 with the mean constraint to write the reconstruction as a sum of the control volume average and derivative terms that have zero mean over the control volume[b]:

$$U_i^R(\vec{x}-\vec{x}_i) = \overline{U}_i + \left(\left.\frac{\partial U}{\partial x}\right|_{\vec{x}_i} ((x-x_i)-\overline{x}_i) + \left.\frac{\partial U}{\partial y}\right|_{\vec{x}_i} ((y-y_i)-\overline{y}_i) \right.$$
$$+ \left.\frac{\partial^2 U}{\partial x^2}\right|_{\vec{x}_i} \left(\frac{(x-x_i)^2}{2} - \frac{\overline{x^2}}{2} \right) + \left.\frac{\partial^2 U}{\partial x \partial y}\right|_{\vec{x}_i} ((x-x_i)(y-y_i)-\overline{xy})$$
$$\left. + \left.\frac{\partial^2 U}{\partial y^2}\right|_{\vec{x}_i} \left(\frac{(y-y_i)^2}{2} - \frac{\overline{y^2}}{2} \right) + \cdots \right). \tag{5}$$

This can be interpreted as meaning that the reconstructed solution at any point is the control volume average plus zero-mean second- and high-order contributions from the reconstruction:

$$U_i^R(\vec{x}-\vec{x}_i) = \overline{U}_i + S(\vec{x}-\vec{x}_i) + H(\vec{x}-\vec{x}_i)$$

where $H(\vec{x}-\vec{x}_i)$ contains only quadratic terms for third-order reconstruction, and both quadratic and cubic terms for fourth-order reconstruction.

Early examples of high-order limiting used a formulation where the limiter value multiplies only the second-order terms while the high-order terms are "switched off" when discontinuities are detected.[7,25] This formulation has the form of

$$U_i^R(\vec{x}-\vec{x}_i, \Phi_i, \sigma_i) = \overline{U}_i + (\Phi_i(1-\sigma_i)+\sigma_i) S(\vec{x}-\vec{x}_i) + \sigma_i H(\vec{x}-\vec{x}_i). \tag{6}$$

The discontinuity detector, σ_i, is zero near discontinuities and one in smooth regions of the flow. However, this approach may violate the monotonicity requirement as our experience shows that the high-order terms — especially the quadratic terms — often contribute to reducing the overshoot in the unlimited reconstruction used in determining the value of Φ_i.

Analogous to the second-order case, no new extrema will be formed when limiting a high-order reconstruction if

$$\delta U_i^{min} \leq S(\vec{x}-\vec{x}_i) + H(\vec{x}-\vec{x}_i) \leq \delta U_i^{max}$$

[b]Even though the zero-mean reconstruction polynomial is used here for clarity of exposition, in practice our implementation computes the non-zero mean polynomials, because these are significantly less expensive to evaluate during flux integration. To conserve the mean when limiting, we must adjust our value of $U_i \equiv \overline{U}_i - \overline{x}_i \left.\frac{\partial U}{\partial x}\right|_{\vec{x}_i} - \overline{y}_i \left.\frac{\partial U}{\partial y}\right|_{\vec{x}_i} - \cdots$ to accommodate changes in the derivatives during limiting.

at the flux quadrature points. Our extension of Barth's approach to high order, then, writes the limited reconstruction as

$$U_i^R(\vec{x} - \vec{x}_i, \Phi_i) = \overline{U}_i + \Phi_i \left(S(\vec{x} - \vec{x}_i) + H_i(\vec{x} - \vec{x}_i) \right). \tag{7}$$

This approach to limiting the reconstruction satisfies monotonicity at solution discontinuities.

3.2. *Accuracy and limiter functions*

Sweby[19] introduced the notion of total-variation diminishing schemes, including showing the conditions under which a limiter function ϕ will preserve second-order accuracy for smooth data on uniform meshes. We will write ϕ as a function of $r \equiv \frac{\delta U_i^{max}}{U_{ik} - \overline{U}_i}$. Using this notation, Sweby's result requires that $\phi\,(r = 2) = 1$, and the Barth–Jespersen limiter computes $\phi\,(r) = \min\,(1, r)$. For a uniform triangular mesh, the flux quadrature point is located midway between two control volume centroids. Therefore, for any smooth solution function, $r_{ik} = 2 \pm \mathcal{O}\,(x_i - x_{ik})$, and $\Phi_{ik} \equiv \phi(r_{ik}) = 1 \pm \mathcal{O}\,(x_i - x_{ik})$ for any smooth ϕ. Using this limiter value to modify the gradient for a second-order scheme introduces an error in the reconstruction that is on the order of truncation error for smooth flows on uniform grids. Many smooth limiter functions satisfy this condition; perhaps the most commonly used in computational aerodynamics is Venkatakrishnan's limiter,[23] which can be written as

$$\phi(r) = \frac{r^2 + 2r}{r^2 + r + 2}, \tag{8}$$

neglecting terms to improve behavior for nearly uniform data.

From an accuracy point of view, however, the requirements for general unstructured grids are more stringent than for uniform meshes, because flux quadrature points can be located at a distance $\mathcal{O}\,(x_i - x_{ik})$ away from the midpoint between centroids. As we shall see, this leads to limiter values that do not necessarily approach one with mesh refinement for Venkatakrishnan's limiter, resulting in only first-order accuracy, regardless of the accuracy of the original reconstruction.

Just as a limiter can maintain second-order accuracy as long as $|\phi - 1| \sim \mathcal{O}\,(\Delta x)$, a third- or fourth-order scheme requires a limiter with $|\phi - 1| \sim \mathcal{O}\,(\Delta x^2)$ or $|\phi - 1| \sim \mathcal{O}\,(\Delta x^3)$, respectively, for the effect of the limiter in smooth regions to be on the order of truncation error. Venkatakrishnan's limiter will not provide sufficient accuracy even in smooth regions on uniform meshes. The Barth–Jespersen limiter would be sufficiently accurate, but its lack of differentiability will still make it impossible to achieve a steady-state solution. Therefore, we seek a new approximation for $\min(1,r)$ used in step 3 of the limiting procedure, which we will call $\widetilde{\min}(1,r)$. Like Venkatakrishnan's function, given in Eq. 8, we require that it be differentiable at all points. We also require this new limiting function to have a value of exactly 1 for $r \geq r_t$ where $r_t < 2$ represents a threshold value. For this function, we propose the form

$$\widetilde{\min}(1,r) = \begin{cases} P(r) & 0 \leq r < r_t \\ 1 & r \geq r_t \end{cases}$$

where $P(r)$ is a polynomial that is tangent to $\min(1,r)$ at both $r=0$ and $r=r_t$. That is, $P(r)$ must satisfy

$$P|_0 = 0 \qquad \left.\frac{dP}{dr}\right|_0 = 1 \quad P|_{r_t} = 1 \qquad \left.\frac{dP}{dr}\right|_{r_t} = 0$$
$$P(r) \leq \min(1,r) \ \ \forall r \in [0, r_t]$$

By design, this function preserves high-order accuracy on uniform grids by satisfying $\left|\widetilde{\min}(1,r) - 1\right| \leq \mathcal{O}\left(\Delta x^3\right)$ near $r=2$. Additionally, this function is also effective in maintaining high-order accuracy in regions of mild mesh non-uniformity. The degree of non-uniformity that can be accommodated is dictated by the choice of the threshold value r_t. Smaller values of r_t are less likely to unduly activate the limiter on non-uniform meshes but result in a limiter that approaches non-differentiability. In addition, when $P(r)$ is a cubic polynomial, values of $r_t < 1.5$ result in a limiter function that falls outside the TVD region. For the results presented in this work, we use $r_t = 1.5$ which yields the following cubic polynomial

$$P(r) = r - \frac{4}{27}r^3.$$

Figure 2 shows a comparison of this new limiter function to those of Barth and Jespersen and of Venkatakrishnan.

3.3. *Preventing unnecessary limiter activation*

Venkatakrishnan[23] showed that a limiter value of $\phi = 1$ for nearly uniform flows on uniform meshes is necessary for good convergence behavior. Also, to maintain high-order accuracy near smooth extrema, the limiter must allow small violations of monotonicity there. Thus, we seek to entirely shut off the limiter when the local solution variation is $\mathcal{O}\left(\Delta x^2\right)$ or smaller.

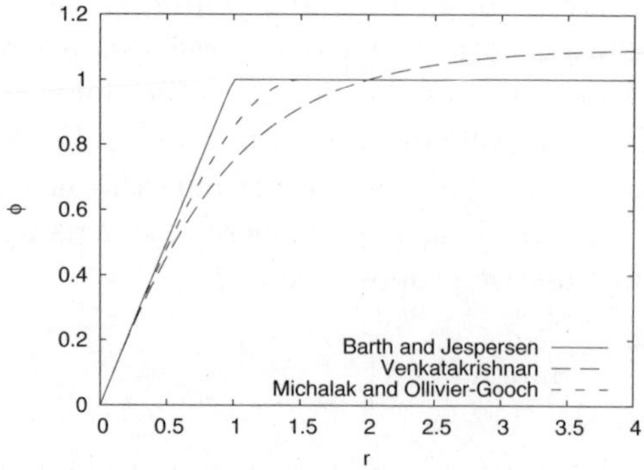

Fig. 2. Comparison of the limiters of Barth and Jespersen, Venkatakrishnan, and Michalak and Ollivier-Gooch.

Specifically, we propose to disable the limiter when

$$\delta U \equiv (\delta U_i^{max} - \delta U_i^{min}) < (K\Delta x)^{\frac{3}{2}}$$

where K is a tunable parameter. To maintain differentiability, we compute a modified limiter value:

$$\widetilde{\Phi}_i = \widetilde{\sigma}_i + (1 - \widetilde{\sigma}_i)\Phi_i \tag{9}$$

where Φ_i is the limiter value as calculated in step 4 of the procedure in Section 3.1 and $\widetilde{\sigma}_i$ is the following function:

$$\widetilde{\sigma}_i = \begin{cases} 1 & \delta U^2 \leq (K\Delta x)^3 \\ s\left(\frac{\delta U^2 - (K\Delta x)^3}{(K\Delta x)^3}\right) & (K\Delta x)^3 < \delta U^2 < 2(K\Delta x)^3 \\ 0 & \delta U^2 \geq 2(K\Delta x)^3 \end{cases} \tag{10}$$

where the smooth transition function s is defined by

$$s(y) = 2y^3 - 3y^2 + 1. \tag{11}$$

The limited reconstruction is then computed for each Gauss point by evaluating $U_i^R(\vec{x} - \vec{x}_i, \widetilde{\Phi}_i)$.

Although this two stage limiting procedure is somewhat more computationally expensive than Venkatakrishnan's limiter in the general case, some "short circuiting" is possible in uniform regions of flow. Since $\widetilde{\sigma}_i$ depends only on neighboring control volume averages, unlike Φ_i which also depends on an evaluation of the unconstrained reconstruction at each Gauss point, it is relatively inexpensive to compute. When $\widetilde{\sigma}_i$ evaluates to 1, computational effort can be saved by not computing Φ_i.

Although the threshold below which we consider the solution to be flat has the same form as the parameter in Venkatakrishnan's limiter that addresses the same issue, the two approaches differ significantly in their action. Venkatakrishnan's ϵ^2 term modifies the limiter value for all cases, and increasing the value of K allows a progressively larger overshoot in the solution at shocks. In our case, near shocks the transition function is exactly one, and the basic limiter enforces monotonicity regardless of the value of K. As we have shown elsewhere,[26] our scheme is less sensitive than Venkatakrishnan's to the choice of K.

3.3.1. *Boundary treatment*

Maintaining high-order accuracy near domain boundaries represents a special challenge. Boundary curvature must, of course, be accounted for in applying boundary conditions, though no special care is necessary in the mathematical formulation of the boundary conditions themselves.[13] In addition, the limiter can adversely affect accuracy at boundaries, because the solution will have extremal values at boundaries without having zero gradient. Therefore, the method used in Sec. 3.3 will not be effective in disabling the limiter in these regions. We will focus here on treatment of inviscid flow tangent to a wall, and refer interested readers elsewhere[26] for more detailed information, including an approach for preventing limiting near stagnation points.

For every wall boundary control volume, we will consider a ghost control volume which is a mirror image of the boundary control volume about the boundary. The solution in this ghost control volume is consistent with a shock-free flow and considered only in finding δU_i^{min} and δU_i^{max} in Step 1 of the limiting procedure in Sec. 3.1. The pressure in this ghost is extrapolated from interior data based on the steady momentum equation in the direction normal to the surface, which reduces to

$$\frac{\partial P}{\partial n} = -\frac{\rho V^2}{R}$$

where n is the direction normal to the wall, ρ is density, V is tangential velocity and R is wall radius of curvature, which can be estimated from variations in the wall normal direction. Since the ghost value is only used in computing the limiter, a first-order extrapolation of the pressure is sufficient:

$$\overline{P}_{gi} = \overline{P}_i - 2d \cdot \frac{\overline{\rho}_i \overline{V}_i^2}{R} \tag{12}$$

where $\overline{P}_i$, $\overline{\rho}_i$, and $\overline{V}_i$ are the control volume pressure, density and velocity respectively, and d is the distance of the control volume centroid from the wall in the convex direction (d is negative for concave boundaries). The ghost values of Mach number and density can be obtained by considering the isentropic transformation from the interior control volume state to the ghost value state with a pressure of $\overline{P}_{gi}$

$$\overline{\rho}_{gi} = \overline{\rho}_i \left(\frac{\overline{P}_{gi}}{\overline{P}_i} \right)^{\frac{1}{\gamma}}$$

$$\overline{M}_{gi}^2 = \frac{2}{\gamma - 1} \left(\left(\frac{\overline{P}_{t_i}}{\overline{P}_{gi}} \right)^{\frac{\gamma-1}{\gamma}} - 1 \right)$$

where $\overline{M}_{gi}$ is the Mach number in the ghost control volume and $\overline{P}_{t_i}$ is the total pressure as calculated using the boundary control volume average flow properties. This, together with the assumption that the flow direction remains tangential to the surface, fully establishes the state of the ghost control volume. As we shall see, considering this data in computing Φ greatly reduces limiter activity near boundaries.

4. Efficient Convergence

To accelerate convergence to steady state, we — like many other researchers — use backward Euler time advance with local timesteps that increase as the solution converges. Schematically, this can be written as:

$$\left(I + \Delta t \frac{\partial R}{\partial \overline{U}} \right) \delta U = R\left(\overline{U}\right) \tag{13}$$

where $\overline{U}$ is the global vector of control volume averages and $R(U)$ is the global residual. We describe elsewhere[27] our residual-based local timestepping scheme and line search globalization of the linear system solution; these techniques improve both convergence rate and robustness. Here, we will focus on our approach for solving the linear system itself.

Direct solution of the sparse linear system in Eq. 13 is prohibitively expensive in computational time and memory. Therefore, as is common, we solve the system iteratively using the GMRES method.[28] Solving a linear system with GMRES only requires a means of computing the product of the left-hand-side matrix with an arbitrary vector. Therefore it is common to avoid forming the exact Jacobian matrix and instead approximate the matrix-vector product using Frechet derivatives; this requires one flux evaluation per inner GMRES iteration plus one flux evaluation per outer iteration. However, since the convergence properties of GMRES are highly sensitive to the conditioning of the matrix, the system must be preconditioned using an approximate factorization of a simplified Jacobian matrix. This matrix is usually chosen to be the Jacobian of the first-order scheme, because this matrix is cheap to compute and store; typically, reconstructed Gauss point solution values are used to compute the Jacobian terms rather than control volume averages, as this results in a more accurate linearization at essentially zero additional cost. In the present work we consider both this "matrix-free" method and a method which explicitly forms the exact high-order Jacobian.

4.1. *Forming the high-order Jacobian matrix*

If we compute the exact Jacobian of the high-order scheme explicitly, we can then use it as an improved preconditioner and to avoid the use of Frechet derivatives and their associated residual evaluations. Here, we will describe this computation in detail only for the unlimited case; interested readers may find a full description of how to find the analytic Jacobian in the presence of a limiter elsewhere.[27] The analytic Jacobian can be written explicitly using the chain rule as

$$\frac{\partial R}{\partial \overline{U}} \equiv \frac{\partial FluxInt}{\partial CVars} = \frac{\partial FluxInt}{\partial Flux} \frac{\partial Flux}{\partial RecSol} \frac{\partial RecSol}{\partial RecCoef} \frac{\partial RecCoef}{\partial PVars} \frac{\partial PVars}{\partial CVars} \tag{14}$$

where *FluxInt* is the flux integral, *Flux* are the numerical fluxes, *RecSol* are the reconstructed solutions at Gauss points, *RecCoef* are the reconstruction coefficients, *PVars* are the control volume averages of the primitive variables, and *CVars* are the control volume averages of the conserved variables. To compute the Jacobian, the following procedure is used at each timestep:

(1) $\frac{\partial PVars}{\partial CVars}$ is computed for each control volume and stored, as each of these is used multiple times below. This is the standard Jacobian for the change of variables from $(\rho\ \rho u\ \rho v\ E)^T$ to $(\rho\ u\ v\ P)^T$.

(2) For each Gauss point, do the following for each of the two adjacent control volumes:
   (a) $\frac{\partial RecSol}{\partial PVars} = \frac{\partial RecSol}{\partial RecCoef} \frac{\partial RecCoef}{\partial PVars}$ is computed. The $\frac{\partial RecCoef}{\partial PVars}$ term is simply the pseudoinverse of the reconstruction matrix precomputed in Eq. 4, while the $\frac{\partial RecSol}{\partial RecCoef}$ term is a geometric term that depends on the location of the Gauss quadrature point.
   (b) $\frac{\partial Flux}{\partial RecSol}$, the Jacobian of the Roe flux, is computed.
   (c) The product $\frac{\partial Flux}{\partial PVars} = \frac{\partial Flux}{\partial RecSol} \frac{\partial RecSol}{\partial PVars}$ is computed efficiently by taking advantage of the sparsity of the reconstruction terms that is due to the lack of coupling between solution variables.
   (d) The product $\frac{\partial Flux}{\partial CVars} = \frac{\partial Flux}{\partial PVars} \frac{\partial PVars}{\partial CVars}$ is computed. Since $\frac{\partial Flux}{\partial PVars}$ couples all solution variables in the reconstruction stencil, this step is computationally intensive.
   (e) $\frac{\partial FluxInt}{\partial CVars} = \frac{\partial FluxInt}{\partial Flux} \frac{\partial Flux}{\partial CVars}$ is the contribution to the flux integral due to one side of this Gauss point. This component is computed by using the appropriate Gauss integration weight. The result is added to the total flux Jacobian.

The sparse analytic Jacobian is found once for every Newton iteration and used to produce the matrix-vector products needed by the GMRES solver.

4.2. *Preconditioning*

Since the rate of convergence of the GMRES method is strongly dependent on the condition number of the matrix, preconditioning is used to alter the matrix spectrum and hence accelerate the convergence rate of the iterative technique. Left preconditioning is applied by modifying the linear system to be solved such that

$$Ax = b$$

becomes

$$M^{-1}Ax = M^{-1}b$$

where M^{-1} is an approximate inverse of the preconditioning matrix $M \approx A$. A common approach[29–31] is to use the flux Jacobian of the first-order scheme for M and to use ILU decomposition to form the approximate inverse. This approach has the advantages of being easier to compute and requiring less memory than using the full-order accurate Jacobian. To form the preconditioning matrix M, a procedure similar to that presented in Sec. 4.1 is used except that the terms $\frac{\partial Flux}{\partial RecSol}\frac{\partial RecSol}{\partial RecCoef}$ are eliminated. This method is used for the matrix-free results presented in the present work.

Since the high-order Jacobian already needs to be computed in the matrix-explicit method, its ILU decomposition can easily be used as a preconditioner. The increase in memory use can be partially mitigated by using a lower level of fill; as we will show, even with low levels of fill, the matrix-explicit method is much better conditioned than the matrix-free method.

5. Verification and Results

The results presented in this section were obtained using a vertex-centered finite-volume solver. The solution process consists of two stages. In the pre-iteration stage the linear system resulting from a local timestepping is solved at each iteration. The Jacobian from the first-order accurate scheme is used on the left-hand side and the full-order accurate flux is used on the right-hand side. At each Newton iteration, the linear system is approximately solved using incomplete-lower-upper factorization (ILU) preconditioned GMRES. This approach is used to avoid wasting time forming and solving an expensive linear system when the solution is still so far from convergence that the linearization of the problem does yet not accurately extrapolate to the converged solution. During the second stage, the left-hand side is replaced with the full-order accurate Jacobian. In both stages, we use a residual-based local time stepping scheme and line search globalization.[27]

5.1. *Ringleb's flow*

We begin by considering Ringleb's flow which is transonic but shock-free and has a known exact solution. This will enable us to quantify the effects of Venkatakrishnan's limiter and the new limiter on the accuracy of the solution in smooth regions of flow. We consider four meshes consisting of 369, 1426, 5467 and 20690 control volumes. For Venkatakrishnan's function we use a tuning parameter of $K = 2$ and for the new limiter we use $K = 1$.

The exact solution to Ringleb's flow is given by the streamlines defined by:

$$x = \frac{1}{2}\frac{1}{\rho}\left(\frac{1}{q^2} - \frac{2}{k^2}\right) + \frac{J}{2} \qquad y = \pm\frac{1}{k\rho q}\sqrt{1 - \left(\frac{q}{k}\right)^2}$$

where k is constant along a streamline and

$$
\begin{aligned}
q &= \left|\vec{V}\right| \\
J &= \frac{1}{c} + \frac{1}{3c^3} + \frac{1}{5c^5} - \frac{1}{2}\ln\frac{1+c}{1-c} \\
c &= \sqrt{1 - \frac{\gamma-1}{2}q^2} \\
\rho &= c^{2/(\gamma-1)}
\end{aligned}
$$

The computational domain is constructed using solid walls at the streamlines for $k = 0.55, 1.2$ and placing the inlet and outlet at $q = 0.5$. This results in a geometry that is symmetric about the horizontal axis. The flow is subsonic at the inlet and outlet but is supersonic near the inside wall of the throat. Using this domain, rather than just the upper half, is a more stringent test of the accuracy of the computational scheme since errors produced at the throat are allowed to propagate to the lower half of the domain.

We examine the limiter value Φ for the pressure component at steady state for the fourth-order solution on the 5467 control volume mesh using three limiting schemes in Fig. 3. Venkatakrishnan's limiter applies some limiting to virtually all control volumes. The new limiter successfully avoids limiting in almost all control volumes; the fraction of control volumes where the limiter is active remains about the same with mesh refinement. All the control volumes that are limited have values $\Phi \geq 0.99$, although this bound becomes somewhat lower with mesh refinement. Without the boundary ghost treatment significant limiting occurs near the inner wall boundary.

Next, in Fig. 4, entropy is plotted for the second- and fourth-order schemes on the second-finest grid. The new limiter causes no distinguishable additional entropy production relative to the unlimited scheme for the second-order method, and only a very slight increase for the fourth-order method. On the other hand, Venkatakrishnan's limiter increases entropy by approximately an order of magnitude for the second-order scheme and two to four orders for the fourth-order scheme. Once again we note that when applying Venkatakrishnan's limiter there is no apparent benefit to using the fourth-order scheme over the second-order scheme.

The L_2 norm of error of the converged solution compared to the known exact solution is used to generate Fig. 5. The grid-convergence orders of the error norms using a regression fit of the results from all grids are given in Tab. 1. The unlimited schemes attain their nominal orders of accuracy in L_1

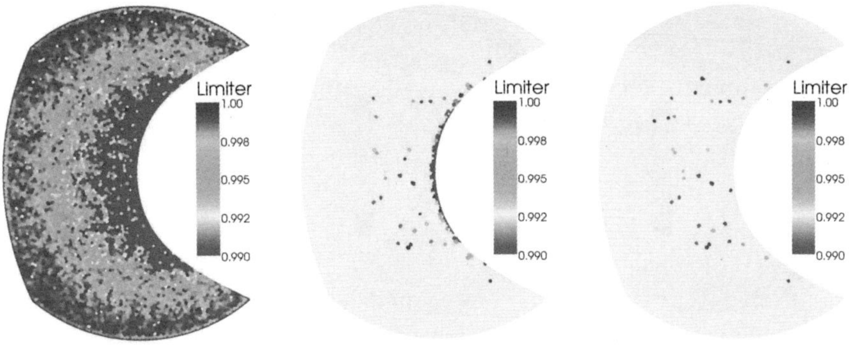

(a) Venkatakrishnan's limiter

(b) New limiter without boundary ghost values

(c) New limiter

Fig. 3. Limiter value for pressure for the fourth-order converged solution on the 5467 control volume mesh. Only values $\Phi \neq 1$ are plotted.

and L_2 norms and one order less than nominal for the L_∞ norm. Regardless of reconstruction order, use of Venkatakrishnan's limiter results in first-order accuracy, though with lower error magnitude than simply using a first-order accurate scheme. The results using the new limiter without the boundary ghost value treatment outperform Venkatakrishnan's limiter only slightly. This indicates that the application of the limiter in even an isolated region has severe implications for global accuracy. The new limiter with boundary ghost values, on the other hand, allows for the second- and third-order schemes to perform virtually identically to the unlimited case and has an adverse effect on the accuracy of the fourth-order scheme only on the finest mesh. The problem here is that, on realistic unstructured meshes, mesh irregularities can induce limiter activity in smooth regions of the flow. To leading order, this alters the reconstructed solution values used in flux integration by $\mathcal{O}\left(|\nabla u|\left(1-\phi\right)h\right)$, where ∇u is the gradient of the solution and h is the characteristic size of the control volume. For coarse meshes and lower orders of accuracy, the impact of these inadvertent limiter firings is lost in the larger discretization error. However, for fourth-order and fine meshes, the discretization error is low enough that this effect comes to dominate the error. While clearly undesirable, this limiter error is still quite small compared with the analogous error introduced by Venkatakrishnan's limiter.

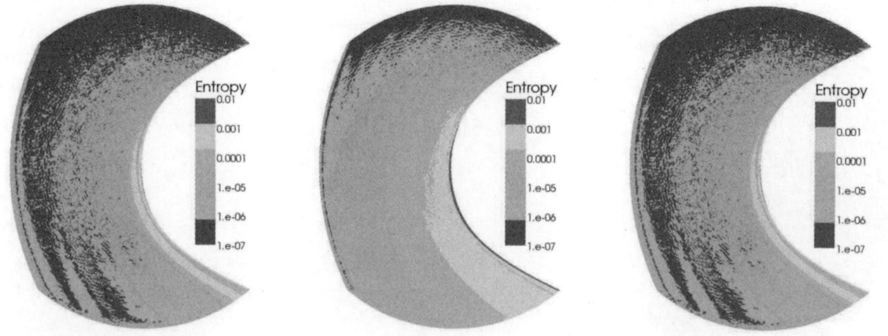

(a) Second order, no limiter (b) Second order, Venkatakrishnan limiter (c) Second order, new limiter

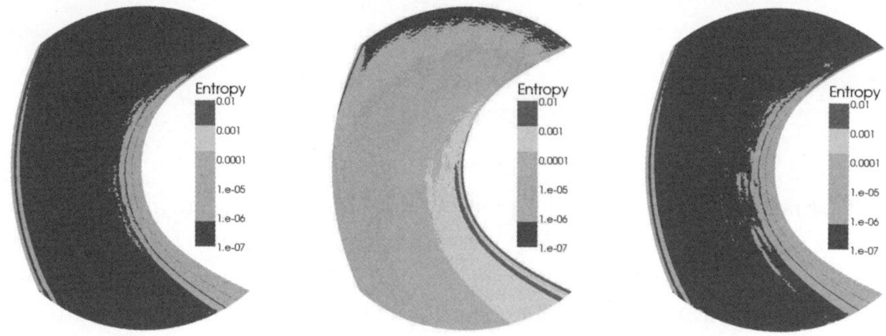

(d) Fourth order, no limiter (e) Fourth order, Venkatakrishnan limiter (f) Fourth order, new limiter

Fig. 4. Difference in dimensionless entropy from the freestream value for Ringleb's flow.

5.2. *Limiter behavior for transonic airfoil Flows*

Next, we present results for transonic flow over a NACA 0012 airfoil at Mach 0.8 and an angle of attack of 1.25 degrees. The computational mesh consists of 4656 control volumes and is shown in Fig. 6. We will consider second- and fourth-order schemes using Venkatakrishnan's limiter, the new limiter, and the new limiter plus a stagnation region fix which disables the limiter at low Mach number.

5.2.1. *Accuracy*

We begin by assessing the quality of the solution upstream of the shock. Entropy near the leading edge of the airfoil is plotted in Fig. 7. The fourth-

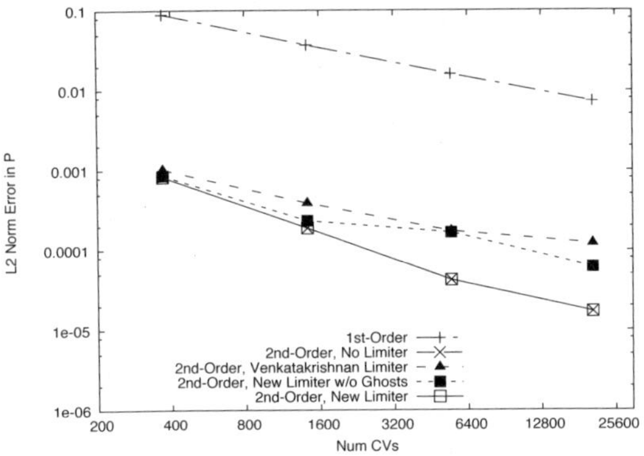

(a) Second-order discretization

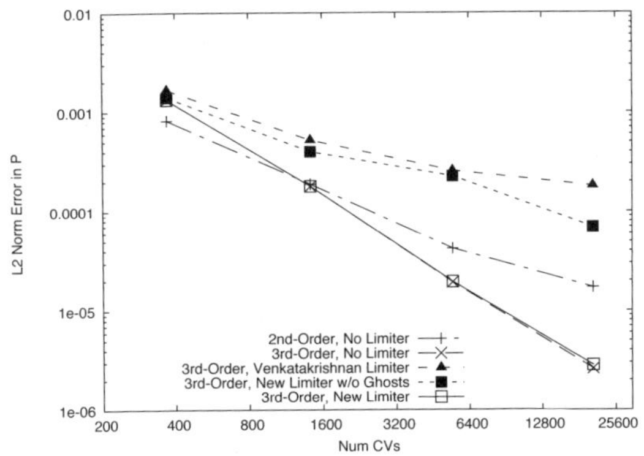

(b) Third-order discretization

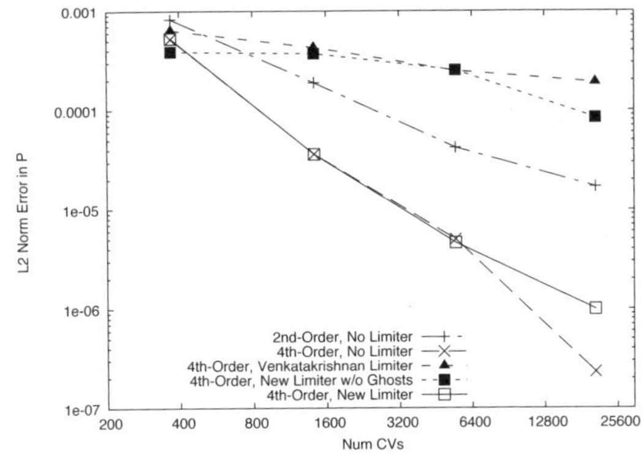

(c) Fourth-order discretization

Fig. 5. Error convergence for pressure in Ringleb's flow.

Table 1. Convergence order of norms of error in pressure for Ringleb's flow computed using regression fit of all mesh results.

Nominal Order	Limiter	L_1 Norm	L_2 Norm	L_∞ Norm
1st	None	1.24	1.24	0.99
2nd	None	2.22	1.96	1.24
2nd	Venkatakrishnan	1.10	1.07	0.47
2nd	New w/o Ghosts	1.71	1.23	0.47
2nd	New	2.22	1.96	1.24
3rd	None	3.18	3.12	2.70
3rd	Venkatakrishnan	1.14	1.10	0.51
3rd	New w/o Ghosts	1.83	1.43	0.59
3rd	New	3.14	3.08	2.40
4th	None	4.07	3.74	3.06
4th	Venkatakrishnan	0.62	0.62	0.12
4th	New w/o Ghosts	1.07	0.74	0.23
4th	New	3.24	3.11	2.08

order solution with Venkatakrishnan's limiter once again fails to outperform its second-order counterpart. The new limiter produces approximately an order of magnitude less entropy than Venkatakrishnan's limiter for the second-order scheme. The new limiter applied to the fourth-order scheme results in even less entropy production. Disabling the stagnation region fix results in a modest increase in entropy production.

The quality of the solutions can also be compared by examining the stagnation pressure ratio along the upper surface of the airfoil shown in Fig. 8. The decrease in total pressure across the shock is comparable for all schemes. However, the schemes limited with the Venkatakrishnan limiter result in substantial stagnation pressure loss upstream of the shock. For the second-order scheme the stagnation region fix has little effect while for the fourth-order scheme applying this step in the limiting procedure results in a further improvement in the conservation of total pressure.

5.2.2. *Shock monotonicity*

The performance of the new limiter in controlling oscillations and overshoots in the solution near the strong shock on the upper surface of the airfoil is demonstrated in the pressure plot of Fig. 9. The new limiter and

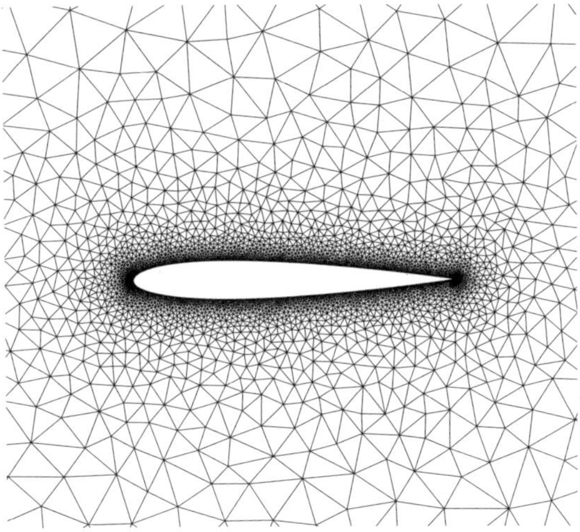

Fig. 6. Mesh consisting of 4656 control volumes used for the NACA 0012 test case.

Venkatakrishnan's limiter are both effective in eliminating any substantial oscillations near the shock. The pressure computed on the upper surface of the airfoil using the different limiters is virtually indistinguishable. Once again the lower dissipation of the new limiter is demonstrated by the sharper profile of the weak shock on the lower surface of the airfoil.

5.2.3. *Convergence*

Next, we consider the residual convergence properties of the new limiting scheme coupled with our Newton-GMRES solver. Figure 10 shows the convergence of the scheme with the new limiter relative to Venkatakrishnan's limiter for second and fourth order accurate schemes. The scheme with the new limiter exhibits a slightly poorer convergence rate than Venkatakrishnan's limiter. This is likely due to the lower dissipation of the scheme which results in poorer conditioning of the nonlinear system of equations. Similarly, we find that the fourth-order scheme requires slightly more iterations to converge than the second-order scheme, and about twice the CPU time.

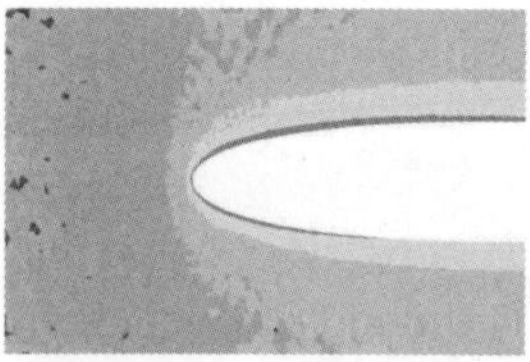

(a) Second-order discretization, Venkatakrishnan limiter

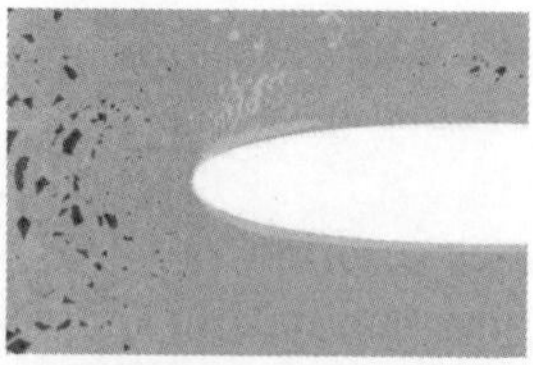

(b) Second-order new limiter

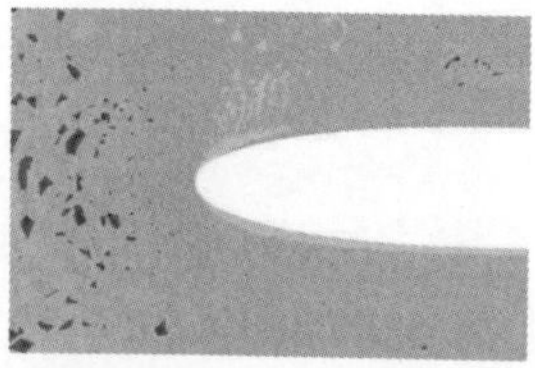

(c) Second-order new limiter, plus fix to turn off limiter near the leading edge stagnation point

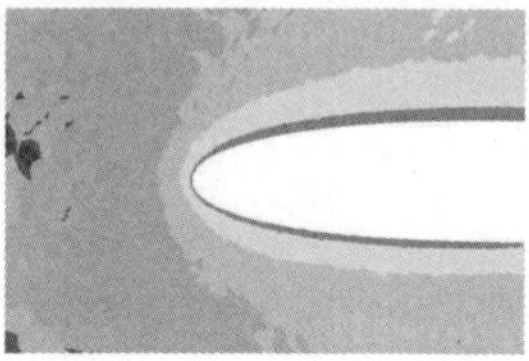

(d) Fourth-order discretization, Venkatakrishnan limiter

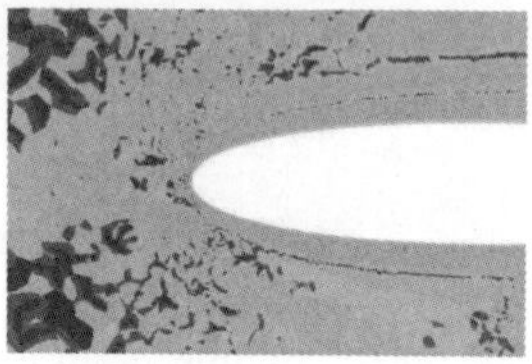

(e) Fourth-order new limiter

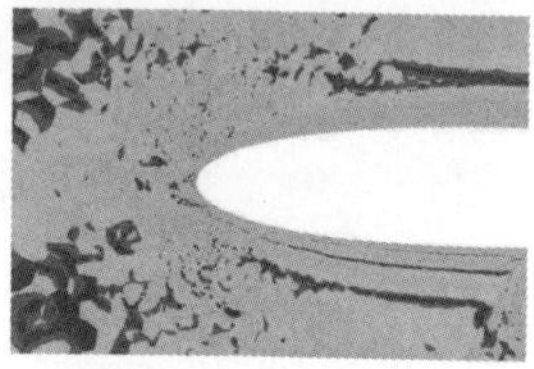

(f) Fourth-order new limiter, plus fix to turn off limiter near the leading edge stagnation point

Fig. 7. Difference in dimensionless entropy from the freestream value for Mach 0.8, $\alpha = 1.25$ flow around a NACA 0012 airfoil.

5.3. *Explicit Jacobian*

We now turn our attention to the impact of Frechet derivatives with a low-order preconditioner versus the explicit high-order Jacobian on efficiency of the Newton-GMRES scheme. We will examine two test cases. Subsonic flow at Mach 0.2 around a multi-element airfoil with zero angle of attack is considered, using a mesh with 4913 vertices. Transonic flow around a NACA 0012 airfoil at Mach 0.8 and an angle of attack of 1.25 degrees is also considered; the mesh for this case has 4656 vertices. Our focus here is on convergence behavior, and we have shown solutions for these cases elsewhere,[27] so we will not repeat the solution results here.

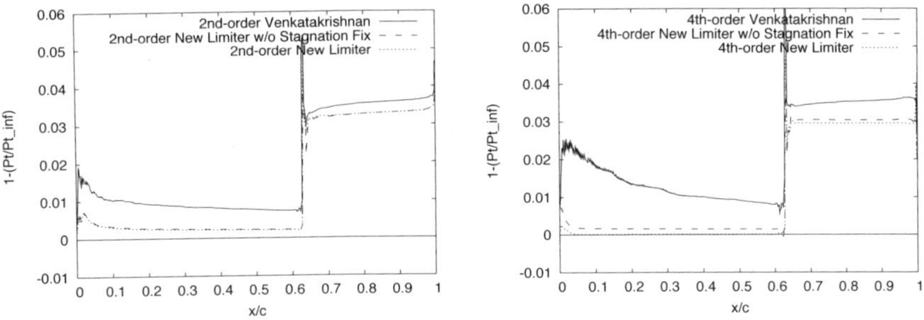

(a) Second order (b) Fourth order

Fig. 8. Decrease in total pressure along the upper surface of the NACA 0012 airfoil at Mach 0.8 $\alpha = 1.25$.

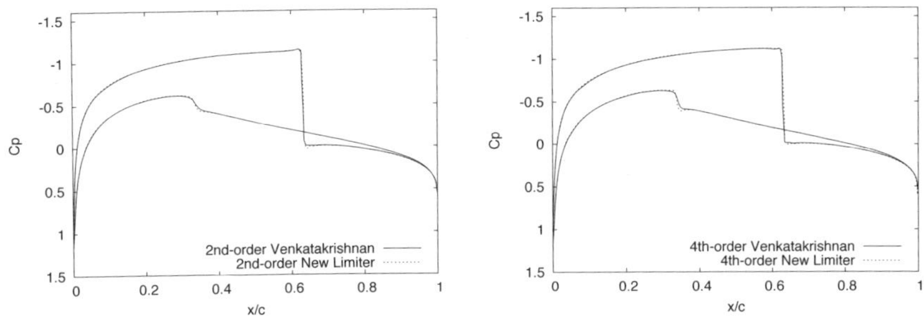

(a) Second order (b) Fourth order

Fig. 9. Surface pressure profiles for transonic flow over a NACA0012 airfoil.

5.3.1. *Cost of evaluating the explicit Jacobian*

We begin by comparing the relative CPU time needed to compute a flux integral, a first-order Jacobian, and a high-order Jacobian. The results are given in Table 2. The large increase in computational time needed for flux and Jacobian evaluations for the third-order scheme relative to the second-order scheme can in large part be attributed to the doubling of the number of Gauss points. Since the fourth-order scheme uses the same number of Gauss points as the third-order scheme, a smaller increase in flux and Jacobian evaluation times is observed. For flux evaluations, this increase is purely due to the cost of the reconstruction procedure. For the Jacobian, the increased cost of matrix-matrix products needed for its assembly is also

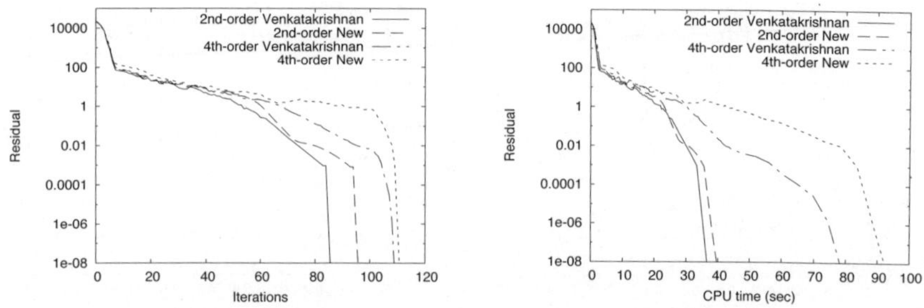

(a) Iteration count (b) CPU time

Fig. 10. Convergence history for transonic airfoil case.

Table 2. Relative computational time of flux and Jacobian evaluation without limiter for subsonic case.

Order	Flux	1st Order Jacobian		Full Order Jacobian	
	seconds	seconds	relative to flux	seconds	relative to flux
		Subsonic case, without limiter			
1	0.0116	0.217	18.7		
2	0.0194	0.217	11.2	0.346	17.8
3	0.0399	0.217	5.4	1.122	28.1
4	0.0568	0.217	3.8	1.219	21.5
		Transonic case, with limiter			
1	0.0109	0.205	18.8		
2	0.0356	0.205	5.8	0.375	10.5
3	0.0733	0.205	2.8	1.623	22.1
4	0.1137	0.205	1.8	2.241	19.7

important. Regardless of order of accuracy, the limiter used with transonic flow significantly increases computational cost.

5.3.2. *Quality of preconditioning*

Next, we compare the relative effectiveness of the ILU decomposition of the first-order and high-order Jacobians. The average number of inner GMRES iterations needed per Newton iteration to obtain a relative residual drop of 10^{-3} is shown in Tab. 3. The results also include for comparison the case where the full LU decomposition of the first-order Jacobian is used to precondition the high-order matrix-free scheme. The results indicate that

Table 3. Average number of inner GMRES iterations per Newton iteration.

Order	First-Order Jacobian (MF)			High-Order Jacobian (ME)	
	ILU(1)	ILU(4)	LU	ILU(0)	ILU(1)
			Subsonic case		
2	85.5	49.7	29.2	54.2	33.7
3	81.3	55.0	38.8	34.0	23.4
4	152.4	105.4	95.7	32.6	23.0
			Transonic case		
2	41.9	28.6	26.8	25.2	13.3
3	40.9	32.9	29.5	10.9	7.1
4	58.9	47.5	44.9	20.6	11.9

the first-order Jacobian is a reasonably good preconditioner for the second-order scheme if ILU with enough levels of fill is used. However, this preconditioner does a poor job for the third- and fourth-order schemes. Even with a full LU decomposition, the number of GMRES iterations required to solve the fourth-order case remains high. Since the matrix-free method must be used when the high-order Jacobian is not available, this large number of GMRES iterations results in a large number of residual evaluations. The costs of these residual evaluations exceed the relative additional cost of computing the high-order Jacobian. Using the full-order Jacobian for the preconditioner, the convergence properties of the high-order schemes are comparable to the second-order scheme. In all cases the transonic test case requires fewer GMRES iterations per Newton iteration indicating that the linear systems are better conditioned. Due to the high non-linearity of the shock, the adaptive timestep method yields smaller timesteps for the transonic case than the subsonic case. This likely explains why the linear system is better conditioned in the transonic case than in the subsonic case. However, since these smaller timesteps also result in a larger number of Newton iterations, the total number of GMRES iterations for the transonic case exceeds that of the subsonic case.

5.3.3. *Overall memory and computational cost*

The major components contributing to the memory requirement for both matrix-explicit and matrix-free methods are:

- *The pseudoinverse of the reconstruction matrix.* To avoid solving the least-squares problem at each flux evaluation, these matrices need to be precomputed and stored for each control volume.

Table 4. Memory use in (average) bytes per control volume along with run time in seconds and equivalent residual evaluations for transonic case.

Scheme	Recon	Jacob	ILU Matrix	Krylov Subsp.	Total	Time (sec)	Time (res eval)
			Second Order				
ME / ILU(0)	94	2378	2378	960	5810	27.5	772
ME / ILU(1)	94	2378	3762	576	6810	27.1	761
MF / ILU(1)	94	870	1141	1728	3833	47.2	1326
MF / ILU(4)	94	870	2334	1152	4450	40.2	1129
			Third Order				
ME / ILU(0)	704	4642	4642	928	10916	184.5	2517
ME / ILU(1)	704	4642	7888	416	13650	156.4	2134
MF / ILU(1)	704	870	1141	2144	4859	179.1	2443
MF / ILU(4)	704	870	2334	1760	5668	155.9	2127
			Fourth Order				
ME / ILU(0)	1434	4934	4934	800	12102	53.5	471
ME / ILU(1)	1434	4934	8587	448	15403	54.8	482
MF / ILU(1)	1434	870	1141	2528	5973	101.6	894
MF / ILU(4)	1434	870	2334	2048	6686	88.1	775

- *The Jacobian.* For the matrix-free scheme, this will always be the first-order Jacobian. For the matrix-explicit scheme, the high-order Jacobian will inevitably have more fill and require more storage.
- *ILU decomposition of the Jacobian.* The memory required for this depends not only on the fill of the Jacobian but also on the additional fill due to the decomposition which increases with n for ILU(n).
- *Krylov subspace.* The maximum number of inner GMRES iterations required to solve a Newton iteration and the number of solution unknowns per control volume determines the memory requirement of the Krylov solver; again, the GMRES iterations are terminated at a relative residual of 10^{-3}.

The breakdown of memory requirements along with the total CPU time is shown in Tab. 4 for the transonic test case. The additional memory required by the matrix-explicit scheme is due to the increased fill of the Jacobian and resulting preconditioning matrix. However, this is partially offset by the lower fill ratio of the ILU decomposition and the reduced memory use of the Krylov solver.

Table 4 also gives CPU time, measured in seconds on a single core of an Intel Core 2 processor and as a multiple of the cost of a residual evaluation. We can not explain with certainty why the third-order scheme is significantly slower than either the second- or fourth-order schemes for the transonic case; our speculation is that the root of the problem may lie in the fact that the leading error for the third-order scheme is dispersive rather than diffusive. Although no attempt was made to hand tune the globalization strategy, the run time in terms of equivalent residual evaluations is competitive with that reported elsewhere: Blanco and Zingg[32] report that the Newton-Krylov algorithm applied to a second-order accurate matrix-dissipation scheme required the equivalent of 660 residual evaluations to converge for the same transonic case. However due to the differences in the discretization scheme and the mesh used, too much emphasis should not be placed on such direct comparisons.

The results indicate that in terms of the memory to computational time trade-off the matrix-explicit method with ILU(0) preconditioning and the matrix-free method with ILU(4) preconditioning represent, in most cases, the best trade-offs. Figure 11 shows the residual as a function of time for these two schemes for the second- and fourth-order schemes. On these plots the start-up procedure is shown with a line while the main stage is shown with a line and a point marker at each iteration.

In all cases the matrix-explicit method outperforms the matrix-free method. The difference is most substantial for the subsonic case where the start-up phase represents a smaller portion of the total time and where the conditioning of the linear system is poorer.

6. Discussion

This chapter has given an overview of our approach to two of the key problems in applying high-order finite-volume methods for inviscid flow simulation: achieving a monotone reconstruction of the solution and accelerating convergence to steady state. Building on well-established TVD methodology, developing a limiter that preserves monotonicity in conjunction with high-order finite-volume methods is relatively straightforward. Maintaining high-order accuracy away from discontinuities is also critically important; our work has shown that this can be achieved by choosing a suitable limiter function and applying the limiter judiciously.

It is not particularly surprising that we could show that a high-order finite-volume scheme (and indeed low-order schemes as well) converges more rapidly in terms of iteration count when the full high-order Jacobian is available.[c] Because the cost of computing the Jacobian is on the order of

[c]From a convergence point of view, using a matrix-free scheme with a full-order preconditioner should also be sufficient, although we have not pursued this option.

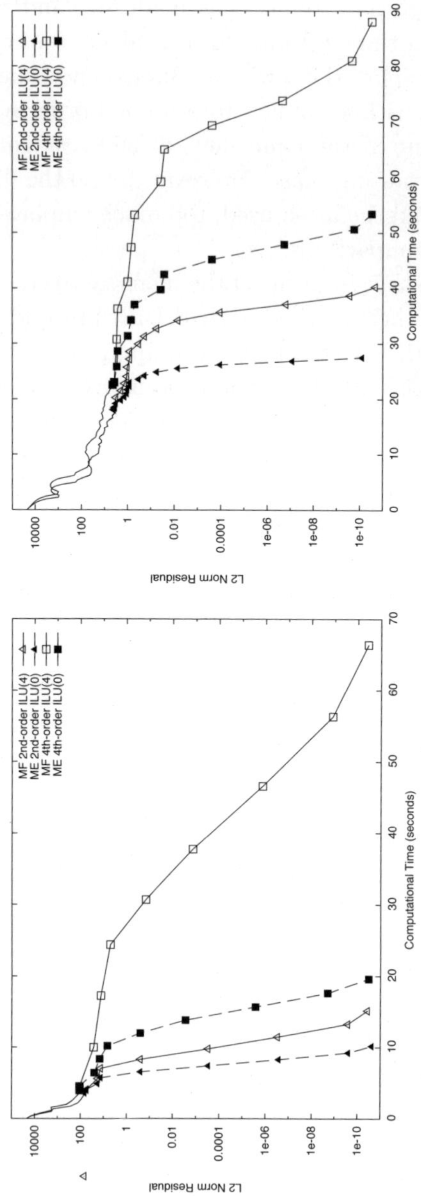

(a) Subsonic test case.

(b) Transonic test case.

Fig. 11. Residual versus computational time for matrix-explicit method and matrix-free method.

10-30 residual evaluations, the more rapid convergence of the GMRES inner iterations when the explicit Jacobian is used more than offsets the cost of computing the Jacobian. When comparing methods of the same order on the same mesh, the explicit Jacobian roughly doubles memory usage. When comparing a fourth-order scheme with an explicit Jacobian matrix to a second-order matrix-free scheme with a first-order preconditioner — a fairly typical example of current production codes in computational aerodynamics — CPU times differ by no more than about 20%, and the fourth-order scheme uses only about four times as much memory. For a given level of accuracy, however, the second-order scheme will require at least one more level of mesh refinement, making the memory comparison quite close. It remains to be seen how this trade-off will work out for large problems in three dimensions.

6.1. *Directions for future development*

There are at least two issues in solution reconstruction that we believe still merit attention. First, it should be possible to identify smooth parts of the solution based on the residual of the least-squares problem. If successful, this could lead to a single approach to shutting off the limiter anywhere the solution is smooth while not affecting its behavior at discontinuities, providing better accuracy and convergence at lower cost without affecting monotonicity; importantly, this new approach would be strictly mathematical, making it easier to apply to other types of problems. Second, sharp corners — most notably, in aerodynamics, the trailing edge — produce singularities in the solution which cannot be resolved using polynomial reconstruction. Adding an additional term to the reconstruction that captures such discontinuities would dramatically improve solution representation at sharp corners, and reduce the errors currently created there commensurately.

We are working to exploit our ability to compute the full high-order Jacobian in other areas. One obvious application area is computing the adjoint solution, which reduces to solving a linear system when the Jacobian is available. We are currently using this approach in aerodynamic optimization. The adjoint solution can also be used directly to produce adaptation indicators[33–35] and compute bounds on errors in functional outputs.[36]

Even with the full Jacobian available, the convergence rate of our scheme degrades for very large meshes; this effect becomes apparent as mesh size approaches 30,000 vertices and becomes severe as mesh size approaches 100,000 vertices in two dimensions. We intend to attack this problem using multigrid methods, both h- and hp-multigrid. Rapid convergence on large meshes will pave the way for extension to complex three-dimensional turbulent flows.

Acknowledgments

This work has been supported by the Canadian Natural Sciences and Engineering Research Council under Discovery Grant OPG-0194467.

References

1. S. E. Rogers, D. Kwak, and C. Kiris, Steady and unsteady solutions of the incompressible Navier-Stokes equations, *American Institute of Aeronautics and Astronautics Journal.* **29**(4), 603–610 (Apr., 1991).
2. D. Zingg, S. De Rango, M. Nemec, and T. Pulliam, Comparison of several spatial discretizations for the Navier-Stokes equations, *Journal of Computational Physics.* **160**, 683–704, (2000).
3. S. De Rango and D. W. Zingg, Higher-order spatial discretization for turbulent aerodynamic computations, *American Institute of Aeronautics and Astronautics Journal.* **39**(7), 1296–1304 (July, 2001).
4. T. J. Barth and P. O. Frederickson. Higher order solution of the Euler equations on unstructured grids using quadratic reconstruction. AIAA paper 90-0013 (Jan., 1990).
5. T. J. Barth. Recent developments in high order k-exact reconstruction on unstructured meshes. AIAA paper 93-0668 (Jan., 1993).
6. C. F. Ollivier-Gooch. High-order ENO schemes for unstructured meshes based on least-squares reconstruction. AIAA paper 97-0540 (Jan., 1997).
7. M. Delanaye and J. A. Essers, Quadratic-reconstruction finite volume scheme for compressible flows on unstructured adaptive grids, *American Institute of Aeronautics and Astronautics Journal.* **35**(4), 631–639 (Apr., 1997).
8. P. Geuzaine, M. Delanaye, and J.-A. Essers. Computation of high Reynolds number flows with an implicit quadratic reconstruction scheme on unstructured grids. In *Proceedings of the Thirteenth AIAA Computational Fluid Dynamics Conference*, pp. 610–619. American Institute of Aeronautics and Astronautics, (1997).
9. R. Abgrall, On essentially non-oscillatory schemes on unstructured meshes: Analysis and implementation, *Journal of Computational Physics.* **114**(1), 45–58, (1994).
10. O. Friedrich, Weighted essentially non-oscillatory schemes for the interpolation of mean values on unstructured grids, *Journal of Computational Physics.* **144**(1), 194–212 (July, 1998).
11. C. Q. Hu and C. W. Shu, Weighted essentially non-oscillatory schemes on triangular meshes, *Journal of Computational Physics.* **150**(1), 97–127 (Mar., 1999).

12. C. F. Ollivier-Gooch and M. Van Altena, A high-order accurate unstructured mesh finite-volume scheme for the advection-diffusion equation, *Journal of Computational Physics*. **181**(2), 729–752, (2002).
13. C. Ollivier-Gooch, A. Nejat, and C. Michalak, On obtaining and verifying high-order finite-volume solutions to the Euler equations on unstructured meshes, *American Institute of Aeronautics and Astronautics Journal*. **47**(9), 2105–2120, (2009). doi: 10.2514/1.40585.
14. A. Harten and S. Osher, Uniformly high-order accurate nonoscillatory schemes, *SIAM Journal on Numerical Analysis*. **24**(2), 279–309 (Apr., 1987).
15. C.-W. Shu and S. Osher, Efficient implementation of essentially non-oscillatory shock-capturing schemes, *Journal of Computational Physics*. **77**, 439–471, (1988).
16. C.-W. Shu and S. Osher, Efficient implementation of essentially non-oscillatory shock-capturing schemes, II, *Journal of Computational Physics*. **83**, 32–78, (1989).
17. X.-D. Liu, S. Osher, and T. Chan, Weighted essentially non-oscillatory schemes, *Journal of Computational Physics*. **115**, 200–212, (1994).
18. W. R. Wolf and J. L. F. Azevedo, High-order ENO and WENO schemes for unstructured grids, *International Journal for Numerical Methods in Fluids*. **55**(10), 917–943, (2007).
19. P. K. Sweby, High resolution schemes using flux limiters for hyperbolic conservation laws, *SIAM Journal on Numerical Analysis*. **21**(5), 995–1011 (Oct., 1984). doi: http://dx.doi.org/10.1137/0721062.
20. B. van Leer, Towards the ultimate conservative difference scheme. V. A second-order sequel to Godunov's method, *Journal of Computational Physics*. **32**, 101–136, (1979).
21. G. D. van Albada, B. van Leer, and W. W. Roberts, Jr., A comparative study of computational methods in cosmic gas dynamics, *Astronomy and Astrophysics*. **108**, 76–84 (Apr., 1982).
22. T. J. Barth and D. C. Jespersen. The design and application of upwind schemes on unstructured meshes. AIAA paper 89-0366 (Jan., 1989).
23. V. Venkatakrishnan, Convergence to steady-state solutions of the Euler equations on unstructured grids with limiters, *Journal of Computational Physics*. **118**, 120–130, (1995).
24. M. J. Berger and M. J. Aftosmis. Analysis of slope limiters on irregular grids. AIAA paper 2005-0490 (Jan., 2005).
25. A. Nejat and C. Ollivier-Gooch, A high-order accurate unstructured finite volume Newton-Krylov algorithm for inviscid compressible flows, *Journal of Computational Physics*. **227**(4), 2592–2609, (2008). doi: 10.1016/j.jcp.2007.11.011.
26. C. Michalak and C. Ollivier-Gooch, Accuracy preserving limiter for the high-order accurate solution of the Euler equations, *Journal of Computational Physics*. **228**(23), 8693–8711, (2009). doi: 10.1016/j.jcp.2009.08.021.
27. C. Michalak and C. Ollivier-Gooch, Globalized matrix-explicit Newton-GMRES for the high-order accurate solution of the Euler equations, *Computers and Fluids*. **39**, 1156–1167, (2010). doi: 10.1016/j.compfluid.2010.02.008.

28. Y. Saad and M. H. Schultz, GMRES: A generalized minimal residual algorithm for solving nonsymmetric linear systems, *SIAM Journal of Scientific and Statistical Computing.* **7**(3), 856–869 (July, 1986).
29. T. J. Barth and S. W. Linton. An unstructured mesh Newton solver for compressible fluid flow and its parallel implementation. AIAA paper 95-0221 (Jan., 1995).
30. A. Pueyo and D. W. Zingg. Improvements to a Newton-Krylov solver for aerodynamic flows. Thirty-sixth AIAA Aerospace Sciences Meeting (Jan., 1998). AIAA Paper 98-0619.
31. A. Nejat and C. Ollivier-Gooch, Effect of discretization order on preconditioning and convergence of a high-order unstructured Newton-GMRES solver for the Euler equations, *Journal of Computational Physics.* **227**(4), 2366–2386, (2008). doi: 10.1016/j.jcp.2007.10.024.
32. M. Blanco and D. W. Zingg. A Newton-Krylov algorithm with a loosely-coupled turbulence model for aerodynamic flows. AIAA Paper 2006-0691. Presented at the 44th AIAA Aerospace Sciences Meeting (Jan., 2006).
33. D. A. Venditti and D. L. Darmofal, Adjoint error estimation and grid adaptation for functional outputs: Application to quasi-one-dimensional flow, *Journal of Computational Physics.* **164**(1), 204–227 (Oct., 2000).
34. D. A. Venditti and D. L. Darmofal, Grid adaptation for functional outputs: Application to two-dimensional inviscid flows, *Journal of Computational Physics.* **175**(1), 40–69 (Feb., 2002).
35. D. A. Venditti and D. L. Darmofal, Grid adaptation for functional outputs: Application to two-dimensional viscous flows, *Journal of Computational Physics.* **187**, 22–46, (2003).
36. N. Pierce and M. Giles, Adjoint recovery of superconvergent functionals from PDE approximations, *SIAM Review.* **42**(2), 247–264 (June, 2000).

CHAPTER 10

A BIASED SHORT REVIEW OF RESIDUAL DISTRIBUTION SCHEMES FOR HYPERBOLIC PROBLEMS

Rémi Abgrall

Team Bacchus and Institut de Mathématiques
INRIA and Université de Bordeaux
341 cours de la Libération, 33 405 Talence cedex, France
remi.abgrall@inria.fr

We describe and review (non oscillatory) residual distribution schemes that are rather natural extension of high order finite volume schemes when a special emphasis is put on the structure of the computational stencil. We provide their connections with standard stabilized finite element and discontinuous Galerkin schemes, show that they are really non oscillatory and provide some research perspective.

1. Introduction

The numerical simulation of compressible flow problems, or more generally speaking, of partial differential equations (PDEs) of hyperbolic nature, has been the topic of a huge literature since the seminal work of von Neumann[33] in the 40's. Among the "hot" topics of the field has been, since the works of Lax,[21] Wendroff,[22] Godunov,[12] van Leer,[32] Roe,[30] Harten,[13] Osher,[25] Yee,[35] to give a few names, the development of robust, parameter free and accurate schemes. This field has had a very rapid and broad development since the work of MacCormack[24] and van Leer. Among the most successful methods one may quote van Leer's MUSCL method[32] and modified flux approach of Roe. These techniques are only second order accurate. The accuracy can be improved via the ENO/WENO methods[14,18,23,31] by Harten, Shu and others.

With the emergence of modern parallels computers, another concern has emerged: what about accuracy and efficiency ? Indeed, it is now important to develop robust algorithms that scale correctly on parallel architecture. This can be achieved more or less easily if the stencil of the numerical scheme is as compact as possible. Good candidates are the schemes relying on finite element technology, such as the Discontinuous Galerkin methods[11] or the stabilized continuous finite element methods.[17,19] In these methods, the numerical stencil is the most compact one.

2. Reinterpretation of Finite Volume Schemes

Let us start with a simple example. We consider the following problem

$$\frac{\partial u}{\partial t} + \frac{\partial f(u)}{\partial x} = 0 \tag{1}$$

with initial and boundary conditions that we do not specify for the moment. Using a regular mesh ($x_j = j\Delta x$), this problem is discretised by a simple finite volume scheme

$$\Delta x \, \frac{u_i^{n+1} - u_i^n}{\Delta t} + F_{i+1/2} - F_{i-1/2} = 0 \tag{2}$$

where $F_{j+1/2}$ is the numerical flux at the cell interface $x_{j+1/2} = \frac{x_j + x_{j+1}}{2}$ which depends on the local cell averages of the solution $\{u_l\}_{l=j-p}^{j+p}$, where

$$u_j \approx \frac{\int_{x_{j-1/2}}^{x_{j+1/2}} u(x,t)dx}{\Delta x}.$$

We can rewrite (2) as

$$\Delta x \, \frac{u_i^{n+1} - u_i^n}{\Delta t} + \phi_{i+1/2}^- + \phi_{i-1/2}^+ = 0 \tag{3}$$

where we have set

$$\phi_{i+1/2}^- = F_{i+1/2} - f(u_i), \qquad \phi_{i-1/2}^+ = f(u_i) - F_{i-1/2}.$$

In each interval $[x_i, x_{i+1}]$, we have introduced the "residuals"

$$\phi_{i+1/2}^- = F_{i+1/2} - f(u_i), \qquad \phi_{i+1/2}^+ = f(u_{i+1}) - F_{i+1/2}. \tag{4}$$

The two formulations (2) and (3) are of course equivalent.

If the numerical scheme $F_{j+1/2}$ is consistent with the continuous one, and depends continuously of its argument, assuming in addition some stability assumptions, the Lax Wendroff theorem states that the solution of (2) converges to a weak solution of (1). In the proof of this theorem, the key algebraic argument is that $F_{i+1/2} - F_{i-1/2}$ is a difference of flux. Considering now (4), this argument is translated into the relation

$$\phi_{i+1/2}^- + \phi_{i+1/2}^+ = f(u_{i+1}) - f(u_i) = \int_{x_i}^{x_{i+1}} \frac{\partial f(u^{\Delta x})}{\partial x} dx \tag{5}$$

where $u^{\Delta x} = u_i \frac{x_{i+1}-x}{\Delta x} + u_{i+1}\frac{x_i - x}{\Delta x}$. On can show, see for example[6] for a more complex case, that under the assumptions of the Lax Wendroff theorem (stability assumptions, and continuous dependency of the residuals with respect to their arguments), that the solution of (3) converges to a weak solution of (1).

The goal is to construct schemes of the type (3)-(5) that have the most possible compact stencil with the maximum accuracy. For example, second order accuracy can be obtained with a 3 point stencil (instead of 5 for a standard high order scheme). This is done in two steps. We first consider the steady version of (1) and then extend the method to the unsteady case. Of course the steady version of (1) is trivial, but is rather enlightening to consider the following problem

$$\begin{aligned} u' &= \lambda u \qquad x \in]0,1] \\ u(0) &= 1. \end{aligned} \tag{6}$$

The solution of (6) is $u(x) = e^{\lambda x}$.

If one wishes to approximate (6) by an upwind finite volume, a natural formulation is

$$F_{i+1/2} - F_{i-1/2} = \lambda \int_{x_{i-1/2}}^{x_{i+1/2}} u(x)dx \approx \Delta x \lambda u_i.$$

Note that the source term is approximated with second order accuracy, and $F_{i+1/2} = u_i$, $F_{i-1/2} = u_{i-1}$. The scheme is

$$u_i - u_{i-1} = \Delta x \lambda u_i, \qquad u_0 = 1. \tag{7}$$

We obtain $u_i = \left(1 - \lambda \frac{x_i}{i}\right)^i$: if i is chosen so that $x_i = i\Delta x$ is *fixed*, $u_i - u(x_i) = \frac{1}{2} e^{\lambda x_i} x_i \lambda^2 \Delta x + O(\Delta x^2)$: the convergence is only first order.

Consider now the scheme

$$u_i - u_{i-1} - \frac{\lambda \Delta x}{2}(u_i + u_{i-1}) = 0 \qquad \text{with } u_0 = 1, \tag{8}$$

we get

$$u_i = \left(\frac{1 + \frac{\lambda \Delta x}{2}}{1 - \frac{\lambda \Delta x}{2}} \right)^i$$

so that when i is chosen with $x_i = i\Delta x$ *fixed*,

$$u_i - u(x_i) = \frac{1}{12} e^{\lambda x_i} \lambda^3 x_i \Delta x^2 + O(\Delta x^4)$$

which shows so the convergence is second order. The scheme (8) can be interpreted in the residual distribution framework. To do that, we define the total residual by

$$\phi_{i+1/2} = \int_{x_i}^{x_{i+1}} (u_h' - \lambda u_h)dx = u_{i+1} - u_i - \lambda \frac{u_i + u_{i+1}}{2}$$

and the sub-residuals by

$$\phi^-_{i+1/2} = \phi_{i+1/2}, \qquad \phi^+_{i+1/2} = 0$$

i.e., we distribute on the downwind vertex of the cell $[x_i, x_{i+1}]$.

This simple example shows that one can maximize accuracy with the smallest stencil. This is precisely the philosophy that is pursued by the Residual Distribution schemes (RD schemes for short), with the goal of deriving non oscillatory schemes.

3. Residual Distribution Schemes

3.1. *Case of scalar problems*

The model problem. We first consider the steady problem

$$\text{div } f(u) = 0 \text{ in } \Omega \subset \mathbb{R}^d \tag{9a}$$

subjected to Dirichlet boundary conditions on the inflow part Γ^- of $\Gamma = \partial\Omega$,

$$u = g \text{ in } \Gamma^-. \tag{9b}$$

If $M \in \Gamma$ and $\vec{n}$ is the outward unit vector at M of Γ, the inflow boundary is defined as

$$\Gamma^- = \{M \in \Gamma, \nabla_u f(u(M)) \cdot \vec{n} < 0\}.$$

Approximation space. The domain Ω is triangulated by a conformal mesh as in Figure 1, the triangulation is denoted by $\mathcal{T}_h$. The elements of this triangulation are triangles and quads in 2D, or tetrahedrons in 3D. Other types of elements could certainly be tackled, but this has not yet been done. The elements of $\mathcal{T}_h$ are denoted by $\{K_i\}_{i=1,n_e}$ and the vertices are denoted by $\{M_i\}_{i=1,n_s}$. In most cases, we deal with one generic element K; since there is no ambiguity, the vertices are denoted by $i = 1, n_K$ where n_K is the number of vertices in K.

The approximate solution of (9) will be sought for in the space

$$V_h = \{u \text{ continuous in } \Omega_h, \text{ for any } u_{|K} \text{ is polynomial of degree r.}\}.$$

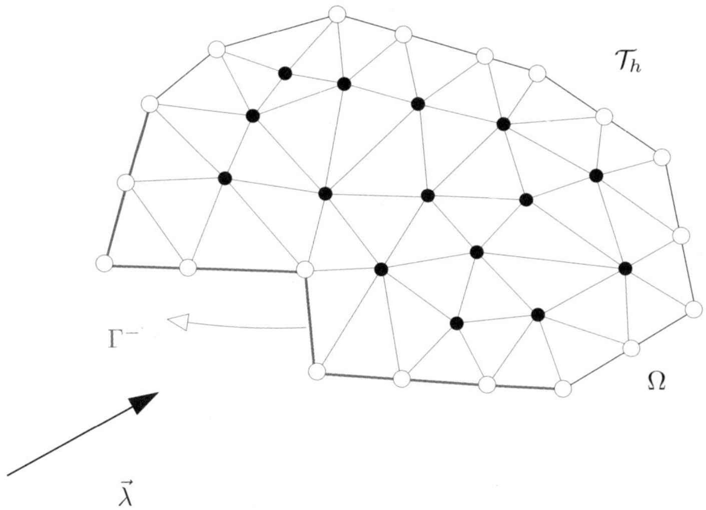

Fig. 1. A typical mesh.

In d dimensions, a polynomial of degree r is defined by $n_r^d = C_{d+r}^d$ coefficients, i.e. $\frac{(r+1)(r+2)}{2}$ in 2D and $\frac{(r+1)(r+2)(r+3)}{6}$ in 3D. This means that a polynomial is uniquely defined if a unisolvant set of points of cardinal n_r^d is given. In the case of triangles/tetrahedrons, the standard Lagrange points , for example by their barycentric coordinates $(\frac{i}{r}, \frac{j}{r}, \frac{k}{r})_{i,j,r\geq 0, i+j+r=d}$ in the case of a triangle and $(\frac{i}{r}, \frac{j}{r})_{i,j\geq 0, i+j\leq r}$ for a quad when it is mapped onto $[0,1]^2$. The Lagrange points are the degrees of freedom at which an approximation of u is sought for.

The class of triangulations that we consider are regular in the finite element meaning, i.e. there is a constant $C_{\mathcal{T}}$ such that if ρ_K is the ratio of the outer diameter of K to the inner diameter of K (so $\rho_K \geq 1$),

$$\max_{K\in\mathcal{T}_h} \rho_K \leq C_{\mathcal{T}}. \tag{10}$$

As classical, the parameter "h" in $\mathcal{T}_h$ refers to the maximum of the diameters of the elements contained in $\mathcal{T}_h$.

Numerical discretisation. The approximation is done in two steps. For any element K, we define the total residual

$$\phi_K(u^h) := \int_{\partial K} f(u^h)\cdot \vec{n} d\sigma \tag{11a}$$

which is splitted into the sub-residuals ϕ_j^K, one for each degree of freedom σ_j in K. Since there is no ambiguity, we denote these degrees of freedom by $j, j = 1, \dots, n_K$. The sub-residuals are constrained by the *conservation* relation

$$\sum_{\sigma_j \in K} \phi_{\sigma_j}^K(u^h) = \phi^K(u^h). \tag{11b}$$

The scheme writes : find $u^h \in V_h$ such that for any degree of freedom $\sigma \ni \Gamma^-$,

$$\sum_{K, \sigma \in K} \phi_\sigma^K(u^h) = 0 \tag{11c}$$

while on the boundary Γ^- we set

$$u^h(\sigma) = g(\sigma). \tag{11d}$$

Of course the problem (11) is in general a (very) non linear problem, which is in practice solved by an iterative technique. We later rapidly come back to this point.

There are a couple of general results which explain the type of structure the residuals and sub-residuals should have in order to guaranty accuracy and convergence to a weak solution of (9), if the method converges.

Convergence to a weak solution. We have the following result that has been shown in Ref. 6.

Proposition 10.1. *We consider a family of triangulations that satisfy* (10) *such that* $h \to 0$. *Assume that the sub-residuals depend continuously on* u^h, *that there exists a constant* C *independent of* h *such that*

$$\max_{\sigma \in \mathcal{T}_h} |u^h(\sigma)| \leq C$$

and a function $v \in L^2(\Omega)$ *such that a sub sequence* $u^{h_{n_k}} \to v$ *in* $L^2(\Omega)$ *when* $k \to +\infty$. *Then* v *is a weak solution of* (9).

The key argument is the conservation relation (11b). In (11a), the integral is generally obtained by numerical quadrature and the result is independent of the numerical quadrature, provided that on any edge/face of $\mathcal{T}_h$, the set of quadrature points only depend on the edge/face and not on the particular element this edge/face is part of.

Accuracy constraints. On an unstructured mesh, it is difficult, if not hopeless, to derive an error analysis via Taylor expansions, because the mesh has in general non geometrical symmetries. Hence, it is better to rely on a weak form of the truncation error. Consider φ a C^1 function on Ω with $||\varphi||_\infty \leq 1$. This inequality is set up for scaling purpose only. We define the truncation error for w^h, the interpolate of the exact solution of (9) (assuming it is smooth enough) by

$$\mathcal{E}_h(u) = \max_{\varphi\in C^1(\Omega), ||\varphi||_\infty \leq 1, ||\nabla\varphi||_\infty\leq 1} \left(\sum_{\sigma\in\mathcal{T}_h} \varphi(\sigma) \Big(\sum_{K,\sigma\in K} \phi_\sigma^K \Big) \right) \tag{12a}$$

and the scheme is p-th order accurate if there exists a constant C independent of h such that

$$\mathcal{E}_h(u) \leq Ch^p. \tag{12b}$$

There is a simple construction that permits, formally at least, to fulfill (12b), it relies on the use of the structure of (9) : it is a steady problem. The case of time dependent problem will be considered later. The key remark is that if for any σ and K, the sub-residuals (evaluated for an interpolation w^h of the exact solution, assuming it is smooth enough) satisfies

$$|\phi_\sigma^K(w^h)| \leq Ch^{p+d} \tag{13}$$

where C is independent of $\mathcal{T}_h$ satisfying (10), then (12b) holds. Again the proof is given in Ref. 6, and we recall it shortly. We introduce the Galerkin residuals,

$$\phi_\sigma^{T,c} = \int_K \psi_\sigma \text{ div } f(u^h)dx$$

and we have

$$\begin{aligned}\sum_{\sigma\in\mathcal{T}_h} \varphi(\sigma)\Big(\sum_{K,\sigma\in K}\phi_\sigma^K\Big) &= \sum_K \sum_{\sigma\in K} \varphi(\sigma)\phi_\sigma^T \\ &= \int_\Omega \varphi^h \text{ div } f(u^h)dx + \sum_K \left(\sum_{\sigma\in K} \varphi(\sigma)(\phi_\sigma^K - \phi_\sigma^{K,c}) \right)\end{aligned}$$

The next step is to see that $\sum_{\sigma\in K} \phi_\sigma^{T,c} = \phi^K$ so that we have

$$\sum_{\sigma\in K} \varphi(\sigma)(\phi_\sigma^K - \phi_\sigma^{K,c}) = \frac{1}{n_K!} \sum_{\sigma,\sigma'\in K} (\phi(\sigma') - \phi(\sigma))(\phi_\sigma^K - \phi_\sigma^{K,c}).$$

Then, we make the following remark: if the exact solution of *steady* version (9) is smooth enough, then for any σ and K

$$\phi_\sigma^K(w^h) = O(h^{k+d}) \text{ and } \phi^K(w^h) = O(h^{k+d}).$$

Let us look at the first relation. The second one is the sum of the first ones. We have

$$\begin{aligned}
\phi_\sigma^K(w^h) &= \int_K \psi_\sigma \text{ div } f(w^h)dx \\
&= \int_K \psi_\sigma \text{ div } \Big(f(w^h) - f(u) \Big) dx \\
&= \int_{\partial K} \psi_\sigma \Big(f(w^h) - f(u) \Big) \cdot \vec{n} dx - \int_K \nabla \psi_\sigma \Big(f(w^h) - f(u) \Big) \\
&= O(h^{d-1}) \times O(h^{k+1}) + O(h^d) \times O(\frac{1}{h}) \times O(h^{k+1}) \\
&= O(h^{k+d}).
\end{aligned}$$

To get the second line, we explicitly use the fact that $\int_K \psi_\sigma \text{ div } f(u)dx = 0$ because the problem is steady, the second line comes from the Gauss theorem, the third line use that fact that f is Lipschitz continuous, $w^h - u = O(h^{k+1})$ and the regularity assumption of the mesh.

Thanks to this, we see that

$$\int_\Omega \varphi^h \text{div } f(w^h) = O(h^{k+1})$$

and

$$\sum_K \Big(\sum_{\sigma \in K} \varphi(\sigma)(\phi_\sigma^K - \phi_\sigma^{K,c}) \Big) = O(h^{k+1})$$

if $\phi_\sigma^K = O(h^{k+d})$ again if the mesh is regular. Indeed, we have

$$\begin{aligned}
\mathcal{E}' = \sum_K \Big(\sum_{\sigma \in K} & \varphi(\sigma)(\phi_\sigma^K - \phi_\sigma^{K,c}) \Big) \\
&= N_K \times n_K \times O(\nabla \varphi) \times h\big(\times O(\phi_\sigma^K) + O(\phi_\sigma^{K,c}) \big).
\end{aligned}$$

The mesh is regular so that the number N_K of elements is $O(h^{-d})$, n_K is fixed, $O(\phi_\sigma^{K,c}) = O(h^{k+d})$ and $\phi_\sigma^K = O(h^{k+d})$, hence

$$\mathcal{E}' = O(h^{-d}) \times O(h) \times O(h^{k+d}) = O(h^{k+1}).$$

This shows that if $\phi_\sigma^K = O(h^{k+d})$ then

$$\mathcal{E}_h(u) \leq C h^{k+1}$$

and the scheme is (formally) $k+1$-th order accurate.

This analysis leads to residuals of the form

$$\phi_\sigma^K = \beta_\sigma^K \phi_K \tag{14}$$

where the family $\{\beta_\sigma^K\}_{\sigma,K}$ is uniformly bounded when $h \to 0$. In the next paragraph, we discuss the construction of scheme of the form (14) that are both formally high order accurate *and* L^∞ stable. In many cases, one can see experimentally that the schemes (14) are over-compressive. This can be cured if one adds dissipation. This can be done without violating (13) by adding a selected form of dissipation, namely

$$\phi_\sigma^K = \beta_\sigma^K \phi_K + \theta h_K \int_K \big(\nabla_u f(u^h) \cdot \nabla \psi_\sigma\big) \tau \big(\nabla_u f(u^h) \cdot \nabla u^h\big) dx \qquad (15)$$

where θ is a positive parameter. This form of dissipation is reminiscent of the stabilization term of the SUPG scheme[17] but here, as shown later, it plays the role of filter. In practice, the positive parameter τ is set to

$$\tau = \left(\sum_{\text{i: vertices of K}} \max(\nabla_u f(\overline{u^h}) \nabla \Psi_i, 0) \right)^{-1}. \qquad (16)$$

In (16), we only consider the vertices of K, and the Ψ_is are the lowest order finite element constructed on K : linear polynomials for triangles and tets, Q^1 for quads and hex, etc. Last $\overline{u^h}$ is the arithmetic average on the degrees of freedom in K.

Remark 10.1 (About the effective accuracy). *In practice, we are never able to exactly satisfy* (11c) *for any degree of freedom, but*

$$\sum_{K,\sigma \in K} \phi_\sigma^K(u^h) = \varepsilon_\sigma. \qquad (17)$$

The previous truncation error analysis leads to

$$\mathcal{E}_h(u) \le C h^{k+1} + \sum_K \varepsilon_\sigma$$

and the same analysis shows that we need to have

$$\max_\sigma |\varepsilon_\sigma| = O(h^{k+d+1}). \qquad (18)$$

Getting both accuracy and stability. All the known RD schemes have the form

$$\phi_\sigma^K = \sum_{\sigma' \in K} c_{\sigma\sigma'}(u_\sigma - u'_\sigma).$$

It is well known that if $c_{\sigma\sigma'} \ge 0$, and if a solution of (11) exists, it satisfies a maximum principle. Hence, we are going to construct schemes of the form (14) with positive $c_{\sigma\sigma'}$. This is done in two steps.

- First step. We construct a family of sub-residuals that ensures first order accuracy and stability in L^∞. The simplest choice is an extension of the local Lax Friedrichs scheme:

$$\phi_\sigma^{K,LxF} = \frac{\phi^K}{n_K} + \alpha\big(u_\sigma - \overline{u}_K\big) \tag{19}$$

with

$$\overline{u} = \frac{\sum_{\sigma\in K} u_\sigma}{n_K} \text{ and } \alpha \geq \max_K ||\nabla_u f(u^h)||.$$

These choices guaranty that the scheme is L^∞ stable, and more precisely we have

$$c_{\sigma\sigma'} = \frac{1}{n_K}\bigg(\int_T \nabla_u f(u^h)\cdot\nabla\psi_{\sigma'}dx + \alpha\bigg).$$

- We define $(\phi_\sigma^K)^\star = \beta_\sigma^K \phi^K$ with

$$\beta_\sigma^K = \frac{\max(0, \frac{\phi_\sigma^{K,LxF}}{\phi_K})}{\sum\limits_{\sigma'\in K} \max(0, \frac{\phi_{\sigma'}^{K,LxF}}{\phi_K})}. \tag{20}$$

This is one of the many choices that guaranty $(\phi_\sigma^K)^\star = \sum_{\sigma'\in K} \widetilde{c}_{\sigma\sigma'}(u_\sigma - u_{\sigma'})$ with $\widetilde{c}_{\sigma\sigma'} \geq 0$. It is constructed following:

$$\begin{aligned}(\phi_\sigma^K)^\star &= \frac{(\phi_\sigma^K)^\star}{\phi_\sigma^{K,LxF}}\phi_\sigma^{K,LxF}\\ &\sum_{\sigma'\in K}\frac{(\phi_\sigma^K)^\star}{\phi_\sigma^{K,LxF}} c_{\sigma\sigma'}^{LxF}(u_\sigma - u_{\sigma'}).\end{aligned}$$

Then we set $\widetilde{c}_{\sigma\sigma'} = \dfrac{(\phi_\sigma^K)^\star}{\phi_\sigma^{K,LxF}} c_{\sigma\sigma'}^{LxF}$ which is positive if (and only if) $(\phi_\sigma^K)^\star \times \phi_\sigma^{K,LxF} \geq 0$ that is,

$$\beta_\sigma^K \frac{\phi_\sigma^{K,LxF}}{\phi_K} \geq 0.$$

The relation (20) is obtained by satisfying these relations for any $\sigma \in K$.

3.2. *Extension to systems*

In the case of systems,

$$\text{div } \mathbf{f}(\mathbf{u}^h) = 0$$

the generalization is straightforward: no modification is needed except the way the coefficients β_σ^K is evaluated. We first note that since ϕ_K and ϕ_σ^{LxF} are vectors, the construction (20) is meaningless. This is why we rely on a characteristic decomposition of the total and sub-residuals. More precisely, we consider a direction $\vec{d}$, the left and right eigen-vectors of $K_{\vec{d}} := \nabla \mathbf{f}(\overline{\mathbf{u}}) \cdot \vec{d}$. They are denoted, respectively, by $\{\mathbf{r}_\xi\}_{\xi \text{ eigenvalues of A}}$ and $\{\ell_\xi\}_{\xi \text{ eigenvalues of } K_{\vec{d}}}$. By construction, we have $\ell_\xi(\mathbf{r}_{\xi'}) = \delta_\xi^{\xi'}$.

Consider a set of first order residuals $\{\phi_\sigma^{K,L}\}_{\sigma \in K}$ with

$$\sum_{\sigma \in K} \phi_\sigma^{K,L} = \phi_K = \int_{\partial K} \mathbf{f}(\mathbf{u}^h) \cdot \vec{n} dl.$$

An example is given by the Lax Friedrichs residuals,

$$\phi_\sigma^{K,LxF} = \frac{1}{n_K}\phi_K + \alpha_K(u_\sigma - \overline{u}) \qquad \text{with } \overline{u} = \frac{1}{n_K}\sum_{\sigma \in K} u_\sigma.$$

We decompose the residuals $\phi_\sigma^{K,L}$ onto the eigen-basis,

$$\phi_\sigma^{K,L} = \sum_{\xi \text{ eigenvalues of } K_{\vec{d}}} \ell_\xi(\phi_\sigma^{K,L})\mathbf{r}_\xi. \tag{21a}$$

By construction, we have, for any ξ,

$$\sum_{\sigma \in K} \ell_\xi(\phi_\sigma^{K,L}) = \ell_\xi(\phi_K) \tag{21b}$$

and we remark that the characteristic $\ell_\xi(\phi_\sigma^{K,L})$ are scalar quantities. We can apply the same technique as in the scalar case to them. For example, using (20), we define

$$\ell_\xi(\phi_\sigma^{K,L})^\star = \beta_\sigma^{K,\xi}\ell_\xi(\phi_K) \tag{21c}$$

and then the high order residuals are

$$\phi_\sigma^{K,\star} = \sum_{\xi \text{ eigenvalues of } K_{\vec{d}}} \ell_\xi(\phi_\sigma^{K,L})^\star \mathbf{r}_\xi. \tag{21d}$$

The last step, as in the scalar case, is to add a dissipation term, as in (15). By analogy, the final residual is

$$\phi_\sigma^K = \phi_\sigma^{K,\star} + \theta h_K \int_K \nabla\psi_\sigma \tau \nabla_u f(u^h) \cdot \nabla u^h dx. \tag{21e}$$

The matrix τ is constructed by analogy to (16), namely

$$\tau = \left(\sum_{\text{i: vertices of K}} \max(\nabla_u f(\overline{u^h}) \nabla \Psi_i, 0) \right)^{-1} \tag{21f}$$

and again $\overline{u^h}$ is the arithmetic average of the solution over the degrees of freedom and $\max(A, 0)$ is the positive part of the matrix A which is assumed to be diagonalisable in $\mathbb{R}$ with real eigenvalues.

We have left unclear the choice of $\vec{d}$. In practice, we choose $\vec{d} = \vec{u}/||\vec{u}||$ and an arbitrary direction if $\vec{u} = 0$. The many experiments we have conducted shows that the non oscillatory behavior of the scheme is independent of the choice of $\vec{d}$. Of course for any direction choice will correspond a particular scheme, but all have the same non oscillatory behavior. The specific choice is motivated by keeping the rotational invariance of the scheme.

Boundary conditions. We have used a simplified version of the boundary conditions. If an element K has an edge, Γ_K, on the boundary, we need to add to the degrees of freedom on Γ_K a boundary residual. We denote it by $\Phi_\sigma^{\Gamma_K}$. These residuals should satisfy the conservation relation

$$\sum_{\sigma \in \Gamma_K} \Phi_\sigma^{\Gamma_K} = \int_{\Gamma_\sigma} (\mathcal{F}_n(u^h) - f(u^h) \cdot \vec{n}) dl$$

where $\mathcal{F}_n$ is a boundary flux. In the examples of this chapter, two types of boundary are considered:

- *Wall boundary conditions* . The condition $\vec{u} \cdot \vec{n} = 0$ is weakly imposed so that

$$\mathcal{F}_n(u^h) = \begin{pmatrix} 0 \\ p(u^h) n_x \\ p(u^h) n_y \\ 0 \end{pmatrix}$$

- *Inflow/outflow boundary conditions.* The state at infinity is U_∞ and we take here the modified Steger-Warming flux

$$\mathcal{F}_n(u^h) = \big(A(u^h) \cdot \vec{n}\big)^+ u^h + \big(A(u^h) \cdot \vec{n}\big)^+ u_\infty.$$

By analogy with what is done in,[1] we have chosen a 'centered' version of the boundary residuals, namely

$$\Phi_\sigma^{\Gamma_K} = \int_{\Gamma_K} (\mathcal{F}_n(u^h) - f(u^h) \cdot \vec{n}) \psi_\sigma(x) dl$$

where again ψ_σ is the Lagrange basis function defined in K for σ. This is approximated by a quadrature formula with positive weights. The quadrature formula should be of order $k + d - 1$, i.e. 3 for a third order scheme in 2D. The actual residual is

$$\Phi_\sigma^{\Gamma_K} = |\Gamma_K| \sum_{\text{quadrature points}} \omega_{quad}\big(\mathcal{F}_n(u^h) - f(u^h)\big)(x_{\text{quad}}) \cdot \vec{n}. \tag{22}$$

In the case of interest ($\mathbb{P}^2/\mathbb{Q}^2$ interpolation), we approximate these relation with Simpson's formula : only one term appears in the sum and it corresponds to σ.

All the meshes we have used are made of triangles or quadrangles. We have used two type of boundary representation. In the first one we adopt a piecewise linear representation of the boundary but we might be quite far from the true geometry. In the second representation, we use a quadratic representation of the geometry. In principle, the situation should be better, but one has to be aware of two difficulties. First, the "numerical" representation of the boundary is not C^1 in general, even if the boundary is C^∞. An example is provided in Figure 2 where we approximate the boundary of a NACA012 airfoil near the symmetry axis. The second problem is that even very simple geometries, such as circle, will not be represented exactly.

The second drawback could be solved by using NURBS representation of the boundary, see section 5, the first one is here solved as follows: instead of trying to interpolate *exactly* in each boundary segment the boundary curve, we use a Bézier representation which amount to interpolate at the boundary points and respect the tangents at these points. We get an approximate

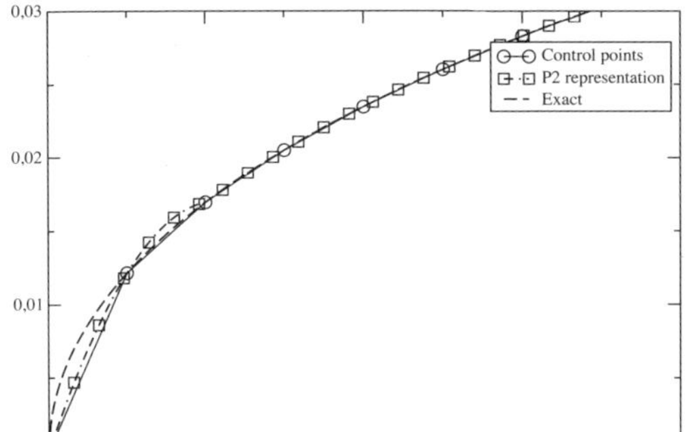

Fig. 2. Comparison with the true geometry between the two boundary representation methods used in this paper. The degrees of freedom are represented by circles.

quadratic representation of the boundary. This is the method we have used in practice.

In order to simplify the coding, we have used an isoparametric representation of each element, even for the interior elements. The filtering operator is adapted to this context : we need a exact evaluation of the gradient and divergence operators.

4. Numerical Examples

We illustrate the previous paragraph by numerical examples.

4.1. *Role of the filtering parameter*

We start with the advection problem with initial states and advection speeds defined by

$$\vec{\lambda} = (1,2)^T \quad \text{and } u(x,y) = \begin{cases} 1 & \text{if } x = 0 \text{ and } y > 0 \\ 0 & \text{if } y = 0 \text{ and } x > 0. \end{cases} \tag{23}$$

The second problem is obtained by setting

$$\vec{\lambda} = (y, -x)^T \quad \text{and} \quad u(x,y) = \begin{cases} \varphi_0(x) & \text{if } y = 0 \\ 0 & \text{otherwise} \end{cases} \tag{24}$$

where

$$\varphi_0(x) = \begin{cases} \cos^2(2\pi x) & \text{if } x \in [0.25, 0.75] \\ 0 & \text{else.} \end{cases}$$

The meshes are made of triangles, but this is not essential in the discussion. Figure 4 show the solution obtained for (23) and (24) by the scheme using $\mathbb{P}^2$ element *without* the term (15), while Figure 5 show the same results with (15). The problem (23) is well resolved without the τ term as it can be seen in Figure 4, top-left, but the cross-section (top-right) shows that the solution looks wiggly in the discontinuity. This is not an instability mechanism, since we can show that the scheme is perfectly stable in the L^∞ norm. The same comments can be done on the solution of problem (24), which, a priori, should be simpler : it is a smooth solution. In fact the situation looks even worse. We emphasis again on the fact that these “wiggles” are not a manifestation of an instability mechanisms. In fact, the scheme appears too compressive, and in Ref. 5, we give an heuristic explanation of the cause of this phenomenon.

Let us come back to the numerical examples, and in particular the results of Figure 5 where (15) has been added. One concern is that when adding (15), the scheme do not any longer preserve the maximum principle. The left picture in Figure 5 shows that, for the discontinuous solution of problem (23), we do not get any spurious oscillations. The right picture instead shows, for problem (24), the positive effect of the extra term in smoothing the contours that now are perfectly circular. We have also run a grid refinement study on this problem using $\mathbb{P}^2$ and $\mathbb{P}^3$ approximations. The results are summarized on table 1. The slope are obtained by least squares fitting, this confirms the expected convergence rates.

To better visualize the improvement in the solution when going from $\mathbb{P}^1$ to $\mathbb{P}^2$ spatial interpolation, we consider, on the spatial domain $[0,2]\times[0,1]$, the solid body rotation of the inlet profile $u(x) = \sin(10\pi x)$. In this case, the advection speed is set to $\vec{\lambda} = (y, 1-x)$. Note that the $\mathbb{P}^1$ run has been performed on the mesh obtained by sub-triangulating the $\mathbb{P}^2$ mesh so that exactly the same number of DOF is used in the two cases. The dramatic improvement brought by the $\mathbb{P}^2$ approximation is clearly visible in the contour plots, and also in the outlet profiles reported in Figure 3.

We test further the behavior of (15)-(20) by solving the 2D Burger's problem

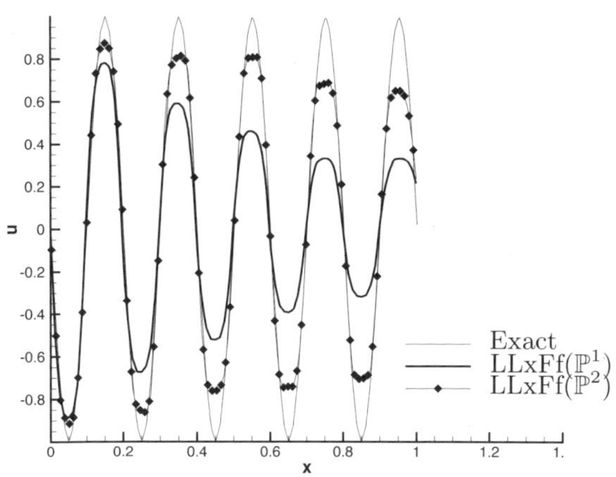

Fig. 3. Rotation of the smooth profile: $u_{\text{in}} = \sin(10\pi x)$. Computed outlet profile. All computations run on the same number of degrees of freedom. Reference mesh size $h = 1/80$.

$$\frac{\partial u}{\partial y} + \frac{1}{2}\frac{\partial u^2}{\partial x} = 0 \quad \text{if } x \in [0,1]^2$$
$$u(x,y) = 1.5 - 2x \quad \text{on } y = 0.$$

The exact solution consists in a fan that merges into a shock which foot is located at $(x,y) = (3/4, 1/2)$. More precisely, the exact solution is

$$u(x,y) = \begin{cases} \text{if } y \geq 0.5 \begin{cases} -0.5 \text{ if } & -2(x-3/4) + (y-1/2) \geq 0 \\ 1.5 & \text{else} \end{cases} \\ \text{else} \qquad \max\Big(-0.5, \min\Big(1, 5, \dfrac{x-3/4}{y-1/2}\Big)\Big). \end{cases}$$

The results obtained on the mesh in Figure 4 are displayed in Figure 6. For the sake of comparison, we give the second and third order results on the same mesh (hence the $\mathbb{P}^2$ results have more degrees of freedom).

We note that there are no spurious oscillation across the shock. We had the same conclusions on all the test cases we have run, even in the non convex case. This indicates that though the term (15) prevent a formal maximum principle, its role is very different to what it is in a SUPG like scheme: it only filters spurious modes, has no role in the stability and helps to converge the iterative scheme so that the error in (17) really behaves like (18).

h	$\epsilon_{L^2}(P^1)$	$\epsilon_{L^2}(P^2)$	$\epsilon_{L^2}(P^3)$
1/25	0.50493E-02	0.32612E-04	0.12071E-05
1/50	0.14684E-02	0.48741E-05	0.90642E-07
1/75	0.74684E-03	0.13334E-05	0.16245E-07
1/100	0.41019E-03	0.66019E-06	0.53860E-08
	$\mathcal{O}^{\text{ls}}_{L^2}$ =1.790	$\mathcal{O}^{\text{ls}}_{L^2}$ =2.848	$\mathcal{O}^{\text{ls}}_{L^2}$ =3.920

4.2. *Compressible flow examples*

We have run many test cases ranging from low subsonic, subsonic, transonic to supersonic flows. We only select two cases: one subsonic flow where we show the behavior of the scheme depending on the mesh structure and a supersonic one. In the latter case, the concern is not in the accuracy but

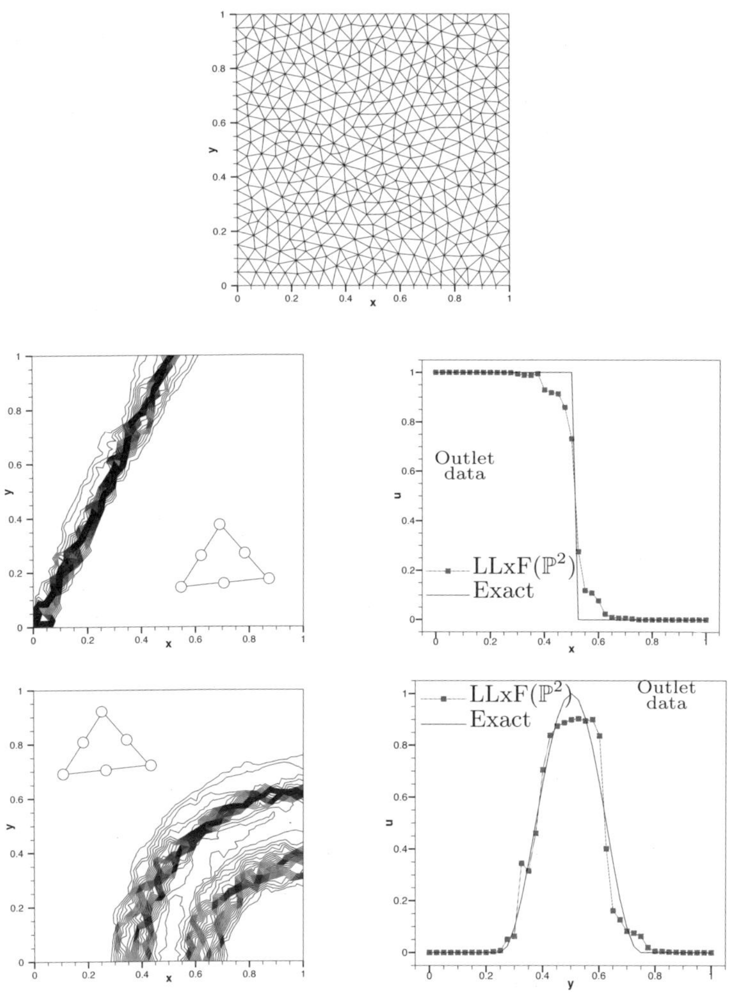

Fig. 4. Convection problem : Results obtained with scheme (14)–(20) for $\mathbb{P}^2$ interpolation. Top : mesh. Middle : result for problem (23). Bottom : results for problem (24). The first order scheme is (19).

on the robustness of the scheme since the solution presents very complex waves interactions.

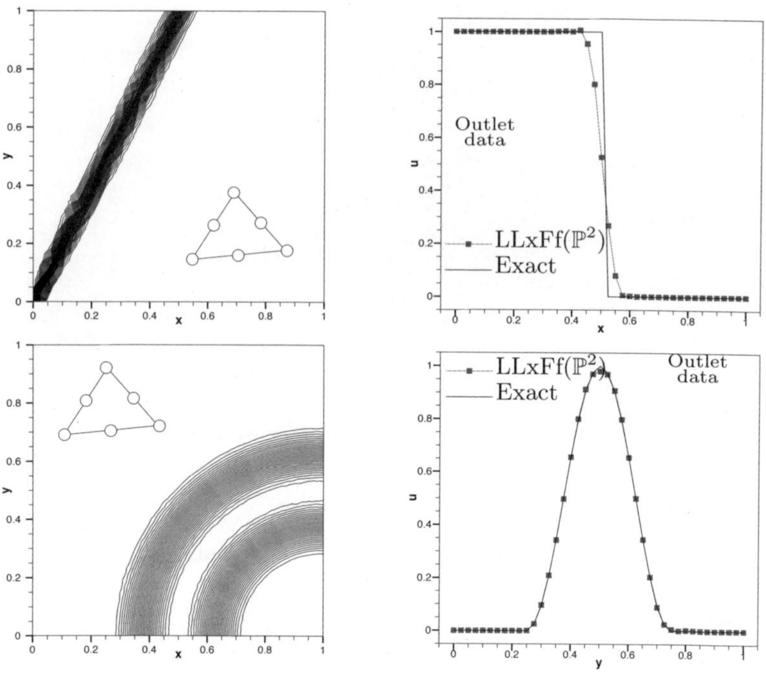

Fig. 5. Rotation problem : Results obtained with the scheme (15)–(20) for $\mathbb{P}^2$ interpolation. Top : result for problem (23) (min $= -1.0094$, max $= 1.01$). Bottom : results for problem (24) (min $= -0.1735\ 10^{-4}$). The first order scheme is (19).

4.2.1. *Subsonic flows*

We have run the case of a flow at $M_\infty = 0.35$ over a sphere. In that case, the flow is symmetric with respect to the x–axis of the domain, but also with respect to the y axis. The flow stays subsonic, so that an easy accuracy criteria is the behavior of the entropy. We have run this case with a second order scheme, a third order scheme, and again the second order scheme on the mesh that has the same degrees of freedom as those of the $\mathbb{P}^2$ scheme. In other words, we subdivide each triangle into 4 smaller triangles which vertices are those of the large triangle and the mid–edges points. The initial mesh has 2719 nodes, 5308 elements and 100 nodes on cylinder. It is displayed in Figure 7.

We see in Figure 8 which displays the pressure coefficient isolines the improvement of the solution quality when the scheme is upgraded from

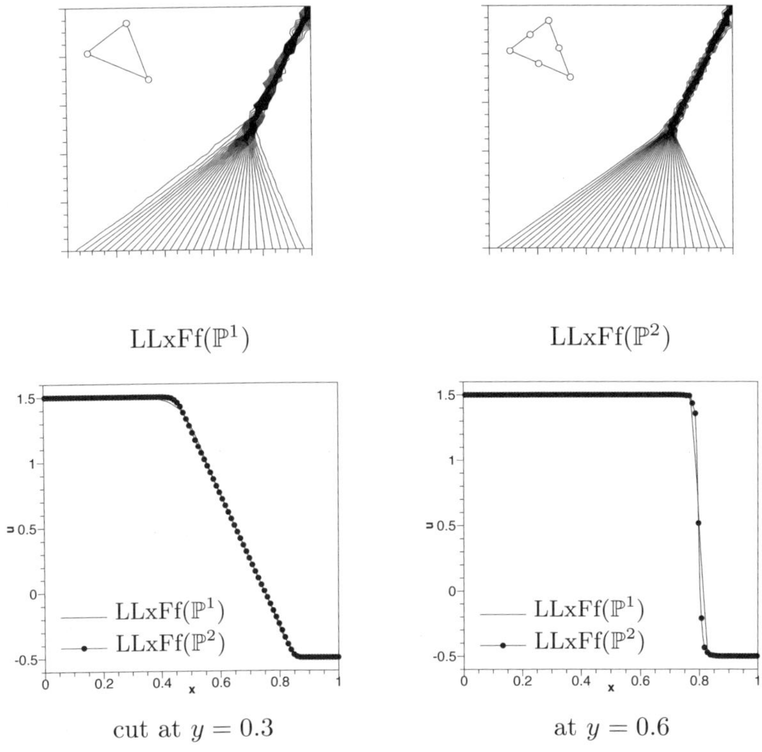

Fig. 6. Burger equation, solution obtained with a $\mathbb{P}^1$ and $\mathbb{P}^2$ Lagrange interpolated and the scheme (15)-(20).

second order to third order. More important, the same Figure indicates clearly that the second order scheme on the refined mesh gives less accurate results than the third order one. Note that we have the same degrees of freedom in both cases.

This result is confirmed by Figure 9 which displays the entropy variation along the boundary. Except at the forefront stagnation point, the entropy deviation of the third order scheme is much closer than the exact one.

We have re-run this test case on an hybrid mesh using the second order and the third order schemes. In both cases, the same degrees of freedom are used (i.e. we use the dofs of the sub-triangulation for the second order scheme). The results are shown in Figure 10. The mesh use 81 points on the sphere. We get the same conclusions as before.

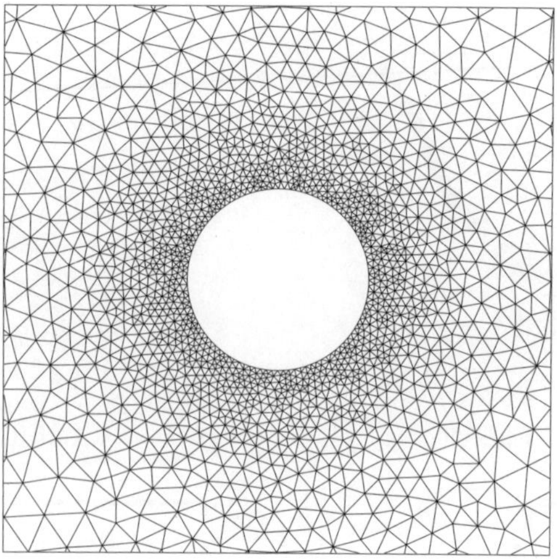

Fig. 7. Subsonic sphere problem : Zoom of the mesh for the sphere problem. The mesh has no symetry.

4.2.2. *Scramjet*

We have run the same scheme on a scramjet–like configuration using an hybrid mesh as shown in Figure 11. The inflow mach number is set to 3.5. The geometry is such that many waves coexist and interact in very complex flow patterns. This situation is particularly clear on the upper part of the internal body where shocks, fans and their reflection due to wall interact. Again, in both cases, the same number of degrees of freedom have been used. Once again, the scheme has been run starting from a uniform flow configuration. Figure 12 shows the Mach number isolines. As expected, there is no real difference between the solutions since the flow is basically made of shock, fans, slip lines and constant states : this is not an accuracy case, but a case that shows that, despite the flow complexity, the third order scheme is robust.

However, one can see a small difference between the solutions : the slip line created by the interaction of two shocks after the blade is a little bit more twisted for the third order scheme than the second order one. We also see that the resolution of the discontinuities is in both cases approximately one cell wide.

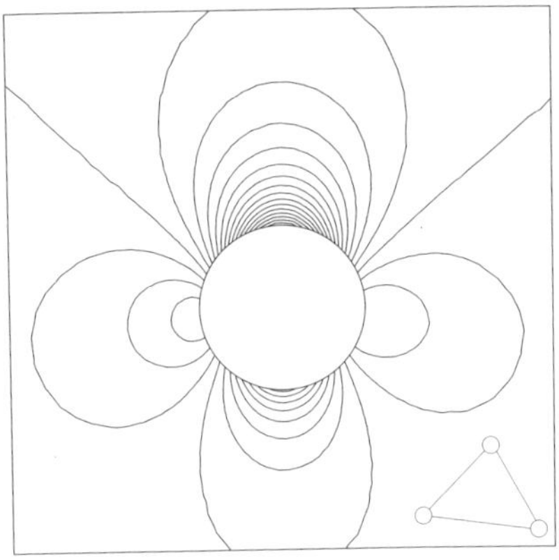

Second order

Second order using the P^2 dofs

third order scheme

Fig. 8. Subsonic sphere problem : Isolines of the pressure coefficient. We have the same isolines in each figure.

5. Extensions

This method can be extended along several directions: unsteady problems, a more complex model such as the (laminar) Navier Stokes equations, different models such as the Shallow water system (see[28] for an extension of the second order scheme for problems including dry beds), or the MHD equations.[4] We quickly cover the unsteady case and the viscous problems.

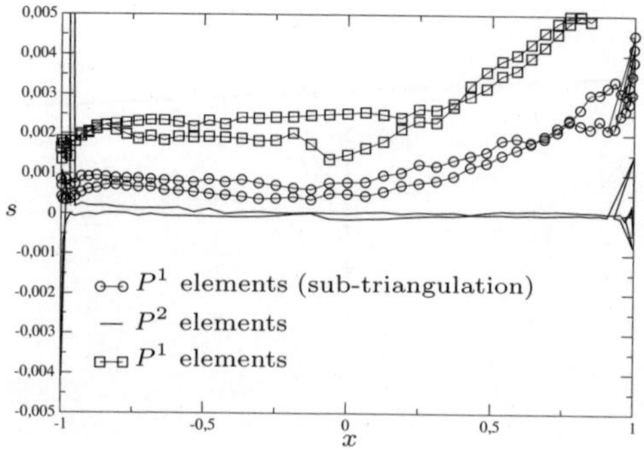

Fig. 9. Subsonic sphere problem : Entropy variation along the boundary.

5.1. *Unsteady problems*

As seen above, the main reason why the schemes can reach arbitrary order of accuracy is because the residual behave, in the case of a smooth enough solution, like

$$\phi_\sigma^K(w^h) = O(h^{k+d})$$

where d is the physical dimension and k the expected order of accuracy. To get this behavior, there are two key ingredients

- the interpolation of the smooth solution of the problem is of order $k+1$,
- we run a steady problem: the fact that

  $$\int_{\partial K} f^h(u^h) \cdot \vec{n} dl = 0$$

  for any element plays a central role.

Because of that, one cannot extend these schemes to unsteady problems via a time/space splitting approach. If this is done, one only get first order accuracy : we need to introduce the structure of the PDE, div $f(u^h) = 0$ somewhere, somehow, in the numerics.

The first natural idea is to consider the time/space problem

$$\frac{\partial u}{\partial t} + \text{div } f(u^h) = 0$$

as a whole. In the RD approach, this has been done by Refs. 10, 15 and 29 to give a few examples. This leads to implicit schemes with possibly stability

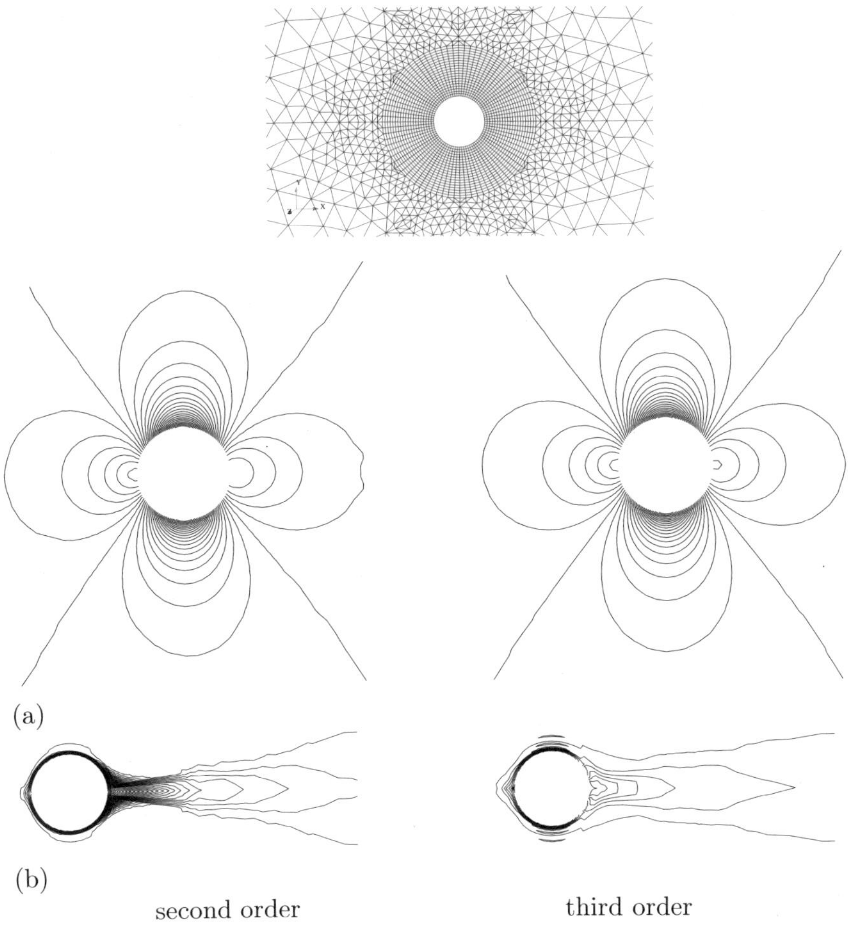

Fig. 10. Subsonic sphere problem, hybrid mesh: (a) Pressure coefficient and (b) entropy variation on an hybrid mesh, $M_\infty = 0.35$.

constraints. These stability constraints can be removed by a "two-layers" technique, see Ref. 29 and then Ref. 10 for details. A simpler method is described in Ref. 15, it uses discontinuous in time finite elements.

The second natural idea is to "pre-discretise" in time, as it is standard in finite element methods. For example, second order accuracy can be reached either by starting from a Crank-Nicholson scheme

$$\frac{u^{n+1} - u^n}{\Delta t} + \frac{1}{2}\Big(\text{ div } f(u^n) + \text{ div } f(u^{n+1})\Big) = 0$$

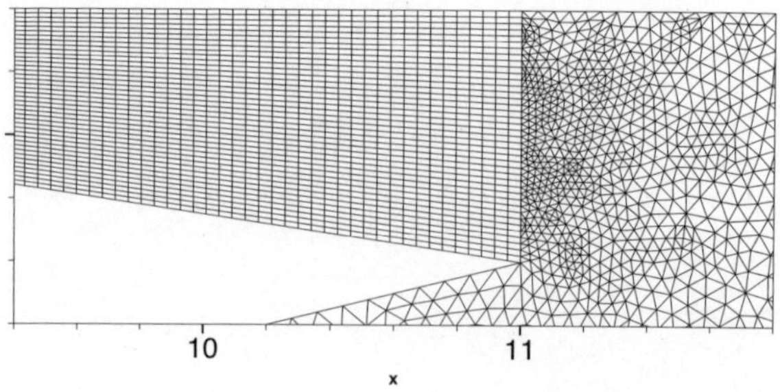

Fig. 11. Zoom of the mesh for the Scramjet problem.

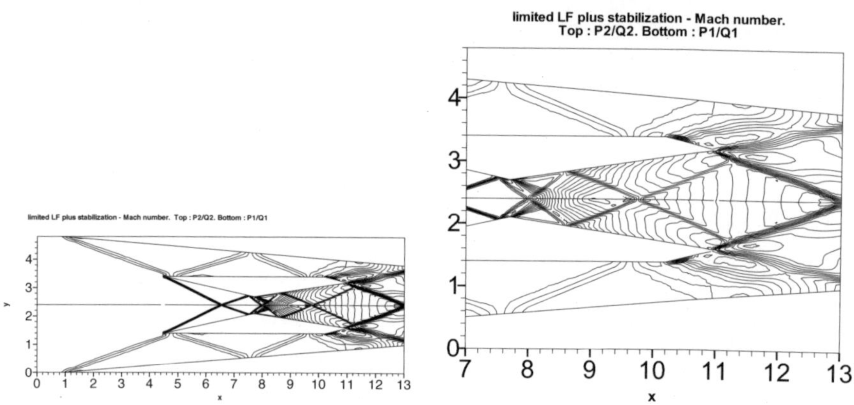

zoom

Fig. 12. Scramjet problem. Mach number distribution. Top : the third order solution, bottom the second order solution. The same isolines are plotted.

or a BDF-like approach

$$\frac{3}{2}\frac{u^{n+1}-u^n}{\Delta t}-\frac{1}{2}\frac{u^n-u^{n-1}}{\Delta t}+\text{ div } f(u^{n+1})=0.$$

In both cases, we end up to solving a problem of the form

$$\alpha v+\text{ div } f(v)-S(x)=0$$

where $v:=u^{n+1}$, $\alpha=1/\Delta t$ in the Cranck Nicholson case and $\alpha=\frac{3}{2\Delta t}$ in the BDF case. The only difference with the previous case is the definition of the total residual. It is naturally

$$\phi^K = \int_K (\alpha v - S(x))dx + \int_{\partial K} f(u^h) \cdot \vec{n} dx.$$

The inclusion of the source term in the total residual is dictated again by accuracy considerations. This approach has been considered in Ref. 3, then extended to flow problems (unpublished).

A much more interesting approach, because it is explicit and very cheap, as well as needing very little modifications of the computer code has been proposed in Ref. 26, only for second order space-time schemes so far with triangular meshes. One example of such scheme for

$$\frac{\partial u}{\partial t} + \text{ div } f(u^h) = 0$$

is: Starting from $v^0 := u^n$

(1) First step. One evaluates v^1 by the scheme

$$|C_\sigma| \frac{v^1_\sigma - v^0_\sigma}{\Delta t} + \sum_{\sigma \in T} \psi^T_\sigma(v^0) = 0$$

with

$$\psi^T_\sigma(v^0) = \beta^T_\sigma(v^0) \int_{\partial T} f(v^0) \cdot \vec{n} dl.$$

We note that in the $\mathbb{P}^1$ case, the filtering term can also be written as $\gamma^T_\sigma \int_{\partial T} f(v^0) \cdot \vec{n} dl$, this is why the previous relation can cover all cases.

(2) Second step. Knowing v^0 and v^1, we define v^2 as

$$|C_\sigma| \frac{v^2 - v^1}{\Delta t} + \sum_{\sigma \in T} \beta^T_\sigma(v_0, v^1) \psi^T_\sigma(v^0, v^1) = 0$$

with

$$\psi^T_\sigma(v^0, v^1) = \int_K \frac{v^1 - v^0}{\Delta t} + \int_{\partial K} f(u^h) \cdot \vec{n} dl.$$

The scheme is fully explicit. In Ref. 26, a full analysis is conducted, other schemes are presented. We pick out one result, that of the Mach 10 DMR test case,[34] to illustrate the results, see Figure 13.

5.2. *Viscous problems*

This topic is also the subject of current active research. Let us write the (steady) system as

$$\text{div } (F_e - F_v) = 0 \qquad (25)$$

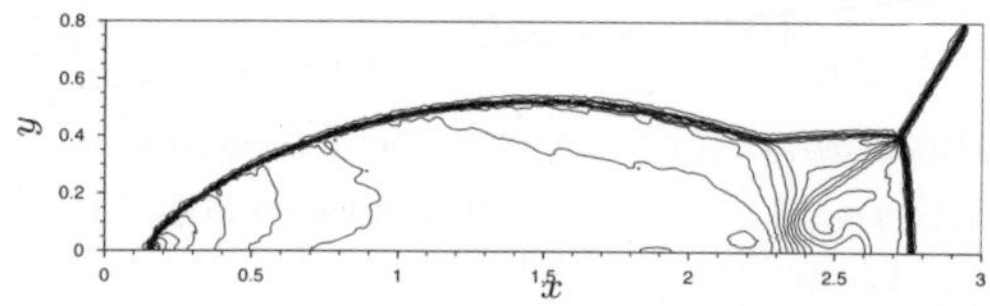

Fig. 13. Double Mach reflection. Density contours. 30 equally spaced contours from 1 to 24. Taken from Ref. 26.

with standard boundary conditions. F_e are the standard Euler fluxes and F_v the viscous ones. In A. Larat's PhD thesis,[20] the system (25) has been discretised in two steps. In the first step, the Euler fluxes are approximated using the method of section 3, and in the second one the viscous fluxes are approximated by a Galerkin variational formulation.

This strategy has already been used in previous works on viscous RD schemes with some refinements when the Peclet becomes small since the viscous effects are predominant, see Ref. 27. A formal justification of the method, in the $\mathbb{P}^1$ case, can be found in Ref. 9.

The approach of Larat[20] is working rather fine (except there is no real theoretical background to this positive result ...). To show this we take a viscous NACA012 airfoil with 0° of incidence, the Mach number at infinity is 0.5 and the Reynolds number is 500. Figure 14 represents the isolines of density colored by the x- component of the velocity. Figure 15 provide the convergence history for the lift. The meshes range from 609 to 230×10^3 vertices. The slope -3 is also represented.

The results are encouraging but a better and more motivated approach is needed. We are currently working on this.

6. Conclusions and Perspectives

We have presented the basic elements that enable us to construct non oscillatory residual distribution schemes on hybrid meshes, for steady and unsteady problems. These schemes have been tested in 2 and 3 space dimensions with excellent results. These schemes have also been extended to different physical problems, such as the Shallow Water equations and the ideal MHD ones. We refer to the references indicated in the text for further details. It is also possible to adapt the method to discontinuous elements, see Refs. 2, 7 and 16 for different versions. The idea, as shown in Ref. 7 can be adapted to Discontinuous Galerkin schemes.

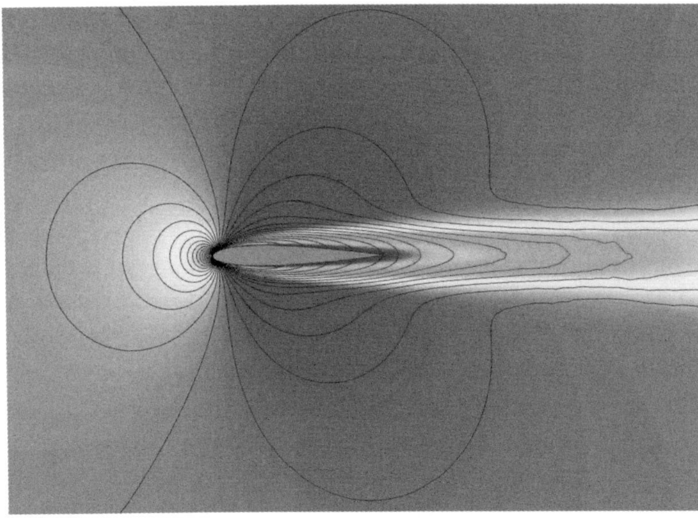

Fig. 14. Third order solution on the finest mesh for the steady viscous NACA012 test case. x-velocity in color and isolines of the density component.

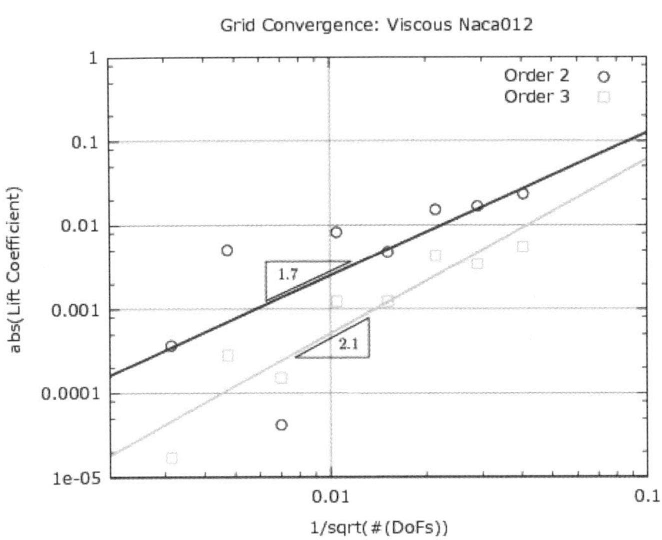

Fig. 15. Convergence of the lift coefficient with respect to the mesh characteristic size $h = \sqrt{\#\{\text{dofs}\}}$ for second and third order simulation of the viscous NACA012 problem.

There is still a lot to be done. Our main efforts are currently on the approximation of the Navier Stokes equations and the use of non Lagrange element to further increase the robustness of the scheme, for example for very strong shocks, see Ref. 8.

Acknowledgments

The author has been supported by the FP7 Advanced Grant # 226316 "ADDECCO". The author would like to acknowledge the contributions of Mario Ricchiuto and Adam Larat (the figures of the viscous problems are taken from his PhD).

References

1. R. Abgrall. Toward the ultimate conservative scheme: Following the quest. *J. Comput. Phys.*, 167(2):277–315, 2001.
2. R. Abgrall. A residual distribution method using discontinuous elements for the computation of possibly non smooth flows. *Adv. Appl. Math. Mech.*, 2(1):32–44, 2010.
3. R. Abgrall, N. Andrianov, and M. Mezine. Towards very high-order accurate schemes for unsteady convection problems on unstructured meshes. *Int. J. Numer. Methods Fluids*, 47(8-9):679–691, 2005.
4. R. Abgrall, R. Huart, and M. Ricchiuto. Approximation of the ideal mhd equations using residual distribution methods. *in preparation*, 2010.
5. R. Abgrall, A. Larat, M. Ricchiuto, and C. Tavé. A simple construction of very high order non-oscillatory compact schemes on unstructured meshes. *Computers and Fluids*, 38(7):1314–1323, 2009.
6. R. Abgrall and P. L. Roe. High-order fluctuation schemes on triangular meshes. *J. Sci. Comput.*, 19(1-3):3–36, 2003.
7. R. Abgrall and C.W. Shu. Development of residual distribution schemes for the discontinuous galerkin method: The scalar case with linear elements. *Communication in Computational Physics*, 5:376–390, 2009.
8. R. Abgrall and J. Treflick. An example of high order residual distribution scheme using non lagrange elements. *Journal of Scientific Computing*, 2010. in press.
9. Rémi Abgrall. Residual distribution schemes: current status and future trends. *Comput. Fluids*, 35(7):641–669, 2006.
10. Rémi Abgrall and Mohamed Mezine. Construction of second order accurate monotone and stable residual distribution schemes for unsteady flow problems. *J. Comput. Phys.*, 188(1):16–55, 2003.
11. Bernardo Cockburn and Chi-Wang Shu. The local discontinuous Galerkin method for time-dependent convection-diffusion systems. *SIAM J. Numer. Anal.*, 35(6):2440–2463, 1998.

12. S.K. Godunov. Über die Eindeutigkeit der Lösung der Gleichungen der Hydrodynamik. 1956.
13. Ami Harten. On a class of high resolution total-variational-stable finite-difference schemes (with appendix by Peter D. Lax). *SIAM J. Numer. Anal.*, 21:1–23, 1984.
14. Ami Harten, Bjorn Engquist, Stanley Osher, and Sukumar Chakravarthy. Uniformly high order accurate essentially non-oscillatory schemes. III. *J. Comput. Phys.*, 71:231–303, 1987.
15. M Hubabrd and M Ricchiuto. Discontinuous upwind residual distribution: A route to unconditional positivity and high order accuracy. *Computers and Fluids*, submitted, 2010.
16. Matthew Hubbard. Discontinuous fluctuation distribution. *J. Comput. Phys.*, 227(24):10125–10147, 2008.
17. Thomas J.R. Hughes and Michel Mallet. A new finite element formulation for computational fluid dynamics. IV: A discontinuity-capturing operator for multidimensional advective-diffusive systems. *Comput. Methods Appl. Mech. Eng.*, 58:329–336, 1986.
18. Guang-Shan Jiang and Chi-Wang Shu. Efficient implementation of weighted ENO schemes. *J. Comput. Phys.*, 126(1):202–228, 1996.
19. Claes Johnson, Uno Nävert, and Juhani Pitkäranta. Finite element methods for linear hyperbolic problems. *Comput. Methods Appl. Mech. Eng.*, 45:285–312, 1984.
20. A Larat. *Conception et Analyse de Schémas Distribuant le Résidu d'Ordre Trés Élevé. Application à la Mécanique des Fluides.* PhD thesis, Université de Bordeaux, 2009. http://tel.archives-ouvertes.fr/tel-00502429/fr/.
21. Peter D. Lax. Hyperbolic systems of conservation laws. II. *Commun. Pure Appl. Math.*, 10:537–566, 1957.
22. Peter D. Lax and B. Wendroff. Systems of conservation laws. *Commun. Pure Appl. Math.*, 13:217–237, 1960.
23. Xu-Dong Liu, Stanley Osher, and Tony Chan. Weighted essentially non-oscillatory schemes. *J. Comput. Phys.*, 115(1):200–212, 1994.
24. R. W. MacCormack. The effect of viscosity in hypervelocity impact cratering. AIAA Paper, 69-354, 1969.
25. Stanley Osher and Fred Solomon. Upwind difference schemes for hyperbolic systems of conservation laws. *Math. Comput.*, 38:339–374, 1982.
26. M Ricchiuto and R. Abgrall. Explicit runge-kutta residual distribution schemes for time dependent problems: Second order case. *J. Comput. Phys.*, 229(16):5653–5691, 1ugust 2010.
27. M. Ricchiuto, N. Villedieu, R. Abgrall, and H. Deconinck. On uniformly high-order accurate residual distribution schemes for advection-diffusion. *J. Comput. Appl. Math.*, 215(2):547–556, 2008.
28. Mario Ricchiuto and Andreas Bollermann. Stabilized residual distribution for shallow water simulations. *J. Comput. Phys.*, 228(4):1071–1115, 2009.
29. Mario Ricchiuto, Árpád Csík, and Herman Deconinck. Residual distribution for general time-dependent conservation laws. *J. Comput. Phys.*, 209(1):249–289, 2005.

30. P.L. Roe. Approximate riemann solver, parameter vectors and difference schemes. *J. Comput. Phys.*, 43:357–372, 1981.
31. Chi-Wang Shu and Stanley Osher. Efficient implementation of essentially nonoscillatory shock-capturing schemes. II. *J. Comput. Phys.*, 83(1):32–78, 1989.
32. Bram van Leer. Towards the ultimate conservative difference scheme. IV: A new approach to numerical convection. *J. comput. Phys.*, 23:276–299, 1977.
33. J. von Neumann and R.D. Richtmeyer. A method for the numerical calculation of hydrodynamic shocks. *J. Appl. Phys.*, 21:232–237, 1950. first released as an unpublished report in 1942.
34. Paul Woodward and Phillip Colella. The numerical simulation of two-dimensional fluid flow with strong shocks. *J. Comput. Phys.*, 54:115–173, 1984.
35. H.C. Yee. Construction of explicit and implicit symmetric TVD schemes and their applications. *J. Comput. Phys.*, 68:151–179, 1987.

CHAPTER 11

RADIAL BASIS FUNCTION-BASED DIFFERENTIAL QUADRATURE (RBF-DQ) METHOD AND ITS APPLICATIONS

Chang Shu

Department of Mechanical Engineering,
National University of Singapore, 10 Kent Ridge Crescent,
Singapore 119260
mpeshuc@nus.edu.sg

In this chapter, we will present an efficient mesh-free method, which combines the derivative approximation by the differential quadrature (DQ) method, and the function approximation by the radial basis functions (RBFs). For simplicity, this combination is termed as RBF-DQ method, which can be used to directly approximate the derivatives of dependent variables on a scattered set of nodes. In particular, when the multiquadrics (MQ) is used to approximate the function, the resultant approach is called MQ-DQ method. The details of MQ-DQ method and its applications to simulate incompressible and compressible flows as well as its accuracy analysis will be shown in this chapter.

1. Introduction

Radial basis functions (RBFs) have been under intensive research as a technique for multivariate data and function interpolation in the past decades, especially in multi-dimensional applications.[1-5] Their performance demonstrates that RBFs constitute a powerful framework for interpolating or approximating data on non-uniform grids. RBFs are attractive for pre-wavelet construction due to their exceptional rates of convergence and infinite differentiability. Since RBFs have excellent performance for function approximation, many researchers turn to explore their ability for solving partial differential equations (PDEs). The first trial of such exploration was made by Kansa.[6,7] After that, a number of researchers[8–15] successfully applied RBFs to solve various problems governed by PDEs. As shown by Kansa,[6,7] using RBFs as a meshless collocation method to solve PDEs possesses the following advantages:

(1) first of all, it is a truly mesh-free method, and is independent of spatial dimension in the sense that the convergence order is of $O(h^{d+1})$ where h is the density of the collocation points and d is the spatial dimension;

(2) in the context of scattered data interpolation, it is known that some RBFs have spectral convergence. In other words, as the spatial dimension of the problem increases, the convergence order also increases, and hence, much fewer scattered collocation points will be needed to maintain the same accuracy as compared with conventional finite difference, finite element and finite volume methods. This shows the advantage of the RBFs for solving multi-dimensional problems.

It should be indicated that although some excellent results were obtained, most of previous works related to the application of RBFs for the numerical solution of PDEs are actually based on the function approximation instead of derivative approximation. In other words, these works directly substitute the expression of function approximation by RBFs into a PDE, and then change the dependent variables into the coefficients of function approximation. The process is very complicated, especially for nonlinear problems. For the nonlinear case, some special techniques such as numerical continuation and bifurcation approach have to be used to solve the resultant nonlinear equations. Since the techniques are very complicated, it is not easy to apply them for solving practical problems such as fluid dynamics, which usually require a large number of mesh points for accurate solution.

On the other hand, it was found that differential quadrature (DQ) method is a global approach for derivative approximation. It was firstly presented by Richard Bellman and his associates[16,17] in the early of 1970's, following the idea of integral quadrature. The basic idea of the DQ method is that any derivative at a mesh point can be approximated by a weighted linear sum of all the functional values along a mesh line. The key procedure in its application is the determination of weighting coefficients. As shown by Shu and Richards,[18] when the solution of a PDE is approximated by a high order polynomial, the weighting coefficients can be computed by a simple algebraic formulation or by a recurrence relationship. Shu and Chew[19] also showed that when the solution of the PDE is approximated by a Fourier series expansion, the weighting coefficients of the first and second order derivatives can be computed explicitly by algebraic formulations. The details of the polynomial-based and Fourier series expansion-based DQ methods can be found in the book of Shu.[20] Note that the above DQ method is only applicable along a straight mesh line. For complex geometry, one has to rely on the coordinate transformation technique to map the irregular domain in the physical space to a regular domain in the computational space first. Then the PDEs and their associated boundary conditions are transformed into relevant forms in the computational space. The numerical discretization is subsequently made in the computational space. The whole process is very tedious. To remove this difficulty, we can combine the DQ method for derivative approximation and the RBF method for function approximation. The combined approach is termed as RBF-DQ method. It not only has the mesh-free feature but also approximates the derivatives directly. Its solution process for a PDE is exactly the same as the conventional DQ method and finite difference schemes. Moreover, it can be consistently well applied to linear and nonlinear problems. In the following, we will describe this method in details.

2. Radial Basis Functions (RBFs) for Function Approximation

In this section, we will briefly describe the radial basis functions and their application for function approximation.

A radial basis function, denoted by $\varphi(\|\mathbf{x}-\mathbf{x}_j\|_2)$, is a continuous spline which depends on the separation distances of a subset of scattered points $\mathbf{x} \in \Omega \subset \Re^d$, d =1, 2, or 3 denotes the spatial dimension. The "radial" is named due to RBFs' spherical symmetry about the centre point $\mathbf{x}_j$. The distances are usually taken to be the Euclidean metric. There are many RBFs (expression of φ) available. The most commonly used RBFs are

$$\text{Multiquadrics (MQ): } \varphi(r) = \sqrt{r^2 + c^2} \tag{1a}$$

$$\text{Thin-plate splines (TPS): } \varphi(r) = r^2 \log(r) \tag{1b}$$

$$\text{Gaussians: } \varphi(r) = e^{-cr^2} \tag{1c}$$

$$\text{Inverse multiquadrics: } \varphi(r) = \frac{1}{\sqrt{r^2 + c^2}} \tag{1d}$$

where $r = \left\| \mathbf{x} - \mathbf{x}_j \right\|_2$ and shape parameter c is a positive constant. Among above popular radial basis functions, the Gaussian and the inverse MQ are positive definite functions, while the TPS and the MQ are conditionally positive definite functions. In recent years, the theory of radial basis function has undergone intensive research and enjoyed considerable success as a technique for interpolating multivariable data and functions. Simply, the RBF interpolation technique can be described as follows: if the values of a function $f(\mathbf{x})$ are known on a set of scattered points $\mathbf{x} \in \Omega \subset \Re^d$, the approximation of $f(\mathbf{x})$ can be written as a linear combination of N radial basis functions,

$$f(\mathbf{x}) \cong \sum_{j=1}^{N} \lambda_j \varphi(\left\| \mathbf{x} - \mathbf{x}_j \right\|_2) + \psi(\mathbf{x}) \tag{2}$$

where N is the number of centers or sometimes called *knots* $\mathbf{x}$, $\mathbf{x} = (x_1, \ x_2, \ \ldots, \ x_d)$, d is the dimension of the problem, λ's are coefficients to be determined and φ is the radial basis function. Equation (2) can be written without the additional polynomial ψ. If Ψ_q^d denotes the space of d-variate polynomials of order not exceeding q, and letting the polynomials $P_1, \ldots, P_m$ be the basis of Ψ_q^d in $\Re^d$, then the polynomial $\psi(\mathbf{x})$, in equation (2), is usually written in the following form:

$$\psi(\mathbf{x}) = \sum_{i=1}^{m} \zeta_i P_i(\mathbf{x}) \tag{3}$$

where $m=(q-1+d)!/(d!(q-1)!)$. To determine the coefficients $(\lambda_1, \ \ldots, \ \lambda_N)$ and $(\zeta_1, \ \ldots, \ \zeta_m)$, extra m equations are required in addition to the N equations resulting from the collocating equation (2) at the N knots. This is insured by the m conditions for equation (2), viz

$$\sum_{j=1}^{N} \lambda_j P_i(\mathbf{x}_j) = 0, \quad i=1, \ldots, M, \tag{4}$$

The matrix form of equations (2) and (4) can be expressed as $\mathbf{Ax} = \mathbf{b}$ with the known function value on the scattered points as the components of vector **b**, and

$$\mathbf{A} = \begin{pmatrix} \varphi & P_m \\ P_m^T & 0 \end{pmatrix}$$

$$\mathbf{x} = (\lambda, \zeta)^T \tag{5}$$

It has been proven that for the case when the nodes are all distinct, the matrix resulting from the above radial basis function interpolation is always non-singular. In 1982, Franke[2] published a review article to evaluate the interpolation methods for scattered data available at that time. Among the methods tested, RBFs outperformed all other methods in terms of accuracy, stability, efficiency, memory requirement, and simplicity of implementation. Among the RBFs tested by Franke,[2] Hardy's multiquadrics (MQ) were ranked the best in accuracy, followed by thin plate splines (TPS).

Though TPS radial basis functions have been considered as optimal functions for multivariate data interpolation, they only converge linearly. Comparatively, the MQ functions converge exponentially and always produce a minimal semi-norm error. However, despite MQ's excellent performance, it contains a shape parameter c, which is given by the end-user to control the surface shape of basis functions. When the value of shape parameter c is small, the resultant interpolating surface forms a cone-like basis functions. As the value of shape parameter c increases, the peak of the cone gradually flattens. The choice of the value of c can greatly affect the accuracy of the approximation. It was found that by increasing c, the root-mean-square error of the goodness-of-fit dropped to a minimum value and then grew rapidly thereafter. This is due to the fact that the MQ coefficient matrix becomes ill-conditioned when $c^2 >> r^2$. How to choose the optimal shape parameter c still remains an open problem. Similar difficulties are also encountered in choosing the shape parameter for the inverse MQ and Gaussian radial basis functions.

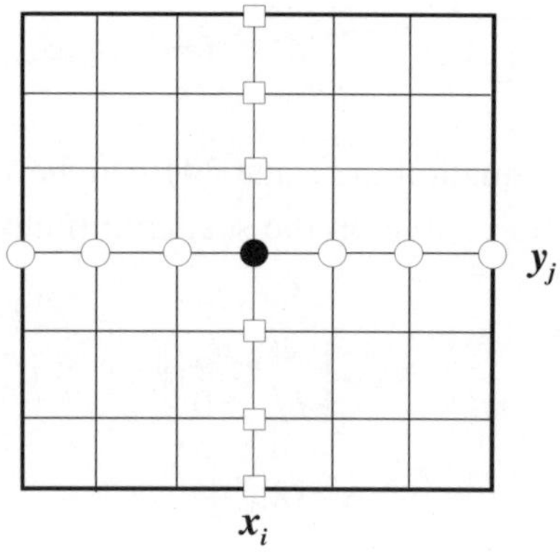

Fig. 1. A Structured mesh for a two-dimensional problem.

3. Differential Quadrature (DQ) Method for Derivative Approximation

In this section, we will briefly describe the derivative approximation by the differential quadrature (DQ) method. The development of DQ method is actually based on the integral quadrature.

It is well known that any integral over a closed domain can be approximated by a linear weighted sum of all the functional values in the integral domain. Following this idea, Bellman *et al.*[16,17] suggested that the partial derivative of a function with respect to an independent variable can be approximated by a linear weighted sum of functional values at all mesh points in that direction. As shown in Fig. 1, DQ approximates the derivative of a function with respect to x at a mesh point (x_i, y_j) (represented by the symbol ●) by all the functional values along the mesh line of $y = y_j$ (represented by the symbol ○), and the derivative of the function with respect to y by all the functional values along the mesh line of $x = x_i$ (represented by the symbol □). Mathematically, the DQ approximation of the nth order derivative with respect to x, $f_x^{(n)}$, and the mth order derivative with respect to y, $f_y^{(m)}$, at (x_i, y_j) can be written as

$$f_x^{(n)}(x_i, y_j) = \sum_{k=1}^{N} w_{i,k}^{(n)} f(x_k, y_j) \tag{6a}$$

$$f_y^{(m)}(x_i, y_j) = \sum_{k=1}^{M} \overline{w}_{j,k}^{(m)} f(x_i, y_k) \tag{6b}$$

where *N*, *M* are respectively the number of mesh points in the *x* and *y* direction, $w_{i,k}^{(n)}$, $\overline{w}_{j,k}^{(m)}$ are the DQ weighting coefficients in the *x* and *y* directions. As shown by Shu,[20] $w_{i,k}^{(n)}$ depends on the approximation of the one-dimensional function $f(x, y_j)$ (*x* is the variable), while $\overline{w}_{j,k}^{(m)}$ depends on the approximation of the one-dimensional function $f(x_i, y)$ (*y* is the variable). When $f(x, y_j)$ or $f(x_i, y)$ is approximated by a high order polynomial, Shu and Richards[18] derived a simple algebraic formulation and a recurrence relationship to compute $w_{i,k}^{(n)}$ and $\overline{w}_{j,k}^{(m)}$. For simplicity, we only give the formulations to compute the weighting coefficients $w_{i,k}^{(n)}$ in the *x* direction as follows

$$w_{i,j}^{(1)} = \frac{M^{(1)}(x_i)}{(x_i - x_j) \cdot M^{(1)}(x_j)}, \quad \text{for } j \neq i, \tag{7a}$$

$$w_{i,i}^{(1)} = -\sum_{k=1, k\neq i}^{N} w_{i,k}^{(1)} \tag{7b}$$

$$w_{i,j}^{(n)} = n \cdot \left(w_{i,j}^{(1)} \cdot w_{i,i}^{(n-1)} - \frac{w_{i,j}^{(n-1)}}{x_i - x_j}\right), \quad \text{for } j \neq i, \tag{7c}$$

$$w_{i,i}^{(n)} = -\sum_{k=1, k\neq i}^{N} w_{i,k}^{(n)} \tag{7d}$$

where $M^{(1)}(x_i) = \prod_{k=1, k\neq i}^{N} (x_i - x_k)$

When the function is approximated by a Fourier series expansion, Shu and Chew[19] also derived simple algebraic formulations to compute the weighting coefficients of the first and second order derivatives as

$$w_{i,j}^{(1)} = \frac{1}{2} \cdot \frac{P(x_i)}{\sin\frac{x_i - x_j}{2} \cdot P(x_j)}, \quad \textit{when } j \neq i \tag{8a}$$

$$w_{i,i}^{(1)} = -\sum_{k=1, k\neq i}^{N} w_{i,k}^{(1)} \tag{8b}$$

$$w_{i,j}^{(2)} = w_{i,j}^{(1)}\left[2w_{i,i}^{(1)} - ctg\frac{x_i - x_j}{2}\right], \quad when \ \ j \neq i \tag{8c}$$

$$w_{i,i}^{(2)} = -\sum_{k=1,k\neq i}^{N} w_{i,k}^{(2)} \tag{8d}$$

where $P(x_i) = \prod_{k=0,k\neq i}^{N} \sin\frac{x_i - x_k}{2}$

For simple geometry, the above DQ approach can obtain very accurate results by using a considerably small number of mesh points. However, for complex geometry, the above scheme cannot be applied directly. The coordinate transformation technique must be introduced. To remove this drawback, we need to develop a more efficient approach.

The basic idea of the DQ method is that any derivative can be approximated by a linear weighted sum of functional values at some mesh points. We can keep this idea but release the choice of functional values along a mesh line in the conventional DQ approximation. In other words, for a two-dimensional problem, any spatial derivative is approximated by a linear weighted sum of all the functional values in the whole two-dimensional domain. In this approximation, a mesh point in the two-dimensional domain is represented by one index, *k*, while in the conventional DQ approximation like equation (6), the mesh point is represented by two indexes *i*, *j*. If the mesh is structured, it is easy to establish the relationship between *i*, *j* and *k*. For the example shown in Fig. 1, *k* can be written as $k = (i-1)M + j, \ \ i = 1,2,...,N; j = 1,2,...,M$. Clearly, when *i* is changed from 1 to *N* and *j* is changed from 1 to *M*, *k* is changed from 1 to $NM = N \times M$. The new DQ approximation for the *n*th order derivative with respect to *x*, $f_x^{(n)}$, and the *m*th order derivative with respect to *y*, $f_y^{(m)}$, at (x_k, y_k) can be written as

$$f_x^{(n)}(x_k, y_k) = \sum_{k1=1}^{NM} w_{k,k1}^{(n)} f(x_{k1}, y_{k1}) \tag{9a}$$

$$f_y^{(m)}(x_k, y_k) = \sum_{k1=1}^{NM} \overline{w}_{k,k1}^{(m)} f(x_{k1}, y_{k1}) \tag{9b}$$

In the following, we will show that the weighting coefficients in equation (9) can be determined by the function approximation of RBFs and the analysis of a linear vector space.

4. Global Radial Basis Function-based Differential Quadrature (RBF-DQ) Method

In this section, we will show in detail the global radial basis function-based differential quadrature (RBF-DQ) method. The development of this method is motivated by our desire to design a numerical scheme that is as simple to implement as traditional finite difference schemes while at the same time keeping the "truly" mesh-free nature.

Among the four RBFs shown in Section 2, MQ, which was first presented by Hardy,[1] is used extensively. Franke[2] did a comprehensive study on various RBFs, and found that MQ generally performs better for the interpolation of two-dimensional scattered data. Therefore, we will concentrate on MQ radial basis functions.

Fig. 2. Point distribution in a two-dimensional domain.

In the following, the MQ RBFs are used as basis functions to determine the weighting coefficients in the DQ approximation of derivatives for a two-dimensional problem. We have to indicate that the method can be easily extended to the case with other RBFs as basis functions or three-dimensional problems.

Consider a two-dimensional problem as shown in Fig. 2. There are N knots randomly distributed in the whole computational domain. Suppose that the solution of a PDE is continuous, which can be approximated by MQ RBFs, and only a constant is included in the polynomial term $\psi(\mathbf{x})$. Then, the function in the domain can be approximated by MQ RBFs as

$$f(x,y) = \sum_{j=1}^{N} \lambda_j \sqrt{(x-x_j)^2 + (y-y_j)^2 + c_j^2} + \lambda_{N+1} \tag{10}$$

To make the problem be well-posed, one more equation is required. From equation (4), we have

$$\sum_{j=1}^{N} \lambda_j = 0 \quad \Rightarrow \quad \lambda_i = -\sum_{j=1, j\neq i}^{N} \lambda_j \tag{11}$$

Substituting equation (11) into equation (10) gives

$$f(x,y) = \sum_{j=1, j\neq i}^{N} \lambda_j g_j(x,y) + \lambda_{N+1} \tag{12}$$

where

$$g_j(x,y) = \sqrt{(x-x_j)^2 + (y-y_j)^2 + c_j^2} - \sqrt{(x-x_i)^2 + (y-y_i)^2 + c_i^2} \tag{13}$$

The number of unknowns in equation (12) is N. As no confusion rises, λ_{N+1} can be replaced by λ_i, and equation (12) can be written as

$$f(x,y) = \sum_{j=1, j\neq i}^{N} \lambda_j g_j(x,y) + \lambda_i \tag{14}$$

It is easy to see that $f(x,y)$ in equation (14) constitutes N-dimensional linear vector space $\mathbf{V}^N$ with respect to the operation of addition and multiplication. From the concept of linear independence, the bases of a vector space can be considered as linearly independent subset that spans the entire space. In the space $\mathbf{V}^N$, one set of base vectors is $g_i(x,y) = 1$, and $g_j(x,y)$, $j = 1,\ldots,N$ but $j \neq i$ given by equation (13).

From the property of a linear vector space, if all the base functions satisfy the linear equation (9), so does any function in the space $\mathbf{V}^N$ represented by equation (14). There is an interesting feature. From equation (14), while all the base functions are given, the function $f(x, y)$ is still unknown since the coefficients λ_i are unknown. However, when all the base functions satisfy equation (9), we can guarantee that $f(x, y)$ also satisfies equation (9). In other words, we can guarantee that the solution of a PDE approximated by the radial basis function satisfies equation (9). Thus, when the weighting coefficients of DQ approximation are determined by all the base functions, they can be used to discretize the derivatives in a PDE. That is the essence of the RBF-DQ method. Substituting all the base functions into equation (9a) as an example, we can obtain

$$0 = \sum_{k=1}^{N} w_{i,k}^{(n)} \tag{15a}$$

$$\frac{\partial^n g_j(x_i, y_i)}{\partial x^n} = \sum_{k=1}^{N} w_{i,k}^{(n)} g_j(x_k, y_k), \quad j = 1,2,\ldots, N, \text{but} \quad j \neq i \tag{15b}$$

For the given i, equation system (15) has N unknowns with N equations. So, solving this equation system can obtain the weighting coefficients $w_{i,k}^{(n)}$. From equation (13), one can easily obtain the first order derivative of $g_j(x, y)$ as

$$\frac{\partial g_j(x, y)}{\partial x} = \frac{x - x_j}{\sqrt{(x - x_j)^2 + (y - y_j)^2 + c_j^2}} - \frac{x - x_i}{\sqrt{(x - x_i)^2 + (y - y_i)^2 + c_i^2}} \tag{16}$$

In the matrix form, the weighting coefficient matrix of the x-derivative can then be determined by

$$[G][W^n]^T = \{G_x\} \tag{17}$$

where $[W^n]^T$ is the transpose of the weighting coefficient matrix $[W^n]$, and

$$[W^n] = \begin{bmatrix} w_{1,1}^{(n)} & w_{1,2}^{(n)} & \cdots & w_{1,N}^{(n)} \\ w_{2,1}^{(n)} & w_{2,2}^{(n)} & \cdots & w_{2,N}^{(n)} \\ \vdots & \vdots & \ddots & \vdots \\ w_{N,1}^{(n)} & w_{N,2}^{(n)} & \cdots & w_{N,N}^{(n)} \end{bmatrix}$$

$$[G] = \begin{bmatrix} 1 & 1 & \cdots & 1 \\ g_1(x_1, y_1) & g_1(x_2, y_2) & \cdots & g_1(x_N, y_N) \\ \vdots & \vdots & \ddots & \vdots \\ g_N(x_1, y_1) & g_N(x_2, y_2) & \cdots & g_N(x_N, y_N) \end{bmatrix}$$

$$[G_x] = \begin{bmatrix} 0 & 0 & \cdots & 0 \\ g_x^n(1,1) & g_x^n(1,2) & \cdots & g_x^n(1,N) \\ \vdots & \vdots & \ddots & \vdots \\ g_x^n(N,1) & g_x^n(N,2) & \cdots & g_x^n(N,N) \end{bmatrix}$$

With the known matrices $[G]$ and $[G_x]$, the weighting coefficient matrix $[W^n]$ can be obtained by using a direct method of LU decomposition. The weighting coefficient matrix of the y-derivative can be obtained in a similar manner. Using these weighting coefficients, we can discretize the spatial derivatives, and transform the governing equations into a system of algebraic equations, which can be solved by iterative or direct method. The details of global RBF-DQ method can be found in Ding *et al.*[21]

One of the most attractive properties in the RBF-DQ method is that the weighting coefficients are only related to the basis functions and the position of the knots. This character is very appealing when we deal with the nonlinear problems. Since the derivatives are directly discretized, the method can be consistently well applied to linear and nonlinear problems. Another attractive property of RBF-DQ method is that it is naturally mesh-free, i.e., numerical discretization is only based on the information of knot coordinates.

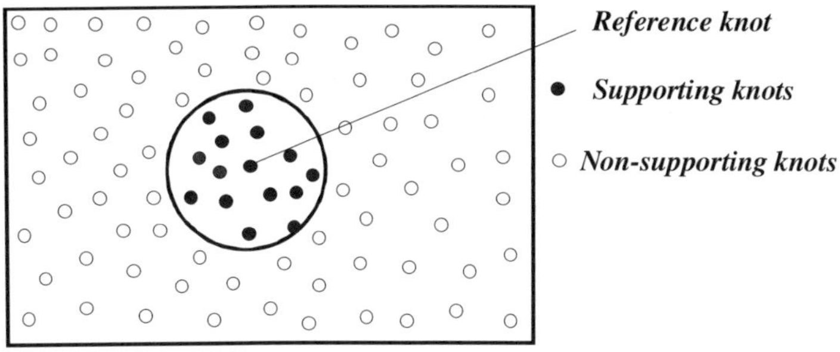

Fig. 3. Supporting knots around a centered knot.

5. Local Radial Basis Function-based Differential Quadrature (RBF-DQ) Method

The RBF-DQ method presented in the last section is a global approach. In other words, the function approximation form (12) uses all the knots in the computational domain. When the number of knots, N, is large, the matrix $[G]$ may be ill-conditioned. This limits its application. To improve it, we developed the local RBF-DQ method.[22–29] To do this, at every knot in the domain, we construct a local support region. As shown in Fig. 3, at any knot, there is a supporting region, in which there are N knots randomly distributed. So, equation (12) is only applied in the local support. That is the only difference between the local RBF-DQ method and the global RBF-DQ method. All the related formulations are the same for these two versions of RBF-DQ method.

As shown in the previous section, the MQ approximation of the function contains a shape parameter c that could be knot-dependent and must be determined by the user. It is well known that the value of c strongly influences the accuracy of MQ approximation, which is used to approximate the solution of PDEs. Thus, there exists a question of how to select a "good" value of c so that the numerical solution of PDEs can achieve satisfactory accuracy. In general, there are three main factors that could affect the optimal shape parameter c for giving the most accurate results. These three factors are the scale of supporting region, the number of supporting knots, and the distribution of supporting knots. Among the three factors, the effect of knot distribution is the most difficult part to be studied since there are infinite kinds of knot distribution. In this section, we will mainly discuss how to minimize the effect of two factors, that is, the scale of supporting region and the number of supporting knots, on the shape parameter c.

In the local MQ-DQ method, the number of supporting knots is usually fixed for an application. Since the knots are randomly generated, the scale of supporting region for each reference knot could be different, and the optimal shape parameter c for accurate numerical results may also be different. Usually, it is very difficult to assign different values of c at different knots. However, this difficulty can be removed from the normalization of scale in the supporting region. The idea is actually motivated from the finite element method, where each element is usually mapped into a regular shape in the computational space. The essence of this idea is to transform the local support into a unit square for the two dimensional case or a unit box for the three dimensional case. So, the discussion about the optimal shape parameter c is now confined to the MQ approximation of functions in the unit square or box. The coordinate transformation has the form

$$\bar{x} = \frac{x}{D_i}, \quad \bar{y} = \frac{y}{D_i} \tag{18}$$

where (x, y) represents the coordinates of supporting region in the physical space, $(\bar{x}, \bar{y})$ denotes the coordinates in the unit square, D_i is the diameter of the minimal circle enclosing all knots in the supporting region for the knot i. The corresponding MQ RBFs in the local support now become

$$\varphi = \sqrt{\left(\bar{x} - \frac{x_i}{D_i}\right)^2 + \left(\bar{y} - \frac{y_i}{D_i}\right)^2 + \bar{c}^2}\ , \quad i = 1, \ldots, N\ , \tag{19}$$

where N is the total number of knots in the support. As compared with traditional MQ-RBF, we can find that the shape parameter c is equivalent to $\bar{c}D_i$. The coordinate transformation (18) also changes the formulation of the weighting coefficients in the local MQ-DQ approximation. For example, by using the differential chain rule, the first order partial derivative with respect to x can be written as

$$\frac{\partial f}{\partial x}=\frac{\partial f}{\partial \bar{x}}\frac{d\bar{x}}{dx}=\frac{1}{D_i}\frac{\partial f}{\partial \bar{x}}=\frac{1}{D_i}\sum_{j=1}^{N} w_j^{(1x)} f_j=\sum_{j=1}^{N}\frac{w_j^{(1x)}}{D_i} f_j \tag{20}$$

where $w_j^{(1x)}$ are the weighting coefficients computed in the unit square, $w_j^{(1x)} / D_i$ are the actual weighting coefficients in the physical domain. Clearly, when D_i is changed, the equivalent c in the physical space is automatically changed. In our application, $\bar{c}$ is chosen as a constant. Its optimal value depends on the number of supporting knots.

6. Numerical Accuracy Analysis for Local MQ-DQ Method

For the conventional numerical schemes which are based on the polynomial approximation, we can easily access their accuracy by using truncated Taylor series expansion. This way cannot be used to access the accuracy of the RBF-DQ method. This is because in the RBF approximation form, every term is equally important. In fact, so far, there is no any theoretical way to analyze the accuracy of RBF-DQ method. On the other hand, in the practical applications, we do need to know the information about the accuracy of RBF-DQ method. In this section, we will show how to use numerical experiments to determine the accuracy of the local MQ-DQ method.

We will use the two-dimensional Poisson equation as an example to illustrate how to access the accuracy of the local MQ-DQ method for approximation of the second order derivatives.[30] The 2D Poisson equation can be written as

$$\frac{\partial^2 u}{\partial x^2}+\frac{\partial^2 u}{\partial y^2}=f(x,y) \tag{21}$$

For simplicity, we consider the solution of equation (21) in the unit square, that is, $0 \le x \le 1, 0 \le y \le 1$. To easily carry out our analysis, we assume that the exact solution of equation (21), u_{exact}, is given. u_{exact} can be used to specify the boundary conditions and the source term $f(x, y)$ as well as the numerical error defined below,

$$Error=\sqrt{\sum_{i=1}^{N}(u_{numeical}-u_{exact})^2} \Big/ \sqrt{\sum_{i=1}^{N}(u_{exact})^2} \tag{22}$$

When we change the mesh spacing h, the numerical error will also be changed. If the numerical error can be written as

$$Error = O(h^m) = C \cdot h^m \tag{23}$$

Then we can say that the scheme has the mth order of accuracy. Equation (23) can also be written as

$$\log(Error) = \log(C) + m\log(h) \tag{24}$$

This means that when the numerical error versus h is plotted in the log-log scale, it could be a straight line and the slope of the line is the order of accuracy m. Indeed, when we choose the exact solution as the following form,

$$u_{exact} = 0.75\exp\left(-\frac{(9x-2)^2+(9y-2)^2}{4}\right)+0.75\exp\left(-\frac{(9x+1)^2}{49}-\frac{9y+1}{10}\right), \tag{25}$$
$$+0.5\exp\left(-\frac{(9x-7)^2+(9y-3)^2}{4}\right)-0.2\exp(-(9x-4)^2-(9x-7)^2)$$

we found that log($Error$) and log(h) does form a straight line. This can be seen clearly in the following figure, where the shape parameter c is taken as 0.12.

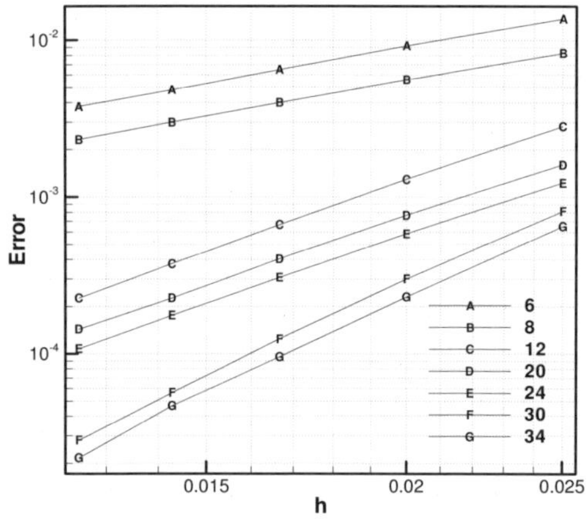

Fig. 4. Numerical error versus mesh size for various number of supporting points (2D case).

It can be observed from the above figure that the convergence lines can be classified into three basic groups by the value of slope, with the number of supporting points ranging from 6 to 34. Specifically, the convergence rate is approximately 1.9 for the scheme with 6 and 8 supporting points, 3.6 for the scheme with 12, 20, and 24 supporting points, and 4.9 for the scheme with 30 and 34 supporting points. Therefore, the accuracy of the local MQ-DQ method for approximation of the second order derivatives in the two-dimensional case with different number of supporting points can be written as[30]

$$Error \sim O(h^m) \text{ and } m \approx \begin{cases} 1.9 & for \quad 6 \le n_s \le 9 \\ 3.6 & for \quad 9 < n_s \le 27 \\ 4.9 & for \quad 27 < n_s \le 34 \end{cases} \tag{26}$$

The above results show the same feature as other polynomial-based numerical schemes. As compared with the traditional finite element method, we can see that the number of supporting points in the local MQ-DQ method plays a similar role as the collocation points in the finite element method. In the finite element method, the use of more collocation points means implementation of higher order polynomials for function approximation. In the local MQ-DQ method, the number of supporting points equals to the number of MQ RBFs used for function approximation. It is known that a polynomial interpolant of degree k requires $(k+1)(k+2)/2$ collocation points in the two-dimensional function approximation, and achieves an accuracy of $O(h^{k-(m+n)+1})$ for a partial derivative $\partial^{m+n}u/\partial x^m \partial y^n$. Taking the second order derivative as an example, we can see that the second order of accuracy (*k=4, m=2, n=0 or k=4, m=0, n=2*) requires 15 collocation points, while the third order of accuracy (*k=5, m=2, n=0 or k=5, m=0, n=2*) requires 21 collocation points. This implies that when the number of collocation points is increased from 15 to 20, the order of accuracy for the numerical results cannot be improved, which keeps the second-order. Only when the number of collocation points is increased to 21, the accuracy of numerical results can be improved to the third-order. It is interesting to see that such feature also holds for the local MQ-DQ method. The details of this analysis can be found in the work of Ding *et al.*[30]

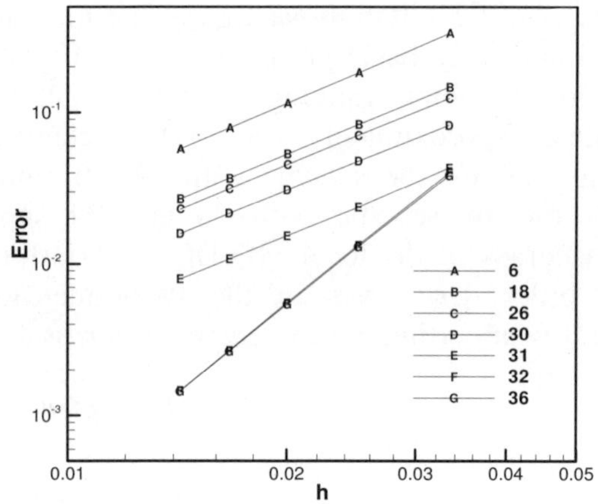

Fig. 5. Numerical error versus mesh size for various number of supporting points (3D case).

The above analysis for the two-dimensional case has been extended to the three-dimensional case.[31] Using 3D Poisson equation as an example, where the exact solution is taken as $u_{exact} = \sin(\pi x)\sin(\pi y)\sin(\pi z)$, Fig. 5 shows the numerical error versus the mesh spacing with different number of supporting points. Clearly, for all the cases, the log(*Error*) and Log(*h*) has a linear relationship. From the slope of the straight line, we can roughly determine the accuracy of the local MQ-DQ method for the 3D second order derivatives as

$$Error \sim O(h^m) \text{ and } m \approx \begin{cases} 2.0 & for\ 6 \le \mathrm{n_s} \le 31 \\ 3.9 & for\ 32 \le \mathrm{n_s} \le 36 \end{cases} \tag{27}$$

7. Application of Local RBF-DQ Method to Simulate Incompressible Flows

When the number of knots is large, the matrix involved in the global RBF-DQ method will be highly ill-conditioned. On the other hand, for the practical flow problems, we usually need to use a considerably large number of mesh points to capture the thin boundary layers or shock waves. In this sense, the global RBF-DQ method is not applicable to the real flow problems. In the present and next sections, we will show how to apply the local RBF-DQ method to simulate both incompressible and compressible flow problems.

The application of local RBF-DQ method to simulate incompressible flows is quite straightforward. Its solution procedure is exactly the same as conventional finite difference schemes. That is, it directly discretizes the derivatives in the governing equations. For the incompressible viscous flows, the difficulty in the numerical simulation is the coupling between the velocity field and the pressure field. For the 2D case, this difficulty can be easily removed by using the vorticity-stream function formulation. For example, for the 2D natural convection problem, we can use the following non-dimensional equations in terms of stream function ψ, vorticity ω and temperature T[26]:

$$\frac{\partial \omega}{\partial t}+u\frac{\partial \omega}{\partial x}+v\frac{\partial \omega}{\partial y}=\mathbf{Pr}\left(\frac{\partial^2 \omega}{\partial x^2}+\frac{\partial^2 \omega}{\partial y^2}\right)-Ra\,\mathbf{Pr}\frac{\partial T}{\partial x} \tag{28}$$

$$\frac{\partial^2 \psi}{\partial x^2}+\frac{\partial^2 \psi}{\partial y^2}=\omega \tag{29}$$

$$\frac{\partial T}{\partial t}+u\frac{\partial T}{\partial x}+v\frac{\partial T}{\partial y}=\frac{\partial^2 T}{\partial x^2}+\frac{\partial^2 T}{\partial y^2} \tag{30}$$

where Pr and Ra are the Prandtl and Rayleigh numbers respectively. The u, v denote the components of velocity in the x and y direction, which can be calculated from the stream function by

$$u=\frac{\partial \psi}{\partial y}, \quad v=-\frac{\partial \psi}{\partial x} \tag{31}$$

Equations (28)-(30) are subjected to proper initial and boundary conditions.

In the numerical simulation, we need to firstly generate a set of knots (randomly distributed or regularly distributed), at which dependent field variables are defined. Only a single index i is required to enumerate the knots. At each knot, we need to find its supporting knots in terms of distance. After that, we can calculate the weighting coefficients of local RBF-DQ method by using equation (17) or its variants. With the weighting coefficients, the governing equations (28)-(30) can be discretized by

$$\frac{d\omega_i}{dt}+u_i\sum_{k=1}^{n_i}w_{i,k}^{(1x)}\omega_i^k+v_i\sum_{k=1}^{n_i}w_{i,k}^{(1y)}\omega_i^k=$$
$$\mathbf{Pr}\left(\sum_{k=1}^{n_i}w_{i,k}^{(2x)}\omega_i^k+\sum_{k=1}^{n_i}w_{i,k}^{(2y)}\omega_i^k\right)-Ra\,\mathbf{Pr}\sum_{k=1}^{n_i}w_{i,k}^{(1x)}T_i^k \tag{32}$$

$$\sum_{k=1}^{n_i}w_{i,k}^{(2x)}\psi_i^k+\sum_{k=1}^{n_i}w_{i,k}^{(2y)}\psi_i^k=\omega_i \tag{33}$$

$$\frac{dT_i}{dt}+u_i\sum_{k=1}^{n_i}w_{i,k}^{(1x)}T_i^k+v_i\sum_{k=1}^{n_i}w_{i,k}^{(1y)}T_i^k=\sum_{k=1}^{n_i}w_{i,k}^{(2x)}T_i^k+\sum_{k=1}^{n_i}w_{i,k}^{(2y)}T_i^k \tag{34}$$

$$u_i=\sum_{k=1}^{n_i}w_{i,k}^{(1y)}\psi_i^k \text{ and } v_i=-\sum_{k=1}^{n_i}w_{i,k}^{(1x)}\psi_i^k \tag{35}$$

where F_i represents the function value at knot i, F_i^k represents the function value at the kth supporting knot of knot i. $w_{i,k}^{(1x)}$, $w_{i,k}^{(1y)}$, $w_{i,k}^{(2x)}$ and $w_{i,k}^{(2y)}$ represent the computed weighting coefficients in the local RBF-DQ approximation for the first and second order derivatives in the x and y direction, respectively. The resultant equations (32) and (34) are ordinary differential equations which can be easily solved by well established explicit or implicit methods. Equation system (33) is a set of algebraic equations, which can be solved by SOR iterative method.

For the 3D case, the vorticity-stream function form involves more dependent variables and differential equations than the original primitive variable form. Thus, the following Navier-Stokes equations in terms of primitive variables are usually adopted for the 3D case,[32]

$$\text{Continuity equation: } \nabla\cdot\mathbf{u}=0 \tag{36}$$

$$\text{Momentum equation: } \frac{\partial\mathbf{u}}{\partial t}+\mathbf{u}\cdot\nabla\mathbf{u}=-\nabla p+\frac{1}{\mathrm{Re}}\Delta\mathbf{u} \tag{37}$$

where Re is the Reynolds number. The solution of above equations confronts difficulties like the lack of an independent equation for the pressure and non-existence of a dominant variable in the continuity equation. One way to circumvent these difficulties is to decouple the pressure computation from the momentum equations and then construct a pressure field to enforce the satisfaction of continuity equation. This method is usually termed as pressure correction or projection method.[33] In the following, we will give a brief description on this method.

For a time increment $\Delta t = t^{n+1} - t^n$, the method consists of two steps. Firstly, an intermediate velocity $\mathbf{u}^*$ is predicted by the advection-diffusion equation, which drops the pressure term. That is, for each interior node in the domain, the intermediate velocity $\mathbf{u}^*$ can be calculated by

$$\frac{\mathbf{u}^* - \mathbf{u}^n}{\Delta t} = -\left[\frac{3}{2} H(\mathbf{u}^n) - \frac{1}{2} H(\mathbf{u}^{n-1})\right] + \frac{1}{2\,\mathrm{Re}} L\left(\mathbf{u}^* + \mathbf{u}^n\right) \tag{38}$$

where H denotes the discrete advection operator, L the discrete Laplace operator. Superscripts (n-1), n and (n+1) denote the time levels. Then, the velocity field $\mathbf{u}$ at t^{n+1} is corrected by including the pressure term, given by

$$\frac{\mathbf{u}^{n+1} - \mathbf{u}^*}{\Delta t} = -Gp^{n+1} \tag{39}$$

where G is the discrete gradient operator. The final velocity field is subject to the continuity constraint given by

$$D\mathbf{u}^{n+1} = 0 \tag{40}$$

where D is the discrete divergence operator. Substituting equation (40) into equation (39) leads to the following Poisson equation for pressure

$$Lp^{n+1} = \frac{1}{\Delta t}(D\mathbf{u}^*) \tag{41}$$

Obviously, the velocity $\mathbf{u}^{n+1}$ is updated by the solution of pressure equation (41). An alternative form of the method is to use the known pressure field in the prediction of the intermediate velocity. Thus, pressure difference instead of the pressure field is computed to correct the velocity. Note that the spatial derivatives in the differential operators H, L, G and D are all discretized by the local RBF-DQ method.

8. Application of Local RBF-DQ Method to Simulate Compressible Inviscid Flows

In this section, we will show how to apply the local RBF-DQ method to simulate compressible inviscid flows.[34,35] The two-dimensional time-dependent compressible Euler equations in the conservative form are taken as an example to illustrate the solution process, which can be written as

$$\frac{\partial}{\partial t}\mathbf{U} + \nabla \bullet \mathbf{F}(\mathbf{U}) = 0 \tag{42}$$

with $\mathbf{U} = \begin{pmatrix} \rho \\ \rho u \\ \rho v \\ e \end{pmatrix}$, $\mathbf{F}_1 = \begin{pmatrix} \rho u \\ \rho u^2 + p \\ \rho uv \\ u(e+p) \end{pmatrix}$, $\mathbf{F}_2 = \begin{pmatrix} \rho v \\ \rho uv \\ \rho v^2 + p \\ v(e+p) \end{pmatrix}$ and $\mathbf{F} = [\mathbf{F}_1, \ \mathbf{F}_2]$

where the dependent variable $\mathbf{U}$ is the vector of conservative variables, and $(\rho, u, v, p)^{\mathrm{T}}$ is the vector of primitive variables. $\vec{\mathbf{m}} = (\rho u, \rho v)^T$ is the momentum vector and $\mathbf{u} = (u, v)^T$ is the velocity vector. $e = \rho[\varepsilon + (u^2 + v^2)/2]$ is the total energy and ε is the specific internal energy. For a thermally perfect gas, the pressure p can be computed by the equation of state

$$p = (\gamma - 1)(e - \rho\frac{u^2}{2}) \tag{43}$$

It is well known that in the compressible flows, there may exist shock waves that are discontinuity in terms of velocity and density. The discontinuity may cause numerical instability in the simulation. To remove this difficulty, one usually adds numerical diffusion when the flux $\boldsymbol{F}$ is evaluated at certain positions. With this in mind, we cannot directly approximate the flux derivatives in equation (42) by the local RBF-DQ method using the reference node and its supporting nodes. This is because at the nodes, the flow variables are defined, and therefore, the fluxes can be directly computed. During the process, there is no way to add numerical diffusion. To remove this drawback, the supporting nodes are defined to locate at the mid-points between the reference node and its supporting nodes, as shown in Fig. 6.

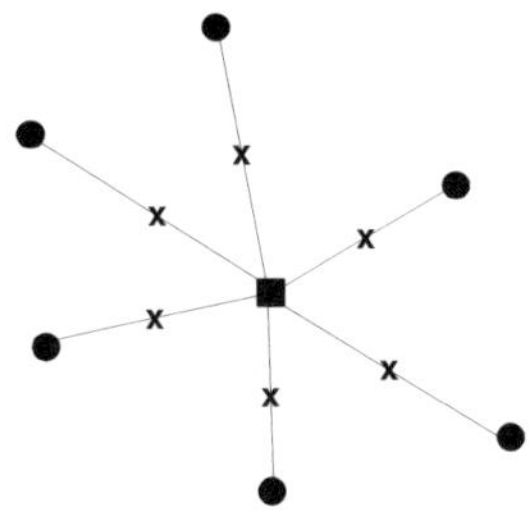

Fig. 6. Reference point, supporting points and mid-points.

After spatial discretization by the local RBF-DQ method, equation (42) can be written as

$$\left.\frac{d\mathbf{U}}{dt}\right|_i = -\sum_{k=0}^{N_I}[w_{i,k}^{(1x)}\mathbf{F}_1(\mathbf{U}_{i,k}) + w_{i,k}^{(1y)}\mathbf{F}_2(\mathbf{U}_{i,k})]^n \tag{44}$$

where $\mathbf{U}_{i,k}$ are the conservative variables at the mid-points between the reference point i and its kth supporting point. $w_{i,k}^{(1x)}$ and $w_{i,k}^{(1y)}$ are the corresponding weighting coefficients for the first-order derivatives in the x and y direction, respectively. N_I denotes the total number of supporting points for the reference point i and $\mathbf{U}_{i,0} = \mathbf{U}_i$.

By observing equation (44), we can find that at each mid-point, a new flux can be defined, based on a unit vector $\vec{1}_w = (\alpha_{i,k}, \beta_{i,k})^T$, which is associated with weighting coefficients of derivative approximation. The new flux can be written as

$$\mathbf{G}_{i,k} = \alpha_{i,k}\mathbf{F}_1(\mathbf{U}_{i,k}) + \beta_{i,k}\mathbf{F}_2(\mathbf{U}_{i,k}) \tag{45}$$

where $\alpha_{i,k} = \dfrac{w_{i,k}^{(1x)}}{\sqrt{(w_{i,k}^{(1x)})^2 + (w_{i,k}^{(1y)})^2}}$ and $\beta_{i,k} = \dfrac{w_{i,k}^{(1y)}}{\sqrt{(w_{i,k}^{(1x)})^2 + (w_{i,k}^{(1y)})^2}}$.

Defining $W_{i,k} = \sqrt{(w_{i,k}^{(1x)})^2 + (w_{i,k}^{(1y)})^2}$, then equation (44) can be simplified as

$$\left.\frac{d\mathbf{U}}{dt}\right|_i = -\sum_{k=0}^{N_I} W_{i,k}\mathbf{G}_{i,k} \tag{46}$$

Equation (46) can be interpreted in such a way that the variation of conservative variables at the reference point can be measured by a linear sum of new fluxes at the reference point and the mid-points. Therefore, how to evaluate the new fluxes effectively and efficiently at the mid-points is a critical issue. Like other upwind schemes, the new fluxes at the mid-points in equation (46) are evaluated by approximate Riemann solvers. One of the most popular schemes in this category is the Roe's approximate Riemann solver.[36] With Roe's scheme, the new flux at the mid-point can be evaluated by

$$\mathbf{G}(U_L,U_R) = \frac{1}{2}[\mathbf{G}(U_L)+\mathbf{G}(U_R)] - \frac{1}{2}\left|\hat{\mathbf{A}}\right|(\mathbf{U}_L - \mathbf{U}_R) \tag{47}$$

where $\mathbf{G}(U_L,U_R)$, $\mathbf{G}(U_L)$, and $\mathbf{G}(U_R)$ denote the new flux at the mid-point, reference point (L) and the supporting point (R), respectively. The symbol $\hat{\mathbf{A}}$ denotes the constant Jacobain matrix, which approximates the Jacobian matrix $\mathbf{A}$ defined by $\partial\mathbf{G}/\partial\mathbf{U}$. Notice that the hat (^) denotes the matrix being constructed with Roe's averaging.[36]

It is noted that Roe's scheme only has the first-order of accuracy. It assumes that the flux between the mid-point and the related node remains a constant. To construct a high-order Roe's approximate Riemann solver, the high-order spatial approximation of the solution must be constructed. For the traditional mesh-based methods, polynomial interpolation is usually employed to do this job. By extrapolating the function values to both sides of the mid-point, the higher order approximation of numerical flux at the mid-point can then be obtained from

$$\mathbf{G}(U_L,U_R) = \frac{1}{2}\left[\mathbf{G}(\mathbf{U}^L)+\mathbf{G}(\mathbf{U}^R)\right] - \frac{1}{2}\left|\mathbf{A}^*(\mathbf{U}^L,\mathbf{U}^R)\right|(\mathbf{U}^L - \mathbf{U}^R) \tag{48}$$

where the superscripts L and R denote the value of flow variables at the mid-point approximated from the side of reference point and supporting point, respectively. $\mathbf{A}^*$ denotes the Roe's approximate Jacobian matrix evaluated at the mid-point with $\mathbf{U}^L$ and $\mathbf{U}^R$. In the conventional FD and FV methods, $\mathbf{U}^L$ and $\mathbf{U}^R$ can be obtained by upwind interpolation using function values at certain mesh points. This way is difficult to be applied in the local RBF-DQ method, in which the nodes may be randomly distributed. On the other hand, we notice that the derivatives at every node can be easily calculated by the local RBF-DQ method. So, in this work, the Taylor series expansion, which only involves the function and its derivatives at the reference node or the supporting node, is used to evaluate $\mathbf{U}^L$ and $\mathbf{U}^R$. Take function f as an example. We suppose that it is approximated from the side of reference node. The interpolation gives

$$f^L = \begin{cases} f_L + s\Delta f, & \text{if } \min\limits_{k\in \sup_i}(f_k - f_L) \le s\Delta f \le \max\limits_{k\in \sup_i}(f_k - f_L) \\ f_L, & \text{otherwise} \end{cases} \tag{49}$$

where

$$\Delta f = f_x \Delta x + f_y \Delta y \tag{50}$$

sup_i denotes the support of point i, and s is the van Albada limiter[37] given by

$$s = \max\left\{0,\ \frac{2\Delta f\cdot(\frac{f_R - f_L}{2}) + \varepsilon}{(\Delta f)^2 + (\frac{f_R - f_L}{2})^2 + \varepsilon}\right\} \tag{51}$$

where ε is a very small number (for example, $\varepsilon = 10^{-6}$), to prevent the division by zero in the uniform flow region, in which the flux difference is very small.

After numerical evaluation of fluxes $\boldsymbol{G}$, equation system (46) can be solved by the four-stage Runge-Kutta scheme.

9. Some Numerical Examples

In the previous sections, we have shown the details of the local MQ-DQ method, its accuracy analysis by numerical experiments, and its implementation for incompressible and compressible flow problems. In this section, we will show four sample applications of the local MQ-DQ method to demonstrate its performance.

9.1 *Poisson equation*

One of interesting features of the local MQ-DQ method is that the accuracy of numerical solution is greatly affected by the optimal shape parameter and the number of supporting knots. We will show this feature through its application to solve the following 2D Poisson equation[23]

$$\frac{\partial^2 u}{\partial x^2}+\frac{\partial^2 u}{\partial y^2}=g(x,y) \tag{52}$$

in the square domain ($0\le x\le 1, 0\le y\le 1$). For simplicity, it is supposed that the exact solution is given by

$$u(x,y)=\frac{\frac{5}{4}+\cos(5.4y)}{6+6(3x-1)^2} \tag{53}$$

Equation (53) will be used to provide the Dirichlet condition on the boundary, the function $g(x,y)$, and to validate the numerical solution. The L_2 norm of relative error is taken to measure the accuracy of numerical results, which is defined as

$$L_2(error)=\sqrt{\sum_{i=1}^{N}\left(\frac{u_{numerical}-u_{analytical}}{u_{analytical}+10^{-8}}\right)^2/N} \tag{54}$$

To conduct numerical experiments, 673 knots are randomly distributed in the domain, and the number of supporting points is respectively taken as 10, 16, 22 and 28. Fig. 7 illustrates the variation of accuracy with different shape parameter and number of supporting knots. It can be seen from the figure that the L_2(error) depends on the value of shape parameter c and the number of supporting knots. It was found that when the number of supporting knots is fixed, with increase of shape parameter c, the accuracy of numerical results is improved. And when the shape parameter c is fixed, with increase of the supporting knots, the accuracy of numerical results is also improved. Another interesting phenomenon is that the shape parameter c with small number of supporting knots is less sensitive than that with large number of supporting knots. In other words, when the number of supporting knots is relatively small, the shape parameter c can be chosen in a wide range to get a convergent solution. But when the number of supporting knots is large, the shape parameter c

can only be selected in a narrow range to get convergent solution. So, one has to balance the good accuracy of numerical results and the sensitivity of the shape parameter c when the number of supporting knots is chosen. From our experiences, 16 supporting knots are a suitable choice.

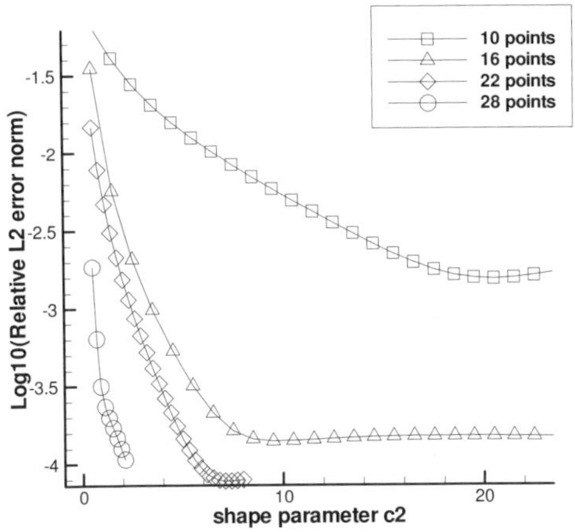

Fig. 7. L_2(error) versus c^2 for solution of 2D Poisson equation.

9.2 *Comparative study for lid-driven cavity flow*

Both the local MQ-DQ method and the least square-based finite difference (LSFD) scheme[38] are mesh-free approaches. It is interesting to compare their performances in terms of accuracy, stability and convergence rate. We will take the lid-driven cavity flow as an example to show the comparison.[39]

The vorticity-stream function formulation as shown in Section 7 is taken as the governing equation. For simplicity, the uniform mesh of 81×81 is chosen for the simulation. For the spatial discretization, the number of supporting point is fixed to 13, and the second order LSFD scheme is adopted. In the local MQ-DQ simulation, the shape parameter is selected as 0.2. Fig. 8 shows the zooming-in view of u-velocity profile along the vertical centerline at Re=1000. Clearly, as compared with the benchmark data of Ghia *et al.*,[40] the local MQ-DQ method is a bit more accurate than the LSFD scheme. On the other hand, it was found[39] that the LSFD method has a much faster convergence rate than the local MQ-DQ method.

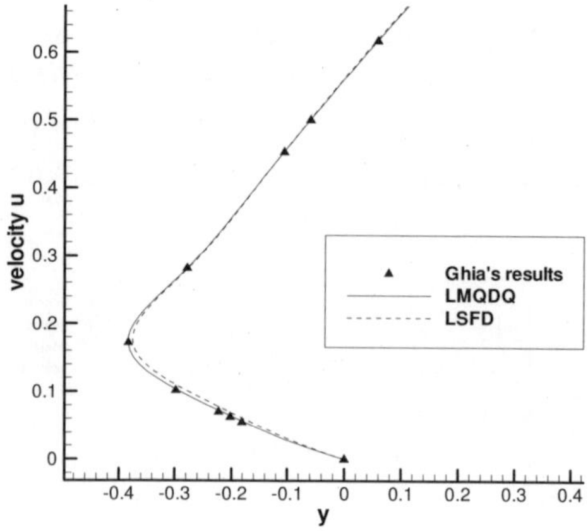

Fig. 8. Zooming-in view of u-velocity profile at Re=1000 on mesh of 81×81.

9.3 *Flow past a sphere*

The problem of flow past a sphere is a standard test case for the 3D simulation. It has been studied by Johnson and Patel[41] in details. They found that when $20 \le Re \le 210$, the flow is separated, steady, axisymmetric and topologically similar. And when $210 \le Re \le 270$, although the flow still remains the steady state, it is non-axisymmetric.

In our 3D simulation,[31] 32 supporting points are employed for every reference node and the shape parameter c is set to be 0.18. Figure 9 shows streamlines in the (x, z) and (x, y) planes, respectively. It is clear from Fig. 9(a) that the flow field is symmetric about the (x, z) plane, which divides the figure across the center. However, as shown in Fig. 9(b), the flow in the (x, y) plane is no longer symmetric. This result is in line with the findings of Johnson and Patel.[41] The recirculating length and drag coefficient computed by the local MQ-DQ method also agree well with the data given by Johnson and Patel.[41]

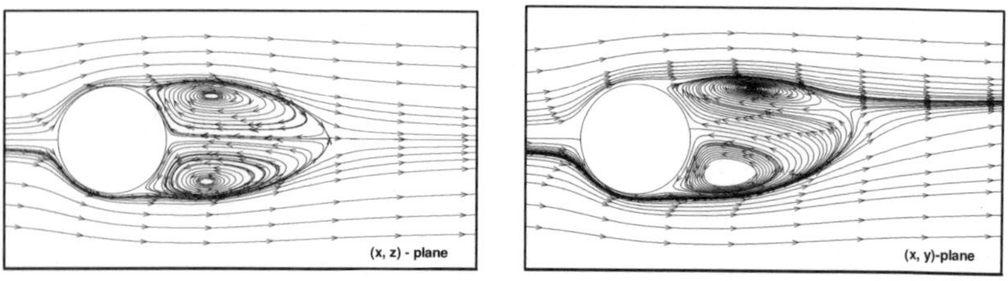

Fig. 9. Streamlines of projected velocity vectors at Re = 250 for flow past a sphere.

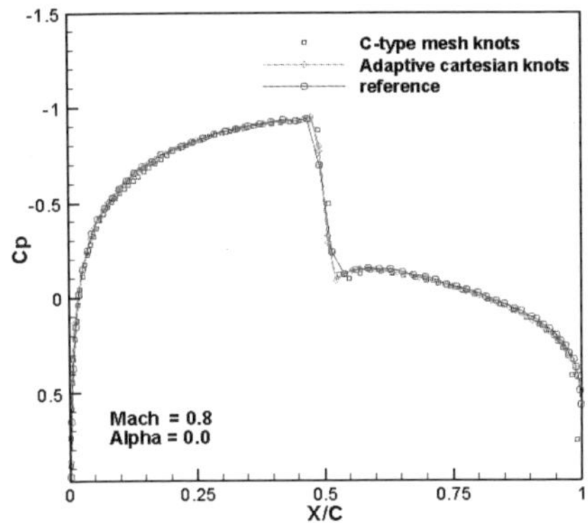

Fig. 10. Pressure coefficient distribution along the airfoil surface.

9.4 *Transonic flow over a NACA 0012*

The transonic flow over a NACA 0012 airfoil is chosen to validate the upwind local MQ-DQ method shown in Section 8. In this study,[35] two types of mesh knots, namely *c*-type mesh knots (8669 points) and adaptive Cartesian knots (7762 points) are used, and 8 supporting points are taken in the local MQ-DQ discretization. Fig. 10 shows the pressure coefficient distribution along the airfoil surface at zero angle of attack and Mach number of 0.8. The numerical results of Jameson[42] are also included in the figure for comparison. Obviously, the present MQ-DQ results agree very well with those of Jameson.[42] From the obtained numerical results, it seems that the physical conservation laws are satisfied by the method since both the shock position and strength are well captured. However, it is very difficult to prove it in mathematics since the knots can be randomly distributed and there are many cases for knot distribution. Hopefully, this problem can be resolved in the future by mathematicians.

10. Conclusions

A mesh-free radial basis function-based differential quadrature (RBF-DQ) method is presented in this chapter. It combines the mesh-free feature of RBF for function approximation and high order property of DQ for derivative approximation. Once the weighting coefficients are computed in advance, the application of RBF-DQ method is as simple as conventional finite difference schemes.

When all the points in the computational domain are used as supporting points in the RBF-DQ approximation, the approach is termed as global RBF-DQ method. Alternatively, when it is applied in a local supporting region, the approach is called local RBF-DQ method. It was found that the global RBF-DQ method is limited by the number of points used due to ill-condition of the matrix in computing the weighting coefficients. Usually, the number of points is limited to a few hundreds. Thus for practical applications, one often uses the local RBF-DQ method.

Among various versions of RBF-DQ method, MQ-DQ method is the most popular approach. In this approach, there is a shape parameter c, which could influence the accuracy of numerical results. How to choose an optimal value of c is still an open problem. To get a convergent solution, it was found that for a small number of supporting points, c can be chosen in a wide range. However, when a large number of supporting points is used, c can only be selected in a narrow range. In a converged range, a larger value of c would provide more accurate numerical results. Currently, there is no theoretical way to access the accuracy of MQ-DQ method. In this work, through numerical experiments, the accuracy of MQ-DQ approximation for the second order derivatives is accessed for the 2D and 3D cases. It was found that the accuracy of MQ-DQ method can be improved by increasing the number of supporting points, but the improvement has a jump when the supporting points exceed a critical value. This feature is in line with conventional polynomial approximation.

Numerical examples show that the local MQ-DQ method can be well applied to simulate incompressible viscous flows and compressible inviscid flows with shock waves.

References

1. R. L. Hardy, *J. Geophys. Res.,* 1905 (1971).
2. R. Franke, *Math. Comp.,* 181 (1982).
3. P. Sosik, *Neural Netw. World,* 221 (1995).
4. X. Li, *Appl. Math. Comput.*, 75 (1998).
5. M. A. Golberg, C. S. Chen, H. Bowman, *Eng. Anal. Bound. Elem.,* 285 (1999).

6. E. J. Kansa, *Computers Math. Applic.*, 127 (1990).
7. E. J. Kansa, *Computers Math. Applic.*, 147 (1990).
8. Z. M. Wu, *Approx. Theory Appl.*, 1 (1992).
9. C. S. Chen, C. A. Brebbia and H. Power, *Comm. Numer. Meths. Eng.*, 137 (1999).
10. Y. C. Hon and Z. M. Wu, *Int. J. Numer. Meth. Eng.*, 1187 (2000).
11. X. Zhang, K. Z. Song and M. W. Liu, *Comput. Mech.*, 333 (2000).
12. B. Fornberg, T. A. Driscoll, G. Wright and R. Charles, *Computers Math. Applic.*, 473 (2002).
13. W. Chen and M. Tanaka, *Comput. Math. Appl.*, 379 (2002).
14. M. A. Golberg, A. S. Muleshkov, C. S. Chen, A. H. D. Cheng, *Numer. Meth. Part. D. E.*, 112 (2003).
15. L. Ling, E. J. Kansa, *Math. Comput. Model.*, 1413 (2004).
16. R. E. Bellman and J. Casti, *J. Math. Anal. Appl.*, 235 (1971).
17. R. E. Bellman, B. G. Kashef, and J. Casti, *J. Comput. Phys.*, 40 (1972).
18. C. Shu and B. E. Richards, *Int. J. Numer. Methods Fluids*, 791 (1992).
19. C. Shu and Y. T. Chew, *Commun. Numer. Methods Eng.*, 643 (1997).
20. C. Shu, *Differential quadrature and its application in engineering* (Springer-Verlag, London, 2000).
21. C. Shu, H. Ding and K. S. Yeo, *Eng. Anal. Bound. Elem.*, 1217 (2004).
22. Y. L. Wu and C. Shu, *Comput. Mech.*, 477 (2002).
23. C. Shu, H. Ding, K. S. Yeo, *Comput. Method. Appl. M.*, 941 (2003).
24. Y. L. Wu, C. Shu, H. Q. Chen and N. Zhao, *Numer. Heat Tr. A-Appl.*, 269 (2004).
25. C. Shu, H. Ding and K. S. Yeo, *CMES-Computer Modeling in Engineering and Sciences*, 195 (2005).
26. H. Ding, C. Shu, K. S. Yeo and Z. L. Lu, *Numer. Heat Tr. A-Appl.*, 291 (2005).
27. C. Shu, Y. Y. Shan and N. Qin, *Int. J. Numer. Methods Fluids*, 367 (2007).
28. C. Shu and Y. L. Wu, *Int. J. Numer. Methods Fluids*, 969 (2007).
29. W. X. Wu, C. Shu and C. M. Wang, *J. Sound Vib.*, 252 (2007).
30. H. Ding, C. Shu and D. B. Tang, *Int. J. Numer. Meth. Eng.*, 1513 (2005).
31. Y. Y. Shan, C. Shu and Z. L. Lu, *CMES-Computer Modeling in Engineering & Sciences*, 99 (2008).
32. H. Ding, C. Shu, K. S. Yeo and D. Xu, *Comput. Method. Appl. M.*, 516 (2006).
33. A.J. Chorin, *Math. Comput.*, 745 (1968).
34. C. Shu, H. Ding, H. Q. Chen and T. G. Wang, *Comput. Method. Appl. M.*, 2001 (2005).
35. H. Q. Chen and C. Shu, *Int. J. Mod. Phys. C*, 439 (2005).
36. P. L. Roe, *J. Comput. Phys.*, 357 (1981).
37. P. K. Sweby, *SIAM J. Numer. Anal.*, 995 (1984).
38. H. Ding, C. Shu, K. S. Yeo and D. Xu, *Comput. Fluids*, 137 (2004).
39. C. Shu, H. Ding and N. Zhao, *Comput. Math. Appl.*, 1297 (2006).
40. U. Ghia, K. N. Ghia and H. B. Keller, *J. Comput. Phys.* 387 (1982).
41. T. A. Johnson, V. C. Patel, *J. Fluid Mech.*, 19 (1999).
42. A. Jameson, *Appl. Math. Comput.*, 327 (1983).

CHAPTER 12

STABILITY AND ACCURACY ANALYSIS OF SPATIAL DISCRETIZATIONS

Chris Lacor* and Kris Van den Abeele

Vrije Universiteit Brussel, Department Mechanical Engineering, Research Group Fluid Mechanics and Thermodynamics, Pleinlaan 2, 1050, Brussels

**chris.lacor@vub.ac.be*

In this chapter, the stability of the Spectral Volume (SV) and Spectral Difference (SD) methods for linear problems is analyzed.

These two methods were proposed a few years ago as alternatives to the popular DG method. The DG method has been under development since the 1980s, and consequently has reached a certain level of maturity. It enjoys a firm mathematical basis and many interesting properties, such as general nonlinear stability for arbitrary cell shapes and superconvergence properties of certain functionals of its numerical solution. However, its formulation is rather complicated, making it difficult to interpret physically, and also quite expensive, due to the numerical evaluations of surface and volume integrals that are required. The formulation of the SV method is based on the total sum of fluxes through the enclosing surface of a control volume (CV), like the FV method. Consequently, it has a clear physical interpretation and requires only the evaluation of surface integrals. The SD method directly computes the divergence of the flux vectors in certain solution points, like the finite difference method. Thus, the SD method is also easily physically interpretable and requires no numerical evaluation of any integrals. The main disadvantages of the SV and the SD methods are that they do not have as firm a mathematical basis as the DG method -yet- and that they are not uniquely defined. For the SV method, partitions of the cells into CVs have to be chosen, while for the SD method, solution and flux point distributions have to be selected. These CV partitions and point distributions have a certain number of identifying parameters, depending on the order of accuracy, which must be specified to define the SV or SD schemes. The stability and accuracy properties of both methods depend strongly on these parameters and consequently, a suitable choice for them is of paramount importance. The proper definition of CV partitions for the SV method and of solution and flux point distributions for the SD method, is the main focus of the present thesis.

For SV schemes of which the partition into CVs has one or more free parameters, as well as for SD schemes that have flux point distributions with free parameters, a stability analysis is used as a tool to identify parameters that result in stable and accurate schemes. The used methodology is based on an analysis of the wave propagation properties of the schemes and is applied here to the 2D SV and SD schemes. This analysis allows to assess both the stability and the accuracy of the schemes.

For details about the methodology of the analysis technique, the reader is refered to Ref. 1. The results that are discussed here are also published in Van den Abeele *et al.*[2–5]

1. Wave Propagation Analysis of 2D Schemes

The wave propagation analysis is based on the 2D linear advection equation

$$\frac{\partial q}{\partial t} + \frac{\partial (qa\cos\psi)}{\partial x} + \frac{\partial (qa\sin\psi)}{\partial y} = 0, \tag{1}$$

where ψ is the direction of the wave propagation. A 2D plane Fourier wave

$$q(t,x) = \check{q}\exp\left[Ik(x\cos\theta + y\sin\theta) + \vartheta t\right], \tag{2}$$

with θ the orientation of the wave and $\vartheta = \vartheta_{\mathcal{R}} - I\omega$, is a solution of this equation if the following exact dispersion relation is satisfied:

$$\vartheta_{\mathcal{R}} = 0 \qquad \text{and} \qquad \omega = ak\cos(\psi - \theta). \tag{3}$$

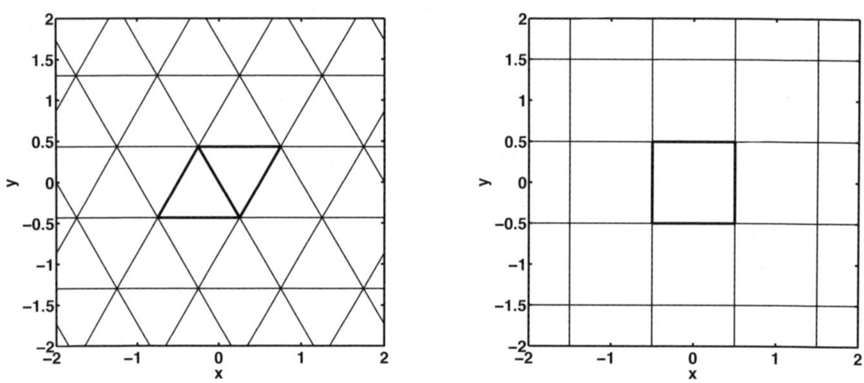

(a) Equilateral triangles grid. (b) Squares grid.

Fig. 1. Grids used for the 2D analysis.

The modified dispersion relation corresponding to a discretization of the linear advection equation on a uniform equilateral triangle or square cells grid, as shown in Figure 1, with the SV or the SD method, should be as close as possible to the exact dispersion relation for accuracy. For stability, the modified dissipation rate $\tilde{\vartheta}_{\mathcal{R}}$ should always be nonpositive. All quantities in this section are non-dimensionalized, using the edge length of the equilateral triangles or the squares as the reference length scale.

A more elaborate discussion on the analysis methodology for 2D can be found in Ref. 1.

It has been shown[6] that in 1D the SV and SD methods are equivalent. In 2D, this is not the case anymore. In the following sections, SV schemes for triangular cells, SD schemes for triangular cells and SD schemes for quadrilateral cells are therefore discussed separately.

1.1. *SV schemes for triangular cells*

The stability and accuracy of second-, third- and fourth-order SV schemes with an upwind Riemann flux and for triangular cells are discussed in the following sections. These results were published in Van den Abeele *et al.*[3]

1.1.1. *Second-order schemes*

The uniquely defined second-order partition of a triangular cell into CVs is shown in Figure 2[a] The modified dispersion relation is a polynomial equation of degree six in the dimensionless complex eigenvalue $\tilde{\Theta}$. For the propagation direction ψ and the wave orientation θ both equal to $\frac{\pi}{6}$, this equation is periodic in the dimensionless wave number K, with the period equal to $\frac{4\pi}{\sqrt{3}}$. The six values of $\tilde{\Theta}$ corresponding to this choice of ψ and θ are plotted versus K in Figure 3.

There are no eigenvalues with positive real components, which shows that the scheme is stable for $\psi = \theta = \frac{\pi}{6}$. This is also the case for any other combination of ψ and θ and thus, the scheme is always stable. Three of the resulting curves for $\psi = \theta = \frac{\pi}{6}$, namely those rendered with the plus symbol (+), the square (□) and the circle (○), have a straightforward physical interpretation. The other three curves (◇, △, ▽) are strongly damped, and thus correspond to solution eigenmodes that do not play any

[a] A partitioning connecting nodes of the triangle with the centroid leads only to first-order accurate schemes and is therefore not considered. The reason is that for such partitioning there is only one control volume on the external faces of the spectral volume to communicate with neighbouring spectral volumes, see Ref. 1.

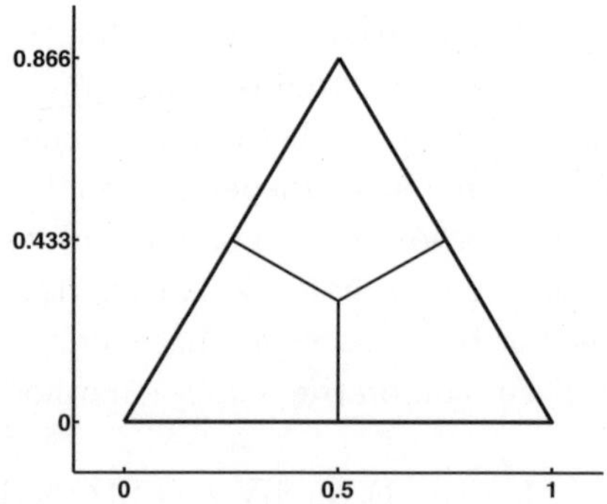

Fig. 2. Second-order SV partition of a triangular cell.

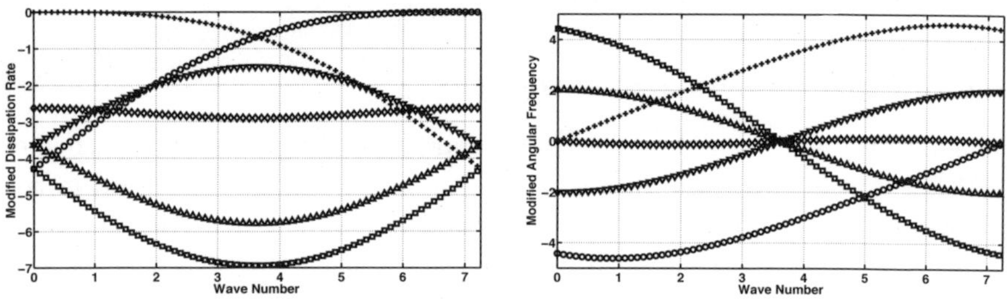

(a) $\tilde{\Theta}_{\mathcal{R}}$ vs. K.

(b) $-\tilde{\Theta}_{\mathcal{I}} = \tilde{\Omega}$ vs. K.

Fig. 3. Eigenvalues for second-order SV scheme with upwind Riemann flux, for $\psi = \theta = \frac{\pi}{6}$. $\tilde{\Theta}_{\mathcal{R}}$ and $\tilde{\Omega}$ versus K.

role of significance. The real dimensionless wave number to which an eigenvalue corresponds can be determined by examining the eigenmode solution shapes, see Ref. 1. This real wave number is K plus a whole multiple of $\frac{4\pi}{\sqrt{3}}$.

The diffusion and dispersion curves of the second-order SV scheme for $\psi = \theta = \frac{\pi}{6}$ are then shown in Figure 4. These curves are similar to those that were obtained for the second-order SV scheme in 1D. The modified dispersion relation follows the exact relation closely for dimensionless wave numbers up to $K \approx 1$.

The dependence of the wave propagation properties on the propagation direction ψ, for plane waves that are oriented in the same direction $\psi = \theta$, is illustrated in Figure 5, for three different dimensionless wave numbers K. In accordance with the symmetry of the equilateral triangles grid that is considered, $\tilde{\Theta}_{\mathcal{R}}$ and $\tilde{\Omega}$ are periodic in $\psi = \theta$, with a period of $\frac{\pi}{3}$. The dependence on $\psi = \theta$ is small for low wave numbers K, but becomes more

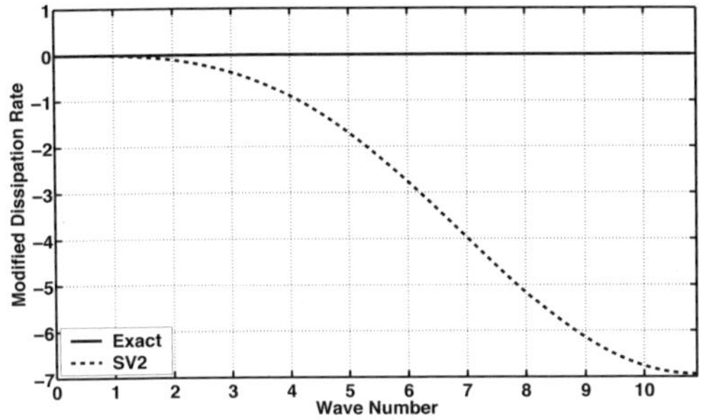

(a) Diffusive properties ($\tilde{\Theta}_{\mathcal{R}}$ vs. K).

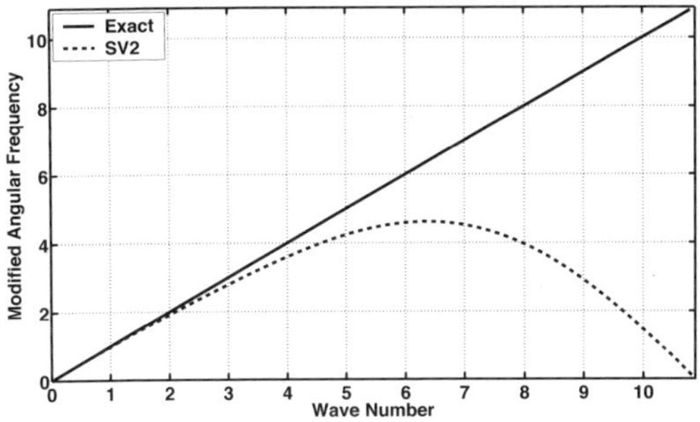

(b) Dispersive properties ($\tilde{\Omega}$ vs. K).

Fig. 4. Diffusive and dispersive properties of second-order 2D SV scheme with upwind Riemann flux, for $\psi = \theta = \frac{\pi}{6}$. $\tilde{\Theta}_{\mathcal{R}}$ and $\tilde{\Omega}$ versus K.

significant for larger K. The scheme is the most accurate for angles $\theta = \psi = \frac{\pi}{6} + l\frac{\pi}{3}$ and the least accurate for $\theta = \psi = l\frac{\pi}{3}$, with l an integer number.

1.1.2. *Third-order schemes*

The third-order partition of a triangular cell is plotted in Figure 6. It has two DOFs, which are defined as

$$\alpha_3 = \frac{|AC|}{|AB|} \in \left]0, \frac{1}{2}\right[\quad \text{and} \quad \beta_3 = \frac{|AE|}{|AD|} \in \left]0, \frac{2}{3}\right], \tag{4}$$

where the points A, B, C, D and E are shown in Fig 6.

By examining the resulting wave propagation properties, appropriate values for α_3 and β_3 can be selected. For the third-order partitions under consideration here, these parameters are summarized in Table 1, along with the corresponding Lebesgue constants $\|\Gamma_{\Pi}\|$.

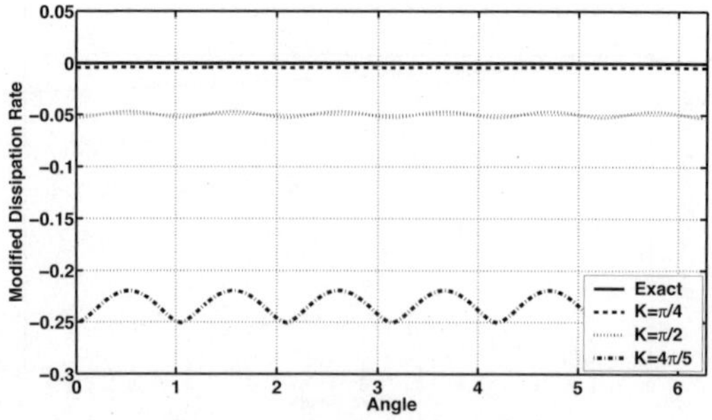

(a) Diffusive properties ($\tilde{\Theta}_{\mathcal{R}}$ vs. K).

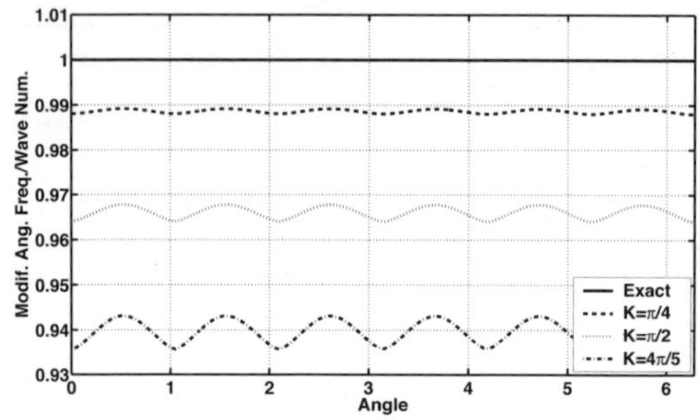

(b) Dispersive properties ($\frac{\tilde{\Omega}}{K}$ vs. K).

Fig. 5. Diffusive and dispersive properties of second-order 2D SV scheme with upwind Riemann flux, for K equal to 0.25π, 0.5π and 0.8π. $\tilde{\Theta}_{\mathcal{R}}$ and $\frac{\tilde{\Omega}}{K}$ versus $\psi = \theta$.

Table 1. Parameters of 2D third-order SV partitions.

Partition	α_3	β_3	$\|\Gamma_\Pi\|$
SV3W	$\frac{1}{4}$	$\frac{2}{3}$	8.0000
SV3L	$\frac{1}{4}$	$\frac{1}{4}$	3.6000
SV3Wb	$\frac{1}{4}$	$\frac{1}{3}$	3.9643
SV3C	0.1093621117	0.1730022492	3.0630
SV3P	0.091	0.18	3.0705

The modified dispersion relation is a polynomial equation of degree twelve in the present case, with twelve eigenvalues $\tilde{\Theta}$ for each combination of K, ψ and θ.

Firstly, three unstable partitions are discussed. The first partition was used in Wang and Liu[7] and corresponds to $\alpha_3 = \frac{1}{4}$ and $\beta_3 = \frac{2}{3}$. This choice

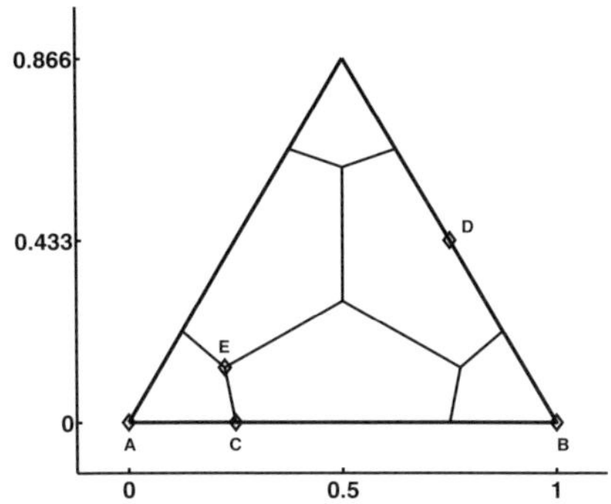

Fig. 6. Third-order SV partition of a triangular cell.

for β_3 leads to a partition in which three interior faces, between the CVs that lie at the center of the cell-faces, vanish. Consequently, it is cheaper to evaluate the residuals, since less flux computations are required, but the direct communication between the CVs that lie at the center of the cell-faces is lost. This partition is labeled SV3W. The second partition was presented in Liu *et al.*[8] and corresponds to $\alpha_3 = \frac{1}{4}$ and $\beta_3 = \frac{1}{4}$. This partition is labeled SV3L. Selecting the same values for α_3 and β_3 results in a partition in which the corner CVs reduce to triangles. The third partition was used in Wang *et al.*[9] and has $\alpha_3 = \frac{1}{4}$ and $\beta_3 = \frac{1}{3}$ as parameter values. It is labeled SV3Wb. These three partitions are plotted in Figure 7. Details, near the imaginary axis, of the Fourier footprints corresponding to these partitions are included in Figure 8. Note, that in these plots, all advection angles are considered with a well chosen discrete step, not too small so that individual points can still be distinguished. This shows that the first partition results in a scheme which suffers from a relatively strong instability. In fact, no stable partitions without interior faces between the CVs at the cell-faces ($\beta_3 = \frac{2}{3}$) exist. The other two partitions lead to schemes which are only weakly unstable.

The dependence of the stability of third-order SV schemes for triangular cells on the partition parameters is illustrated in more detail in Figure 9, where the logarithm in base ten of the maximum real eigenvalue, $\log_{10}\left(\max\tilde{\Theta}_{\mathcal{R}}\right)$, is plotted versus the parameters α_3 and β_3. Since this maximum must be zero for a scheme to be stable, the logarithm should be minus infinity. The zone of partition parameters that result in stable schemes thus corresponds to the white regions within the bold rectangle, which bounds the range of partitions that were investigated. It is clear that

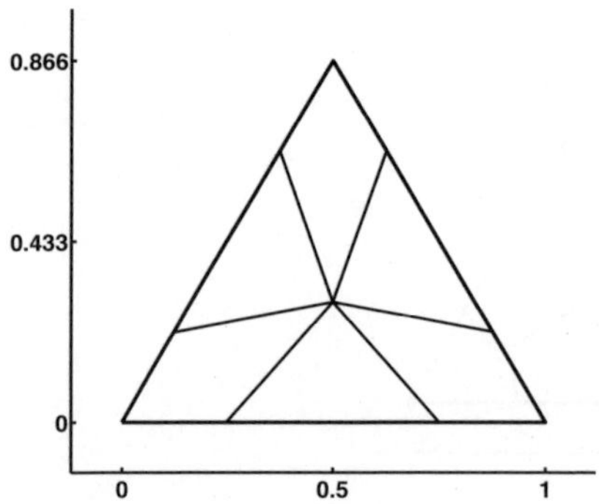

(a) SV3W,.[7]

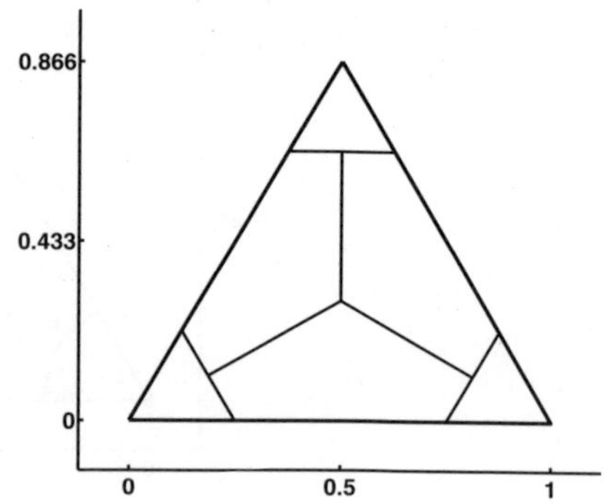

(b) SV3L,.[8]

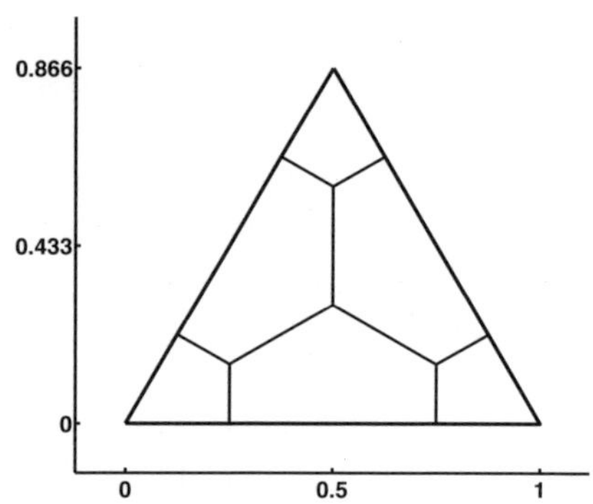

(c) SV3Wb,.[9]

Fig. 7. Examples of unstable third-order SV partitions of a triangular cell.

the partitions that were discussed above lie outside this stable zone.

Two partitions that were used in literature lie inside the stable zone. The first partition is named SV3C here, and was obtained by Chen,[10] using a systematic technique based on the Voronoi diagram and its variants. It is defined by $\alpha_3 = 0.1093621117$ and $\beta_3 = 0.1730022492$. The other partition is labeled SV3P, where P stands for 'present'. Based on the present analysis, it was designed to have good wave propagation properties. Its parameters are $\alpha_3 = 0.091$, $\beta_3 = 0.18$. Figure 10 shows the diffusion and dispersion curves versus K, with $\psi = \theta = \frac{\pi}{6}$, for these two partitions. The SV3P-scheme is slightly more accurate than the SV3C-scheme for these values of ψ and θ. The variation of $\tilde{\Theta}_{\mathcal{R}}$ and $\tilde{\Omega}$ versus $\psi = \theta$ for $K = \frac{\pi}{2}$ and $K = \pi$ is illustrated in Figure 11. The modified dissipation rate curves of the two

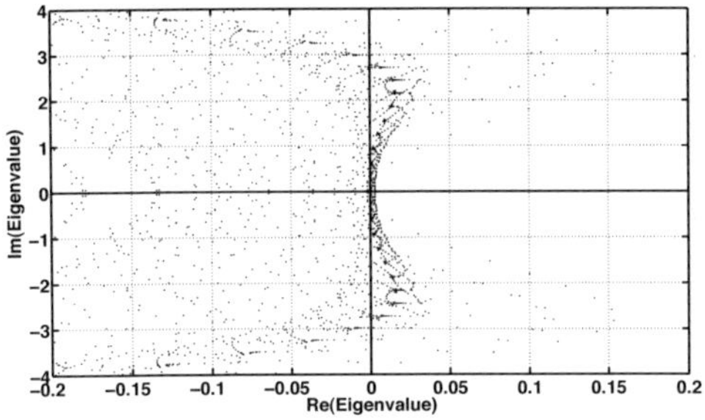

(a) SV3W,.[7]

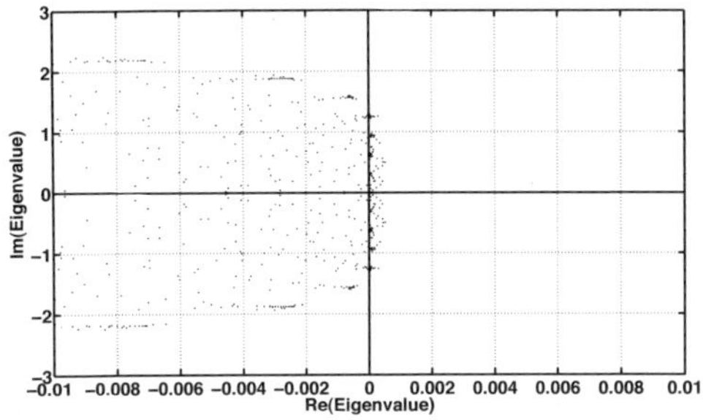

(b) SV3L,.[8]

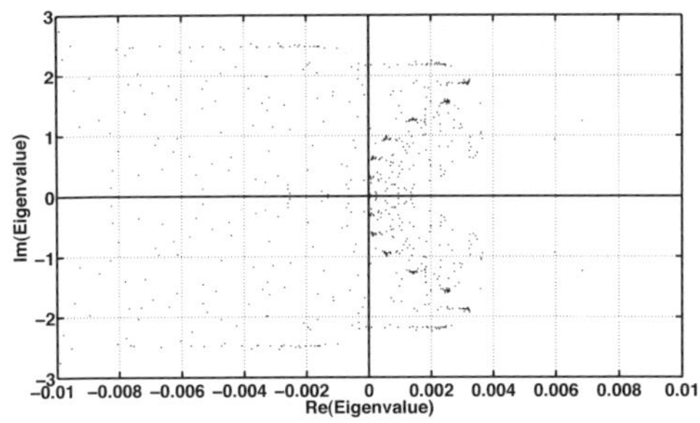

(c) SV3Wb,.[9]

Fig. 8. Detail near imaginary axis of Fourier footprints of unstable third-order SV schemes for triangular cells.

schemes are nearly indistinguishable. The ratio $\tilde{\Omega}/K$ of the SV3P-scheme is closer to the ideal value of one however. Notice the negative peaks in the dispersion curves for both schemes at angles $\theta = \psi = l\frac{\pi}{6}$, with l an integer number. This shows that the schemes are significantly less accurate for these angles. The loss of accuracy is expected to be less for the SV3P-

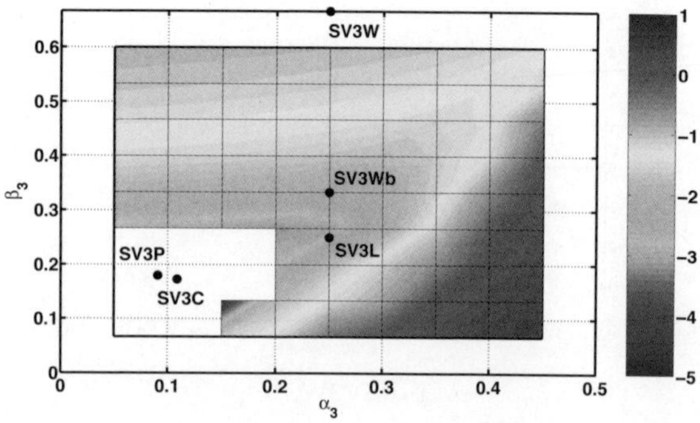

(a) Whole range of α_3 and β_3.

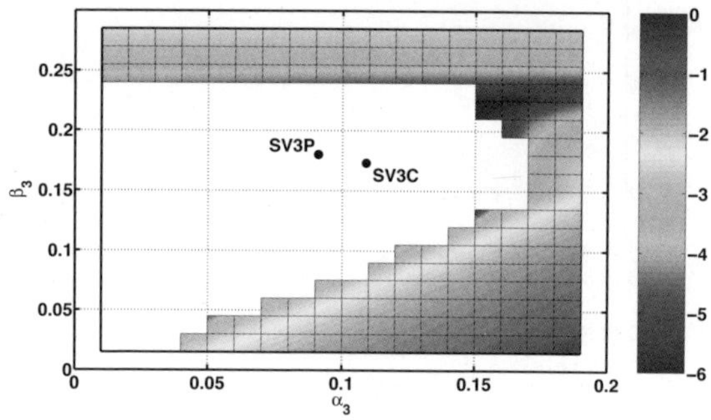

(b) Detail of the stable zone.

Fig. 9. Dependence of stability of third-order SV schemes for triangular cells on the partition: $\log_{10}\left(\max\left(\tilde{\Theta}_{\mathcal{R}}\right)\right)$ versus α_3 and β_3.

scheme, since the negative peaks are less pronounced.

1.1.3. *Fourth-order schemes*

The general fourth-order partition is included in Figure 12. Here, there are four DOFs:

$$\alpha_4 = \frac{|AC|}{|AB|} \in \left]0, \frac{1}{2}\right[\quad , \quad \beta_4 = \frac{|AE|}{|AD|} \in \left]0, \frac{2}{3}\right[,$$
$$\gamma_4 = \frac{|GD|}{|AD|} \in \left]0, \frac{1}{3}\right[\text{ and } \delta_4 = \frac{|AF|}{|AD|} \in \left[\beta_4, \frac{2}{3}\right[, \tag{5}$$

with the involved points again shown in the figure.

Using the present wave propagation analysis, these four parameters can

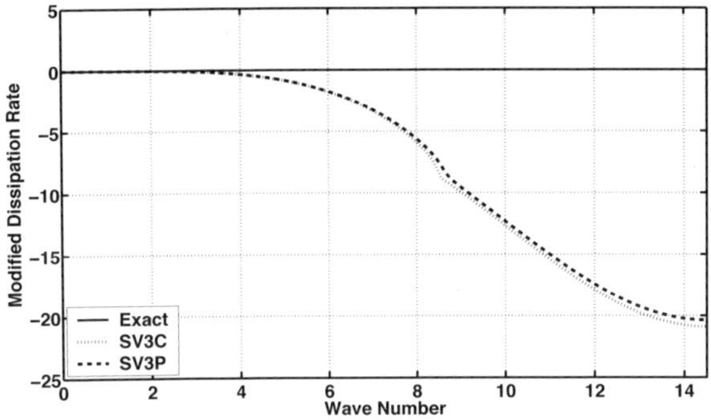

(a) Diffusive properties ($\tilde{\Theta}_{\mathcal{R}}$ vs. K).

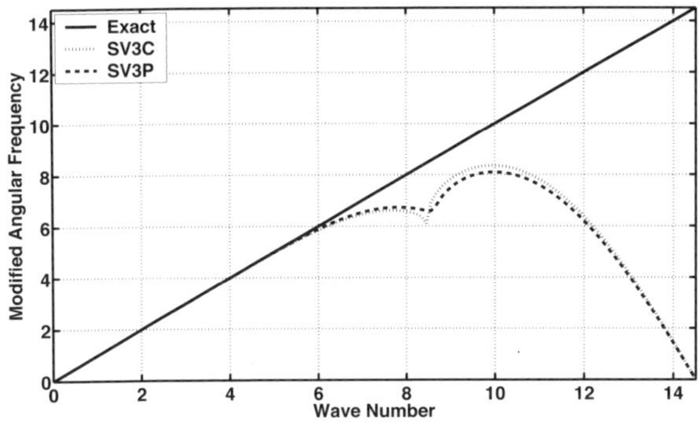

(b) Dispersive properties ($\tilde{\Omega}$ vs. K).

Fig. 10. Diffusive and dispersive properties of third-order 2D SV schemes with upwind Riemann flux, for $\psi = \theta = \frac{\pi}{6}$. $\tilde{\Theta}_{\mathcal{R}}$ and $\tilde{\Omega}$ versus K.

again be selected such that the resulting schemes are stable and accurate. The fourth-order partitions that are considered in the present section are summarized in Table 2, along with their Lebesgue constants $\|\Gamma_{\Pi}\|$. The modified dispersion relation is now a polynomial equation of degree twenty, with twenty eigenvalues $\tilde{\Theta}$ for each combination of K, ψ and θ.

Two of the partitions listed in Table 2 lead to weakly unstable schemes. The fourth-order partition labeled SV4W was first proposed in Wang and Liu.[7] The other partition labeled SV4H was presented in Harris and Wang.[11] The SV4H partition results in a scheme with very good wave propagation properties for $\psi = \theta = \frac{\pi}{6}$. The instability of this scheme occurs for propagation angles that are about zero. Details of the Fourier footprints of these two schemes can be seen in Figure 13. All advection angles are

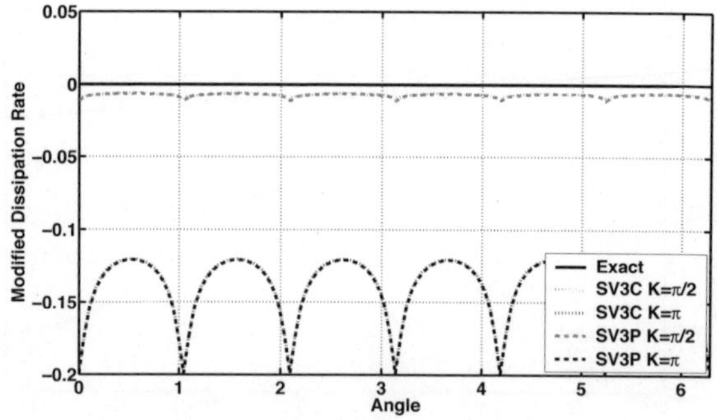

(a) Diffusive properties ($\tilde{\Theta}_{\mathcal{R}}$ vs. K).

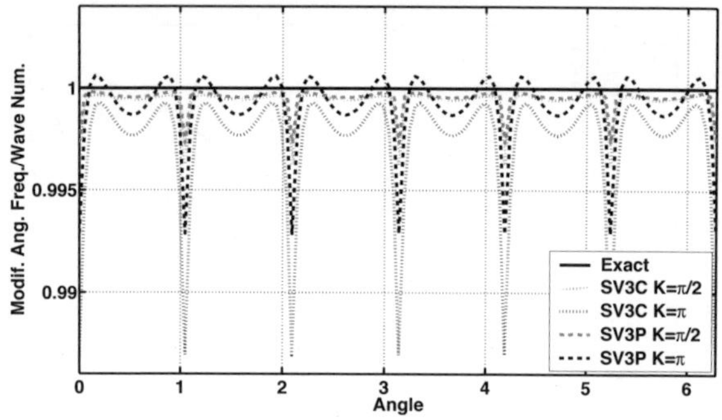

(b) Dispersive properties ($\frac{\tilde{\Omega}}{K}$ vs. K).

Fig. 11. Diffusive and dispersive properties of third-order 2D SV schemes with upwind Riemann flux, for K equal to $\frac{\pi}{2}$ and π. $\tilde{\Theta}_{\mathcal{R}}$ and $\frac{\tilde{\Omega}}{K}$ versus $\psi = \theta$.

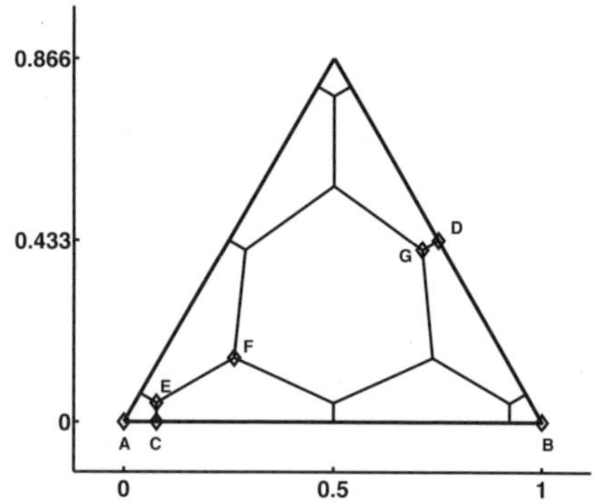

Fig. 12. Fourth-order ($p = 3$) triangular SV cell.

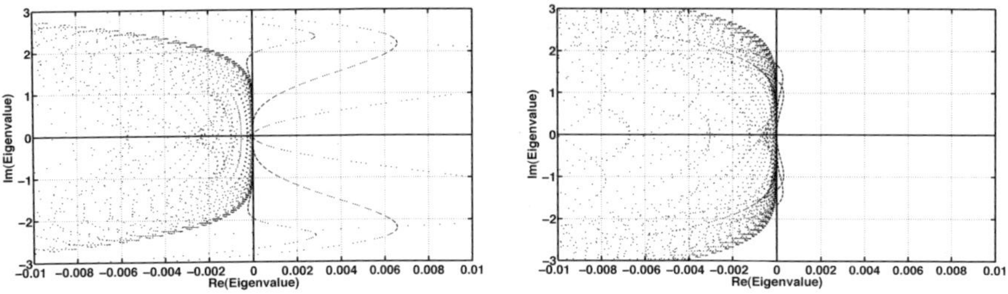

(a) SV4W,.[7] (b) SV4H,.[11]

Fig. 13. Detail near imaginary axis of Fourier footprints of unstable third-order SV schemes for triangular cells.

considered with a well chosen discrete step, not too small so that individual points can still be distinguished in the plot. Some of the eigenvalues lie in the right half of the complex plane in both cases. A general discussion about the dependence of stability on the partition parameters for fourth-order SV schemes is very complicated, because the parameter space of these partitions is four-dimensional.

Two stable partitions are known in literature. The first, labeled SV4C, see Table 2, was obtained by Chen,[10] using the same technique that lead to the third-order SV3C partition. The second was presented by the present authors, and was designed to have good wave propagation properties. It is referred to as SV4P, see Table 2. These two partitions are illustrated in Figure 14. The modified dissipation rate $\tilde{\Theta}_{\mathcal{R}}$ and angular frequency $\tilde{\Omega}$ are plotted versus K for $\psi = \theta = \frac{\pi}{6}$ in Figure 15. The influence of the angle $\psi = \theta$ is illustrated in Figure 16. It is clear that the SV4P-scheme is superior to the SV4C-scheme.

Table 2. Parameters of 2D fourth-order SV partitions.

Partition	α_4	β_4	γ_4	δ_4	$\|\Gamma_\Pi\|$
SV4W	$\frac{1}{15}$	$\frac{2}{15}$	$\frac{1}{15}$	$\frac{2}{15}$	3.4448
SV4C	0.0326228301	0.0425080882	0.0504398911	0.1562524902	3.2129
SV4P	0.078	0.104	0.052	0.351	4.2446
SV4H	0.12061033	0.12129456	0.066666667	0.312260947	4.0529

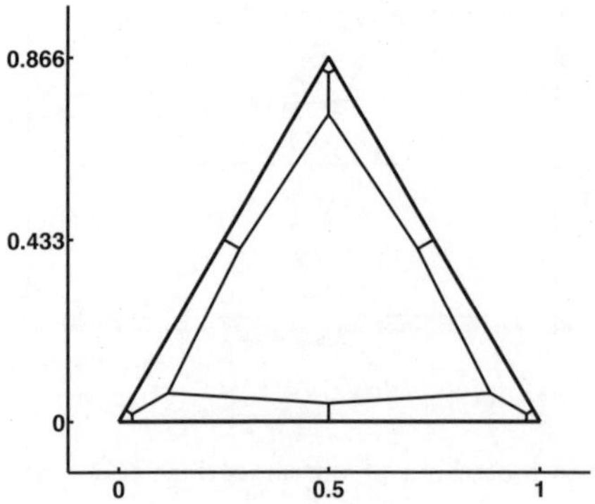

(a) SV4C,.[10]

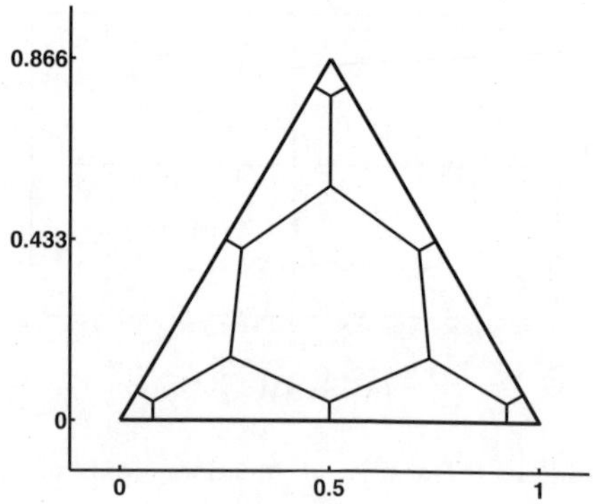

(b) SV4P,.[3]

Fig. 14. Two stable fourth-order SV partitions of a triangular cell.

1.1.4. *Illustration*

The results of the analysis above are verified with a test case that is governed by the 2D linear advection equation (1), with $a = 1$. The initial solution is a Gaussian pulse

$$q^0(x,y) = \exp\left[-\frac{(x-x_0)^2 + (y-y_0)^2}{b^2}\right], \tag{6}$$

with $b = 0.1$, and x_0 and y_0 the initial coordinates of the center of the pulse.

The stability or instability of the third- and fourth-order accurate schemes that were discussed above is verified first. A uniform grid that consists of equilateral triangles with edge length 0.1, as illustrated in Figure 1(a), is considered.

A four-stage fourth-order accurate R-K scheme is used for time marching.

For the third-order SV schemes, the propagation angle ψ is equal to $\frac{\pi}{2}$ and a time step Δt of 0.005 is used. Figure 17(a) shows the obtained residual histories. As predicted by the analysis, the SV3W-, SV3L- and SV3Wb-partitions result in unstable schemes. Notice that the SV3W-scheme diverges very quickly. The SV3Wb-scheme diverges after a larger number of iterations and the SV3L-scheme takes almost one hundred thousand iterations before it starts to diverge. This is in accordance with the relative magnitudes of the real components of the eigenvalues of these three schemes, as shown in Figure 8. The SV3C- and SV3P-partitions result in

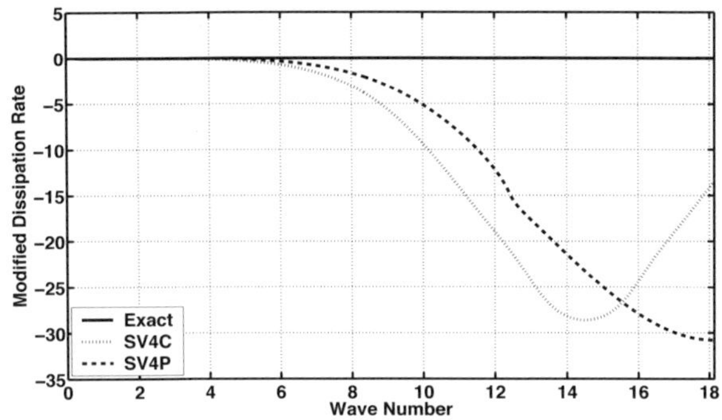

(a) Diffusive properties ($\tilde{\Theta}_{\mathcal{R}}$ vs. K).

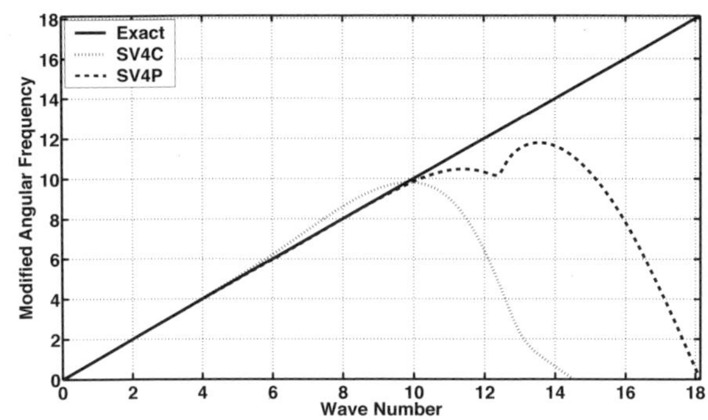

(b) Dispersive properties ($\tilde{\Omega}$ vs. K).

Fig. 15. Diffusive and dispersive properties of fourth-order 2D SV schemes with upwind Riemann flux, for $\psi = \theta = \frac{\pi}{6}$. $\tilde{\Theta}_{\mathcal{R}}$ and $\tilde{\Omega}$ versus K.

stable schemes, in agreement with the results of the analysis. In Figure 17(b), the residual histories corresponding to the fourth-order partitions are shown. For these schemes, the propagation angle ψ is 0 and the time step Δt is 0.001. The instability of the SV4W- and SV4H-schemes is clearly illustrated. The SV4W-scheme diverges before the SV4H-scheme, as predicted by Figure 13. The computations with the other schemes are stable.

A grid convergence study on a sequence of uniform grids consisting of equilateral triangles was performed for the stable schemes. Propagation angles $\psi = 0$ and $\psi = \frac{\pi}{2}$ were considered. These angles correspond to respectively the least accurate and the most accurate direction of the schemes. The same four-stage fourth-order accurate R-K scheme was used for time marching, with a sufficiently small time step Δt to ensure negligible time

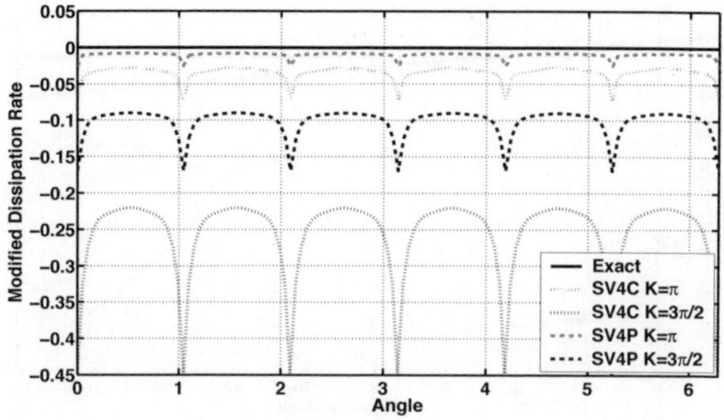

(a) Diffusive properties ($\tilde{\Theta}_{\mathcal{R}}$ vs. K).

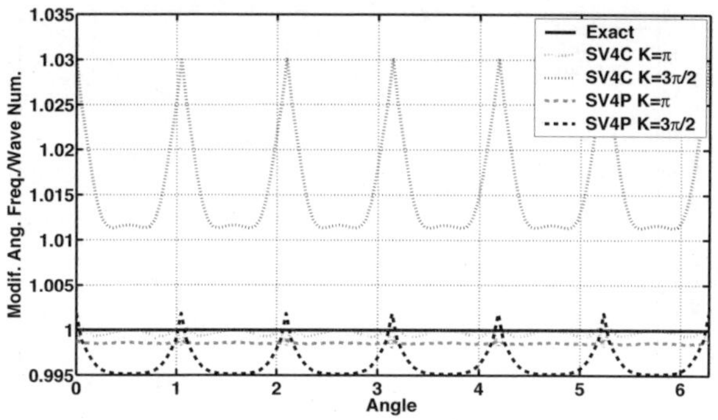

(b) Dispersive properties ($\frac{\tilde{\Omega}}{K}$ vs. K).

Fig. 16. Diffusive and dispersive properties of fourth-order 2D SV schemes with upwind Riemann flux, for K equal to π and $\frac{3\pi}{2}$. $\tilde{\Theta}_{\mathcal{R}}$ and $\frac{\tilde{\Omega}}{K}$ versus $\psi = \theta$.

discretization errors. The resulting errors in the L_1- and the L_∞-norm at $t = 1$ are listed in Table 3 for $\psi = 0$ and in Table 4 for $\psi = \frac{\pi}{2}$.

One immediately notices that all schemes fail to achieve the expected order of accuracy for the case $\psi = 0$. The second-order SV scheme attains only a first-order accuracy. The third- and fourth-order SV schemes perform slightly better, attaining orders of accuracy that are significantly higher than, respectively, two and three. Even though the analysis predicts that $\psi = 0$ is the direction in which the schemes are the least accurate, this does not explain the decrease in the order of accuracy. A possible explanation for this is the fact that, for $\psi = \theta = l\frac{\pi}{3}$, with l an integer number, the modified dispersion relations of all these schemes support multiple zero roots $\tilde{\Theta} = 0$, if $K = 0$. This leads to a solution eigenmode with a polynomial growth,

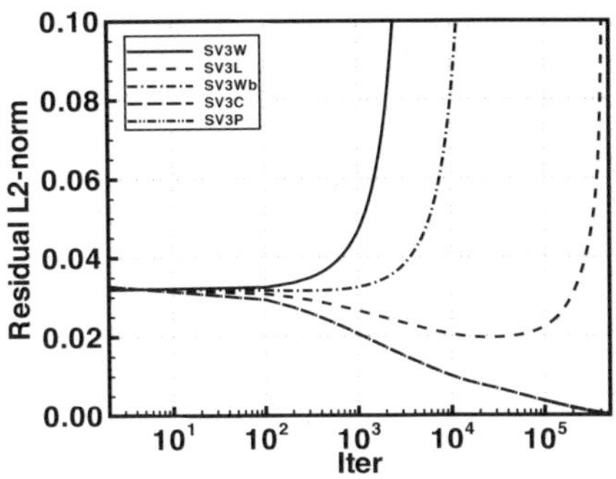

(a) $p = 2$, $\psi = \frac{\pi}{2}$, $\Delta t = 0.005$.

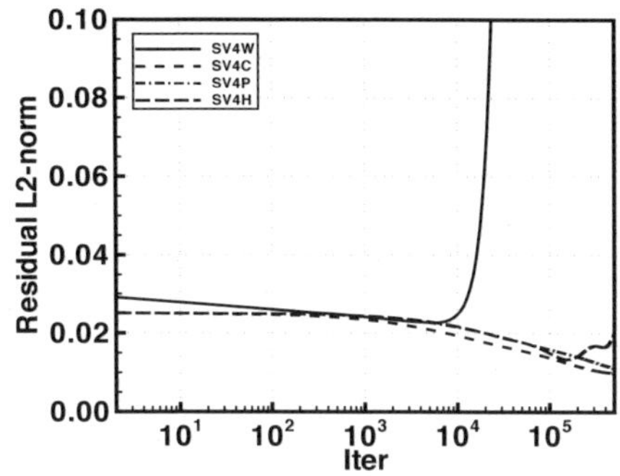

(b) $p = 3$, $\psi = 0$, $\Delta t = 0.001$.

Fig. 17. Residual histories for the linear advection of a 2D Gaussian pulse, obtained with the SV method.

see[1] and thus to a very weak instability, which could cause the loss of the expected order of accuracy that is observed in Table 3. Also notice that for $\psi = 0$, the rows of cells parallel to the propagation direction are uncoupled, since the fluxes between these rows are zero.

The somewhat strange-looking values for the cell size in Table 4 are chosen to obtain an integer number of cells in an interval of length one in the propagation direction. The cell sizes in Table 4 thus correspond to those in Table 3 times $\frac{2}{\sqrt{3}}$. For this case, the expected orders of accuracy are achieved with all the schemes. It is also seen from Tables 3 and 4 that lower errors can be obtained with less DOFs if higher-order schemes are used.

For the third-order accurate schemes and propagation direction $\psi = \frac{\pi}{2}$, there is very little difference between the errors obtained with the SV3C- and the SV3P-schemes, although the errors obtained with the first scheme are systematically smaller. As predicted by the wave propagation analysis, the SV3P-scheme performs better than the SV3C-scheme for $\psi = 0$, as the error levels are always significantly lower and the observed order of accuracy is higher. Regarding the fourth-order accurate schemes, it can be concluded that the SV4P-scheme systematically yields much lower error levels than the SV4C-scheme.

1.2. *SD schemes for triangular cells*

The wave propagation properties of second- and third-order SD schemes for triangular cells are discussed in this section. These results were published

Table 3. Grid convergence study for the 2D linear advection equation using the SV schemes for triangular cells. Wave propagation angle $\psi = 0$.

p	Δx	#DOF	L_1 error	L_1 order	L_∞ error	L_∞ order
1	0.20000	180	$2.52e-2$	–	$2.22e-1$	–
	0.10000	720	$2.00e-2$	0.33	$3.58e-1$	−0.69
	0.05000	2880	$1.05e-2$	0.93	$2.03e-1$	0.82
	0.02500	11520	$5.04e-3$	1.06	$9.87e-2$	1.04
	0.01250	46080	$2.47e-3$	1.03	$4.48e-2$	1.14
	0.00625	184320	$1.24e-3$	0.97	$2.13e-2$	1.04
2	0.20000	360	$1.67e-2$	–	$3.95e-1$	–
SV3C	0.10000	1440	$6.32e-3$	1.40	$1.92e-1$	1.04
	0.05000	5760	$1.36e-3$	2.21	$5.14e-2$	1.90
	0.02500	23040	$2.76e-4$	2.31	$1.16e-2$	2.15
	0.01250	92160	$6.29e-5$	2.13	$2.51e-3$	2.21
2	0.20000	360	$1.65e-2$	–	$3.88e-1$	–
SV3P	0.10000	1440	$6.07e-3$	1.45	$1.83e-1$	1.08
	0.05000	5760	$1.26e-3$	2.27	$4.55e-2$	2.01
	0.02500	23040	$2.26e-4$	2.47	$8.56e-3$	2.41
	0.01250	92160	$4.74e-5$	2.26	$1.64e-3$	2.39
3	0.20000	600	$1.12e-2$	–	$2.20e-1$	–
SV4C	0.10000	2400	$1.92e-3$	2.55	$5.32e-2$	2.05
	0.05000	9600	$1.66e-4$	3.53	$6.56e-3$	3.02
	0.02500	38400	$1.48e-5$	3.48	$5.20e-4$	3.66
3	0.20000	600	$9.26e-3$	–	$1.52e-1$	–
SV4P	0.10000	2400	$1.22e-3$	2.93	$4.76e-2$	1.67
	0.05000	9600	$1.14e-4$	3.41	$5.24e-3$	3.18
	0.02500	38400	$1.03e-5$	3.47	$4.17e-4$	3.65

in Van den Abeele *et al.*[4]

1.2.1. *Second-order schemes*

Consider the second-order triangular SD cell shown in Figure 18.

Different approaches for the Riemann solvers in the flux points on the faces can be used, see Liu *et al.*[12] and Wang *et al.*[13]

With the first approach, d 1D Riemann solvers, with d the dimensionality of the problem, are used at corner flux points to compute the normal flux components between cells that share a face. From these normal flux components, the full flux vector at a corner flux point can be reconstructed for a cell. At face flux points, a 1D Riemann solver is used for the nor-

Table 4. Grid convergence study for the 2D linear advection equation using the SV schemes for triangular cells. Wave propagation angle $\psi = \frac{\pi}{2}$.

p	Δx	#DOF	L_1 error	L_1 order	L_∞ error	L_∞ order
1	0.28868	150	$3.32e-2$	–	$2.54e-1$	–
	0.11547	600	$1.97e-2$	0.57	$4.08e-1$	−0.52
	0.05774	2400	$7.67e-3$	1.36	$2.52e-1$	0.70
	0.02887	9600	$2.31e-3$	1.73	$1.03e-1$	1.29
	0.01443	38400	$6.06e-4$	1.93	$2.96e-2$	1.80
	0.00707	153600	$1.53e-4$	1.93	$7.75e-3$	1.88
2	0.28868	300	$3.70e-2$	–	$6.71e-1$	–
SV3C	0.11547	1200	$5.32e-3$	2.12	$2.03e-1$	1.30
	0.05774	4800	$7.53e-4$	2.82	$4.11e-2$	2.30
	0.02887	19200	$9.43e-5$	3.00	$5.52e-3$	2.90
	0.01443	76800	$1.19e-5$	2.99	$6.85e-4$	3.01
2	0.28868	300	$3.69e-2$	–	$6.71e-1$	–
SV3P	0.11547	1200	$5.32e-3$	2.12	$2.04e-1$	1.30
	0.05774	4800	$7.61e-4$	2.80	$4.17e-2$	2.29
	0.02887	19200	$9.64e-5$	2.98	$5.61e-3$	2.89
	0.01443	76800	$1.22e-5$	2.98	$6.99e-4$	3.01
3	0.28868	500	$3.47e-2$	–	$6.20e-1$	–
SV4C	0.11547	2000	$1.85e-3$	3.20	$8.13e-2$	2.22
	0.05774	8000	$1.04e-4$	4.16	$6.32e-3$	3.69
	0.02887	32000	$5.04e-6$	4.36	$3.75e-4$	4.07
3	0.28868	500	$1.39e-2$	–	$3.13e-1$	–
SV4P	0.11547	2000	$7.68e-4$	3.16	$3.40e-2$	2.42
	0.05774	48000	$4.02e-5$	4.26	$3.17e-3$	3.42
	0.02887	32000	$2.31e-6$	4.12	$2.29e-4$	3.79

mal component of the flux. The tangential component can be the internal component or the average of the tangential components. This first possibility is illustrated for the face (□) and corner (○) flux point of cell C in Figure 19(a). The approach with the internal tangential component is labeled 'semi-upwind' approach and the one with the averaged tangential component 'averaged-upwind' approach.

The second possible treatment for face and corner flux points consists of using multi-dimensional Riemann solvers. The full flux is then evaluated using the solution in the cell where propagating waves are coming from. This treatment is illustrated in Figure 19(b). For the face flux point (□), the full flux vector from within cell D is used, while for the corner flux

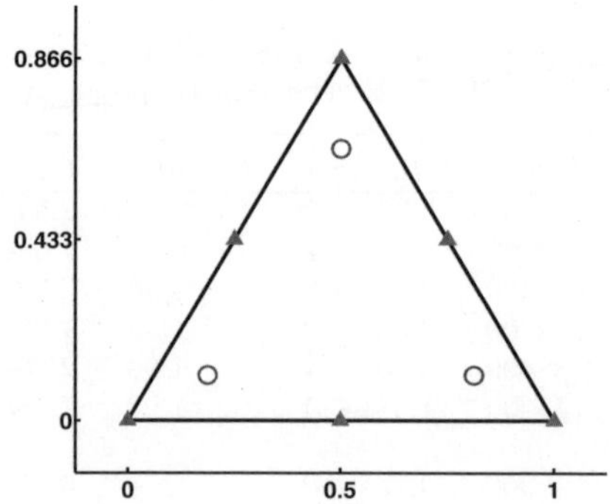

Fig. 18. Second-order triangular SD cell. Solution (○) and flux points (▲).

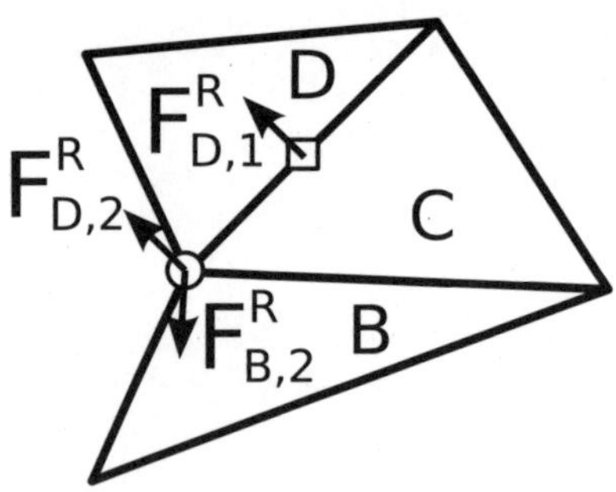

(a) Multiple 1D Riemann solvers.

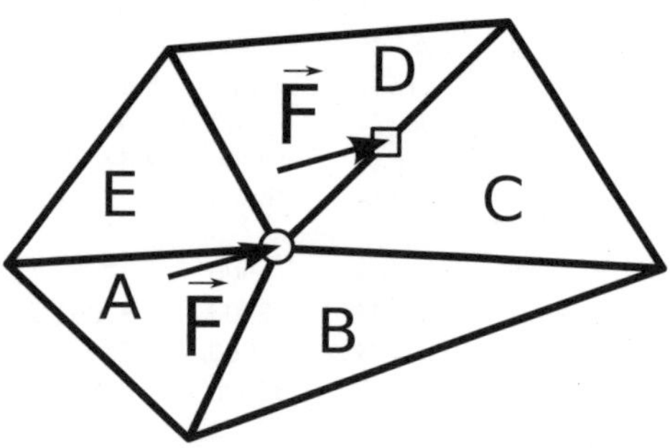

(b) One multi-D Riemann solver.

Fig. 19. Different treatments of face and corner flux points.

point (○), the flux from within cell A is selected. This treatment is labeled 'full-upwind' approach. The problem with this approach is that such multi-dimensional Riemann solvers are only available when the physics of the problem is a simple unidirectional wave propagation.

Combining the averaged-upwind Riemann flux approach with this SD cell, an unstable scheme is obtained. This is illustrated in Figure 20, which shows the scheme's Fourier footprint. The other two Riemann flux approaches, the semi-upwind and the full-upwind approach, lead to stable schemes. The corresponding Fourier footprints are shown in Figure 21. As before, the Fourier plots contain all advection angles with a well chosen discrete step, not too small so that individual points can still be distinguished. Notice that the full-upwind approach leads to a more compact footprint than the semi-upwind approach. Consequently, the former will generally allow larger time steps than the latter, if an explicit time marching scheme is used.

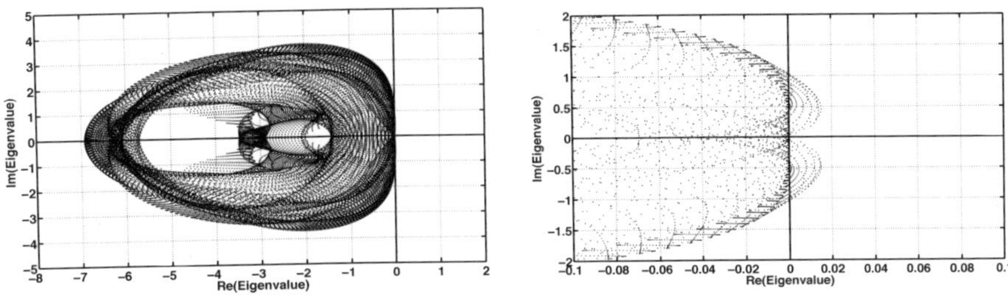

(a) Full footprint. (b) Detail near the imaginary axis.

Fig. 20. Fourier footprint of second-order SD scheme for triangular cells, with averaged-upwind Riemann flux approach.

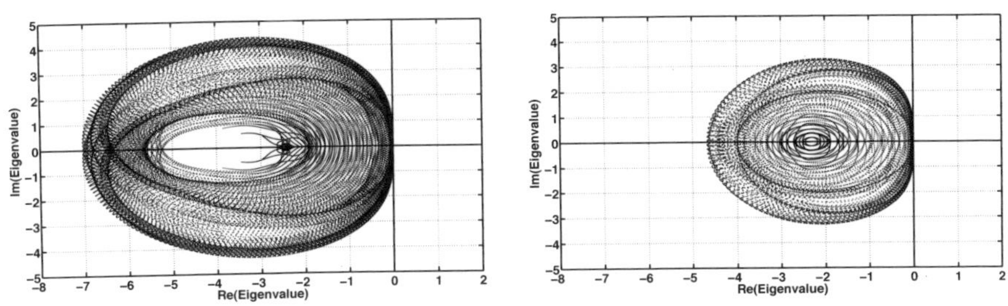

(a) Semi-upwind Riemann flux approach. (b) Full-upwind Riemann flux approach.

Fig. 21. Fourier footprint of second-order SD schemes for triangular cells.

The dependence of the diffusive and dispersive properties upon the angles $\psi = \theta$, for various wave numbers K, is illustrated in Figure 22 for the semi-upwind Riemann flux approach and in Figure 23 for the full-upwind Riemann flux approach. Due to the symmetry of the equilateral triangle grid, they are periodic in $\psi = \theta$ with a period equal to $\frac{\pi}{3}$. Both approaches lead to schemes that are the most accurate for $\psi = \theta = \frac{\pi}{6} + l\frac{\pi}{3}$ and the least accurate for $\psi = \theta = l\frac{\pi}{3}$, with l an integer number. The scheme with the semi-upwind approach is significantly less diffusive, and consequently more accurate, than the one with the full-upwind approach.

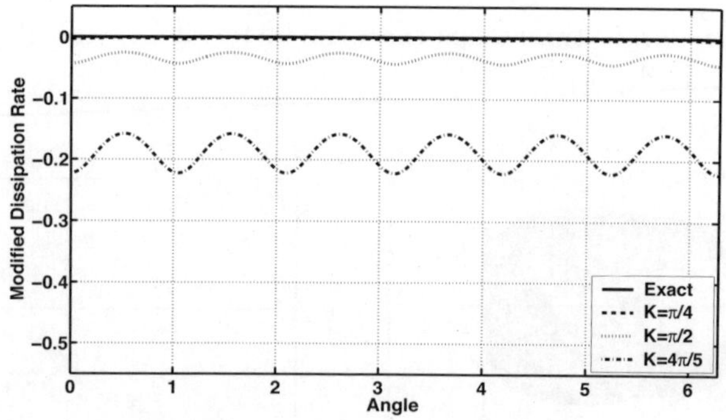

(a) Diffusive properties ($\tilde{\Theta}_{\mathcal{R}}$ vs. K).

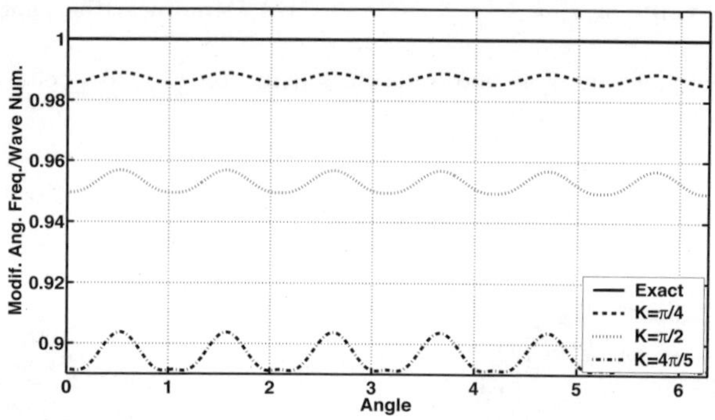

(b) Dispersive properties ($\frac{\tilde{\Omega}}{K}$ vs. K).

Fig. 22. Diffusive and dispersive properties of second-order 2D SD schemes for triangular cells, with semi-upwind Riemann flux approach, for $K = \frac{\pi}{4}$, $K = \frac{\pi}{2}$ and $K = \frac{4\pi}{5}$. $\tilde{\Theta}_{\mathcal{R}}$ and $\frac{\tilde{\Omega}}{K}$ versus $\psi = \theta$.

1.2.2. *Third-order schemes*

Figure 24 shows the two possible general triangular third-order SD cells, with cubic flux polynomial distributions with at least three points at each face. The dependence of the maximum real component of the eigenvalues $\max\left(\tilde{\Theta}_{\mathcal{R}}\right)$ upon the flux point distribution parameter α_3 is illustrated in Figure 25 for the cell with corner flux points, shown in Figure 24(a). Similar plots for the cell without corner flux points, see Figure 24(b), are included in Figure 26. It is seen that neither cell leads to a stable scheme, with neither the semi-upwind nor the full-upwind Riemann flux approach, for any value of α_3.

The third-order schemes shown in Figure 24 include all possible symmetric flux point distributions for an order-complete cubic flux polynomial

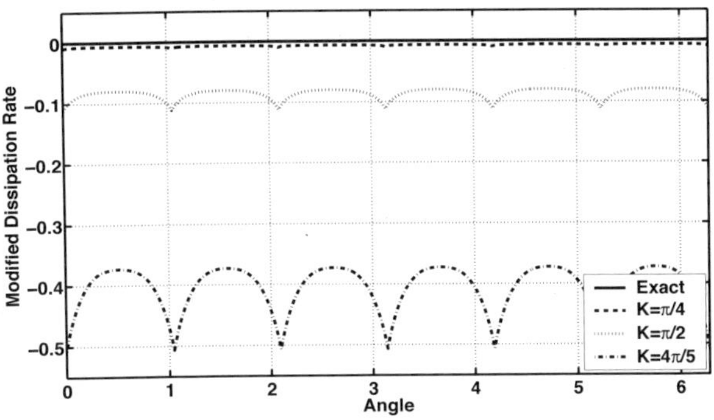

(a) Diffusive properties ($\tilde{\Theta}_{\mathcal{R}}$ vs. K).

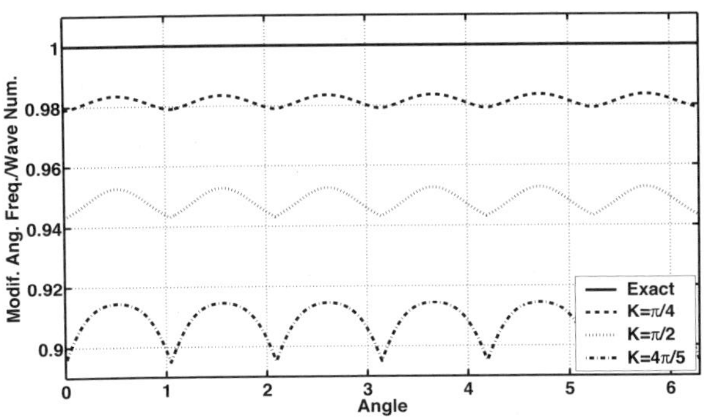

(b) Dispersive properties ($\frac{\tilde{\Omega}}{K}$ vs. K).

Fig. 23. Diffusive and dispersive properties of second-order 2D SD schemes for triangular cells, with full-upwind Riemann flux approach, for $K = \frac{\pi}{4}$, $K = \frac{\pi}{2}$ and $K = \frac{4\pi}{5}$. $\tilde{\Theta}_{\mathcal{R}}$ and $\frac{\tilde{\Omega}}{K}$ versus $\psi = \theta$.

in a triangle, with at least three flux points on each face. Consequently, it has to be concluded that no stable third-order accurate SD schemes, with a standard third-order Lagrangian flux polynomial treatment, exist. There is no stable flux point distribution, with any treatment of the corner and face flux points, even though the instability is very small in some cases.

Recently a new formulation based on flux interpolation on Raviart-Thomas elements has been proposed.[14] Third- and fourth-order accurate schemes are formulated that seem to be linearly stable.

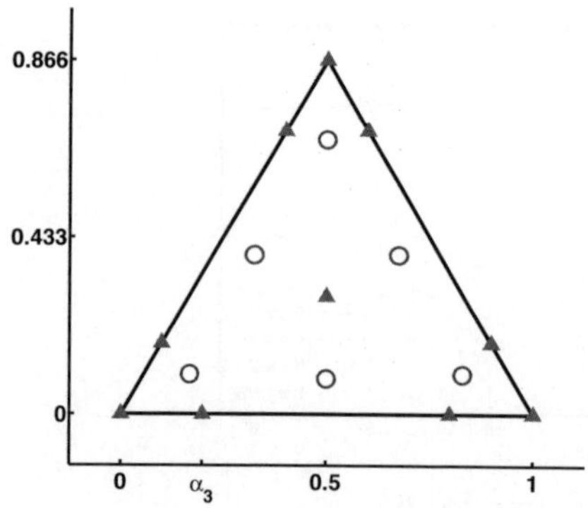

(a) With corner flux points.

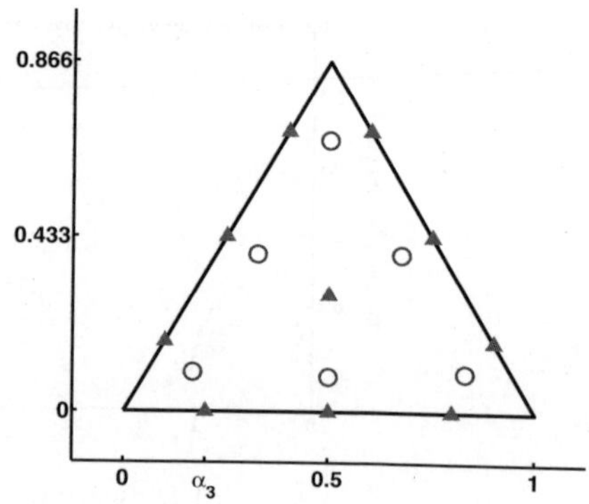

(b) Without corner flux points.

Fig. 24. Third-order triangular SD cells. Solution (○) and flux points (▲).

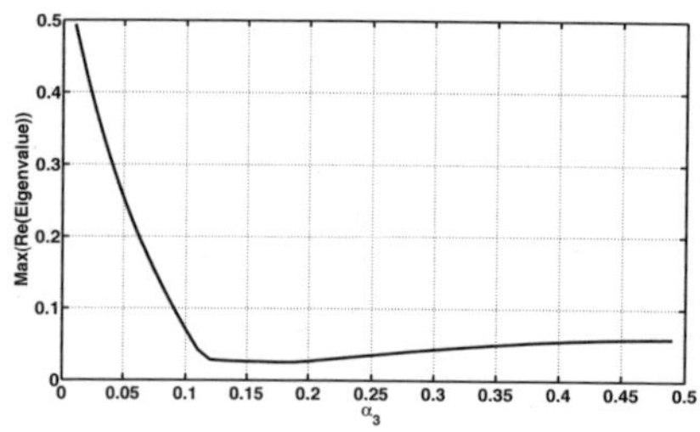

(a) Semi-upwind Riemann flux approach.

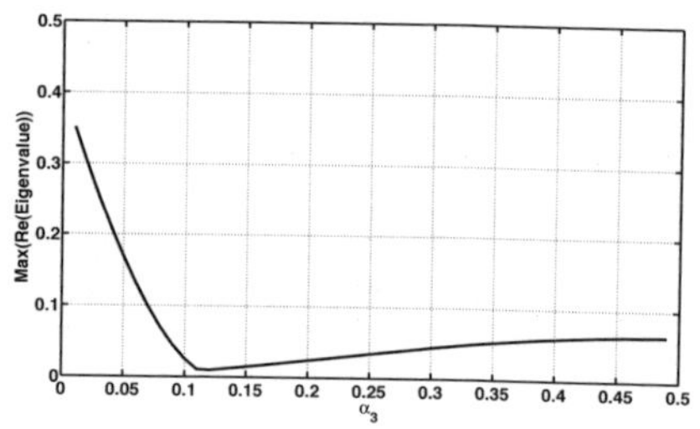

(b) Full-upwind Riemann flux approach.

Fig. 25. Dependence of stability of third-order SD schemes for triangular cells with corner flux points on the flux point distribution: $\max\left(\tilde{\Theta}_{\mathcal{R}}\right)$ versus α_3.

1.3. *SD schemes for quadrilateral cells*

The properties of quadrilateral SD cells are discussed in this section. All results assume a semi-upwind Riemann flux approach. It can be shown, see[1] (Theorems 6.2 to 6.4) that the results for a scheme of a certain order are valid for any variant of this scheme, if the flux point distributions are equivalent. This analysis was also published in Van den Abeele *et al.*[4]

1.3.1. *Second-order schemes*

The second-order SD scheme for quadrilateral cells, as illustrated in Figure 27, is stable. In Figure 28, the wave propagation properties of this scheme are plotted versus the propagation angle ψ, with the orientation of the plane

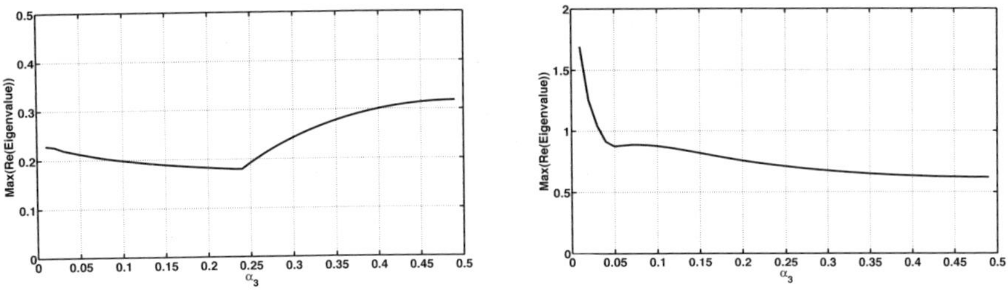

(a) Semi-upwind Riemann flux approach. (b) Full-upwind Riemann flux approach.

Fig. 26. Dependence of stability of third-order SD schemes for triangular cells without corner flux points on the flux point distribution: $\max\left(\tilde{\Theta}_{\mathcal{R}}\right)$ versus α_3.

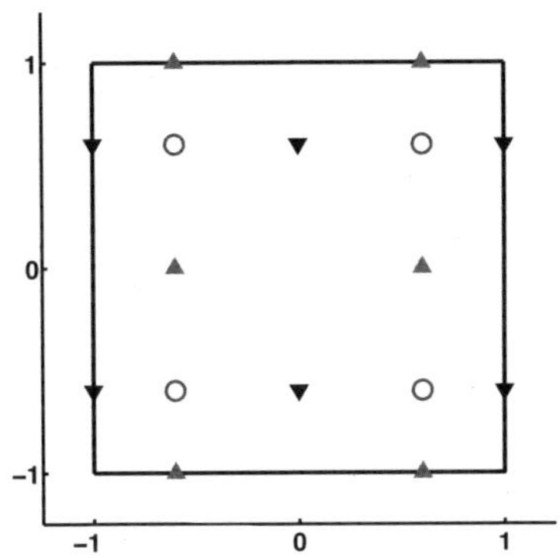

Fig. 27. Second-order quadrilateral SD cell.

Fourier wave θ equal to ψ. It is seen that the scheme is the most accurate for propagation along the diagonals of square cells ($\psi = \theta = \frac{\pi}{4} + l\frac{\pi}{2}$), and the least accurate for propagation along the edges ($\psi = \theta = l\frac{\pi}{2}$), with l an integer number. As expected, for propagation along the edges, the same wave propagation properties as for the 1D SD scheme are found. For larger K, there is a more important dependency on $\psi = \theta$.

Upon comparison of the properties of the second-order SD scheme for quadrilateral cells and the properties of the second-order SD schemes for triangular cells, shown in Figures 22 and 23 for respectively the semi-upwind and the full-upwind Riemann flux approach, it is seen that the triangular SD cells with the semi-upwind approach are the most accurate. It should be noted however that for the second-order quadrilateral SD cells, there are

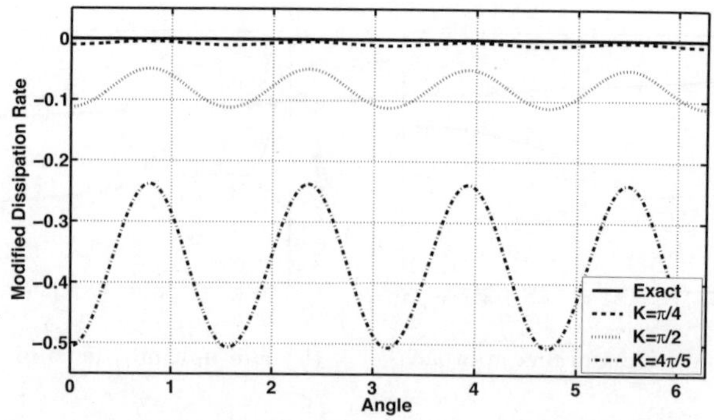

(a) Diffusive properties ($\tilde{\Theta}_{\mathcal{R}}$ vs. K).

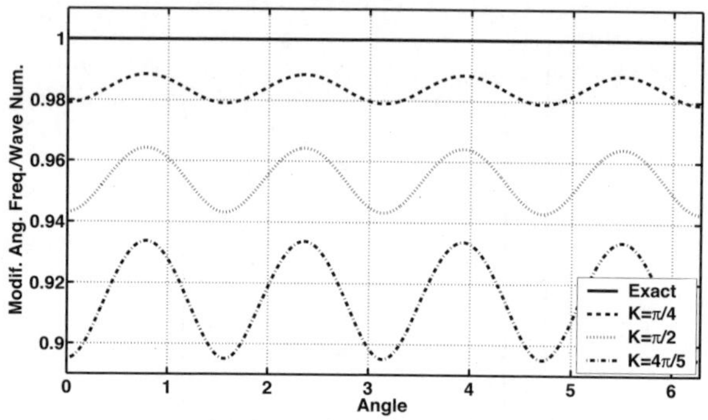

(b) Dispersive properties ($\frac{\tilde{\Omega}}{K}$ vs. K).

Fig. 28. Diffusive and dispersive properties of second-order 2D SD schemes for quadrilateral cells, with full-upwind Riemann flux approach, for $K = \frac{\pi}{4}$, $K = \frac{\pi}{2}$ and $K = \frac{4\pi}{5}$. $\tilde{\Theta}_{\mathcal{R}}$ and $\frac{\tilde{\Omega}}{K}$ versus $\psi = \theta$.

only four solution points in the generating pattern[b], whereas there are six solution points in the one for the triangular cells –three in each cell. In fact, even though they use less solution points, the quadrilateral SD cells yield the same level of accuracy as the triangular SD cells with a full-upwind Riemann flux approach.

1.3.2. *Third-order schemes*

Consider a third-order quadrilateral SD cell, as shown in Figure 29. The flux point distributions have one parameter, α_3, which is the same parameter as

[b]The generating pattern of a uniform grid is the smallest part from which the full grid can be reconstructed by periodically repeating the pattern in all directions.

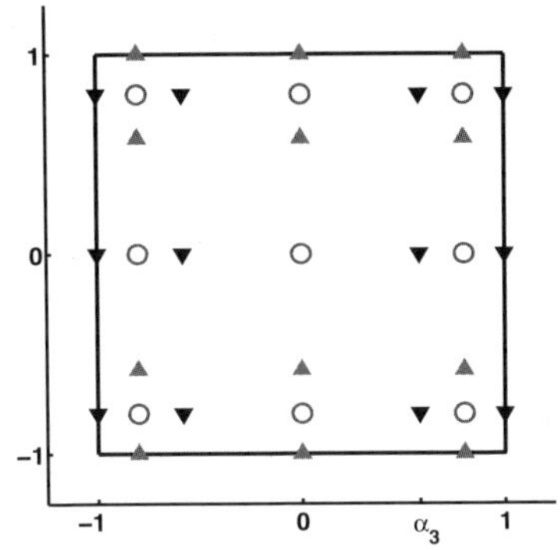

Fig. 29. Third-order quadrilateral SD cell.

with the 1D SD scheme from which the present scheme is derived. For the 1D case, the optimal value of α_3 was 0.58. The corresponding 2D scheme for quadrilaterals is stable, and its wave propagation properties are illustrated in Figure 30. As with the second-order scheme that was discussed in the previous section, the properties for wave propagation along the edges of the square cells ($\psi = \theta = l\frac{\pi}{2}$) are the same as the properties of the 1D scheme. For other angles, the scheme is more accurate and it is the most accurate for wave propagation along the diagonals of the squares ($\psi = \theta = \frac{\pi}{4} + l\frac{\pi}{2}$).

1.3.3. *Higher-order schemes*

Higher-order SD schemes for quadrilateral cells have similar properties as the second- and third-order schemes that were discussed previously. Deriving the schemes from their stable 1D counterparts always leads to a stable scheme. The properties for wave propagation along the edges of square cells are the same as the properties of the 1D schemes. For propagation in other directions, the 2D SD schemes for quadrilaterals are more accurate.

1.3.4. *Illustration*

The accuracy of the SD schemes for quadrilateral cells, based on the 1D SD schemes that use the Legendre-Gauss quadrature points and the end points as flux points, is verified with a grid convergence study for the 2D linear advection equation (1), with $a = 1$. The initial solution is again a Gaussian pulse given by (6), with $b = 0.1$. A sequence of uniform grids consisting of square cells, as illustrated in Figure 1(b), was used. Propagation

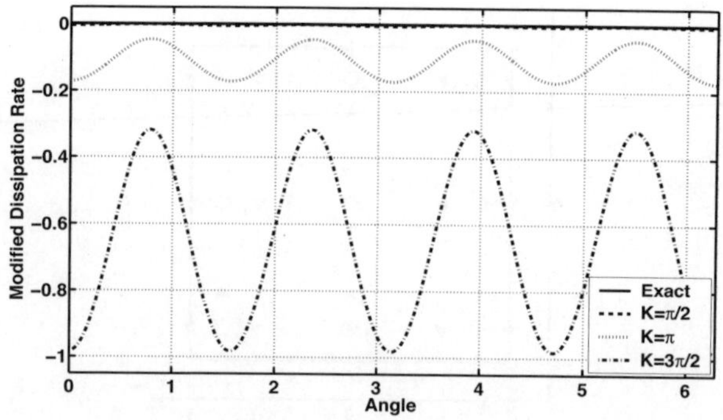

(a) Diffusive properties ($\tilde{\Theta}_{\mathcal{R}}$ vs. K).

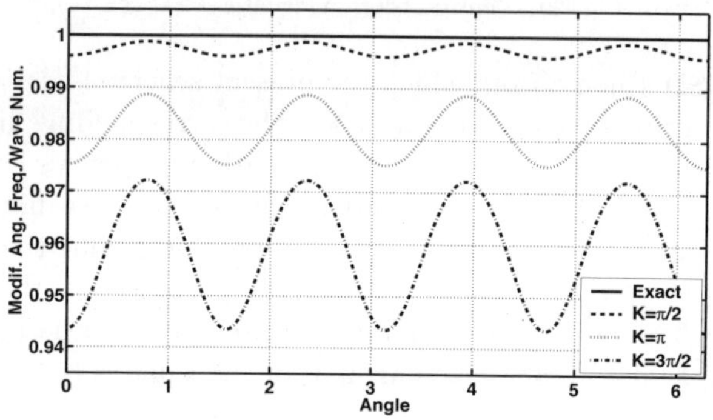

(b) Dispersive properties ($\frac{\tilde{\Omega}}{K}$ vs. K).

Fig. 30. Diffusive and dispersive properties of third-order 2D SD schemes for quadrilateral cells, with $\alpha_3 = 0.58$ and a semi-upwind Riemann flux approach, for $K = \frac{\pi}{2}$, $K = \pi$ and $K = \frac{3\pi}{2}$. $\tilde{\Theta}_{\mathcal{R}}$ and $\frac{\tilde{\Omega}}{K}$ versus $\psi = \theta$.

angles $\psi = 0$ and $\psi = \frac{\pi}{4}$ were considered, corresponding to respectively the least accurate and the most accurate direction of the schemes. A four-stage fourth-order accurate R-K scheme was used for time marching, with a sufficiently small time step Δt.

The resulting errors at $t = 1$ are listed in Table 5 for $\psi = 0$ and in Table 6 for $\psi = \frac{\pi}{4}$. The SD schemes are convergent and the expected order of accuracy is observed in all cases. In the L_1-norm, the errors obtained for $\psi = \frac{\pi}{4}$ are indeed smaller than those obtained for $\psi = 0$, in agreement with the analysis. For the errors in the L_∞-norm, the opposite is true however.

Table 5. Grid convergence study for the 2D linear advection equation using the SD schemes for quadrilateral cells. Wave propagation angle $\psi = 0$.

p	Δx	#DOF	L_1 error	L_1 order	L_∞ error	L_∞ order
1	0.20000	100	$6.50e-03$	–	$6.08e-02$	–
	0.10000	400	$9.50e-03$	−0.55	$4.14e-01$	−2.77
	0.05000	1600	$4.50e-03$	1.08	$2.28e-01$	0.86
	0.02500	6400	$1.60e-03$	1.49	$9.36e-02$	1.28
	0.01250	25600	$4.35e-04$	1.88	$2.67e-02$	1.81
	0.00625	102400	$1.10e-04$	1.98	$6.70e-03$	1.99
2	0.20000	225	$9.30e-03$	–	$2.79e-01$	–
	0.10000	900	$2.40e-03$	1.95	$1.05e-01$	1.41
	0.05000	3600	$2.38e-04$	3.34	$1.81e-02$	2.53
	0.02500	14400	$2.00e-05$	3.57	$1.50e-03$	3.59
	0.01250	57600	$1.98e-06$	3.34	$1.44e-04$	3.38
3	0.20000	400	$3.30e-03$	–	$1.41e-01$	–
	0.10000	1600	$2.80e-04$	3.56	$1.63e-02$	3.11
	0.05000	6400	$1.19e-05$	4.55	$1.30e-03$	3.65
	0.02500	25600	$5.95e-07$	4.32	$1.05e-04$	3.63
	0.01250	102400	$3.61e-08$	4.05	$7.11e-06$	3.89
4	0.20000	625	$9.73e-04$	–	$3.90e-02$	–
	0.10000	2500	$4.03e-05$	4.60	$4.50e-03$	3.12
	0.05000	10000	$7.39e-07$	5.77	$1.31e-04$	5.10
	0.02500	40000	$2.11e-08$	5.13	$4.19e-06$	4.97
5	0.20000	900	$3.38e-04$	–	$1.94e-02$	–
	0.10000	3600	$1.93e-06$	7.46	$2.73e-04$	6.15
	0.05000	14400	$4.63e-08$	5.38	$8.13e-06$	5.07
	0.02500	57600	$9.16e-10$	5.66	$1.98e-07$	5.36

2. Conclusions

Both the SV and the SD methods are not uniquely defined for orders of accuracy higher than two. They have a certain number of parameters that must be specified, which increases with the order of accuracy. In the case of the SV method, these parameters define the partition of cells into CVs, while for the SD method, they define the flux point distribution. The influence of these parameters on the stability and accuracy of 2D SV and SD schemes has been investigated by means of an *analysis of the wave propagation properties* of the methods. The most important results of this analysis can be summarized as follows.

Table 6. Grid convergence study for the 2D linear advection equation using the SD schemes for quadrilateral cells. Wave propagation angle $\psi = \frac{\pi}{4}$.

p	Δx	#DOF	L_1 error	L_1 order	L_∞ error	L_∞ order
1	0.20000	100	$1.34e-02$	–	$8.83e-01$	–
	0.10000	400	$6.60e-03$	1.02	$5.84e-01$	0.60
	0.05000	1600	$2.90e-03$	1.19	$2.89e-01$	1.02
	0.02500	6400	$8.60e-04$	1.75	$1.20e-01$	1.27
	0.01250	25600	$2.20e-04$	1.97	$3.34e-02$	1.85
	0.00625	102400	$5.47e-05$	2.01	$8.40e-03$	1.99
2	0.20000	225	$5.30e-03$	–	$3.83e-01$	–
	0.10000	900	$1.40e-03$	1.92	$1.54e-01$	1.31
	0.05000	3600	$1.42e-04$	3.30	$2.58e-02$	2.58
	0.02500	14400	$1.25e-05$	3.51	$2.20e-03$	3.55
	0.01250	57600	$1.35e-06$	3.20	$2.64e-04$	3.06
3	0.20000	400	$2.40e-03$	–	$2.48e-01$	–
	0.10000	1600	$1.69e-04$	3.83	$1.79e-02$	3.79
	0.05000	6400	$7.84e-06$	4.43	$2.50e-03$	2.84
	0.02500	25600	$4.44e-07$	4.14	$1.94e-04$	3.68
	0.01250	102400	$2.76e-08$	4.01	$1.26e-05$	3.95
4	0.20000	625	$7.44e-04$	–	$7.46e-02$	–
	0.10000	2500	$2.75e-05$	4.76	$7.70e-03$	3.28
	0.05000	10000	$5.34e-07$	5.69	$1.90e-04$	5.34
	0.02500	40000	$1.53e-08$	5.13	$6.81e-06$	4.80
5	0.20000	900	$2.33e-04$	–	$4.24e-02$	–
	0.10000	3600	$1.12e-06$	7.70	$2.87e-04$	7.21
	0.05000	14400	$3.23e-08$	5.12	$1.41e-05$	4.35
	0.02500	57600	$5.80e-10$	5.80	$3.18e-07$	5.46

- **2D SV schemes for triangular cells:** The uniquely defined second-order SV scheme has been confirmed to be stable by the wave propagation analysis. Weak instabilities in several third- and fourth-order SV schemes that are used in the literature have been identified analytically and verified numerically. Stable and accurate third- and fourth-order schemes have been proposed and tested. They were found to be more accurate than previously proposed SV schemes.
- **2D SD schemes for triangular cells:** Different Riemann flux approaches for the second-order SD scheme, the flux point distribution of which is uniquely defined, have been examined using the wave propagation analysis. Two approaches, namely the semi-upwind and the full-upwind approaches, result in a stable scheme.

A third approach, named the averaged-upwind approach, does not. The wave propagation analysis of the third-order SD schemes for triangular cells indicates that no stable flux point distribution for such schemes exists, with neither the semi-upwind nor the full-upwind Riemann flux approach. The correctness of the analysis was verified with numerical tests.

- **2D SD schemes for quadrilateral cells:** The wave propagation analysis of 2D SD schemes for quadrilateral cells confirmed that stable schemes are obtained if a tensor-product formulation based on a stable 1D scheme is used. The expected high-order accuracy of these schemes was observed numerically.

References

1. K. Van den Abeele. *Development of high-order accurate schemes for unstructured grids.* PhD thesis, Vrije Universiteit Brussel (May, 2009).
2. K. Van den Abeele, T. Broeckhoven, and C. Lacor, Dispersion and dissipation properties of the 1D spectral volume method and application to a p-multigrid algorithm, *J. Comput. Phys.* **224**(2), 616–636, (2007).
3. K. Van den Abeele and C. Lacor, An accuracy and stability study of the 2D spectral volume method, *J. Comput. Phys.* **226**(1), 1007–1026, (2007).
4. K. Van den Abeele, C. Lacor, and Z. J. Wang, On the stability and accuracy of the spectral difference method, *J. Sci. Comput.* **37**(2), 162–188, (2008).
5. K. Van den Abeele, G. Ghorbaniasl, M. Parsani, and C. Lacor, A stability analysis for the spectral volume method on tetrahedral grids, *J. Comput. Phys.* **228**, 257–265, (2009).
6. K. Van den Abeele, C. Lacor, and Z. J. Wang, On the connection between the spectral volume and the spectral difference method, *J. Comput. Phys.* **227**(2), 877–885, (2007).
7. Z. J. Wang and Y. Liu, Spectral (finite) volume method for conservation laws on unstructured grids II: Extension to two-dimensional scalar equation, *J. Comput. Phys.* **179**, 665–697, (2002).
8. Y. Liu, M. Vinokur, and Z. J. Wang, Spectral (finite) volume method for conservation laws on unstructured grids V: Extension to three-dimensional systems, *J. Comput. Phys.* **212**, 454–472, (2006).
9. Z. J. Wang and Y. Liu, Extension of the spectral volume method to high-order boundary representation, *J. Comput. Phys.* **211**, 154–178, (2006).
10. Q.-Y. Chen, Partitions of a simplex leading to accurate spectral (finite) volume reconstruction, *SIAM J. Sci. Comput.* **27**(4), 1458–1470, (2006).
11. R. Harris and Z. J. Wang, Partition Design and Optimization for High-Order Spectral Volume Schemes, *AIAA paper.* **2009-1333**, 1–9, (2009).
12. Y. Liu, M. Vinokur, and Z. J. Wang, Spectral difference method for unstructured grids I: Basic formulation, *J. Comput. Phys.* **216**, 780–801, (2006).

13. Z. J. Wang, Y. Liu, G. May, and A. Jameson, Spectral difference method for unstructured grids II: Extension to the Euler equations, *J. Sci. Comput.* **32** (1), 45–71, (2006).
14. G. May and J. Schöberl. Analysis of a spectral difference scheme with flux interpolation on raviart-thomas elements. Technical report, AICES technical report 2010-04-8, (2010).

CHAPTER 13

EFFICIENT RELAXATION METHODS FOR HIGH-ORDER DISCRETIZATION OF STEADY PROBLEMS

Georg May

Graduate School AICES, RWTH Aachen University,
Schinkelstr 2., 52056 Aachen, Germany
may@aices.rwth-aachen.de

Antony Jameson

Department of Aeronautics & Astronautics,Stanford University,
Durand Building, Stanford, CA 94305, USA
jameson@baboon.stanford.edu

We review the current status of solution methods for nonlinear systems arising from high-order discretization of steady compressible flow problems. In this context, many of the difficulties that one faces are similar to, but more pronounced than, those that have always been present in industrial-strength CFD computations. We highlight similarities and differences between the high-order paradigm and the mature solver technology of lower oder discretization methods, such as second order finite-volume schemes.

1. Introduction

Many have anticipated the arrival of high-order discretization as the CFD method of choice for compressible fluid flow. However, for industrial applications in external aerodynamics lower order methods, such as finite-volume schemes, are still far more popular. Numerical schemes of third or higher spatial order are often not efficient enough for high-throughput CFD computations to engineering levels of accuracy. Among the reasons for this is the fact that for established CFD methodologies tailor-suited convergence acceleration techniques have emerged over the past decades [Jameson (1983); Jameson and Yoon (1987); Pierce and Giles (1997); Jameson and

Caughey (2001); Mavriplis (2002)]. High-order solvers thus compete with very mature technology, and consequently novel discretization techniques have to be augmented by extremely efficient solution algorithms.

We present an overview of relaxation methods for steady compressible flow problems. This is to be understood in the sense that by virtue of spatial discretization the steady-state governing equations are converted to a nonlinear algebraic system of equations, which has to be solved. No time accuracy is required in this context, but time-accurate computations may also fall under this relaxation paradigm. For instance when implicit time discretization is employed, the solution of such an algebraic system of equations is required at each time instance.

One may argue that, in principle, the same relaxation methods and the same convergence theory may be applied to high-order discretization and low-order discretization. After all, a nonlinear algebraic system of equations is the result of spatial discretization in both cases. It is nevertheless true that the circumstances change when the order of accuracy is increased. As an example, consider two very popular paradigms in CFD computations for compressible flow, namely nonlinear multigrid methods with explicit multistage schemes, and implicit relaxation methods. Stability restrictions become a major concern for multigrid methods using explicit multistage relaxation, even on non-stretched meshes, as permissible CFL numbers of high-order methods typically behave as CFL $\propto m^{-2}$, where m is the polynomial degree of approximation [Hesthaven and Gottlieb (1999)]. Furthermore, the direct extension of multigrid methods to higher order schemes via the multi$-p$ approach is not entirely straight forward. On the other hand, implicit relaxation methods, such as Newton-Krylov methods, suffer from drawbacks as well, such as excessive storage requirements for high orders of approximation. We present an overview of viable relaxation methods with particular emphasis on constraints imposed by high-order spatial discretization, emphasizing such methods that are applicable to general unstructured grids.

2. Discretization Methods

The current state-of-the art in CFD focuses on solving the Euler or Navier-Stokes equations, the latter with suitable turbulence modeling. We write these equations generically as a system of conservation laws

$$\frac{\partial w}{\partial t} + \nabla \cdot f(w) = S(w) \; , \tag{1}$$

where w is the vector of conserved variables, f is the flux vector, including inviscid and viscous contributions from the governing equations, and possibly a turbulence model. The right-hand side may include a source term that often comes from a turbulence model.

For example, for the two-dimensional Euler equations for inviscid rotational fluid flow, w and f are written as

$$w = \begin{pmatrix} \rho \\ \rho u_1 \\ \rho u_2 \\ E \end{pmatrix} , \qquad f_j = \begin{pmatrix} \rho u_j \\ \rho u_j u_1 + p\delta_{j1} \\ \rho u_j u_2 + p\delta_{j2} \\ \rho u_j H \end{pmatrix} , \qquad j = 1, 2 \; . \tag{2}$$

Here ρ is the density, p is the pressure, E is the energy, and $H = (E+p)/\rho$ is the enthalpy. The fluid velocity vector is given by $u = (u_1, u_2)^T$. For a thermally and calorically perfect gas, one closes the equations by the equation of state

$$p = (\gamma - 1)\left(E - \frac{1}{2}\rho||u||^2\right) , \tag{3}$$

where γ is the ratio of specific heats.

There are a wide variety of high-order discretization methods for conservation laws, such as high-order finite-volume schemes [Barth (1993)], WENO schemes of finite difference or finite volume type [Shu (2003)], residual-distribution schemes [Abgrall and Roe (2003)], or hp finite-element methods [Karniadakis and Sherwin (2005)]. A very popular paradigm in high-order discretization is given by schemes based on piecewise polynomial representation, i.e. such schemes that, for a partition of the computational domain $\mathcal{T}_h = \{T\}$, approximate the solution of (1) as $w \approx w_h \in \mathcal{V}_h^p$, where $\mathcal{V}_h^p$ is the space of functions that are polynomials of degree p in each element, but are discontinuous across elements. Examples are the Discontinuous Galerkin (DG) method [Cockburn and Shu (2001)] or the Spectral Difference method [Liu *et al.* (2006); Wang *et al.* (2007)]. Attempts have been made to put some of these discretization approaches into a unified setting, such as Huynh's flux reconstruction approach [Huynh (2007)], the newly established Lifting Collocation Penalty method [Wang and Gao (2009)], or P_nP_m schemes [Dumbser *et al.* (2008); Dumbser (2010)].

We shall not be overly concerned with discretization methods here, as the focus is very much on relaxation methods for steady problems, which is generally the task of solving the nonlinear algebraic set of equations resulting from the spatial discretization. However, we do emphasize such schemes that are based on local polynomial approximation on unstructured grids.

For example, omitting any limiting or shock capturing terms, a simple DG discretization for a steady hyperbolic conservation law of the type (1) without source term leads to the problem of finding $w_h \in \mathcal{V}_h^p$ such that

$$R(w_h; v_h) := \sum_{T \in \mathcal{T}_h} \left\{ - \int_T f(w_h) \nabla v_h \, dx + \int_{\partial T} g(w_h^+, w_h^-; n) v_h^+ \, ds \right\} = 0 \,, \tag{4}$$

for all $v_h \in \mathcal{V}_h^p$. The function g is a numerical flux, which defines the flux on element boundaries, where the solution is discontinuous, as a function of the solution u_h^+ in element T, and u_h^-, the solution in its neighbor. See e.g. [Roe (1981); Jameson (1995)] for the case of the Euler equations. While usually the DG discretization is formulated using the semilinear form (4), it is clear that once the basis and test functions are chosen, the residual is a function of the solution coefficients for w_h only, and we may suppress the test function v_h in the notation. Another example is the Spectral Difference scheme for which one seeks $w_h \in \mathcal{V}_h^p$, using a nodal (Lagrange) basis, such that

$$R(w_h) := \nabla \cdot f_h(w_h) = 0 \,, \tag{5}$$

where f_h is a global interpolant of the nonlinear flux function f, which is continuous in normal direction across element interfaces by virtue of using numerical flux functions in the interpolation in a suitable manner, see [Kopriva and Kolias (1996); Liu *et al.* (2006); Wang *et al.* (2007)].

Since we only deal with the numerical solution w_h, we drop the subscript by default, and use it only when reference to a characteristic mesh length h is deemed necessary. Note that in Eq. (4) and Eq. (5) we use w_h to denote the assembled polynomial solution. Naturally, enforcing these equations means solving for discrete degrees of freedom, such as the modal coefficients or the collocation values, that together with corresponding basis functions define the numerical solution. In the following we shall associate w with the vector of discrete degrees of freedom. Likewise R corresponds to the pertaining residual evaluations. Thus we are left with a vector-valued nonlinear system of algebraic equations

$$R(w) = 0 \,, \tag{6}$$

where $R(w)$ is the residual vector.

The core of the present exposition is a pseudo time-dependent relaxation, marching the field equations to a steady state in a method of lines approach. This means one considers the system of nonlinear ODE

$$\frac{dw}{d\tau} + R(w) = 0 \,. \tag{7}$$

Obviously no time accuracy is required if one wishes to iterate toward the steady state, allowing such convergence acceleration techniques as local time stepping and multigrid methods. The advantage of this approach is that a wide variety of methods may be formulated in this framework.

3. Explicit Multistage Methods

In the early development of DG methods, multistage time-stepping schemes have been very popular. Early publications introduced the Runge-Kutta Discontinuous Galerkin (RKDG) method [Cockburn and Shu (1988)], presenting spatial discretization and multistage time stepping as a combined scheme. While much of this classical work is devoted to presenting and analyzing the method for time-dependent problems, Runge-Kutta methods have since become popular for steady problems as well [Bassi and Rebay (1997); May *et al.* (2010)]. Runge-Kutta methods are easy to implement and parallelize, and have low memory requirements.

Consider the pseudo-time ODE Eq. (7). An M-stage multistage temporal discretization may be written

$$\begin{aligned} w^{(0)} &= w^n \ , \\ w^{(k)} &= \sum_{l=0}^{k-1} \left\{ \alpha_{kl} w^{(l)} - \Delta\tau \beta_{kl} R^{(l)} \right\} \ , \qquad k = 1, \ldots, M \ , \\ w^{n+1} &= w^{(M)} \ , \end{aligned} \tag{8}$$

where w^n is the nth iterate of the solution, and $R^{(l)} := R(w^{(l)})$. Given a discretization that is TVD [Harten (1983)] with forward Euler time stepping, Shu proposed high order multistage schemes [Shu and Osher (1988)], which preserve the TVD property at high CFL numbers. These concepts have since been generalized under the paradigm of strong stability preserving (SSP) Runge-Kutta schemes [Gottlieb *et al.* (2001)]. TVD properties have been shown for Discontinuous Galerkin and Spectral Difference Schemes using standard limiting methods [Cockburn and Shu (1988); May (2006)]. The coefficients of the popular Shu RK3 scheme [Shu and Osher (1988)] may be written, arranged in matrix form, as

$$\alpha = \begin{pmatrix} 1 & & \\ \frac{3}{4} & \frac{1}{4} & \\ \frac{1}{3} & 0 & \frac{2}{3} \end{pmatrix} \ , \qquad \beta = \begin{pmatrix} 1 & & \\ 0 & \frac{1}{4} & \\ 0 & 0 & \frac{2}{3} \end{pmatrix} \ . \tag{9}$$

It should be noted that this scheme allows preservation of TVD properties only at the same CFL number as a forward Euler time stepping scheme [Shu and Osher (1988)]. For time dependent problems this still may lead to superior efficiency due to high order accuracy in time. For steady problems however, temporal order of accuracy is immaterial, and the use of this scheme is merely justified by the fact that simpler schemes, such as forward Euler or the 2-stage TVD RK scheme [Shu and Osher (1988)] are not linearly stable with DG or Spectral Difference methods, which may lead to overactive limiters in the TVD discretization, and hence to compromised accuracy.

An alternative are low order but high-CFL number schemes, such as TVD / SSP schmes [Shu (1988); Gottlieb *et al.* (2001)] or Jameson's high-CFL number multistage schemes [Jameson (1983, 1993, 2004)] , which have been very popular in standard finite-volume CFD computations. These latter schemes have been designed using Fourier analysis for a linear model equation with the aim to maximize the stability region and at the same time provide good high-frequency error damping properties, which improves performance within multigrid algorithms.

The success of such multistage schemes for steady problems depends to a large extent on convergence acceleration techniques. Certainly the use of local time stepping methods is mandatory if no time accuracy is required. Time steps are adjusted so that they are always close to the local stability limit. If the mesh size increases, the time step, which is proportional to the local characteristic mesh length, will also increase, producing an effect comparable to that of an increasing wave speed. Furthermore, the combination of multistage schemes with multigrid, which we address in section 5.1, is one of the classic paradigms in compressible flow simulation. It should not be overlooked that the success of multistage methods in classical CFD methods have also relied on other convergence acceleration methods, such as implicit residual smoothing and related methods [Jameson (1988); Swanson *et al.* (2007)], which have not found a straight forward extension in the realm of higher order discretization methods.

While explicit relaxation methods are attractive due to ease of implementation and parallelization, stability restrictions are a concern. Often spectra of the (linearized) discrete advection operators are investigated to infer stability properties [Karniadakis and Sherwin (2005)]. In the context of nodal DG schemes, or Spectral Difference schemes, such analysis has revealed that the spectral radius is proportional to m^2, where m is the polynomial order of approximation [Hesthaven and Gottlieb (1999); May (2006)], which suggests that stability for explicit methods necessitates $CFL \sim m^{-2}$. As an example, consider the one-dimensional linear advection equation and Discontinuous Galerkin or Spectral Difference Discretization, where permissible CFL numbers with respect to linear L_2 stability have been explicitly computed [May (2006)], see Table 1. The measure CFL · DOF used in Table 1, where DOF is the number of local degrees of freedom, is appropriate when making comparison with standard finite difference schemes using the same number of total degrees of freedom. The Spectral Difference schemes in Table 1 use Gauss-Legendre quadrature points, augmented with cell interval end-points, which has recently been confirmed to be a stable choice by means of numerical eigenvalue analysis [Huynh (2007)] as well as rigorous proof [Jameson (2009)].

The rapid asymptotic decrease of permissible CFL numbers poses a severe challenge, certainly if the problem is exacerbated by numerical stiffness induced by stretched meshes. It remains to be seen if explicit relaxation methods will remain popular for practical high order viscous CFD computations. This depends to a large extent on whether convergence acceleration techniques such as multigrid methods can be incorporated succesfully.

The popular focus on nonlinear TVD stability theory has to some extent led to negligence of linear stability analysis for high-order schemes. It has to be stressed that many combinations of explicit time integration methods with higher order schemes, such as the Spectral Difference scheme, or standard RKDG schemes [Cockburn and Shu (2001)] are not unconditionally linearly stable [May (2006)]. While nonlinear stability results may still hold, flux limiters or artificial viscosity techniques are needed for stabilization. These may locally degrade the accuracy, even in smooth regions, if oscillations are generated by linear instability. For example, in the case of the 1D Spectral Difference Scheme, the popular Chebyshev-Lobatto nodes are not unconditionally stable [May (2006); Van den Abeele *et al.* (2008)] (and by extension tensor-products thereof). Linear instability has also been shown for the Spectral Difference scheme using different nodal sets for multivariate

Table 1. Linear stability limits for Spectral Difference and DG Schemes with the Shu RK3 scheme, and Jameson's four-stage scheme [Jameson (1983)].

	SD / Jameson RK4		SD / Shu RK3		DG / Shu RK3	
Pol. Order	CFL	CFL · DOF	CFL	CFL · DOF	CFL	CFL · DOF
0	0.696	1.392	0.595	1.190	0.409	0.818
1	0.363	1.089	0.322	0.966	0.209	0.627
2	0.226	0.904	0.201	0.804	0.130	0.520
3	0.156	0.780	0.139	0.695	0.089	0.445
4	0.115	0.690	0.103	0.618	0.066	0.396
5	0.089	0.623	0.079	0.559	0.051	0.357

interpolation on triangular meshes [Van den Abeele *et al.* (2008)], although more recently a new formulation of the Spectral Difference scheme has been proposed, based on interpolation in Raviart-Thomas spaces, which seems to be linearly stable [May and Schöberl (2010)].

High-order DG or spectral methods for nonlinear equations can be stabilized quite effectively with filtering methods [Hesthaven and Warburton (2007)], meaning attenuation of higher order modes. While it has been demonstrated that even for fixed (intermediate) order schemes such an approach may be viable without significantly degrading the accuracy [Hesthaven and Warburton (2007)], this has not been explored too much for CFD applications.

Regardless of the stabilization method of choice, restrictive stability conditions of the type shown in Tab. 1 always apply for explicit relaxation methods, which makes it absolutely necessary to combine them with convergence acceleration techniques for steady problems.

4. Implicit Relaxation Methods

A linearized backward Euler temporal discretization of Eq. (7) may be written

$$(I + \Delta\tau A(w^n))\,\Delta w^n = -\Delta\tau R(w^n)\ , \tag{10}$$

where $\Delta w^n = w^{n+1} - w^n$ and A is the Jacobian matrix of the residual vector, i.e. the differentiation of the residual vector R with respect to the state vector w. For $\Delta\tau \to \infty$ one obtains a Newton iteration, while finite time steps may be interpreted as damped Newton iterations. For simplicity we shall often use the symbol $\mathbb{M}$ to denote the entire left-hand-side matrix in Eq. (10).

The hallmark of implicit methods is that a large sparse linear system, given by Eq. (10), has to be solved at each nonlinear iteration step n. For most practical applications direct solution of the system is out of the question, so one has to resort to iterative methods. The key parameters in the implementation are

- Approximation and assembly of the Jacobian matrix A
- The iterative solution method for the linear system
- Preconditioning of the system

Finding the best overall approach is not trivial if the time to solution is to be minimized. A variety of different approaches have been proposed even for standard low-order CFD methods. Nevertherless, two approaches may be identified that are particularly popular: Newton-Krylov methods, corresponding to infinite-time steps in Eq. (10), and finite-time-step implicit methods using classical iterative solvers with convergence acceleration methods.

4.1. *Newton-Krylov methods*

The Newton iteration potentially achieves quadratic convergence, provided the exact Jacobian matrix is available, and the linear systems arising at each iteration are solved to high precision. Newton's method is often combined with Krylov methods to solve the linearized equations at each iteration. Krylov methods (with a good preconditioner) are often advantageous if solution of the linear system to high precision is desired, and the system by itself is not necessarily well conditioned. This is usually the case if the time step in Eq. (10) is increased to infinity. Methods that rely on diagonal dominance, as many classical iteration methods do [Hackbusch (1994)], may not be a good choice for this case. Among the Krylov methods for non-symmetric systems that arise in CFD applications the GMRES method [Saad and Schultz (1986)] is quite popular. GMRES is very robust, in the sense that it cannot break down, unless the exact solution of the linear system is reached, and furthermore guarantees that the (linear) residual 2-norm is non-increasing. On the other hand, the method is quite expensive due to long recurrences of Krylov vectors, and usually requires good preconditioning to attain acceptable rates of convergence. We defer the issue of preconditioning to section 4.4.

In practice it is very difficult, if not impossible, to quantify *a priori* the region of attraction that must be reached, to attain convergence of the Newton iteration. Therefore some kind of globalization must be added to the method to allow convergence from an initial guess that may be far away from the converged solution. For CFD applications a simple time-step control of the implicit temporal discretization, based on the size of the residual is usually fairly robust. As an example of a Newton-Krylov method applied to a compressible flow problem consider the test case depicted in Fig. 1.

The convergence using a DG method with polynomials of degree $m = 2$ and $m = 4$ in terms of the norm $||R(w)||_\infty$ is shown in Fig. 2 and Fig. 3, respectively. Here and in the following NDOF is the number of degrees of freedom in the numerical approximation, i.e. $NDOF = N_e \times N_m \times N_w$, where N_e is the number of mesh elements, N_m is the number of local degrees

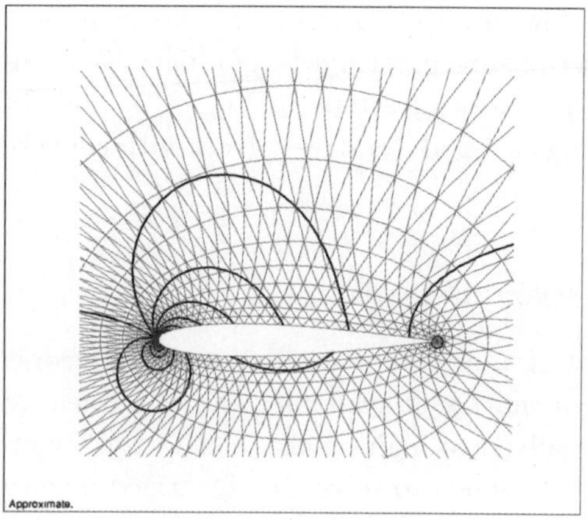

Fig. 1. Mach contours for inviscid flow around the NACA0012 profile. Free-stream Mach number $M_\infty = 0.4$, Angle of Attack $\alpha = 5^o$.

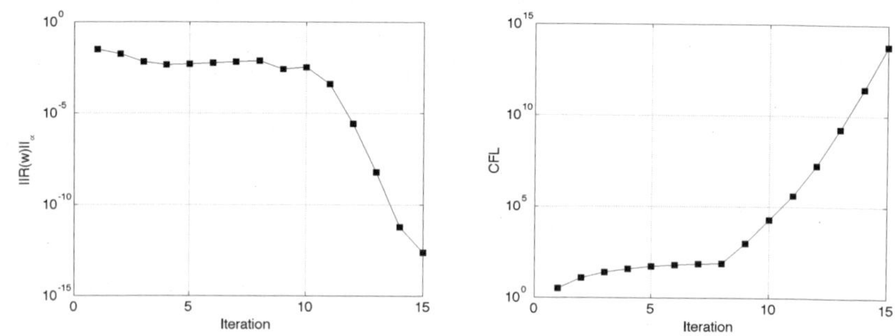

Fig. 2. Inviscid flow around the NACA0012 profile at $M_\infty = 0.4$, $\alpha = 5°$. Degrees of freedom: NDOF $= 61,440$. DG method with polynomial degree $m = 2$. Left: Convergence of the residual against nonlinear iterations. Right: Adaptive CFL number against nonlinear iterations.

of freedom, and N_w is the number of equations. It can be seen that very rapid convergence is attained, once the asymptotic region is reached.

To summarize, Newton-Krylov methods imply increased cost per iteration by requiring

(1) A good approximation of the Jacobian
(2) Solution of the linear system to relatively high precision (at least in the asymptotic region)
(3) A good preconditioner

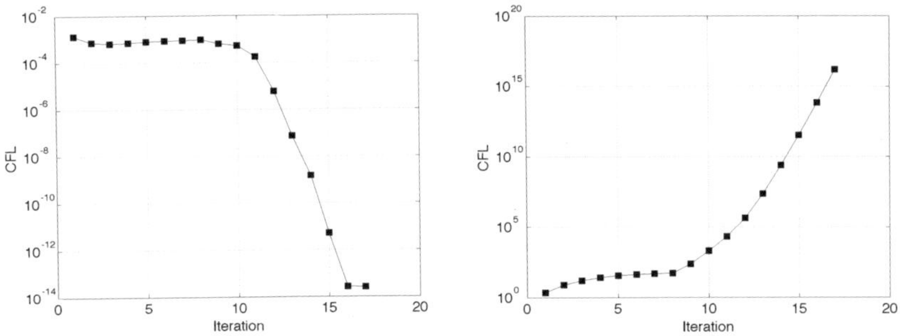

Fig. 3. Inviscid flow around the NACA0012 profile at $M_\infty = 0.4$, $\alpha = 5°$. Degrees of freedom: NDOF $= 61,440$. DG method with polynomial degree $m = 4$. Left: Convergence of the residual against nonlinear iterations. Right: Adaptive CFL number against nonlinear iterations.

4.2. *Implicit Schemes with Finite Time Steps*

Newton-Krylov methods imply a high cost per iteration, but at the same time a very low nonlinear iteration count. The opposite approach may also lead to success. One may use finite time steps in Eq. (10), resulting in a linear system with (relatively) high diagonal dominance, so that classical iteration techniques may be used. If a rather inexact solution of Eq. (10) is accepted, i.e. solving the system to relatively high residual levels and perhaps using a crude approximation of the Jacobian, the result is a significantly higher nonlinear iteration count, but also a dramatically reduced cost per iteration. The viability of the concept depends on its successful combination with convergence acceleration techniques, such as multigrid, for which classical iterative linear solvers can be quite effective smoothers.

An example for this approach is given by relaxation methods of the Gauss-Seidel type. Consider a splitting $\mathbb{M} = D + L + U$, where D is the (block) diagonal, while L and U denote the strictly lower and upper triangular submatrices, respectively. A standard Gauss-Seidel method may be written upon setting $\Delta w^{n,0} = 0$,

$$(D^n + L^n)\Delta w^{n,k+1} = -R(w^n) - U^n \Delta w^{n,k}\ , \qquad k = 0, 1, 2, \dots\ . \tag{11}$$

Very often symmetric Gauss Seidel methods are used, that basically concatenate a forward and backward solve:

$$(D^n + L^n)\Delta w^{n,k+\frac{1}{2}} = -R(w^n) - U^n \Delta w^{n,k}\ , \tag{12}$$

$$(D^n + U^n)\Delta w^{n,k+1} = -R(w^n) - L^n \Delta w^{n,k+\frac{1}{2}}\ . \tag{13}$$

Particularly popular is the so-called LU-SGS method [Jameson and Yoon (1987); Yoon and Jameson (1988)], which is basically a one-step symmetric Gauss-Seidel method with zero initial guess. It is a matter of straightforward computation to show that this corresponds to a splitting

$$\Delta w^n = -(D^n + U^n)^{-1} D^n (D^n + L^n)^{-1} R(w^n) \, . \tag{14}$$

It is quite obvious that a small number of Gauss-Seidel sweeps does not solve the linear system to high precision. Nevertheless, such schemes have been applied to high-order discretizations by numerous researachers in combination with multilevel convergence acceleration techniques [Luo *et al.* (2006); Nastase and Mavriplis (2006)], making this approach a good example for the trade-off considered above: The cost per iteration is extremely low, so that a higher nonlinear iteration count may be tolerated. Furthermore, the quality of the sweeps can be controlled by appropriately selecting the ordering of the state vector. Many examples exist in the literature where for both classical CFD computations and higher order methods such reorderings have been constructed to reflect lines of strong coupling of the equations [Mavriplis (1998); Fidkowski *et al.* (2005)], while in regions of generally weak coupling the relaxation may even be reduced to a Jacobi iteration. In order to reduce memory overhead, a nonlinear variant of the LU-SGS scheme, similar to that used by Jameson and Caughey in a finite-volume context [Jameson and Caughey (2001)], is sometimes applied to high-order discretization [Sun and Wang (2007); Premasuthan (2010); Parsani *et al.* (2010)].

Whether or not this approach is superior to a Newton iteration is highly problem dependent, and often also depends on the measure of convergence: Since the asymptotic quadratic convergence of Newton's method is very hard to beat, the solution of the nonlinear problem to machine accuracy is often most efficiently done with a good Newton solver. On the other hand, convergence of output functionals, such as lift or drag coefficients, is often very efficiently achieved to engineering levels of accuracy by multigrid-accelerated classical smoothing techniques.

4.3. *Matrix-free methods*

For higher order methods based on local polynomial approximations, a major difficulty may be already encountered in the assembly of the Jacobian matrix. Let $\mathcal{P}^m$ be the space of polynomials of degree m. In 2D, $\dim(\mathcal{P}^m) \propto m^2$, while in 3D one has $\dim(\mathcal{P}^m) \propto m^3$. Since all local degrees of freedom are coupled in each cell, the overall storage requirements grow with the fourth power of the polyomial order m in 2D, and with the sixth power in 3D.

Storing the Jacobian matrix may not be feasible in some cases, forcing one either to resort to explicit relaxation methods, or to consider matrix-free formulations of implicit methods. For the latter, Krylov methods are particularly suited, since they require, neglecting preconditioning for the moment, the system matrix $\mathbb{M}$ only in the action on Krylov vectors, i.e. matrix-vector products of the form

$$z = \mathbb{M}v \ . \tag{15}$$

Note that the nontrivial part of this operation involves the Jacobian of the residual vector $R(w^n)$, namely in the computation of $A(w^n)v$, which is a projection of the derivative of the residual onto the Krylov vector v. One may generate a numerical approximation to first order accuracy in a small parameter ε by writing

$$A(w^n)v \approx \frac{R(w^n + \varepsilon v) - R(w^n)}{\varepsilon} \ . \tag{16}$$

Here the cost of applying the matrix-vector product is the same as one residual evaluation. There is some freedom in the choice of ε. Several methods have been proposed in the literature to estimate the step size [Knoll and Keyes (2004)]. A simple choice, supposing normalized Krylov vectors, is:

$$\varepsilon = \sqrt{1 + ||w||}\varepsilon_{rel} \ , \tag{17}$$

where the parameter ε_{rel} should roughly represent the square-root of machine accuracy.

4.4. *Preconditioning*

For Newton-Krylov methods it is usually the preconditioning that is most problem-dependent, the rest being a rather generic procedure. In case of the GMRES method, unfortunately the eigenvalue spectrum does not completely specify convergence properties, complicating the process of enabling fast convergence through good preconditioning. Pathological examples with extremely benign spectrum of the matrix, yet extremely poor GMRES convergence, may be constructed [Van der Vorst (2009)]. Preconditioning methods that reflect the physics and numerics of certain problems are often proposed, CFD applications being no exception [Persson and Peraire (2008)]. That being said, standard preconditioners based on incomplete LU factorizations (ILU) [Saad (2003)] are also often used with good results [May *et al.* (2010)].

If the preconditioner is explicitly assembled as a sparse matrix it is normally independent of the Krylov iteration index, i.e. does not change while the linear system is being solved. On the other hand, explicit storage of the preconditioner may not always be possible any more than storage of the matrix itself. For matrix-free preconditioners alternative formulations of the GMRES method, such as the flexible GMRES method [Soulaimani *et al.* (2002)] facilitate the implementation by allowing the preconditioner to depend on the linear iteration index. One generates a preconditioned Krylov vector by solving the linear system

$$P_j \widetilde{w}_j = z_j \ , \tag{18}$$

where $z_j = Av_j$. Since the preconditioning matrix is allowed to depend on the GMRES iteration index, one may use iterative solvers as preconditioners. Noting that the preconditioning matrix P should approximate A one may apply to equation Eq. (18) a few iteration steps with the same GMRES method that is used to solve Eq. (10), i.e. apply GMRES recursively. In particular, this may be done using the same matrix-free approximation. This method is denoted "squared preconditioning", due to the recursive application of the linear GMRES solver [May *et al.* (2010)]. In principle this algorithm could be recursively applied even further.

It should thus be pointed out that the method is completely matrix-free. This means that storage requirements grow linearly in the degrees of freedom, as opposed to quadratically, which led to the extreme asymptotic storage requirements outlined in section 4.3. For the matrix-free variant memory requirements are now dominated by storage of the Krylov vectors, which in 3D is certainly still considerable, but manageable with modern computer architectures.

As an example of a computation using matrix-free implicit relaxation with squared preconditioning, consider the flow depicted in Fig. 4. Figure 5 shows the convergence in terms of the density residual for different orders of approximation using a Spectral Difference method at constant CFL number (CFL=550). It may be seen that the convergence in terms of linear iterations, i.e. cumulative number of Krylov vectors (excluding preconditioning iterations), deteriorates with higher polynomial orders. This is due to the fact that the condition of the system matrix also deteriorates, and it is usually not advisable to increase the number of preconditioning iterations (inner iterations) too much as a countermeasure, since the matrix-free preconditioner is not as effective as matrix-based ones, and thus net runtime may increase despite fewer linear iterations.

5. Multilevel Methods

5.1. *Geometric multigrid*

Multigrid is certainly one of the most popular paradigms within the applied CFD community. One may distinguish between linear multigrid methods, that may be used as preconditioners in the context of implicit relaxation methods, and nonlinear methods, under the paradigm of the Full Approximation Scheme (FAS) [Brandt (1977)]. The latter is very popular in the combination with explicit multistage methods, following the (now classic) approach in [Jameson (1983)].

The FAS approach has traditionally been associated with geometric multigrid methods, which we consider first. Since it is standard practice to use only first order accurate solution methods on coarse-grid approximations, we first consider the special case $w_h|_T \in \mathcal{P}^0$ for all mesh elements T, and discuss the extension to higher order approximations afterwards.

Assume that the equations have been iterated n steps on a given mesh of characteristic length h, the "fine" mesh, by an explicit multistage scheme, as in section 3. This results in an approximation w_h^n, and residual $R_h^n = R_h(w_h^n)$. Using a suitable coarser mesh of characteristic length H,

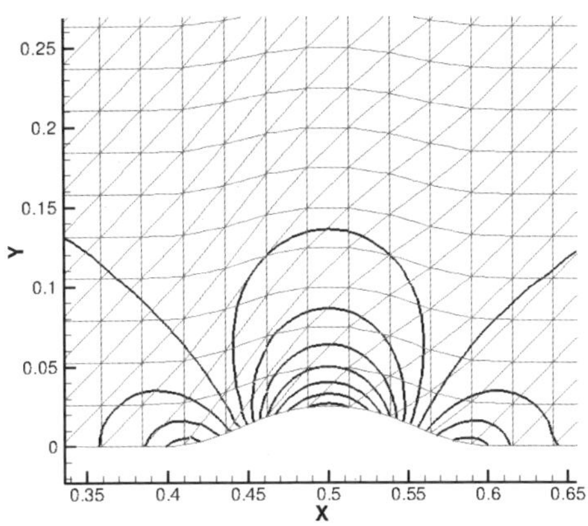

Fig. 4. Mach contours for inviscid flow around a smooth bump. Free-stream Mach number $M_\infty = 0.3$.

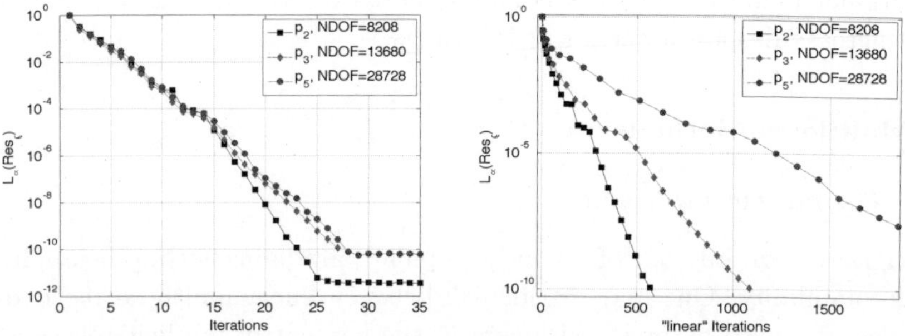

Fig. 5. Convergence of the matrix-free method for smooth inviscid flow around a bump at free-stream Mach number $M_\infty = 0.3$. Spectral Difference computation with polynomial degrees of 2,3, and 5. Left: Convergence against outer, nonlinear iterations. Right: Convergence against linear iterations, i.e. number of generated Krylov vectors.

and defining appropriate restriction operators for the solution and residual, $I_h^H w_h^n$, and $\widetilde{I}_h^H R_h^n$, respectively, one may advance the solution on a coarse grid by the modified multistage scheme

$$w_H^{(k)} = \sum_{l=0}^{k-1} \left\{ \alpha_{kl} w_H^{(l)} + \Delta t \beta_{kl} \left(R_H^{(l)} + S_H \right) \right\} , \qquad k = 1, \ldots, M , \tag{19}$$

where the additional defect correction term

$$S_H = \widetilde{I}_h^H R_h - R_H^{(0)} \tag{20}$$

appears on the right-hand side [Jameson (1983); Mavriplis (2002)]. After iterating on the coarse mesh for n_c iterations the corrected solution on the fine grid is computed as

$$w_h^+ = w_h^n + I_H^h (w_H^{n_c} - w_H^0) , \tag{21}$$

where I_H^h is an interpolation operator, and optionally additional smoothing on the fine mesh may be applied, before the updated solution is declared the new iterate at $n+1$.

One uses recursive application of this concept to extend the method to more than two meshes. Good results are usually obtained using W cycles, following standard nomenclature, see e.g. [Jameson (2004)]. These are defined by allowing transfer to the next higher level only if the solution has been advanced twice on the current mesh. Figure 6 shows a schematic depiction of a 4-level W-cycle. This approach has proved particularly effective when combined with a nonlinear variant of the LU-SGS scheme [Jameson and Caughey (2001)].

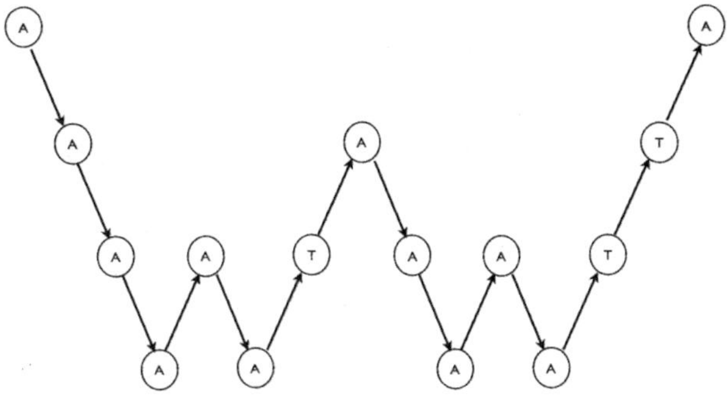

Fig. 6. W-Cycle for a 4-mesh sequence. The letter A stands for advancing the flow solution on a particular level, while the letter T stands for transfer of the solution to the next higher level.

The geometric approach introduces some overhead, since the availability of mesh sequences is implied. Often those are generated by automatic coarsening procedures, such as agglomeration methods [Mavriplis and Venkatakrishan (1995)].

5.2. *Multi-p methods*

In direct analogy to geometric multigrid we may define multi-p methods in broad terms as computing an approximation for the error of the current solution, $w_h|_T \in \mathcal{P}^m$, using a lower polynomial degree m_c.[a] However, just as for geometric multigrd methods the optimal mesh coarsening ratio is not always *a priori* clear, with multi$-p$ methods the same question applies to the lower polynomial degree one ought to use. While some Fourier analysis for linear model equations has been carried out to assess convergence factors for multi$-p$ methods [Atkins and Helenbrook (2005); Fidkowski *et al.* (2005)], the issue of how many polynomial levels one ought to include for high m is still open in general, and likely problem and discretization dependent. We shall simply assume that in a two-grid cycle we use polynomial levels m and m_c with $0 \leq m_c < m$.

One may use multi$-p$ in a similar fashion as described in the previous section for geometric multigrid methods. Assume that the equations have

[a]In our nomenclature we prefer to use multi-p in place of the somewhat more popular, but misleading, term p-multigrid.

been iterated n steps using a discretization of local polynomial degree m, and mesh of characteristic length h, resulting in an approximation w_m^n, and residual $R(w_m^n)$. Note that here the mesh index has been suppressed, as it will not change in the multi$-p$ iteration, and instead the subscript m has been added. Defining appropriate transfer operators for the solution and residual, $I_m^{m_c} w_m^n$, and $\widetilde{I}_m^{m_c} R(w_m^n)$, respectively, one may solve the equation

$$R(w_{m_c}) + S_{m_c} = 0 \tag{22}$$

where the additional defect correction term

$$S_{m_c} = \widetilde{I}_m^{m_c} R(w_m^n) - R(w_{m_c}^0) \tag{23}$$

appears on the right-hand side. After relaxing on the polynomial level m_c for n_c iterations the corrected solution on the level m may be computed, using the prolongation operator $I_{m_c}^m$, as

$$w_m^+ = w_m^n + I_{m_c}^m (w_{m_c}^{n_c} - w_{m_c}^0) , \tag{24}$$

which may be declared the next solution iterate, upon optionally applying some further smoothing, as in the geometric multigrid case. Such an approach has been used, for example, in [Premasuthan (2010)] with Spectral Difference discretization and a Runge-Kutta smoother, and in [Fidkowski *et al.* (2005)] with DG discretization and implicit smoothers of the block Jacobi and line relaxation type.

For multi$-p$ methods it has been, however, at least as popular to use a linear multilevel paradigm with implicit relaxation schemes, instead of the FAS approach. Applying multilevel techniques when solving Eq. (10) allows a straightforward interpretation as a preconditioner for linear systems. Using a suitable smoother for Eq. (10), one transfers the residual

$$r_m = -R(w_m^n) - \mathbb{M}_m \Delta w_m^n \tag{25}$$

to a lower order approximation, i.e. $r_{m_c} = \mathcal{I}_m^{m_c} r_m$. Subsequently one solves the error equations directly, i.e.

$$\mathbb{M}_{m_c} e_{m_c} = r_{m_c} \tag{26}$$

which may be done recursively using yet more levels. The corrected solution is then obtained as $w_m^+ = w_m^n + \mathcal{I}_{m_c}^m e_{m_c}$. In [Nastase and Mavriplis (2006)] the linear approach applied to a DG discretization was found superior in terms of runtime. A major advantage is certainly the reduced number of nonlinear residual evaluations, which are particularly costly in a higher order context. (Keep in mind that during a multi$-p$ iteration the mesh is fixed, so that the cost of evaluating the residual does not decrease as dramatically as with geometric multgigrid.)

For best results one ought to combine multi$-p$ and geometric multgrid. Recall that for nonlinear convection-dominated problems geometric multigrid aids through two mechanisms: firstly, the elimination of high-frequency error modes on successively coarser meshes; secondly, the propagation of error modes, and expulsion from the computational domain [Pierce and Giles (1997)]. While asymptotic convergence rates are dominated by high-frequency smoothing, early convergence is dominated by convection. Often one observes effectively converged output functionals, such as lift and drag coefficients at relatively high residual levels, before asymptotic convergence rates are reached. In this phase, geometric multigrid may be viewed primarily as an increase of the effective wave speed propagating error modes, which is, however, dependent on global coarsening. Since multi$-p$ methods do not provide such global coarsening, it is likely that best-practice multilevel solvers will still have to include geometric multigrid.

5.3. *Hybrid multilevel schemes*

It is certainly possible to use multigrid with different relaxation schemes on different mesh levels or levels of polynomial approximation. This leads to hybrid multilevel schemes. Depending on the constraints deemed important one may find very different "optimal" combinations. For example, in [Luo *et al.* (2006)] a multi$-p$ DG scheme is proposed that combines Shu's three-stage Runge-Kutta method, cf. Eq. (9), for polynomial levels of approximation $m > 0$, with implicit LU-SGS solves, cf. Eq. (14), for $m = 0$, with the primary concern being storage requirements.

A different method was proposed in [May *et al.* (2010)], where a damped Newton/GMRES implicit method is used for the highest level of polynomial approximation $m > 0$. Storage concerns are addressed with an optional matrix-free formulation. A geometric multigrid method with explicit multistage smoothing is used for the volume averages (i.e. for $m = 0$) between Newton iterations to accelerate the convection of the volume-averaged large-scale error modes. The smoothed volume averages replace the volume averages of the high-order relaxation. The rationale behind this is that experience indicates that error convection and expulsion is the primary mode of convergence, when considering integrated quantities, such as force coefficients. When using geometric multigrid methods, which accelerate the effective wave speed for error convection and expulsion, force coefficients are often essentially converged at rather high residual levels, when high-frequency errors still persist.

The method in [May *et al.* (2010)] is completed by a full multigrid (FMG) finite-volume start-up procedure. Algorithm 5.1 gives an example of a practical implementation of the overall approach. Let $RKMG(w;l,n)$ denote the application of n iterations of Runge-Kutta smoothing with $l-$level geometric multigrid. The mesh levels are identified by an indexed characteristic length h_l, where the coarsest mesh is indexed with $l=1$, while the finest available mesh is indexed $l=L$. For easy reference we have denoted volume averages by an overbar, i.e. $\overline{w}_h$, while w_h indicates the solution at the current high-order polynomial level m. The implicit solves are denoted by $NK(w;m,n)$, where again, n is the number of iterations, and m is the level of polynomial approximation.

Algorithm 5.1. Hybrid Multilevel with Full Multigrid

(1) Initialize $\overline{w}^0_{h_1}$ with free stream conditions
(2) For $l=1,\ldots,L$, Do
(3) $\quad \overline{w}^n_{h_l} = \mathrm{RKMG}(\overline{w}^0_{h_l}; l, n_l)$
(4) $\quad$ if (l = L) exit
(5) $\quad \overline{w}^0_{h_{l+1}} = I^{h_{l+1}}_{h_l}\overline{w}^n_{h_l}$
(6) EndDo
(7) $w^0_{h_L} = Inject(\overline{w}^n_{h_L}; m)$
(8) For $n=0,\ldots,N_{cyc}$, Do
(9) $\quad w^+_{h_L} = \mathrm{NK}(w^n_{h_L}; m, 1)$
(10) $\quad$ if (converged) exit
(11) $\quad \overline{w}^0_{h_L} = \mathcal{V}(w^+_{h_L})$
(12) $\quad \overline{w}^+_{h_L} = \mathrm{RKMG}(\overline{w}^0_{h_L}; L, n_{RK})$
(13) $\quad w^{n+1}_{h_L} = w^+_{h_L} - Inject(\overline{w}^0_{h_L} - \overline{w}^+_{h_L}; m)$
(14) EndDo

The first loop over the meshes defines the startup procedure. The computation starts on the coarsest grid using a finite-volume method with explicit multistage relaxation, and then proceeds up to the next finer grid when a sufficiently good approximation to the solution has been achieved. This is applied recursively, re-using all available coarser meshes with FAS multigrid, until the finest mesh is reached. The number of multigrid cycles n_l should be enough to attain reasonable convergence of integrated quantities, such as lift and drag, on each mesh level.

The result of the finite-volume relaxation procedure is used as initial guess for the damped Newton iteration acting on the high-order discretization in Algorithm 5.1, step 7. We define the injection operator $Inject(\overline{w}; m)$, which injects the volume average for the current approximation order m. Obviously the definition of this operator depends on the chosen basis and degrees of freedom. This operator is also used in the subsequent multilevel relaxation procedure. Because of the good start value provided by the initial full multigrid relaxation, only very mild damping for a few iterations has to be used to avoid start-up problems in the Newton-Krylov method on the highest level.

The main loop is over the combined Newton / GMRES and explicit smoothing operators. First the implicit iterator $NK(w; m, n)$ is applied. Note that usually $n = 1$, as shown in Algorithm 5.1, step 9. Subsequently the volume averages are extracted in step 11, where the operator is denoted $\mathcal{V}$. This is particularly easy for hierarchical bases that are often used with DG methods, e.g. [Dubiner (1991)]. It is easily accomplished also for the non-hierarchical Spectral Difference basis by (exact) numerical quadrature based on the solution nodes. Finally the explicit multigrid iterations are performed for the volume averages, which produces updated values that replace the previous ones. Typically around $n_{RK} = 20$ iterations are used for the additional $RKMG$ smoothing between Newton iterations. Intermediate polynomial levels $0 < m_c < m$ are not used in the nonlinear multigrid cycles, but may be used within this framework under the linear multigrid paradigm, i.e. as a preconditioner for the linear systems, although incomplete LU factorizations also work effectively.

As a computational example, consider the inviscid flow test case summarized in Fig. 7 using the Spectral Difference Scheme with $m = 2$. Figure 8 shows the convergence of the hybrid method, Algorithm 5.1, under mesh refinement in terms of the drag coefficient. Here the CFL number has been kept constant at CFL=550 to highlight the mesh independent convergence. Both nonlinear iterations, and linear iterations are shown, where the latter refers to the cumulative number of generated Krylov vectors. It can be seen that convergence of both degrades very severely for the single-grid method, while the convergence is nearly mesh independent for the hybrid mulitlevel method. The mesh sequence used for these computations is shown in Table 2 and Table 3. More precisely, in Table 2 the three meshes used in the refinement study are summarized, while in Table 3 the multilevel data for the coarsest of these meshes is shown. For the finer meshes in Table 2 it should be understood that all previously defined coarser meshes are used

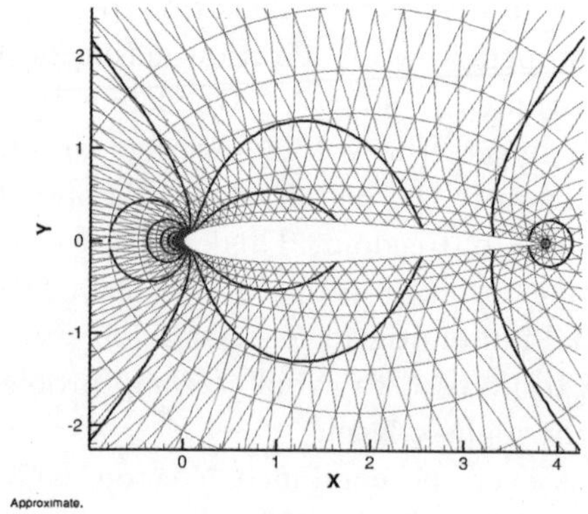

Fig. 7. Mach contours for inviscid flow around the NACA0012 profile. Free-stream Mach number $M_\infty = 0.3$, Angle of Attack $\alpha = 0^o$.

Table 2. Meshes used in h-refinement study.

Level	$NDOF$	Elements
fine	983,040	40,960
medium	245,760	10,240
coarse	61,440	2,560

Table 3. Meshes and Degrees of freedom used with the hybrid multilevel method on the coarsest mesh of Table 2.

Hybrid Multilevel					
Level	$NDOF$	m	Cells	CFL	Smoothing
4	61,440	2	2,560	550	Implicit
3	10,240	0	2,560	6	Explicit
2	2,560	0	640	6	Explicit
1	640	0	160	6	Explicit

recursively (with finite volume approximation). Thus the medium mesh uses a 4-level strategy, while the finest mesh uses 5 levels. A similar test case has been computed for the flow conditions $M_\infty = 0.4$ and $\alpha = 5^o$, i.e.

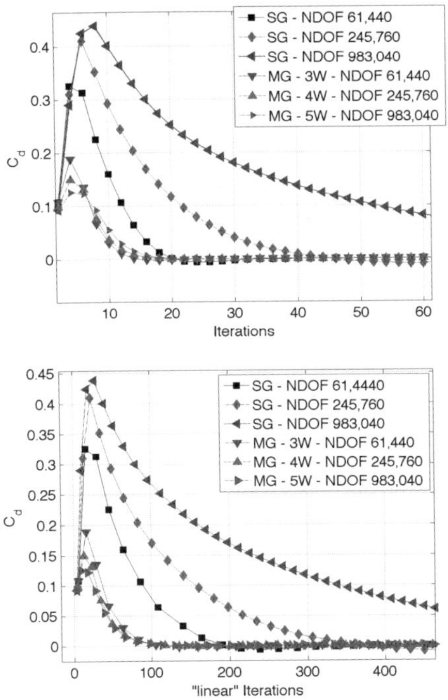

Fig. 8. Inviscid flow around the NACA0012 profile at free-stream Mach number $M_\infty = 0.3$, angle-of-attack $\alpha = 0^o$. Convergence of the single-grid method (SG), and the hybrid multilevel method, Algorithm 5.1, using 20 finite-volume multigrid cycles between Newton iterations. Krylov solver: GMRES(30) with ILU(2) preconditioning. Left: Convergence against outer, nonlinear iterations. Right: Convergence against linear iterations, i.e. number of generated Krylov vectors.

the case depicted in Fig. 1. The convergence in terms of the lift coefficient, plottet against CPU time, is shown in Fig. 9 for the three different meshes in Table 2.

6. Conclusion

We reviewed approaches to the solution of nonlinear systems arising from high-order spatial discretization in a CFD context. It is not advisable to end such a review with a clear recommendation on what method ought to be generally preferred, as such a choice is always problem-dependent. Readers familiar with best-practice low-order CFD methods will recognize the

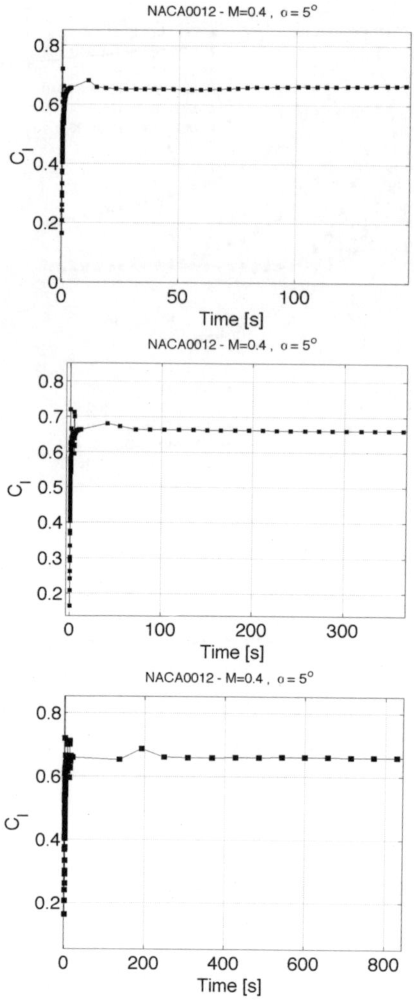

Fig. 9. Mesh refinement study. Inviscid flow around the NACA0012 profile at $M_\infty = 0.4$, $\alpha = 5°$. Convergence of the lift coefficient. Degrees of freedom: NDOF $= 61,440$ (left) , NDOF $= 245,760$ (middle) , NDOF $= 983,040$ (right).

same trade-offs that have always existed: Restrictive stability restrictions with explicit methods that require fine-tuned convergence acceleration techniques, high memory requirement with implicit relaxation methods, and the problem of adequate preconditioning.

In the context of higher order discretization methods, however, these trade-offs are often more pronounced. Efficient solution needs to be defined in terms of available resources and objective of the calculation, which may inform the decision as to what relaxation scheme should be used. For example, solving the nonlinear equations to machine accuracy is a very different task compared with high-throughput computations that focus on convergence of output quantities to engineering levels of accuracy. Available resources, such as computer memory, may just dictate the choice of relaxation method. For example, the enormous storage requirements of Newton-Krylov methods for 3D computations may sometimes be prohibitive.

It must be said, that efficient solution of steady compressible flow problems to relatively modest levels of accuracy is still a domain dominated by the mature technology of standard best-practice lower order methods. However, solution methods that are well adapted to the unique environment of high-order discretization are an area of very active research, and the transition from model problems to more realistic applications is well underway. It is entirely possible that well designed hp-adaptive solvers will be able to challenge the current status-quo in the near future.

References

Abgrall, R. and Roe, P. L. (2003). High order fluctuation schemes on triangular meshes, *J. Sci. Comp.* **19**, 1-3, pp. 3–36.

Atkins, H. L. and Helenbrook, B. T. (2005). Numerical evaluation of p-multigrid for the solution of discontinuous galerkin discretizations of diffusive equations, AIAA Paper 05-5110, American Institute of Aeronautics and Astronautics.

Barth, T. J. (1993). Recent developments in high order k-exact reconstruction on unstructured meshes, AIAA paper 93-0668, American Institute of Aeronautics and Astronautics.

Bassi, F. and Rebay, S. (1997). A high-order accurate discontinuous finite-element method for the numerical solution of the compressible Navier-Stokes equations, *J. Comp. Phys.* **131**, pp. 267–279.

Brandt, A. (1977). Multi-level adaptive solutions to boundary-value problems, *Math. Comp.* **31**, 138, pp. 333–390.

Cockburn, B. and Shu, C. W. (1988). TVB Runge-Kutta local projection Discontinuous Galerkin finite element method for conservation laws II: General framework, *Math. Comp.* **52**, 186, pp. 411–435.

Cockburn, B. and Shu, C. W. (2001). Runge-Kutta Discontinuous Galerkin methods for convection-dominated problems, *J. Sci. Comp.* **16**, 3, pp. 173–261.

Dubiner, M. (1991). Spectral methods on triangles and other domains, *J. Sci. Comput.* **6**, 4, pp. 345–390.

Dumbser, M. (2010). Arbitrary high order pnpm schemes on unstructured meshes for the compressible Navier-Stokes equations, *Computers and Fluids* **39**, 1, pp. 60–76.

Dumbser, M., Balsara, D. S., Toro, E. F. and Munz, C.-D. (2008). A unified framework for the construction of one-step finite volume and discontinuous Galerkin schemes on unstructured meshes, *J. Comp. Phys.* **227**, 18, pp. 8209–8253.

Fidkowski, K. J., Oliver, T. A., Lu, J. and Darmofal, D. L. (2005). p–Multigrid solution of high-order discontinuous Galerkin discretizations of the compressible Navier–Stokes equations, *J. Comp. Phys.* **207**, pp. 92–113.

Gottlieb, S., Shu, C.-W. and Tadmor, E. (2001). Strong stability-preserving high-order time discretization methods, *SIAM Review* **43**, 1, pp. 89–112.

Hackbusch, W. (1994). *Iterative Solution of Large Sparse Systems of Equations*, *Applied Mathematical Sciences*, Vol. 95 (Springer-Verlag).

Harten, A. (1983). High-resolution schemes for hyperbolic conservation laws, *J. Comp. Phys.* **49**, 3, pp. 357–393.

Hesthaven, J. S. and Gottlieb, D. (1999). Stable spectral methods for conservation laws on triangles with unstructured grids, *Comput. Meth. Appl. Mech. Engrg.* **175**, pp. 361–381.

Hesthaven, J. S. and Warburton, T. (2007). *Nodal Discontinuous Galerkin Methods: Algorithms, Analysis, and Applications*, no. 54 in Texts in Applied Mathematics (Springer Verlag).

Huynh, H. T. (2007). A flux reconstruction approach to high-order schemes including discontinuous galerkin methods, AIAA Paper 07-4079, American Institute of Aeronautics and Astronautics.

Jameson, A. (1983). Solution of the Euler equations for two dimensional transonic flow by a multigrid method, *Appl. Math. Comp.* **13**, pp. 327–356.

Jameson, A. (1988). Computational transonics, *Comm. Pure Appl. Math.* **41**, 5, pp. 507–549.

Jameson, A. (1993). Computational algorithms for aerodynamic analysis and design, *Appl. Numer. Math.* **13**, 5, pp. 383–422.

Jameson, A. (1995). Analysis and design of numerical schemes for gas dynamcics 2: Artificial diffusion and discrete shock structure, *Int. J. Comp. Fluid. Dyn.* **5**, pp. 1–38.

Jameson, A. (2004). Aerodynamics, in E. Stein, R. De Borst and T. J. R. Hughes (eds.), *Encyclopedia of Computational Mechanics*, Vol. 3, chap. 11 (Wiley).

Jameson, A. (2009). A proof of the stability of the spectral difference method for all orders of accuracy, Report ACL 2009-1, Aerospace Computing Laboratory, Stanford University.

Jameson, A. and Caughey, D. A. (2001). How many steps are required to solve the Euler equations of steady compressible flow: In search of a fast solution algorithm, AIAA Paper 01-2673, American Institute of Aeronautics and Astronautics.

Jameson, A. and Yoon, S. (1987). Lower-upper implicit schemes with multiple grids for the Euler equations, *AIAA Journal* **25**, 7, pp. 929–935.

Karniadakis, G. E. and Sherwin, S. (2005). *Spectral/hp Element Methods for Computational Fluid Dynamics*, 2nd edn. (Oxford University Press).

Knoll, D. A. and Keyes, D. E. (2004). Jacobian-free newton–krylov methods: a survey of approaches and applications, *J. Comp. Phys.* **193**, pp. 357–397.

Kopriva, D. A. and Kolias, J. H. (1996). A conservative staggered-grid Chebyshev multidomain method for compressible flows, *J. Comp. Phys.* **125**, pp. 244–261.

Liu, Y., Vinokur, M. and Wang, Z. J. (2006). Spectral Difference method for unstructured grids I: Basic formulation, *J. Comp. Phys.* **216**, 2, pp. 780–801.

Luo, H., Baum, J. D. and Löhner, R. (2006). A *p*-multigrid Discontinuous Galerkin method for the Euler equations on unstructured grids, *J. Comp. Phys.* **211**, pp. 767–783.

Mavriplis, D. J. (1998). Multigrid strategies for viscous flow solvers on anisotropic unstructured meshes, *J. Comp. Phys.* **145**, 1, pp. 141–165.

Mavriplis, D. J. (2002). An assessment of linear versus nonlinear multigrid methods for unstructured mesh solvers, *J. Comp. Phys.* **175**, 1, pp. 302 – 325.

Mavriplis, D. J. and Venkatakrishan, V. (1995). Agglomeration multigrid for two-dimensional viscous flows, *Computers and Fluids* **24**, 5, pp. 553–570.

May, G. (2006). *A Kinetic Scheme for the Navier-Stokes Equations and High-Order Methods for Hyperbolic Conservation Laws*, Ph.D. thesis, Stanford University, Stanford, CA 94305.

May, G., Iacono, F. and Jameson, A. (2010). A hybrid multilevel method for high-order discretization of the Euler equations on unstructured meshes, *J. Comp. Phys.* **229**, 10, pp. 3938–3956.

May, G. and Schöberl, J. (2010). Analysis of a spectral difference scheme with flux interpolation on raviart-thomas elements, AICES Technical Report 2010-04/8, Aachen Institute for Advanced Study in Computational Engineering Science.

Nastase, C. R. and Mavriplis, D. J. (2006). High-order Discontinuous Galerkin methods using an *hp*-multigrid approach, *J. Comp. Phys.* **213**, pp. 330–357.

Parsani, M., Van den Abeele, K., Lacor, C. and Turkel, E. (2010). Implicit LU-SGS algorithm for high-order methods on unstructured grid with p-multigrid strategy for solving the steady Navier-Stokes equations, *J. Comp. Phys.* **229**, 3, pp. 828–850.

Persson, P.-O. and Peraire, J. (2008). Newton-GMRES preconditioning for discontinuous Galerkin discretizations of the Navier-Stokes equations, *SIAM J. Sci. Comput.* **30**, 6, pp. 2709–2733.

Pierce, N. A. and Giles, M. B. (1997). Preconditioned multigrid methods for compressible flow calculations on stretched meshes, *J. Comp. Phys.* **136**, pp. 425–445.

Premasuthan, S. (2010). *Towards an efficient and Robust High Order Accurate Flow Solver for Viscous Compressible Flows*, Ph.D. thesis, Stanford University, Stanford, CA 94305.

Roe, P. L. (1981). Approximate Riemann solvers, parameter vectors, and difference schemes, *J. Comp. Phys.* **43**, pp. 357–372.

Saad, Y. (2003). *Iterative Methods for Sparse Linear Systems*, 2nd edn. (Society for Industrial and Applied Mathematics).

Saad, Y. and Schultz, M. H. (1986). GMRES: A generalized minimal residual algorithm for solving non-symmetric linear systems, *SIAM J. Sci. Stat. Comp.*. **7**, pp. 856–869.

Shu, C.-W. (1988). Total-variation-diminishing time discretizations, *SIAM J. Sci. Stat. Comput.* **9**, 6, pp. 1073–1084.

Shu, C.-W. (2003). High-order finite difference and finite volume WENO schemes and discontinuous Galerkin methods for CFD, *Int. J. Comput. Fluid Dyn.* **17**, 2, pp. 107–118.

Shu, C.-W. and Osher, S. (1988). Efficient implementation of essentially non-oscillatory shock capturing schemes, *J. Comp. Phys.* **77**, pp. 439–471.

Soulaimani, A., Salah, N. B. and Saad, Y. (2002). Enhanced GMRES acceleration techniques for some CFD problems, *Int. J. Comp. Fluid. Dyn.* **16**, 1, pp. 1–20.

Sun, Y. and Wang, Z. J. (2007). Efficient implicit non-linear LU-SGS approach for viscous flow computation using high-order Spectral Difference method, AIAA Paper 07-4322, American Institute of Aeronautics and Astronautics.

Swanson, R. C., Turkel, E. and Rossow, C. C. (2007). Convergence acceleration of Runge-Kutta schemes for solving the Navier-Stokes equations, *J. Comp. Phys.* **224**, 1, pp. 365–388.

Van den Abeele, K., Lacor, C. and Wang, Z. J. (2008). On the stability and accuracy of the spectral difference method, *J. Sci. Comput.* **37**, 2, pp. 162–188.

Van der Vorst, H. A. (2009). *Itertive Krylov Methods for Large Linear Systems* (Cambridge University Press).

Wang, Z. J. and Gao, H. (2009). A unifying lifting collocation penalty formulation including the discontinuous Galerkin, spectral volume/difference methods for conservation laws on mixed grids, *J. Comp. Phys.* **228**, 21, pp. 8161–8186.

Wang, Z. J., Liu, Y., May, G. and Jameson, A. (2007). Spectral Difference method for unstructured grids II: Extension to the Euler equations, *J. Sci. Comput.* **32**, 1, pp. 54–71.

Yoon, S. and Jameson, A. (1988). Lower-upper symmetric-Gauss-Seidel method for the Euler and Navier-Stokes equations, *AIAA Journal* **26**, 9, pp. 1025–1026.

CHAPTER 14

HIGH-ORDER METHODS BY CORRECTION PROCEDURES USING RECONSTRUCTIONS

H. T. Huynh

NASA Glenn Research Center,
Cleveland, Ohio 44135, U.S.A
huynh@grc.nasa.gov

A new approach to high-order accuracy for the numerical solution of conservation laws with the advantages of simplicity and economy is presented. The approach employs the differential form of the equation and accounts for the jumps in flux values at the cell boundaries by a correction procedure based on the concept of reconstruction. Named 'correction procedure via reconstruction' or CPR, the approach results in numerous new schemes with favorable properties. It also unifies several existing methods: with appropriate choices of correction terms, it recovers discontinuous Galerkin, staggered-grid, spectral volume, and spectral difference. The CPR versions are generally more economical than the original ones. Fourier analyses are carried out to determine the accuracy and stability of schemes by this formulation. Tradeoffs between accuracy and time-step sizes are discussed.

1. Introduction

In the field of Computational Fluid Dynamics (CFD), low-order methods are less accurate, but generally robust and reliable; as a result, they are routinely employed in practical calculations. High-order methods have the potential of providing accurate solutions at reasonable cost; however, they are more complicated and less robust. The need to improve and develop new methods with more favorable properties has attracted the interest of many researchers.

For high-order accuracy, a solution is typically approximated by a piecewise polynomial function (a polynomial in each cell or element).

This piecewise polynomial function is either required to be continuous (see, e.g., (Hughes 1987) and chapter 10 by Abgrall) or allowed to be discontinuous across the cell interfaces. While the pros and cons of these two approaches are still debatable, methods with discontinuous solution spaces appear to be more popular in CFD and are the main focus here. For these methods, the interaction of the data among cells takes place in the form of a common flux at each cell interface (shared by the two adjacent cells). Popular schemes of this type include discontinuous Galerkin (Reed and Hill 1973, Cockburn, Karniadakis, and Shu 2000), staggered-grid (Kopriva and Kolias 1996), spectral volume (Wang, Zhang, and Liu 2004), and spectral difference (Liu, Vinokur, and Wang 2006). Among these, discontinuous Galerkin (DG) and spectral volume are formulated via the integral form of the equation, whereas staggered-grid and spectral difference use the differential form.

In a recent paper, the author (Huynh 2007) introduced a new approach to high-order accuracy by solving the equations in differential form. The approach, called flux reconstruction (FR), results in numerous schemes with favorable properties. In addition, it unifies several existing methods: with appropriate choices of correction terms, it recovers discontinuous Galerkin, staggered-grid, spectral volume, and spectral difference. The FR versions are also generally simpler and more economical than the original versions. The framework was applied to solve diffusion problems using quadrilateral meshes in (Huynh 2009). Wang and Gao (2009) extended the FR idea to 2D triangular and mixed meshes with the lifting collocation penalty (LCP) formulation. The method was employed to solve the Euler and Navier-Stokes equations in both two and three dimensions (Gao and Wang, 2009, Haga, Gao, and Wang 2010). Since these two approaches are tightly related, the involved authors combine the names and call them CPR (Correction Procedure or Collocation Penalty via Reconstruction; 'CP' from 'LCP' and 'R' from 'FR'). The CPR formulation requires no numerical integrations; the mass matrix inversion is built-in (and not needed) regardless of the choice of basis functions; therefore, the resulting schemes are generally simpler and more efficient than those by quadrature-based formulations.

In this chapter, the basic theory of the CPR approach is established for the one-dimensional case. This case has the advantage of simplicity, and it contains essentially all of the key new ideas. In one spatial dimension, the derivative of a discontinuous piecewise polynomial function is evaluated in each cell by employing a straightforward derivative estimate using the data within the cell together with correction terms that account for the jumps at the two interfaces. The correction terms, in turn, are derived from the concept of reconstruction, which defines a continuous function approximating the discontinuous piecewise polynomial data. On each cell, the problem reduces to dealing with the jumps at the two interfaces. For the left interface (a reflection yields the solution for the right), it reduces to constructing a correction function g on the interval $I = [-1,1]$ of one degree higher than that of the discontinuous piecewise polynomial data such that $g(-1) = 1$, $g(1) = 0$, and g approximates the zero function. Several correction functions resulting in schemes with favorable properties are introduced. Accuracy and stability are examined by Fourier analyses. Tradeoffs between accuracy and time-step sizes for these schemes are discussed. It is shown that the CPR scheme with correction function defined by the Radau polynomial is identical to the DG scheme. Extension to a quadrilateral mesh is straightforward via tensor products (Huynh 2007). Extension to triangular and hybrid meshes as well as results for the Euler and Navier-Stokes equations can be found in chapter 15 by Wang, Gao, and Haga.

This chapter is self-contained. Section 2 deals with conservation laws. Section 3 introduces the derivative calculation via the CPR approach. Section 4 describes various choices of correction functions. Section 5 contains the proof that the CPR scheme with the Radau polynomial as correction function is identical to the DG scheme. The evaluation of the second derivative for the diffusion equation is presented in Section 6. Fourier analyses are provided in Section 7. Stability and accuracy of various schemes are given in Section 8. Finally, conclusions and discussion can be found in Section 9.

2. Conservation Laws

Consider the conservation law

$$u_t + f_x = 0 \tag{1}$$

with initial condition $u(x,0) = u_{\text{init}}(x)$ where $u_{\text{init}}(x)$ is periodic or of compact support so that boundary conditions are trivial. The flux f is assumed to depend on u. Denoting the signal speed df/du by $a(u)$, the above can be cast in nonconservation form:

$$u_t + a u_x = 0.$$

Let the domain of calculation be divided into (possibly nonuniform) cells or elements E_j, $j = \ldots -1, 0, 1, 2, \ldots$ On each cell, let the solution be approximated by K pieces of data $u_{j,k}$, $k = 1, \ldots, K$, at locations $x_{j,k}$, which are called *solution points*. The K solution points are typically the Gauss or Lobatto points, but equidistant points can also be employed. In fact, the Fourier stability and accuracy results below are independent of the type of points chosen. For convenience, we use the same type of points for all cells. Note that if the Lobatto points are selected, then since they include the two cell interfaces, each interface $x_{j+1/2}$ corresponds to two values of u, namely $u_{j,K}$ and $u_{j+1,1}$, and these left and right interface values are readily available for the upwind flux calculation.

At each solution point, the solution $u_{j,k}(t)$ depends on t. For simplicity, $u_{j,k}(t^n)$ is abbreviated to $u_{j,k}$. At time level n, suppose the data $u_{j,k}$ are known for all j and k. We wish to calculate $du_{j,k}(t)/dt$ at time $t = t^n$ which is abbreviated to $du_{j,k}/dt$ (note that the notation d/dt replaces $\partial/\partial t$ due to the fixed location of the solution points). In other words, we wish to evaluate f_x at the solution points $x_{j,k}$ in terms of the data. Then, we march in time by, say, a Runge-Kutta method.

As is standard in finite element methods (Hughes 2000), instead of dealing with the global element E_j, it is more convenient to deal with the local element, i.e., the interval $I = [-1,1]$. Denote the center of E_j by x_j and its width by h_j. With ξ varying on I and x on E_j, the linear function mapping I onto E_j and its inverse are

$$x(\xi) = x_j + \xi h_j / 2 \quad \text{and} \quad \xi(x) = 2(x - x_j)/h_j.$$

The local solution points on I are denoted by ξ_k, $k = 1, \ldots, K$. They relate to the global solution points on E_j by

$$x_{j,k} = x(\xi_k) = x_j + \xi_k h_j / 2.$$

A function $r_j(x)$ on E_j results in a function on I denoted by $r_j(\xi)$

$$r_j(\xi) = r_j(x(\xi)) = r_j(x).$$

The global and local derivatives are related by the chain rule

$$\frac{dr_j(x)}{dx} = \frac{2}{h_j}\frac{dr_j(\xi)}{d\xi}.$$

Returning to solving the conservation law (1), the first task is to approximate u on each cell by a polynomial of degree $K-1$ interpolating the K pieces of data $u_{j,k}$, $k=1,\ldots,K$. To this end, for each k, let the basis function $\phi_{j,k}$ be the Lagrange polynomial on E_j of degree $K-1$ that takes on value 1 at $x_{j,k}$ and 0 at the other $K-1$ solution points. The global and local basis functions are (see Fig. 1(a)),

$$\phi_{j,k}(x) = \prod_{l=1,\, l\neq k}^{K} \frac{x - x_{j,l}}{x_{j,k} - x_{j,l}} \quad \text{and} \quad \phi_k(\xi) = \prod_{l=1,\, l\neq k}^{K} \frac{\xi - \xi_l}{\xi_k - \xi_l}.$$

On E_j, let $u_j(x)$ be the polynomial of degree $K-1$ interpolating $u_{j,k}$, $k=1,\ldots,K$; $u_j(x)$ is called a *solution polynomial* (Fig. 1(b)),

$$u_j(x) = \sum_{k=1}^{K} u_{j,k}\,\phi_{j,k}(x) \quad \text{and} \quad u_j(\xi) = \sum_{k=1}^{K} u_{j,k}\,\phi_k(\xi).$$

Next, we define the *discontinuous flux function*. Set $f_{j,k} = f(u_{j,k})$. Let $f_j(x)$ be of degree $K-1$ interpolating $f_{j,k}$, $k=1,\ldots,K$,

$$f_j(x) = \sum_{k=1}^{K} f_{j,k}\,\phi_{j,k}(x) \quad \text{and} \quad f_j(\xi) = \sum_{k=1}^{K} f_{j,k}\,\phi_k(\xi).$$

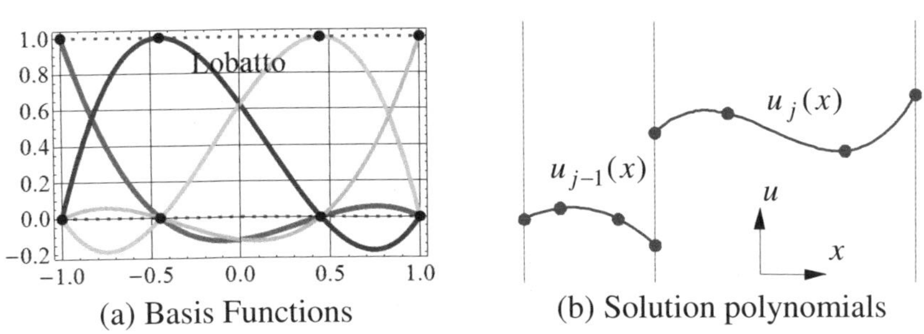

Fig. 1. (a) Cubic basis functions on $I=[-1,1]$ for Lobatto points with $K=4$, (b) Solution polynomials $u_{j-1}(x)$ and $u_j(x)$.

(Note that both f_j and u_j are of degree $K-1$.) The flux polynomials $\{f_j\}$ form a function, which is generally *discontinuous* across cell interfaces and is called the *discontinuous flux function*.

If we employ $(f_j)_x$ to evaluate f_x for the conservation law (1), we obtain erroneous solutions: such a derivative does not include the interaction of the data between adjacent cells. To account for interaction, we will construct a *continuous flux function*, which approximates the discontinuous function in some sense, and then calculate its derivative. The continuous flux function will be obtained by adding a correction to the discontinuous one. As a result, we still need the derivative of the discontinuous function. At each solution point ξ_k, $1 \le k \le K$,

$$(f_\xi)_{j,k} = \sum_{l=1}^{K} d_{kl} f_{j,l}. \tag{2}$$

The coefficients $\{d_{kl}\}$ form a $K \times K$ matrix denoted by $\boldsymbol{D}$ and is called the derivative matrix (in the local description). For each j, set

$$\boldsymbol{f}_j = \{f_{j,k}\}_{k=1}^{K} \quad \text{and} \quad (\boldsymbol{f}_\xi)_j = \{(f_\xi)_{j,k}\}_{k=1}^{K}.$$

Then

$$(\boldsymbol{f}_\xi)_j = \boldsymbol{D}\boldsymbol{f}_j \quad \text{and} \quad (\boldsymbol{f}_x)_j = \frac{2}{h_j}(\boldsymbol{f}_\xi)_j.$$

As an alternative, the chain rule can be employed (Huynh 2007):

$$(f_\xi)_{j,k} = (f_u)_{j,k}(u_\xi)_{j,k} = a(u_{j,k})(u_\xi)_{j,k}. \tag{3}$$

Wang and Gao (2009) found that for nonlinear equations, compared to (2), the chain rule above, in fact, yields more accurate solutions. However, it does not ensure that the scheme is conservative, whereas (2), as will be shown, does.

We now define the various left and right interface values. Appropriate use of these values can assure that the resulting the method is conservative. At $x_{j+1/2}$, let u_L and u_R be given by,

$$u_L = u_{j+1/2,L} = u_j(x_{j+1/2}) \quad \text{and} \quad u_R = u_{j+1/2,R} = u_{j+1}(x_{j+1/2}). \tag{4}$$

Similarly,

$$f_L = f_{j+1/2,L} = f_j(x_{j+1/2}) \quad \text{and} \quad f_R = f_{j+1/2,R} = f_{j+1}(x_{j+1/2}).$$

In general, $u_L \neq u_R$, $f_L \neq f_R$, $f_L \neq f(u_L)$, and $f_R \neq f(u_R)$.

Note that $f(u_L)$ and $f(u_R)$ are employed for the upwind flux in (5) below, whereas to assure conservation, f_L and f_R are employed in the jumps in flux values of (10).

For the common flux at each interface (shared by the two adjacent cells), we use an upwind flux, e.g., Roe's splitting (1986). The common flux, for the Euler and Navier-Stokes equations by finite volume methods, is often called 'approximate Riemann solver' or, for the DG methods, 'numerical flux'. Here, we use the term 'common' due to the critical property of commonality relative to the two adjacent cells; in addition, for diffusion problems, the common flux is often 'centered' and we also need a 'common derivative' quantity. At $x_{j+1/2}$, with u_L and u_R via (4), let $\tilde{u}$ be defined by the mean value theorem

$$a(\tilde{u}) = \begin{cases} (f(u_R) - f(u_L)) / (u_R - u_L) & \text{if} \quad u_L \neq u_R, \\ a(u_L) = a(u_R) & \text{otherwise.} \end{cases}$$

The common (upwind) flux is determined by the sign of $a(\tilde{u})$,

$$f_{\text{com}} = f_{\text{upw}} = f_{j+1/2,\text{upw}} = \begin{cases} f(u_L) & \text{if} \quad a(\tilde{u}) \geq 0, \\ f(u_R) & \text{otherwise.} \end{cases} \tag{5}$$

Equivalently, $f_{\text{upw}} = \frac{1}{2}[f(u_L) + f(u_R)] - \frac{1}{2}|a(\tilde{u})|(u_R - u_L)$. Such an upwind flux is typically employed with an entropy correction, which is beyond the scope of this chapter.

3. Derivative Evaluation via CPR Approach

First, we reconstruct the flux by a continuous function F such that on each cell E_j, F is a polynomial denoted by F_j approximating the discontinuous flux function f_j. To assure continuity across cells, F_j is required to take on the common flux values at the two interfaces:

$$F_j(x_{j-1/2}) = f_{j-1/2,\text{com}} \quad \text{and} \quad F_j(x_{j+1/2}) = f_{j+1/2,\text{com}} .$$

What degree should F_j be? To answer this question, we allow $u_j(x)$ to depend on t, i.e., for each fixed t, $u_j(x,t)$ is a polynomial of degree $K-1$. The conservation law can be approximated by

$$\frac{\partial}{\partial t} u_j(x,t) + (F_j(x))_x = 0.$$

After one time step of size τ (Euler explicit or one stage of Runge-Kutta), denoting the solution at the new time by $u_j^{\text{new}}(x)$, the above implies

$$u_j^{\text{new}}(x) = u_j(x) - \tau(F_j(x))_x.$$

The right hand side is evaluated at time t^n, and $u_j(x)$ is of degree $K-1$. For $u_j^{\text{new}}(x)$ to also be of degree $K-1$, since τ is a constant, $(F_j(x))_x$ must be of degree $K-1$. Therefore, F_j is required to be of degree K. It must also approximate f_j, which is of degree $K-1$.

Thus, on E_j, the polynomial F_j is required to take on the common flux values at the two interfaces, to be of degree K, and to approximate f_j. Instead of defining F_j, we define $F_j - f_j$, which approximates the zero function. Switching to the local description, at $\xi = \pm 1$, the function $F_j - f_j$ takes on the following values, which are called the interface corrections:

$$F_j(-1) - f_j(-1) = f_{j-1/2,\text{com}} - f_j(-1) \tag{6a}$$

and

$$F_j(1) - f_j(1) = f_{j+1/2,\text{com}} - f_j(1). \tag{6b}$$

Therefore, $F_j - f_j$ takes on the above prescribed left and right correction values, is of degree K, and approximates the zero function.

We now separate the prescription of the correction at the left interface from that of the right. This separation plays a critical role in the new approach. Again, using the local coordinate, for the left interface, let g_{LB} be the *correction function* ('LB' for 'left boundary') defined by

$$g_{\text{LB}}(-1) = 1, \quad g_{\text{LB}}(1) = 0, \tag{7a,b}$$

and g_{LB} is of degree K and approximates the zero function in some sense (Fig. 2(a)). As for g_{RB}, by reflection, $g_{\text{RB}}(\xi) = g_{\text{LB}}(-\xi)$. Thus,

$$g_{\text{RB}}(-1) = 0, \quad g_{\text{RB}}(1) = 1. \tag{8a,b}$$

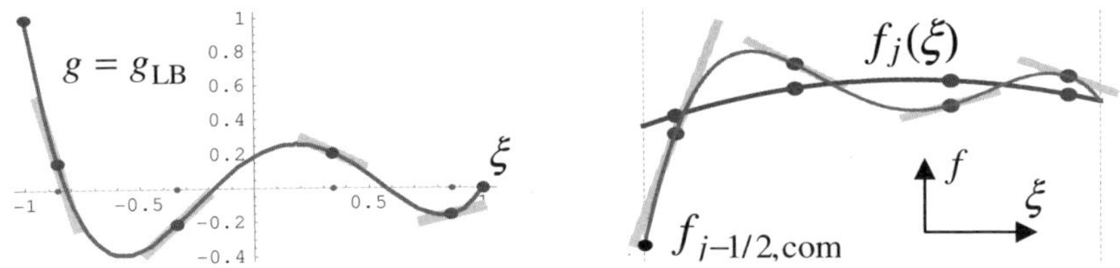

(a) Correction function (b) Correcting for the left boundary

Fig. 2. (a) Correction function g_{LB} for the left boundary with $K = 4$ (the right Radau polynomial); (b) Discontinuous flux function and the polynomial by Eq. (9) accounting for the correction at the left boundary.

Consider the left interface $x_{j-1/2}$, i.e., $\xi = -1$. The polynomial

$$f_j(\xi) + [f_{j-1/2,\text{com}} - f_j(-1)]\, g_{\text{LB}}(\xi) \tag{9}$$

provides a correction for $f_j(\xi)$ by changing the flux value at this interface from $f_j(-1)$ to $f_{j-1/2,\text{com}}$, while leaving the value at the right interface unchanged, namely $f_j(1)$ (Fig. 2b)). Next, the polynomial

$$\begin{aligned} F_j(\xi) &= f_j(\xi) + [f_{j-1/2,\text{com}} - f_j(-1)]\, g_{\text{LB}}(\xi) \\ &\quad + [f_{j+1/2,\text{com}} - f_j(1)]\, g_{\text{RB}}(\xi) \end{aligned} \tag{10}$$

provides corrections to both interfaces: using (7) and (8), one can verify that $F_j(-1) = f_{j-1/2,\text{com}}$ and $F_j(1) = f_{j+1/2,\text{com}}$. Thus, F_j is of degree K, takes on the two common flux values, and approximates f_j in the same sense that g_{LB} and g_{RB} approximate the zero function.

Finally, the derivative of $F_j(\xi)$ at the solution point ξ_k is

$$\begin{aligned} (F_\xi)_{j,k} &= (f_\xi)_{j,k} + [f_{j-1/2,\text{com}} - f_j(-1)]\, g_{\text{LB}}'(\xi_k) \\ &\quad + [f_{j+1/2,\text{com}} - f_j(1)]\, g_{\text{RB}}'(\xi_k) \end{aligned} . \tag{11}$$

In the global coordinate, at the solution point $x_{j,k}$,

$$(F_x)_{j,k} = (2/h_j)\,(F_\xi)_{j,k}\,. \tag{12}$$

The quantities $(F_x)_{j,k}$ above are employed to approximate f_x for the conservation law at $x_{j,k}$:

$$\frac{du_{j,k}}{dt} = \left[\frac{du_{j,k}(t)}{dt}\right]_{t=t^n} = -(F_x)_{j,k}\,. \tag{13}$$

The solution $u_{j,k}$ can then be updated via, say, a Runge-Kutta method.

What is crucial in (11) is that at each solution point, the derivative $(F_\xi)_{j,k}$ of the continuous flux function is obtained by correcting the derivative $(f_\xi)_{j,k}$ of the discontinuous flux function. The correction amount is straightforward once the values $g_{\text{LB}}'(\xi_k)$ and $g_{\text{RB}}'(\xi_k)$ are known for $1 \le k \le K$. These derivative values, in turn, can easily be derived once g_{LB} and g_{RB} are defined on *I*.

We summarize the CPR algorithm below.

Algorithm. At time level *n*, suppose $u_{j,k}$ are known for all *j* and *k*.

(1) At each interface $j+1/2$, if the left and right values of u are not available, calculate them by (4); then estimate and store the common (upwind) fluxes at all interfaces via (5).

(2) In cell *j*, for $k = 1,\ldots,K$, evaluate $f_{j,k} = f(u_{j,k})$; then obtain $(f_\xi)_{j,k}$ of the discontinuous flux function by (2). Alternatively, the chain rule (3) can be employed.

(3) At the two interfaces of E_j, get the corrections $f_{j-1/2,\text{com}} - f_j(-1)$ and $f_{j+1/2,\text{com}} - f_j(1)$ by (6). At the solution points, evaluate $(F_\xi)_{j,k}$ by (11) and $(F_x)_{j,k}$ by (12).

(4) Calculate $du_{j,k}/dt$ via (13) and march in time by a Runge-Kutta method. This completes the algorithm.

Next, we show that the resulting scheme is conservative, that is,

$$\left[\frac{\partial}{\partial t}\int_{E_j} u_j(x,t)\,dx\right]_{t=t^n} = f_{j-1/2,\text{com}} - f_{j+1/2,\text{com}}\,. \tag{14}$$

To this end, consider a quadrature of degree of precision at least $K-1$ using the K solution points, i.e., it is exact for any polynomial of degree $K-1$ or less. Denote the weight at ξ_k by w_k. Then, since $u_j(\xi)$ is of degree $K-1$,

$$\int_{-1}^{1} u_j(\xi)\,d\xi = \sum_{k=1}^{K} w_k\, u_{j,k}\ .$$

Note the exactness of the quadrature. For the global coordinate,

$$\int_{E_j} u_j(x)\,dx = \frac{h_j}{2} \sum_{k=1}^{K} w_k\, u_{j,k}\ . \tag{15}$$

The left hand side of (14) can then be written as

$$\left[\frac{\partial}{\partial t}\int_{E_j} u_j(x,t)\ dx\right]_{t=t^n} = \frac{h_j}{2}\frac{\partial}{\partial t}\sum_{k=1}^{K} w_k\, u_{j,k}$$

$$= \frac{h_j}{2}\sum_{k=1}^{K} w_k \frac{du_{j,k}}{dt} = -\frac{h_j}{2}\sum_{k=1}^{K} w_k\,(F_x)_{j,k}\ .$$

Here, the last equality is a result of (13). Since $(F_j)_x$ is of degree $K-1$, by (15),

$$\int_{E_j} (F_j)_x(x)\,dx = \frac{h_j}{2}\sum_{k=1}^{K} w_k\,(F_x)_{j,k}\ .$$

It follows from the fundamental theorem of calculus that

$$\int_{E_j} (F_j(x))_x\,dx = F_j(x_{j+1/2}) - F_j(x_{j-1/2})$$

$$= f_{j+1/2,\text{com}} - f_{j-1/2,\text{com}}\ .$$

The above three equations imply (14). This completes the proof.

4. Correction Functions

To define the various correction functions, we need some review. Let the L^2 inner product of any two polynomials v and w on E_j be

$$(v,w) = \int_{x_{j-1/2}}^{x_{j+1/2}} v(\xi)w(\xi)\,d\xi\ .$$

For any integer $m \geq 0$, let $\boldsymbol{P}_m$ be the space of polynomials of degree m or less. Then $\boldsymbol{P}_m$ is a vector space of dimension $m+1$. When the domain, say E_j, is important, we use the notation $\boldsymbol{P}_m(E_j)$.

We now focus on $I=[-1,1]$. A polynomial v is orthogonal to $\boldsymbol{P}_m=\boldsymbol{P}_m(I)$ if, for each l, $0\le l\le m$,

$$(v,\xi^l)=\int_{-1}^{1} v(\xi)\xi^l\,d\xi=0\,.$$

The criterion of being orthogonal to $\boldsymbol{P}_m$ provides $m+1$ equations.

For $k=0,1,2,\ldots$, let the Legendre polynomial P_k be defined on I as the unique polynomial of degree k satisfying the $k+1$ conditions of being orthogonal to $\boldsymbol{P}_{k-1}$ and $P_k(1)=1$. The Legendre polynomials are given by a recurrence formula (e.g., Hildebrand 1987):

$$P_0=1,\quad P_1=\xi,$$

and, for $k\ge 2$,

$$P_k(\xi)=\frac{2k-1}{k}\,\xi P_{k-1}(\xi)-\frac{k-1}{k}\,P_{k-2}(\xi)\,.$$

The first few Legendre polynomials are plotted in Fig. 3(a). Useful properties of the Legendre polynomials are listed below. If $k>m$, then P_k is orthogonal to $\boldsymbol{P}_m$. Next, P_k is an even function (involving only even powers of ξ) for even k, and an odd function for odd k. For all k,

$$P_k(-1)=(-1)^k,\qquad P_k(1)=1\,. \tag{16a,b}$$

The derivative values at the end points are

$$P_k'(-1)=(-1)^{k-1}k(k+1)/2\,,\quad P_k'(1)=k(k+1)/2\,. \tag{17a,b}$$

In addition,

$$(P_k,P_k)=2/(2k+1)\,;\ \text{ for } k\ne l,\ (P_k,P_l)=0\,.$$

The zeros of P_k are the k Gauss points on $[-1,1]$.

The right Radau polynomial of degree k ($k\ge 1$) is given by

$$R_{R,k}=\frac{(-1)^k}{2}(P_k-P_{k-1})\,. \tag{18}$$

The above definition is nonstandard so that (19a) below holds. The first few Radau polynomials are plotted in Fig. 3(b). The above definition implies that $R_{R,k}$ is orthogonal to $\boldsymbol{P}_{k-2}$. In addition, by (16),

$$R_{R,k}(-1)=1 \quad \text{and} \quad R_{R,k}(1)=0. \tag{19a,b}$$

It is important to note that $R_{R,k}$, which is of degree *k*, is defined by the above two conditions and the $k-1$ conditions that it is orthogonal to $\boldsymbol{P}_{k-2}$. This definition of the Radau polynomial shows that it approximates the zero function in the sense of least squares and is a natural choice for the correction function. For later use, at the two boundaries, by using (17),

$$R_{R,k}{}'(-1)=-k^2/2, \quad \text{and} \quad R_{R,k}{}'(1)=(-1)^{k-1}k/2. \tag{20a,b}$$

The zeros of the Radau polynomial $R_{R,k}$ are the *k* right Radau points.

The Lobatto polynomial of degree *k* ($k \ge 1$) is defined by

$$\text{Lo}_k = P_k - P_{k-2}.$$

They can be expressed in terms of Radau polynomials via (18):

$$\text{Lo}_k = 2(-1)^k (R_{R,k} - R_{R,k-1}). \tag{21}$$

The zeros of the Lobatto polynomial of degree *k* are the *k* Lobatto points; they include the two boundaries ± 1. As can be observed from Fig. 3(b), consistent with (21), the Lobatto points are also the ξ-coordinates of the intersections of the graphs of $R_{R,k}$ and $R_{R,k-1}$.

Returning to the correction functions, we always choose g_{RB} by reflection: $g_{\text{RB}}(\xi)=g_{\text{LB}}(-\xi)$. Consequently, we only need to focus on g_{LB}. For simplicity of notation, set

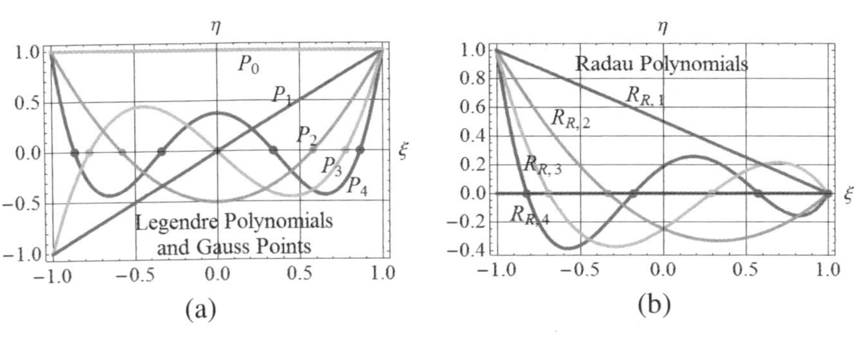

Fig. 3. (a) Legendre polynomials and (b) right Radau polynomials.

$$g = g_{\text{LB}} \,.$$

Since g is of degree K, it is determined by $K+1$ conditions. Two conditions are known, namely

$$g(-1) = 1 \quad \text{and} \quad g(1) = 0. \tag{22}$$

Therefore, $K-1$ additional conditions remain.

Under the CPR approach, the problem of designing high-order schemes reduces to defining g such that the above holds together with $K-1$ additional conditions. These additional conditions are prescribed so that g 'approximates' the zero function. What criteria should we employ? The final criteria are the stability and accuracy of the resulting scheme discussed later. Since there is a trade-off between accuracy and stability, and the property of super-convergence (super-accuracy) does not hold for the general case of the Euler and Navier-Stokes equations, an optimal scheme has not been determined by this author. In fact, such a scheme is likely to be problem dependent. Therefore, in the rest of this section, we discuss three choices for g using approximation theory as well as some guidance from Fourier analysis. For simplicity, a scheme is identified by its correction function, e.g., scheme g_{DG}.

The first choice for g, denoted by g_1 (the right boundary is a zero of multiplicity one), is defined by the criterion of least squares: to approximate the zero function by $K-1$ conditions, we require the projection of g onto $\boldsymbol{P}_{K-2}$ to be 0. This requirement implies that g is the right Radau polynomial of degree K, namely $R_{R,K}$ defined by (18). Note that the correction function for the *left* boundary is the *right* Radau polynomial (vanishing at $\xi = 1$). As will be shown in the next section, the resulting scheme is identical to DG; therefore, g_1 is also denoted by g_{DG}:

$$g_{\text{DG}} = g_1 = R_{R,K} \,.$$

It will shown via Fourier analysis that the scheme is stable and accurate to order $2K-1$. This order of accuracy, which is higher than the expected order of K, is consistent with the super-convergence (super-accuracy) property of the DG method (Adjerid *et al.* 2002).

Loosely put, the current formulation is a finite difference formulation (versus finite element) for DG. It involves no quadratures and has the advantage of simplicity and economy. In addition, regardless of the choice of solution points, no matrix inversion is needed. For example, if we choose the Lobatto points or the equidistant points as solution points, then, since the corresponding basis functions are not orthogonal, the standard DG formulation requires a matrix inversion, whereas the current formulation does not. In other words, the mass matrix inversion is built-in.

For the next two choices of correction functions, in addition to (22), we require g to be orthogonal to $\boldsymbol{P}_{K-3}$ (yielding $K-2$ conditions) together with one additional condition. It can be verified via Fourier analysis that the requirement of g being orthogonal to $\boldsymbol{P}_{K-3}$ gives rise to stable schemes (the converse, as discussed in (Huynh 2007) is not true, however). Since both $R_{R,K}$ and $R_{R,K-1}$ are orthogonal to $\boldsymbol{P}_{K-3}$, the next two correction functions can be written as, with $0<\alpha<1$,

$$g = \alpha R_{R,K} + (1-\alpha) R_{R,K-1}.$$

where, α remains to be determined (Huynh 2009). Note that by (19), the above satisfies (22).

The second choice for g, denoted by g_2 or $g_{\text{Lump, Lo}}$ (for 'lumping for Lobatto points' explained later), is defined as follows. Since a steeper correction function tends to result in a scheme with a smaller CFL limit, we wish to make g less steep. On the other hand, since g approximates the function 0 on $I=[-1,1]$, it seems reasonable to require that all zeros of g lie on I, not outside. To make g less steep, therefore, the extra condition is obtained by pushing one of the zeros to the right boundary, i.e., we require $\xi=1$ to be a zero of multiplicity two (thus, the notation g_2):

$$g_2'(1)=0. \tag{23}$$

Using (20b) with $k=K$ and $k=K-1$ respectively, the above implies $\alpha=(K-1)/(2K-1)$, i.e.,

$$g_2 = g_{\text{Lump, Lo}} = \frac{K-1}{2K-1} R_{R,K} + \frac{K}{2K-1} R_{R,K-1}. \tag{24}$$

The function g_2 has the following remarkable property, which holds true for any K. Among the K Lobatto points, g_2' vanishes at

$K-1$ of them; the exception is the left boundary $\xi=-1$ as can be seen in Fig. 4(a). Therefore, if we employ g_2, it is convenient and economical to select the K Lobatto points as *solution points*. With such a selection, the jump in flux values at the left interface results in a correction to only $(f_\xi)_{j,1}$ and not to any $(f_\xi)_{j,k}$, $k>1$. That is, the correction due to the jump at the left boundary is lumped into that boundary, and the corrections at all other solution points are zero. For this reason, g_2 is also denoted by $g_{\text{Lump, Lo}}$. Employing (20a) with $k=K$ and then with $k=K-1$, we can calculate the correction at the boundary by (24):

$$g_2'(-1) = -K(K-1)/2. \tag{25}$$

Note that the above quantity equals $-w_1$ where w_1 is the weight at the boundary point of the Lobatto quadrature with K points. Thus, (25) is consistent with the fact that the Lobatto quadrature for $\int_{-1}^{1} g_2'(\xi)d\xi$ is exact and equals $g_2(1)-g_2(-1)=-1$.

As will be shown by Fourier analysis, scheme g_2 is stable and accurate to order $2K-2$ as opposed to order $2K-1$ of the DG scheme. To its advantage, the CFL limit is roughly twice as large as that of DG.

The third choice for g, denoted by g_{Ga}, requires that in addition to (22), g vanishes at the $K-1$ Gauss points. (These points are the zeros of the Legendre polynomial P_{K-1} and are completely different from the K Gauss points, which typically are the solution points). It can be verified that (note the reverse in the order of the weights compared to (24))

$$g_{\text{Ga}} = \frac{K}{2K-1} R_{R,K} + \frac{K-1}{2K-1} R_{R,K-1}. \tag{26}$$

As will be shown by Fourier analysis, scheme g_{Ga} is stable and accurate to order $2K-2$. A stability proof using energy estimates and a norm of Sobolev type for this scheme can be found in (Jameson 2010).

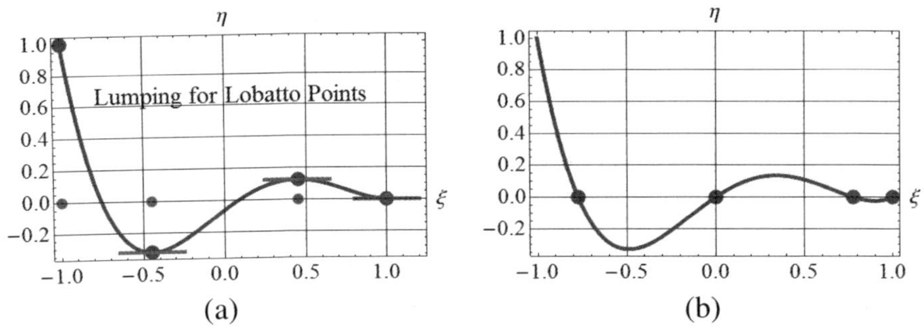

Fig. 4. Correction functions for $K = 4$ (a) $g_2 = g_{\text{Lump,Lo}}$ and (b) g_{Ga}.

Note that if we require g to vanish at all $K+1$ Chebyshev-Lobatto points except the left boundary, then the resulting method is identical to the staggered-grid scheme (Kopriva and Kolias 1996) provided that the solution points are the K Chebyshev-Gauss points. The staggered-grid scheme, however, is mildly unstable. Scheme g_{Ga} above resolves this stability problem and is also more economical (Huynh 2007).

Among the three correction functions, loosely put, g_{DG} is the steepest, and $g_{\text{Lump, Lo}}$, least steep. At the left boundary, by (20a), (23), and (26), in the order of decreasing steepness,

$$g_{\text{DG}}'(-1) = -K^2/2, \; g_{\text{Ga}}'(-1) = -[K(K-1)+1]/2, \text{ and}$$
$$g_{\text{Lump, Lo}}'(-1) = -K(K-1)/2.$$

The schemes in the order of decreasing accuracy as well as increasing time-stepping limit are: g_{DG}, g_{Ga}, and $g_{\text{GLump,Lo}}$. That is, g_{DG} is the most accurate but possesses the smallest time-stepping limit.

The plots of the three correction functions for various K will be shown in Figs. 6–8 of Section 8. Additional correction functions can be found in (Huynh 2007).

A stability proof using energy estimates for a one parameter family of schemes of the form $\alpha R_{R,K} + (1-\alpha) R_{R,K-1}$ was presented by Vincent, Castonguay, and Jameson (2010) where the family was expressed in terms of the Legendre polynomials instead of the Radau polynomials as in this chapter and in (Huynh 2009).

5. CPR and DG Approaches

We now show that, with the Radau polynomial as correction function, the CPR scheme for the conservation law (1) yields a result identical to that by the DG method.

Readers who are not interest in the proof can skip this section with no loss of continuity. To prove the above claim, we first review the standard DG scheme (see, e.g., Hesthaven and Warburton 2008). For simplicity of notation, the subscripts j and E_j are often omitted. On the cell $E = E_j$, recall that

$$(v,w) = (v,w)_E = \int_{x_{j-1/2}}^{x_{j+1/2}} v(\xi)w(\xi)d\xi \,.$$

Let ϕ be a test function. Since ϕ is independent of t, $(u_t,\phi) = (u,\phi)_t$. Formally, we require u to satisfy $(u,\phi)_t + (f_x,\phi) = 0$, or, using integration by parts,

$$(u,\phi)_t + [f\phi]_{\partial E} - (f,\phi_x) = 0 \,. \tag{27}$$

With the DG method, u is replaced by u_j, ϕ by one of the basis functions $\phi_{j,k}$, $k = 1,...,K$, and f by f_j, all of degree $K-1$. Concerning the boundary terms, at say $j+1/2$, for *f*, the upwind flux is employed as the common flux for the two adjacent cells. For ϕ, however, the value from E (no upwinding) is used. The common flux provides coupling between adjacent cells and results in a conservative scheme. Instead of (27), the solution is required to satisfy

$$(u,\phi)_t + [f_{\text{com}}\,\phi]_{\partial E} - (f,\phi_x) = 0 \,. \tag{28}$$

The standard DG scheme evaluates the above numerically. Here, we wish to eliminate ϕ. Applying integration by parts again to (f,ϕ_x) and, since the polynomials f and ϕ are smooth on *E*, denoting by f_{int} the flux value from the *interior* of cell E (no upwinding),

$$(u,\phi)_t + ([f_{\text{com}} - f_{\text{int}}]\phi)_{\partial E} + (f_x,\phi) = 0 \,. \tag{29}$$

Note that $[f_{\text{com}} - f_{\text{int}}]$, which is often denoted by $[f]$, is simply the correction at the interface in (6). The above, sometimes called the 'strong form' (as opposed to (28), the 'weak form'), is the result of integrating by parts twice. It is simply the inner product of the conservation law with the test function except for the correction term $([f]\phi)_{\partial E}$, which accounts for interaction.

Our task is to eliminate the test function. To this end, we first switch to the local coordinate ξ on $I=[-1,1]$. Denote the length of $E=E_j$ by h. Noting that $dx=(h/2)d\xi$ and $df=f_x\,dx=f_\xi\,d\xi$, the above can be written as (again, $u=u_j$, $f=f_j$, and $\phi=\phi_{j,k}$)

$$\frac{h}{2}\frac{\partial}{\partial t}\int_{-1}^{1}u\phi d\xi + [f_{\text{com}}(1)-f(1)]\phi(1) \\ - [f_{\text{com}}(-1)-f(-1)]\,\phi(-1)+\int_{-1}^{1}f_\xi\,\phi d\xi = 0\,. \tag{30}$$

Focusing on the term $-[f_{\text{com}}(-1)-f(-1)]\,\phi(-1)$ at the left boundary, due to the term $\int_{-1}^{1}f_\xi\,\phi d\xi$, to eliminate ϕ, we raise the following question: can we find a polynomial g_{LB} on $I=[-1,1]$ which possesses the property that for any ϕ of degree $K-1$ or less,

$$-\phi(-1)=\int_{-1}^{1}g_{\text{LB}}{}'\,\phi d\xi \tag{31}$$

where $g_{\text{LB}}{}'(\xi)=(g_{\text{LB}})_\xi(\xi)$. From (30), due to the term $\int_{-1}^{1}u\phi d\xi$, we require $g_{\text{LB}}{}'(\xi)$ to have the same degree as $u_j(\xi)$, i.e., of degree $K-1$; as a result, g_{LB} is required to be of degree K. To determine g_{LB}, applying integration by parts to the above right hand side, we have

$$-\phi(-1)=g_{\text{LB}}(1)\phi(1)-g_{\text{LB}}(-1)\phi(-1)-\int_{-1}^{1}g_{\text{LB}}\,\phi'\,d\xi.$$

Thus, (31) holds if g_{LB} satisfies,

$$g_{\text{LB}}(-1)=1,\qquad g_{\text{LB}}(1)=0, \tag{32}$$

and, for all ϕ in $\boldsymbol{P}_{K-1}$,

$$\int_{-1}^{1}g_{\text{LB}}\,\phi'\,d\xi=0\,. \tag{33}$$

Since ϕ is of degree $K-1$, ϕ' is of degree $K-2$; moreover, ϕ' spans $\boldsymbol{P}_{K-2}$ as ϕ spans $\boldsymbol{P}_{K-1}$. The above then implies that g_{LB} is orthogonal to $\boldsymbol{P}_{K-2}$, i.e., for any polynomial φ of degree $K-2$,

$$\int_{-1}^{1}g_{\text{LB}}\,\varphi d\xi=0.$$

The criterion that g_{LB} is orthogonal to $\boldsymbol{P}_{K-2}$ provides $K-1$ conditions; (32) provides the other two. These $K+1$ conditions imply that g_{LB} is the right Radau polynomial defined in (18),

$$g_{\mathrm{LB}}(\xi) = R_{R,K}(\xi).$$

Thus, the answer to the question posed for (31) is positive. Note again that the correction function for the left boundary is the right Radau polynomial (vanishing at $\xi = 1$).

Next, switching to the right boundary, in a manner similar to (32) and (33), let g_{RB} be defined by $g_{\mathrm{RB}}(-1)=0$, $g_{\mathrm{RB}}(1)=1$, and g_{RB} is orthogonal to $\boldsymbol{P}_{K-2}$. Then $g_{\mathrm{RB}} = R_{L,K}$, the left Radau polynomial defined by $R_{L,K}(\xi) = R_{R,K}(-\xi)$. Using (31),

$$\phi(1) = \int_{-1}^{1} g_{\mathrm{RB}}{}' \phi \, d\xi .$$

We now return to our task of eliminating ϕ. By the above and (31), we can write (30) as

$$\begin{aligned} &\frac{h}{2}\frac{\partial}{\partial t}\int_I u\phi \, d\xi + [f_{\mathrm{com}}(1) - f(1)]\int_I g_{\mathrm{RB}}{}' \phi \, d\xi \\ &+ [f_{\mathrm{com}}(-1) - f(-1)]\int_I g_{\mathrm{LB}}{}' \phi \, d\xi + \int_I f_\xi \phi \, d\xi = 0 . \end{aligned} \tag{34}$$

What is crucial here is that ϕ can be factored out. Indeed, with u replaced by u_j and f by f_j, set

$$\begin{aligned} F_j(\xi) = f_j(\xi) &+ [f_{j+1/2,\mathrm{com}} - f_j(1)]\, g_{\mathrm{RB}}(\xi) \\ &+ [f_{j-1/2,\mathrm{com}} - f_j(-1)]\, g_{\mathrm{LB}}(\xi) . \end{aligned}$$

Then, (34) implies

$$\int_I \left[\frac{h_j}{2}(u_j)_t + (F_j)_\xi \right] \phi \, d\xi = 0 .$$

Switching to the global coordinate,

$$\int_{E_j} [(u_j)_t + (F_j)_x] \, \phi \, dx = 0 .$$

Since u_j and $(F_j)_x$ are of degree $K-1$, and the above holds for any ϕ of degree $K-1$,

$$(u_j)_t + (F_j)_x = 0\,.$$

This equation is identical to (13). Thus, the DG method is identical to the CPR scheme using g_{DG} as correction function.

6. The Diffusion Equation

To apply the CPR approach to diffusion problems, on $(-\infty,\infty)$, consider the diffusion equation,

$$u_t = u_{xx}$$

with initial condition

$$u(x,0) = u_0(x)\,.$$

At time level n, assume that the data $u_{j,k}$ are known. We wish to evaluate the second derivative in a manner which takes into account the data interaction among cells. For simplicity and efficiency, the stencil of the scheme is required to remain compact in the sense that the second derivative evaluation in a cell involves the data of only that cell and the two immediate neighbors.

Common values and corrected derivative estimates. The first task is to estimate u_x at the solution points $x_{j,k}$. Since the function $\{u_j\}$ is discontinuous across the interfaces, to estimate u_x, we first reconstruct u by a piecewise polynomial function $\{u_j^C\}$, which is continuous across the cell interfaces and, on each E_j, is of degree K and approximates u_j (the superscript 'C' stands for 'continuous' or 'corrected'). At the solution points, $(u_j^C)_x$ provides a derivative approximation that accounts for the data interaction.

In order for $\{u_j^C\}$ to be continuous at the interfaces, u_j^C and u_{j+1}^C must take on the same value at $x_{j+1/2}$. Thus, at each interface, we need to define a common interface value (or common value). For an advection problem, between the left and right values $u_L = u_{j+1/2,L}$ and $u_R = u_{j+1/2,R}$, the common value is typically an upwind (flux) value. For diffusion problems, we use a centered-type quantity

$$u_{\text{com}} = u_{j+1/2,\text{com}} = \tfrac{1}{2}(u_L + u_R)\,.$$

The above formula was employed by Bassi and Rebay (1997, 2000). A more general formula is the weighted average, with $0 \le \kappa \le 1$,

$$u_{\text{com}} = u_{j+1/2,\text{com}} = \kappa u_L + (1-\kappa) u_R . \quad (35)$$

For $\kappa = 0$ or $\kappa = 1$, we have the one-sided formula used in the local DG or LDG (Cockburn and Shu 1998) as well as the compact DG or CDG methods (Peraire and Persson 2008).

Next, we require $u_j^C(x)$ to take on the common values $u_{j-1/2,\text{com}}$ at $x_{j-1/2}$ and $u_{j+1/2,\text{com}}$ at $x_{j+1/2}$, to be of degree K, and to approximate $u_j(x)$. That is, in the local coordinate,

$$u_j^C(\xi) = u_j(\xi) + [u_{j-1/2,\text{com}} - u_j(-1)]\, g_{\text{LB}}(\xi) + [u_{j+1/2,\text{com}} - u_j(1)]\, g_{\text{RB}}(\xi) .$$

The derivative of $u_j^C(\xi)$ is

$$(u_j^C)_\xi(\xi) = (u_j)_\xi(\xi) + [u_{j-1/2,\text{com}} - u_j(-1)]\, g_{\text{LB}}{}'(\xi) + [u_{j+1/2,\text{com}} - u_j(1)]\, g_{\text{RB}}{}'(\xi) .$$

At the solution point ξ_k, the *corrected derivative* is given by

$$(u_x)_{j,k}^C = (u_x^C)_{j,k} = (2/h_j)(u_j^C)_\xi(\xi_k) . \quad (36)$$

Note that the reconstruction polynomial $u_j^C(\xi)$ clarifies the ideas; in practice, we only need the values of its derivative at the solution points.

Common derivative and corrected second derivative estimates. At each interface, in formula (35) for the common value, with $0 \le \kappa \le 1$, the weight for u_L is κ and that for u_R, $1-\kappa$. To define the common derivative value, we switch the two weights. Loosely put, this switch makes the method unbiased and, therefore, consistent with the centered nature of the diffusion process. Since the corrected derivative $(u_j^C)_x$ is readily available, an obvious choice is

$$(u_x)_{j+1/2,\ \text{com}} = (1-\kappa)(u_j^C)_x(x_{j+1/2}) + \kappa (u_{j+1}^C)_x(x_{j+1/2}) . \quad (37)$$

The function u_j^C involves the data in the three cells $j-1$, j, and $j+1$. Consequently, the above formula has a stencil of four cells, from $j-1$ to $j+2$ (see Fig. 5(a)). Since the calculation of u_{xx} in cell j employs

$(u_x)_{j-1/2,\mathrm{com}}$ and $(u_x)_{j+1/2,\mathrm{com}}$, the corresponding scheme has a five-cell stencil.

We now define a common derivative at $j+1/2$ that involves only the data in the two adjacent cells. A scheme with such a compact stencil is desirable since it is easy to code, the boundary conditions involved are simple, and the resulting implicit version has a sparse and generally invertible matrix. To this end, correcting for the right boundary of cell *j*, set

$$u_j^{\mathrm{RB}}(\xi) = u_j(\xi) + [u_{j+1/2,\mathrm{com}} - u_j(1)]\, g_{\mathrm{RB}}(\xi) \tag{38a}$$

i.e., u_j^{RB} corrects for the right boundary, namely $u_j^{\mathrm{RB}}(1) = u_{j+1/2,\mathrm{com}}$, while leaving the value at the left boundary unchanged, namely $u_j(-1)$. Next, correcting for the left boundary of cell $j+1$, set

$$u_{j+1}^{\mathrm{LB}}(\xi) = u_{j+1}(\xi) + [u_{j+1/2,\mathrm{com}} - u_{j+1}(-1)]\, g_{\mathrm{LB}}(\xi). \tag{38b}$$

Then u_{j+1}^{LB} corrects for the left boundary, $u_{j+1}^{\mathrm{LB}}(-1) = u_{j+1/2,\mathrm{com}}$, while leaving the value at the right boundary unchanged, namely $u_j(1)$.

Finally, for the common derivative at $j+1/2$, set (see Fig. 5(b)),

$$\begin{aligned}(u_x)_{j+1/2,\mathrm{com}} &= (1-\kappa)\frac{2}{h_j}\{(u_j)_\xi(1) + [u_{j+1/2,\mathrm{com}} - u_j(1)]\, g_{\mathrm{RB}}{}'(1)\} \\ &+ \kappa\frac{2}{h_{j+1}}\{(u_{j+1})_\xi(-1) + [u_{j+1/2,\mathrm{com}} - u_{j+1}(-1)]\, g_{\mathrm{LB}}{}'(-1)\}.\end{aligned} \tag{39}$$

Note the dependence only on $u_{j+1/2,\mathrm{com}}$ and the data on E_j and E_{j+1}.

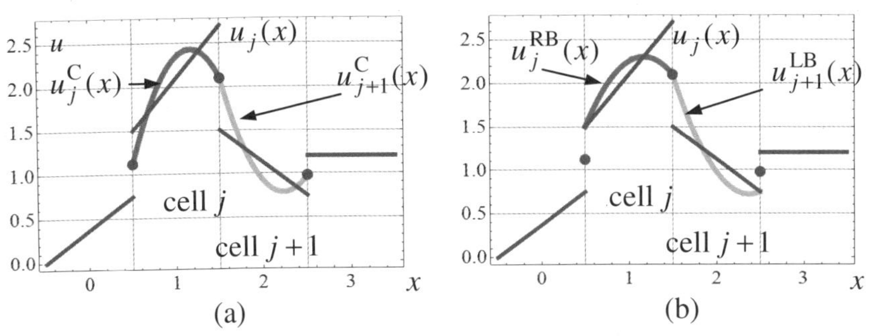

Fig. 5. Centered-type common derivative: (a) via (37) using a four-cell stencil and (b) via (39) using a two-cell stencil. Here, the solution polynomials are linear, and the correction function g_{DG} is parabolic.

With the corrected derivative given by (36) and the common derivative (39), we can obtain the *corrected second derivative* estimates.

The above procedure yields the CPR versions of the BR2 scheme if $\kappa = 1/2$ (Bassi and Rebay, 2000) and the LDG (Cockburn and Shu 1998) or CDG schemes (Peraire and Persson 2008) if $\kappa = 0$ or $\kappa = 1$.

7. Fourier (Von Neumann) Analysis

Fourier analysis provides information on both stability and accuracy. The accuracy criterion (42) is presented below since it is critical and not widely known. On the domain $(-\infty, \infty)$, consider the equations

$$u_t + u_x = 0 \quad \text{or} \quad u_t = u_{xx}.$$

The initial condition is periodic: $u_{\text{init}}(x) = e^{iwx}$ where w is a real number between $-\pi$ and π called a wave number. Low frequency data corresponds to w of small magnitude, high frequency, to w near $\pm\pi$. For the advection case (the diffusion case is similar), the exact solution is $u_{\text{exact}}(x,t) = e^{iw(x-t)}$. At $x = 0$ and $t = 0$,

$$(u_{\text{exact}})_t(0,0) = -iw.$$

The cells are $E_j = [j - 1/2, j + 1/2]$. The solution points on $I = [-1,1]$ are ξ_k, $k = 1, ..., K$. The global solution points are $x_{j,k} = j + \xi_k / 2$. The data are $u_{j,k} = \exp[iw(j + \xi_k / 2)]$. However, it is not the data but their following property, which plays a key role in the calculation of eigenvalues,

$$u_{j-1,k} = e^{-iw} u_{j,k}. \tag{40}$$

To calculate the eigenvalues, the K solution points are grouped together as a vector: with superscript T denoting the transpose, set

$$\boldsymbol{u}_j = (u_{j,1}, ..., u_{j,K})^T.$$

For both the advection and diffusion cases,

$$\frac{d\boldsymbol{u}_j}{dt} = \boldsymbol{C}_{-1}\boldsymbol{u}_{j-1} + \boldsymbol{C}_0\boldsymbol{u}_j + \boldsymbol{C}_1\boldsymbol{u}_{j+1}$$

where $\boldsymbol{C}_{-1}$, $\boldsymbol{C}_0$, and $\boldsymbol{C}_1$ are $K \times K$ matrixes. Using (40), we replace $\boldsymbol{u}_{j-1}$ by $e^{-iw}\boldsymbol{u}_j$ and $\boldsymbol{u}_{j+1}$ by $e^{iw}\boldsymbol{u}_j$. The spatial discretization yields

$$\frac{d\boldsymbol{u}_j}{dt} = \boldsymbol{S}\,\boldsymbol{u}_j \tag{41}$$

where the $K \times K$ matrix $\boldsymbol{S}$ (for 'space' or 'semidiscrete') is given by

$$\boldsymbol{S} = e^{-iw}\boldsymbol{C}_{-1} + \boldsymbol{C}_0 + e^{iw}\boldsymbol{C}_1 .$$

Equation (41) is similar to the differential equation $du/dt = \lambda u$ whose solution is $ce^{\lambda t}$ where c is an arbitrary constant. If $\mathrm{Re}(\lambda) \le 0$, the solution is stable. Here, the eigenvalues of $\boldsymbol{S}$ take the place of λ and, for all schemes discussed here, $\boldsymbol{S}$ has K eigenvalues. For advection, the one approximating $-iw$ is the *principal eigenvalue* denoted by $S(w)$:

$$S(w) \approx -iw .$$

All other eigenvalues are *spurious*. All eigenvalues must lie in the left half of the complex plane for the semidiscretization to be stable. The collection of all eigenvalues forms the *spectrum* of the scheme.

To find the order of accuracy for the advection case, note that with a uniform mesh of width h, the principal eigenvalue S 'approximates' $-h\partial/\partial x$. A scheme is accurate to order m if S 'approximates' $-h\partial/\partial x$ to $O(h^{m+1})$; more precisely, for small w,

$$S(w) = -iw + O(w^{m+1}) .$$

In practice, it is difficult if not impossible to derive an expression for $S(w)$ when the number of solution points is greater than 2. Therefore, we obtain the order of accuracy of a scheme by the following procedure. First, set w to be, say, $\pi/4$. We can estimate the error

$$E(w) = S(w) + iw .$$

Next, by halving the wave number w (which is equivalent to doubling the number of mesh points), the error corresponding to $w/2$ ($=\pi/8$) is

$$E(w/2) = S(w/2) + iw/2 .$$

Since $O((w/2)^{m+1}) \approx (1/2)^{m+1} O(w^{m+1})$, for a scheme to be m-th order accurate, the following condition must hold: $E(w)/E(w/2) \approx 2^{m+1}$. That is, the order of accuracy is given by

$$m \approx \left[\mathrm{Log}\left(\frac{E(w)}{E(w/2)} \right) \Big/ \mathrm{Log}(2) \right] - 1. \tag{42}$$

Similarly, the order of accuracy for the diffusion case is

$$m \approx \left[\mathrm{Log}\left(\frac{E(w)}{E(w/2)} \right) \Big/ \mathrm{Log}(2) \right] - 2. \tag{43}$$

Here, the constant 2 is due to the fact that the principal eigenvalue 'approximates' $h^2 \partial^2 / \partial x^2$.

We conclude this section by the following observation. The eigenvalues of $\boldsymbol{S}$, and thus, the accuracy and stability of CPR schemes by Fourier analysis, are independent of the solution points chosen.

The proof boils down to the fact that a different set of solution points corresponds to a change of basis. Indeed, let $\tilde{\xi}_l$, $l = 1, ..., K$ be another set of solution points, and $\tilde{u}_l$, the interpolated value at $\tilde{\xi}_l$, i.e., $\tilde{u}_l = \sum_{k=1}^{K} u_k \phi_k(\tilde{\xi}_l)$. For $1 \le k, l \le K$, set $m_{lk} = \phi_k(\tilde{\xi}_l)$ and let the $K \times K$ matrix $\boldsymbol{M} = \{m_{lk}\}$. In addition, set $\boldsymbol{u} = \{u_k\}_{k=1}^{K}$ and $\tilde{\boldsymbol{u}} = \{\tilde{u}_l\}_{l=1}^{K}$. Then, $\tilde{\boldsymbol{u}} = \boldsymbol{Mu}$. By (41), $d\boldsymbol{u}/dt = \boldsymbol{S}\,\boldsymbol{u}$. It follows that

$$d(\boldsymbol{M}\,\boldsymbol{u})/dt = \boldsymbol{MSM}^{-1}\boldsymbol{M}\,\boldsymbol{u} = (\boldsymbol{MSM}^{-1})(\boldsymbol{M}\,\boldsymbol{u}).$$

Next, $d\tilde{\boldsymbol{u}}/dt = \tilde{\boldsymbol{S}}\,\tilde{\boldsymbol{u}} = \tilde{\boldsymbol{S}}\,\boldsymbol{Mu}$. These two expressions imply $\tilde{\boldsymbol{S}} = \boldsymbol{MSM}^{-1}$.

8. Stability and Accuracy of CPR Schemes

Note that Fourier analyses of schemes are independent of the solution points chosen: whether the Gauss, Radau, Lobatto, or equidistant points are selected, the results are the same. These results vary only with the choice of correction functions. Again, for brevity, we identify a scheme by its correction function.

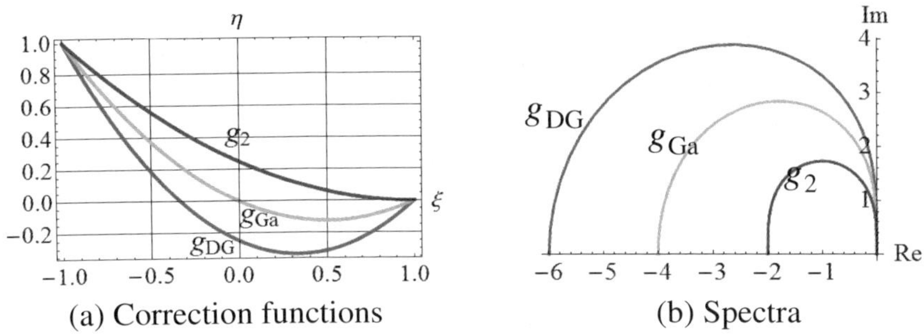

Fig. 6. (a) Correction functions and (b) Spectra for $K = 2$.

The advection case.

The case $K = 1$. There is only one correction function, namely $g = (1-\xi)/2$, and the CPR scheme reduces to the first order upwind scheme $du_j/dt = u_{j-1} - u_j$. The eigenvalue is

$$S(w) = e^{-iw} - 1 = -iw - w^2/2 + O(w^3).$$

As w varies on $[0, 2\pi]$, $S(w)$ varies on a circle of radius 1 centered at $-1+0i$ on the complex plane.

The case $K = 2$. The three correction functions are (see Fig. 6(a))

$$g_{DG} = \frac{3\xi^2}{4} - \frac{\xi}{2} - \frac{1}{4}, \quad g_{Ga} = \frac{\xi^2}{2} - \frac{\xi}{2}, \quad \text{and} \quad g_2 = \frac{\xi^2}{4} - \frac{\xi}{2} + \frac{1}{4}.$$

Figure 6(b) shows the spectra (collection of all eigenvalues) of the corresponding semidiscrete schemes. Due to symmetry, only the top half of the complex plane is shown. Note that each spectrum (Sp) lies to the left of the complex plane; therefore, all three schemes are stable. If the RK2 time stepping is employed, then the CFL limits are: 1/3 for DG; $1/2$ for g_{Ga}, and 1 for $g_2 = g_{\text{Lump,Lo}}$.

Table 1 tabulates the orders of accuracy and errors of the semidiscrete schemes for $K = 2$. Note that all errors have negative real parts—a fact consistent with the stability of the three schemes.

The case $K = 3$. Figure 7(a) shows the three correction functions g_{DG}, g_{Ga}, and g_2. Figure 7(b) shows the spectra. Again, all three schemes are stable. If the RK3 method is employed for time stepping, then the CFL limits are approximately .21, .32, and .45, respectively.

Table 1. Orders of accuracy and errors of schemes for $K = 2$.

Scheme	Ord. acc.	Coarse mesh error, $w = \pi/8$	Fine mesh error, $w = \pi/16$
DG	3	$-3.2\times10^{-4} - 3.3\times10^{-5}i$	$-2.1\times10^{-5} - 1.1\times10^{-6}i$
g_{Ga}	2	$-7.1\times10^{-4} + 2.4\times10^{-3}i$	$-4.6\times10^{-5} + 3.1\times10^{-4}i$
g_2	2	$-2.5\times10^{-3} + 9.\times10^{-3}i$	$-7.1\times10^{-4} + 2.4\times10^{-3}i$

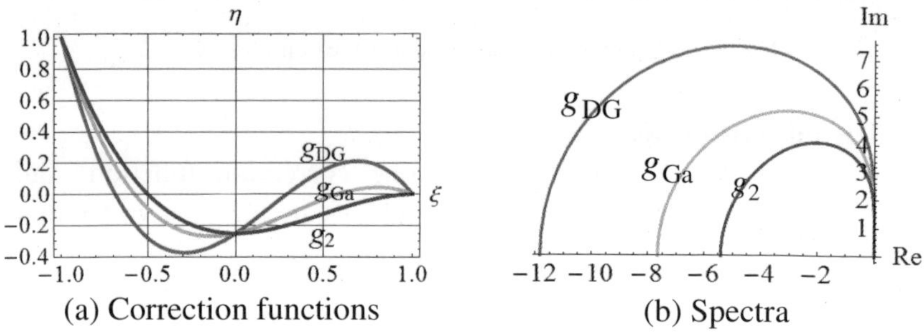

Fig. 7. (a) Correction functions and (b) Spectra for $K = 3$.

Table 2. Orders of accuracy and errors of schemes for $K = 3$.

Scheme	Ord. acc.	Coarse mesh error, $w = \pi/8$	Fine mesh error, $w = \pi/16$
DG	5	$-5.\times10^{-7} - 3.4\times10^{-8}i$	$-7.9\times10^{-9} - 2.7\times10^{-10}i$
g_{Ga}	4	$-1.4\times10^{-6} + 8.5\times10^{-6}i$	$-2.2\times10^{-8} + 2.7\times10^{-7}i$
g_2	4	$-3.2\times10^{-6} + 1.9\times10^{-5}i$	$-5.\times10^{-8} + 6.\times10^{-7}i$

Table 2 tabulates the orders of accuracy and errors of the semi-discrete schemes for $K = 3$. Note that again, all errors have negative real parts.

The case $K = 4$. Figure 8(a) shows the three correction functions g_{DG}, g_{Ga}, and g_2.

Figure 8(b) shows the spectra. Note that all three schemes are stable. The spectra intersect the real axis at $x_{DG} = -19.2$, $x_{Ga} = -12.3$, and $x_2 = -9.6$. If the RK4 method is employed for time stepping, then the CFL limits are approximately .145, .227, and .289, respectively.

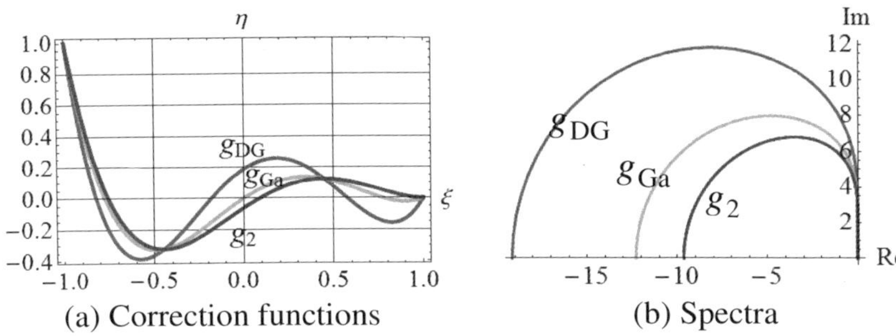

Fig. 8. (a) Correction functions and (b) Spectra for $K = 4$.

Table 3 tabulates the orders of accuracy and errors of these semi-discrete schemes for $K = 4$. Here, the coarse mesh error corresponds to $w = \pi/4$, and fine mesh error, $w = \pi/8$.

The diffusion case. Due to space limitation, only the results for $K = 4$ with the correction function g_{DG} for the BR2 and CDG schemes (CDG is identical to LDG for the 1D case) are shown. Additional results can be found in (Huynh 2009). For this case, each scheme has four eigenvalues, which are real and negative. The plots of these eigenvalues as functions of w are shown in Fig. 9. Note that the minimum eigenvalues are approximately -170 for BR2 and -439 for LDG. As a result, if an explicit time stepping method such as the standard Runge-Kutta scheme is employed, the time step for BR2 is roughly 2.6 times that of LDG for this case.

Table 4 tabulates the orders of accuracy and errors of the two diffusion schemes for $K = 4$. Here, the coarse mesh error corresponds to $w = \pi/8$, and fine mesh error, $w = \pi/16$; the orders of accuracy given by (43) are respectively 6 and 8.

Table 3. Orders of accuracy and errors of schemes for $K = 4$.

Scheme	Ord. acc.	Coarse mesh error, $w = \pi/4$	Fine mesh error, $w = \pi/8$
DG	7	$-1.\times10^{-7} - 1.\times10^{-8} i$	$-4.\times10^{-10} - 2.\times10^{-11} i$
g_{Ga}	6	$-3.1\times10^{-7} + 1.3\times10^{-6} i$	$-1.2\times10^{-9} + 1.1\times10^{-8} i$
g_2	6	$-5.4\times10^{-7} + 2.3\times10^{-6} i$	$-2.2\times10^{-9} + 1.9\times10^{-8} i$

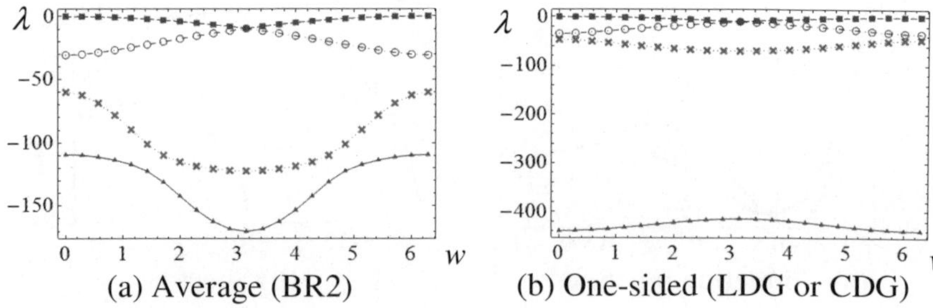

Fig. 9. Eigenvalues as functions of wave numbers for schemes for diffusion; here, $K = 4$; (a) BR2 and (b) LDG scheme.

Table 4. Orders of accuracy and errors of diffusion schemes for $K = 4$.

Scheme	Ord. acc.	Coarse mesh error, $w = \pi/8$	Fine mesh error, $w = \pi/16$
BR2	6	-5.06×10^{-9}	-2.14×10^{-11}
LDG	8	1.96×10^{-12}	1.92×10^{-15}

Remark. Our final remark in this section is that as shown above, via Fourier analysis, all CPR schemes discussed are super-accurate for $K \geq 3$. In practice, since the solution is approximated by a polynomial of degree $K-1$, the degree of accuracy is no higher than K. To observe super-accuracy, besides a uniform mesh and a linear equation, we may need to compare special quantities such as the average of the solution on the cell.

9. Conclusions and Discussion

In summary, a new approach to high-order accuracy for the numerical solution of conservation laws was presented. The approach employs the differential form of the equation and accounts for the jumps in flux values at the cell boundaries by a correction procedure derived from reconstruction (CPR). It results in numerous new schemes and also unifies several existing methods. To determine the accuracy and stability of the CPR schemes, Fourier analysis was carried out. Tradeoffs between accuracy and time-step sizes were discussed.

The correction procedure (using g') has been extended to triangular and hybrid meshes (see chapter 15 by Wang, Gao, and Haga). It turns out that the concept of reconstruction (i.e., g) can be extended to a triangular mesh as well. The author hopes to present these results in the near future.

Acknowledgments

This work was supported by the Fundamental Aeronautics Program of NASA.

References

S. Adjerid, K.D. Devine, J.E. Flaherty and L. Krivodonova (2002). A posteriori error estimation for discontinuous Galerkin solutions of hyperbolic problems, Computer Methods in Applied Mechanics and Engineering, 191, pp. 1097–1112.

F. Bassi and S. Rebay (1997). A high-order accurate discontinuous finite element method for the numerical solution of the compressible Navier-Stokes equations, J. Comput. Phys., 131, pp. 267-279.

F. Bassi and S. Rebay (2000). A high order discontinuous Galerkin method for compressible turbulent flows, in "*Discontinuous Galerkin methods*", eds. B. Cockburn, G. Karniadakis, and C.-W. Shu (Springer), pp. 77–88.

B. Cockburn, G. Karniadakis, and C.-W. Shu, Eds., (2000). *Discontinuous Galerkin methods: Theory, Computation, and Application* (Springer).

B. Cockburn and C.-W. Shu (1998). The local discontinuous Galerkin methods for time-dependent convection diffusion systems, SIAM J. Numer. Anal., 35, 2440–2463.

H. Gao and Z.J. Wang (2009). A High-Order Lifting Collocation Penalty Formulation for the Navier-Stokes Equations on 2-D Mixed Grids, AIAA-2009-3784.

T. Haga, H. Gao, Z.J. Wang (2010). A High-Order Unifying Discontinuous Formulation for 3-D Mixed Grids, AIAA Paper 2010-540.

J.S. Hesthaven and Tim Warburton (2008). *Nodal Discontinuous Galerkin Methods* (Springer).

F.B. Hildebrand (1987). *Introduction to Numerical Analysis* (Dover).

T.J.R. Hughes (1987). Recent progress in the development and understanding of SUPG methods with special reference to the compressible Euler and Navier-Stokes equations. Int. J. Numer. Methods Fluids 7:1261–75.

T.J.R. Hughes (2000). *The finite element method* (Dover).

H.T. Huynh (2007). A flux reconstruction approach to high-order schemes including discontinuous Galerkin methods, AIAA Paper 2007-4079.

H.T. Huynh (2009). A Reconstruction Approach to High-Order Schemes Including Discontinuous Galerkin for Diffusion, AIAA Paper 2009-403.

A. Jameson (2010). A proof of the stability of the spectral difference method for all orders of accuracy, J. Sci. Comput., 45(1–3), 348–358.

D.A. Kopriva and J.H. Kolias (1996). A conservative staggered-grid Chebyshev multidomain method for compressible flows, J. Comput. Phys. 125, 244.

Y. Liu, M. Vinokur, and Z.J. Wang (2006), Discontinuous Spectral Difference Method for Conservation Laws on Unstructured Grids, J. Comput. Phys., 216, 780–801.

J. Peraire and P.-O. Persson (2008). The compact discontinuous Galerkin (CDG) method for elliptic problems, SIAM J. Sci. Comput. 30, No. 4, pp. 1806–1824.

W.H. Reed and T.R. Hill (1973), Triangular mesh methods for the neutron transport equation, Los Alamos Scientific Laboratory Report, LA-UR-73-479.

P.L. Roe (1986). Characteristic-based schemes for the Euler equations, Ann. Rev. Fluid Mech., 18, pp. 337–365

P.E. Vincent, P. Castonguay, and A. Jameson (2010). A New Class of High-Order Energy Stable Flux Reconstruction Schemes, J. Sci. Comput., (to appear).

Z.J. Wang and H. Gao (2009). A Unifying Lifting Collocation Penalty formulation including the discontinuous Galerkin, spectral volume/difference methods for conservation laws on mixed grids, J. Comput. Phys., 228, No. 2, pp. 8161–8186.

Z.J. Wang, L. Zhang and Y. Liu (2004). Spectral (finite) volume method for conservation laws on unstructured grids IV: extension to two-dimensional Euler equations, J. Comput. Phys., 194, No. 2, pp. 716–741.

CHAPTER 15

A UNIFYING DISCONTINUOUS FORMULATION FOR HYBRID MESHES

Z. J. Wang*, H. Gao† and T. Haga‡

Department of Aerospace Engineering and CFD Center,
Iowa State University, Ames, Iowa 50011, USA
**zjw@iastate.edu*
†hgao@iastate.edu
‡thaga@iastate.edu

This chapter describes a differential discontinuous formulation for conservation laws named the Correction Procedure via Reconstruction (CPR) on hybrid meshes. CPR is inspired by several other discontinuous methods such as the discontinuous Galerkin, staggered grid multi-domain, spectral volume and spectral difference methods. In fact, all of them can be unified under the CPR framework, which is relatively simple to implement especially for high-order elements. The extension to viscous flows and to 3D elements is also described. Several benchmark test cases including an accuracy study are presented to demonstrate its capability. Several remaining challenges in adaptive high-order methods are outlined to conclude the chapter.

1. Introduction

The history of discontinuous high-order methods can be traced to the well-known Godunov[15] finite volume method. In fact, all high-order discontinuous methods gracefully reduce to the Godunov method at the lowest order of accuracy, i.e., the first order. An excellent review of these methods has been given in another chapter[12] of this book and thus will not be repeated here. Two other reviews of high-order methods are given in Refs. 13 and 41. In this chapter, we give a brief review of numerical methods which motivated the development of the present work. The lifting collocation penalty (LCP) formulation[42] is directly inspired by the flux reconstruction (FR) method,[20,21] and can be viewed as an extension of the original FR method to simplex elements. Instead of directly reconstructing the flux function, a "correction field" due to interface flux jumps is computed in LCP. Because these two formulations are so tightly related, they have been renamed the Correction Procedure via Reconstruction or CPR (FR+LCP = CPR). From here on, we will use the name CPR to refer to both FR and LCP methods.

The CPR method was developed to improve the efficiency or stability of several well-known high-order methods, including the discontinuous Galerkin (DG)[3,4,7–9,14,25,33], staggered grid multi-domain (SG),[26] spectral volume (SV)[28,40,43] and spectral difference (SD)[29,30,37] methods. As a matter of fact, it unified all these methods into a simple nodal or collocation-type differential formulation. In 1D or multiple dimensions with a tensor-product basis, there is a one to one connection between different formulations and special polynomials. These connections are described in another chapter of this book.[22] This chapter focuses on the development of CPR for simplex and hybrid meshes. As mentioned earlier, all these high-order methods, similar to second and higher order finite volume methods,[2,10,31,38] reduce to the Godunov method at the lowest order.

In the CPR method, the degrees-of-freedom (DOFs) are the state variables at a pre-defined nodal set named solution points (SPs), where the differential form of the governing equation is solved. As a result, explicit surface and volume integrals are avoided. This formulation has the following properties. The framework is easy to understand, and efficient to implement especially for high-order curved elements. The CPR formulation is among the most efficient discontinuous methods in term of the number of operations.

This chapter is organized as follows. The basic CPR formulation is presented in Sec. 2. Section 3 describes the discretization of diffusion/viscous terms. Section 4 is devoted to the 3D implementation. Numerical tests are shown in Sec. 5 for various benchmark and demonstration problems. Conclusions and possible future research directions are given in Sec. 6.

2. Framework of the CPR Formulation

2.1. *Basic idea*

The CPR formulation can be derived from a weighted residual method by transforming the integral formulation into a differential one. First, a hyperbolic conservation law can be written as

$$\frac{\partial Q}{\partial t} + \vec{\nabla} \cdot \vec{F}(Q) = 0, \tag{1}$$

with proper initial and boundary conditions, where Q is the state vector, and $\vec{F} = (F, G)$ is the flux vector. Assume that the computational domain Ω is discretized into N non-overlapping triangular elements $\{V_i\}_{i=1}^N$. Let W be an arbitrary weighting function or test function. The weighted residual formulation of Eq. (1) on element V_i can be expressed as

$$\begin{aligned} &\int_{V_i} \left(\frac{\partial Q}{\partial t} + \vec{\nabla} \cdot \vec{F}(Q) \right) W dV \\ &\quad = \int_{V_i} \frac{\partial Q}{\partial t} W dV + \int_{\partial V_i} W \vec{F}(Q) \cdot \vec{n} dS - \int_{V_i} \vec{\nabla} W \cdot \vec{F}(Q) dV = 0. \end{aligned} \tag{2}$$

Let Q_i be an approximate solution to the analytical solution Q on V_i. On each element, the solution belongs to the space of polynomials of degree k or less, i.e. $Q_i \in P^k(V_i)$, (or P^k if there is no confusion) with no continuity requirement across element interfaces. Let the dimension of P^k be $K = (k+1)(k+2)/2$. In addition, the numerical solution Q_i, for the moment, is required to satisfy Eq. (2)

$$\int_{V_i} \frac{\partial Q_i}{\partial t} W dV + \int_{\partial V_i} W \vec{F}(Q_i) \cdot \vec{n} dS - \int_{V_i} \vec{\nabla} W \cdot \vec{F}(Q_i) dV = 0. \tag{3}$$

Obviously the surface integral is not properly defined because the numerical solution is discontinuous across element interfaces. Following the idea used in the Godunov method, the normal flux term in Eq. (3) is replaced with a common Riemann flux, e.g., in Refs. 23,27,34 and 35

$$F^n(Q_i) \equiv \vec{F}(Q_i) \cdot \vec{n} \approx F^n_{com}(Q_i, Q_{i+}, \vec{n}), \tag{4}$$

where Q_{i+} denotes the solution outside the current element V_i. Instead of Eq. (3), the approximate solution is required to satisfy

$$\int_{V_i} \frac{\partial Q_i}{\partial t} W dV + \int_{\partial V_i} W F^n_{com} dS - \int_{V_i} \vec{\nabla} W \cdot \vec{F}(Q_i) dV = 0. \tag{5}$$

Applying integration by parts again to the last term of the above LHS, we obtain

$$\int_{V_i} \frac{\partial Q_i}{\partial t} W dV + \int_{V_i} W \vec{\nabla} \cdot \vec{F}(Q_i) dV + \int_{\partial V_i} W \left[F^n_{com} - F^n(Q_i) \right] dS = 0. \tag{6}$$

Here, the test space has the same dimension as the solution space, and is chosen in a manner to guarantee the existence and uniqueness of the numerical solution.

Note that the quantity $\vec{\nabla} \cdot \vec{F}(Q_i)$ involves no influence from the data in the neighboring cells. The influence of these data is represented by the above boundary integral, which is also called a "penalty term", penalizing the normal flux differences.

The next step is critical in the elimination of the test function. The boundary integral above is cast as a volume integral via the introduction of a "correction field" on V_i, $\delta_i \in P^k(V_i)$,

$$\int_{V_i} W\delta_i dV = \int_{\partial V_i} W[F^n]dS, \tag{7}$$

where $[F^n] = F^n_{com} - F^n(Q_i)$ is the normal flux difference. The above equation is sometimes referred to as the "lifting operator", which has the normal flux differences on the boundary as input and a member of $P^k(V_i)$ as output. Substituting Eq. (7) into Eq. (6), we obtain

$$\int_{V_i} \left[\frac{\partial Q_i}{\partial t} + \vec{\nabla} \cdot \vec{F}(Q_i) + \delta_i\right] WdV = 0. \tag{8}$$

If the flux vector is a linear function of the state variable, then $\vec{\nabla} \cdot \vec{F}(Q_i) \in P^k$. In this case, the terms inside the square bracket are all elements of P^k. Because the test space is selected to ensure a unique solution, Eq. (8) is equivalent to

$$\frac{\partial Q_i}{\partial t} + \vec{\nabla} \cdot \vec{F}(Q_i) + \delta_i = 0. \tag{9}$$

For nonlinear conservation laws, $\vec{\nabla} \cdot \vec{F}(Q_i)$ is usually not an element of P^k. As a result, Eq. (8) cannot be reduced to Eq. (9). In this case, the most obviously choice is to project $\vec{\nabla} \cdot \vec{F}(Q_i)$ into P^k. Denote $\Pi(\vec{\nabla} \cdot \vec{F}(Q_i))$ a projection of $\vec{\nabla} \cdot \vec{F}(Q_i)$ to P^k. One choice is

$$\int_{V_i} \Pi\left(\vec{\nabla} \cdot \vec{F}(Q_i)\right) WdV = \int_{V_i} \vec{\nabla} \cdot \vec{F}(Q_i)WdV. \tag{10}$$

Then Eq. (8) reduces to

$$\frac{\partial Q_i}{\partial t} + \Pi\left(\vec{\nabla} \cdot \vec{F}(Q_i)\right) + \delta_i = 0. \tag{11}$$

With the introduction of the correction field δ_i, and a projection of $\vec{\nabla} \cdot \vec{F}(Q_i)$ for nonlinear conservation laws, we have reduced the weighted residual formulation to a differential formulation, which involves no integrals. Note that for δ_i defined by Eq. (7), if $W \in P^k$, Eq. (11) is equivalent to the DG formulation, at least for linear conservation laws; if W belongs to another space, the resulting δ_i is different. We obtain a formulation corresponding to a different method such as the SV method.

Next, let the DOFs be the solutions at a set of solution points (SPs) $\{\vec{r}_{ij}\}$ (j varies from 1 to K), as shown in Fig. 1. Then Eq. (11) holds true at the SPs, i.e.,

$$\frac{\partial Q_{i,j}}{\partial t} + \Pi_j\left(\vec{\nabla}\cdot\vec{F}(Q_i)\right) + \delta_{i,j} = 0, \tag{12}$$

where $\Pi_j\left(\vec{\nabla}\cdot\vec{F}(Q_i)\right)$ denotes the values of $\Pi\left(\vec{\nabla}\cdot\vec{F}(Q_i)\right)$ at SP j. The efficiency of the CPR approach hinges on how the correction field δ_i and the projection $\Pi\left(\vec{\nabla}\cdot\vec{F}(Q_i)\right)$ are computed. To compute δ_i, we define $k+1$ points named flux points (FPs) along each interface, where the normal flux differences $[F^n]$ are computed, as shown in Fig. 1. We approximate (for nonlinear conservation laws) the normal flux difference $[F^n]$ with a degree k interpolation polynomial along each interface,

$$[F^n]_f \approx \mathbf{I}_k[F^n]_f \equiv \sum_l [F^n]_{f,l} L_l^{FP}, \tag{13}$$

where f is an face (or edge in 2D) index, and l is the FP index, and L_l^{FP} is the Lagrange interpolation polynomial based on the FPs in a local interface coordinate. For linear triangles with straight edges, once the solution points and flux points are chosen, the correction at the SPs can be written as

$$\delta_{i,j} = \frac{1}{|V_i|}\sum_{f\in\partial V_i}\sum_l \alpha_{j,f,l}[F^n]_{f,l}S_f, \tag{14}$$

where $\alpha_{j,f,l}$ are lifting constants independent of the solution, S_f is the face area, $|V_i|$ is the volume of V_i. Note that the correction for each solution point, namely $\delta_{i,j}$, is a linear combination of all the normal flux differences

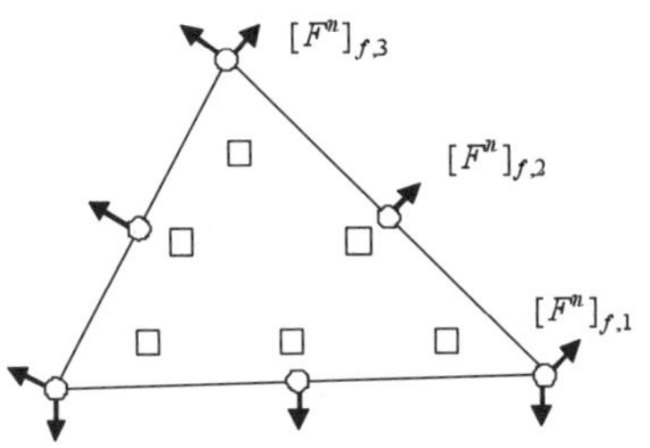

Fig. 1. Solution points (squares) and flux points (circles) for a triangular element of $k = 2$.

on all the faces of the cell. Conversely, a normal flux difference at a flux point on a face, say (f, l) results in a correction at a solution point j of an amount $\alpha_{j,f,l}[F^n]_{f,l}S_f/|V_i|$.

Next, we focus on how to compute $\Pi_j\left(\vec{\nabla}\cdot\vec{F}(Q_i)\right)$ efficiently. A brute-force implementation based on Eq. (10) requires high-order integral quadratures, and is expensive. Two more efficient approaches are developed in Ref. 42, and reviewed here for the sake of completeness.

2.1.1. *Lagrange polynomial (LP) approach*

Based on the solution at a SP, the flux vector at each SP can be computed. Then a degree k Lagrange interpolation polynomial for the flux vector is used to approximate the (nonlinear) flux vector

$$\vec{F}(Q_i) \approx \mathbf{I}_k\left(\vec{F}(Q_i)\right) \equiv \sum_j L_j^{SP}(\vec{r})\vec{F}(Q_{i,j}), \tag{15}$$

where $L_j^{SP}(\vec{r})$ is the Lagrange polynomial based on the solution points $\{\vec{r}_{i,j}\}$. After that, the projection is computed using

$$\Pi\left(\vec{\nabla}\cdot\vec{F}(Q_i)\right) \approx \vec{\nabla}\cdot\mathbf{I}_k\left(\vec{F}(Q_i)\right) = \sum_j \vec{\nabla}L_j^{SP}\cdot\vec{F}(Q_{i,j}). \tag{16}$$

In this case, $\Pi\left(\vec{\nabla}\cdot\vec{F}(Q_i)\right)$ is a degree $k-1$ polynomial, which also belongs to P^k. Numerical experiments indicate that there is a slight loss of accuracy with the LP approach, but it is fully conservative.[42]

2.1.2. *Chain rule (CR) approach*

We recognize that the divergence of the flux vector can be computed analytically given the approximate solution using the chain rule, i.e.,

$$\begin{aligned}\vec{\nabla}\cdot\vec{F}(Q_{i,j}) &= \frac{\partial F(Q_{i,j})}{\partial x} + \frac{\partial G(Q_{i.j})}{\partial y}\\ &= \frac{\partial F(Q_{i,j})}{\partial Q}\frac{\partial Q_{i,j}}{\partial x} + \frac{\partial G(Q_{i.j})}{\partial Q}\frac{\partial Q_{i,j}}{\partial y}\\ &= \frac{\partial \vec{F}(Q_{i,j})}{\partial Q}\cdot\nabla Q_{i,j},\end{aligned} \tag{17}$$

where $\frac{\partial \vec{F}}{\partial Q}$ is composed of the flux Jacobian matrices, which can be computed analytically. Then the projection is approximated by the Lagrange interpolation polynomial of the flux vector divergence at the solution points, i.e.,

$$\Pi\left(\vec{\nabla}\cdot\vec{F}(Q_i)\right) \approx \sum_j L_j^{SP}(\vec{r})\vec{\nabla}\cdot\vec{F}(Q_{i,j}). \tag{18}$$

Numerical experiments indicate that the CR approach is much more accurate than the LP approach, at the expense of full conservation.[42]

Substituting Eq. (14) into Eq. (12) we obtain the following CPR formulation

$$\frac{\partial Q_{i,j}}{\partial t} + \Pi_j\left(\vec{\nabla}\cdot\vec{F}(Q_i)\right) + \frac{1}{|V_i|}\sum_{f\in\partial V_i}\sum_l \alpha_{j,f,l}[F^n]_{f,l}S_f. \tag{19}$$

It can be easily shown that the location of SPs does not affect the numerical scheme for linear conservation laws.[37] For efficiency, therefore, the solution points and flux points are always chosen to include corners of the cell. In addition, the solution points are chosen to coincide with the flux points along cell faces, as shown in Fig. 2(a) to avoid any solution reconstruction. Furthermore, in computations with hybrid meshes, the flux points are always the same for different cell types for ease of interface treatment, as shown in Fig. 2(b). For the 2D cases presented here, the Legendre-Lobatto points along the edges are used as the flux points and also (part of) the solution points for both triangular and quadrilateral cells. Due to the special choice of DOFs, the reconstruction cost in CPR is completely avoided.

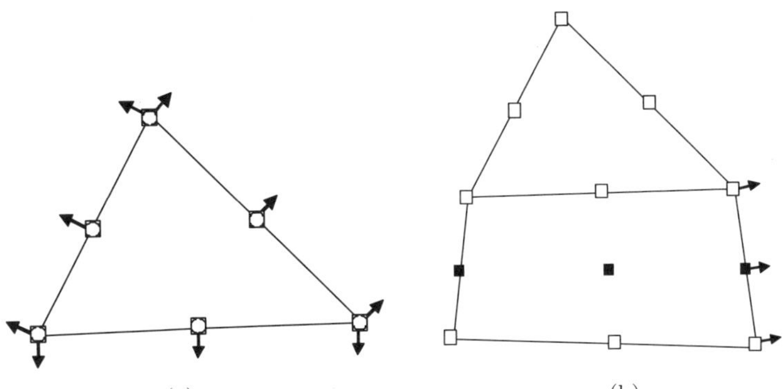

(a) (b)

Fig. 2. Efficient arrangement of solution (squares) and flux points (circles) for $k = 2$.

2.2. *Connection between the CPR, DG, SV and SD methods*

Let's first express the solution and the correction in terms of the values at the SPs, i.e.,

$$Q_i = \sum_j L_j^{SP} Q_{i,j}, \tag{20}$$

$$\delta_i = \sum_j L_j^{SP} \delta_{i,j}. \tag{21}$$

In the DG method, the weighting function W is set to be one of the Lagrange polynomials. Substituting W into Eq. (7), we obtain the following equations

$$\int_{V_i} L_k^{SP} \sum_j L_j^{SP} \delta_{i,j} dV = \sum_{f \in \partial V_i} \int_f L_k^{SP} \sum_l L_l^{FP} [F^n]_{f,l} dS, \; k = 1, ..., K. \tag{22}$$

The unknowns in Eq. (22) $\delta_{i,j}$ can be easily solved in terms of the normal flux jumps at the flux points $[F^n]_{f,l}$, and the coefficients $\alpha_{j,f,l}$ be determined, which are constant for any straight-sided triangles. In the case of $k = 1$, the coefficients for the first solution point are $\{2.5, 0.5, -1.5, -1.5, 0.5, 2.5\}$. Therefore, the formula for the correction is

$$\begin{aligned}\delta_{i,1} = &\frac{1}{|V_i|}\{(2.5[F^n]_{1,1} + 0.5[F^n]_{1,2})S_1 \\ &+ (-1.5[F^n]_{2,1} - 1.5[F^n]_{2,2})S_2 + (0.5[F^n]_{3,1} + 2.5[F^n]_{3,2})S_3\}.\end{aligned} \tag{23}$$

Although all the flux points coincide with the solution points, as shown in Fig. 2, it is necessary to distinguish flux points according to which face they are located on because each face has a different normal direction.

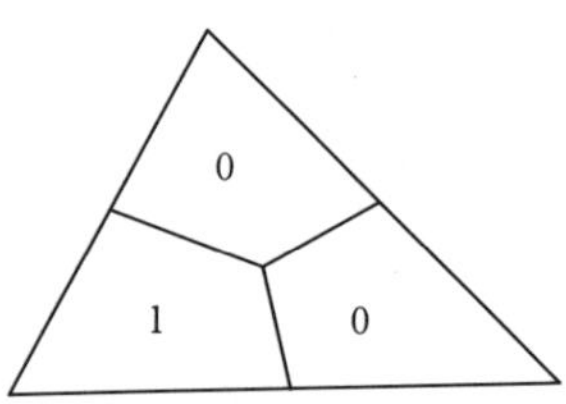

Fig. 3. One of the weighting functions for the spectral volume method, $k = 1$.

In addition, the flux points on each face are numbered independently for easy identification and implementation.

In the SV method, the weighting function is 1 within a partition of the element, and 0 elsewhere, for example, as shown in Fig. 3 in the case of $k = 1$. Repeating the same with all the partitions, we obtain again K equations for K unknowns $\delta_{i,j}$, which can be uniquely solved. The coefficients for the first solution point of the second order SV scheme are $\{2, 0.2, -0.7, -0.7, 0.2, 2\}$, corresponding to the following formula

$$\begin{aligned}\delta_{i,1} = \frac{1}{|V_i|}\{&(2[F^n]_{1,1} + 0.2[F^n]_{1,2})S_1 \\ &+ (-0.7[F^n]_{2,1} - 0.7[F^n]_{2,2})S_2 + (0.2[F^n]_{3,1} + 2[F^n]_{3,2})S_3\}.\end{aligned} \tag{24}$$

In the SD method, the correction field is computed based on the direct differential of a reconstructed flux vector. The derivation is a little more involved than those for the DG and SV methods. We found that only on an equilateral triangular grid can the SD method degenerate into the CPR formulation. This is not surprising because the SD method is generally not only dependent on the normal fluxes at element interfaces, but also on the tangential fluxes. The $k = 1$ linear case has the following coefficients $\alpha_{j,f,l}$ at the first solution point $\{2, 0, -0.5, -0.5, 0, 2\}$, resulting in the following formula

$$\delta_{i,1} = \frac{1}{|V_i|}\{2[F^n]_{1,1}S_1 + (-0.5[F^n]_{2,1} - 0.5[F^n]_{2,2})S_2 + 2[F^n]_{3,2}S_3\}. \tag{25}$$

Note that the coefficients are quite different for the DG, SV and SD methods. These schemes have been numerically verified to be 2nd order accurate.

2.3. *Extension to high order elements and mixed grids*

For the sake of simplicity, we have limited our discussions to linear triangles. However, many of the descriptions carry over directly to arbitrary elements including high-order elements. In this section, we consider general triangular and quadrilateral elements with possible high-order edges. To achieve an efficient implementation, all elements are transformed from the physical domain (x, y) into a standard element in the computational domain (ξ, η). The standard triangle is

$$\mathbf{T} = \left\{\vec{\xi} = (\xi, \eta) | (\xi, \eta) \geq 0; \xi + \eta \leq 1\right\}, \tag{26}$$

and the standard quadrilateral is

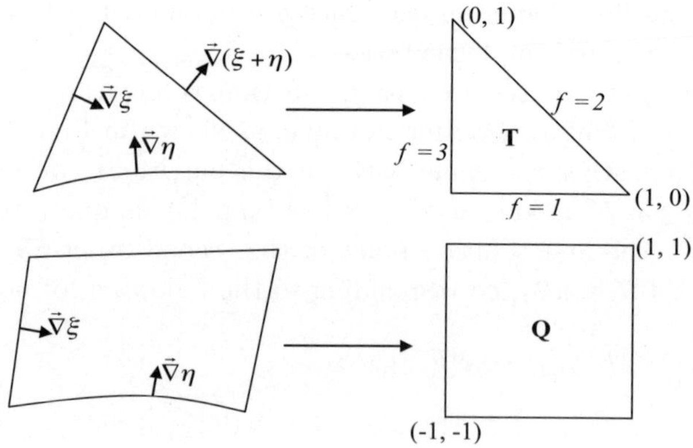

Fig. 4. Transformation of general elements to standard elements.

$$\mathbf{Q} = \left\{ \vec{\xi} = (\xi, \eta) | -1 \le (\xi, \eta) \le 1 \right\}, \tag{27}$$

as shown in Fig. 4. The transformation can be written as

$$\vec{r} = \sum_j M_j(\vec{\xi})\vec{r}_j \tag{28}$$

where $\vec{r}_j$ is the physical coordinates used to define an element, and $M_j(\vec{\xi})$ is the shape function. The Jacobian matrix J takes the following form

$$J = \frac{\partial \vec{r}}{\partial \vec{\xi}} = \begin{bmatrix} x_\xi \; x_\eta \\ y_\xi \; y_\eta \end{bmatrix}. \tag{29}$$

The metrics can be computed according to

$$\xi_x = y_\eta/|J|,\ \xi_y = -x_\eta/|J|,\ \eta_x = -y_\xi/|J|,\ \eta_y = x_\xi/|J|. \tag{30}$$

The transformed equation takes the following form

$$\frac{\partial \tilde{Q}}{\partial t} + \frac{\partial \tilde{F}}{\partial \xi} + \frac{\partial \tilde{G}}{\partial \eta} = 0, \tag{31}$$

where

$$\begin{aligned} \tilde{Q} &= |J|Q \\ \tilde{F} &= |J|(\xi_x F + \xi_y G) \\ \tilde{G} &= |J|(\eta_x F + \eta_y G). \end{aligned} \tag{32}$$

Let $\vec{S}_\xi = |J|\vec{\nabla}\xi$, $\vec{S}_\eta = |J|\vec{\nabla}\eta$, which physically represent the "area vector" of constant ξ and η lines in the physical domain, and obviously

$\tilde{F} = \vec{F} \cdot \vec{S}_\xi$ and $\tilde{G} = \vec{F} \cdot \vec{S}_\eta$. Equation (31) can be cast in the following divergence form

$$\frac{\partial \tilde{Q}}{\partial t} + \vec{\nabla}^\xi \cdot \vec{\tilde{F}} = 0, \tag{33}$$

where $\vec{\tilde{F}} = (\tilde{F}, \tilde{G})$ and $\vec{\nabla}^\xi \cdot$ is the divergence operator in the computational domain. Since the standard element is a linear triangle, the CPR formulation can be directly applied

$$\frac{\partial \tilde{Q}_{i,j}}{\partial t} + \Pi_j \left(\vec{\nabla}^\xi \cdot \vec{\tilde{F}}(\tilde{Q}_i) \right) + \frac{1}{|V_i^\xi|} \sum_{f \in \partial V_i} \sum_l \alpha_{j,f,l} [\tilde{F^n}]^\xi_{f,l} S^\xi_f = 0, \tag{34}$$

where superscript ξ means that the variables or operations are evaluated on the computational domain. For the standard triangle, $|V_i^\xi| = 1/2$. For face 1, $\vec{n}_1^\xi = (0, -1)$, $S_1^\xi = 1$, and

$$\begin{aligned} [\tilde{F^n}]^\xi_{1,l} S^\xi_1 &= -(\tilde{G}_{com} - \tilde{G}(\tilde{Q}_i))_{1,l} \\ &= (F^n_{com} - F^n(Q_i))_{1,l} |\vec{S}_\eta|_{1,l} = [F^n]_{1,l} |\vec{S}_\eta|_{1,l}. \end{aligned} \tag{35}$$

Similar formulas can be obtained for the other 2 faces. Taking into account that

$$\frac{1}{|J|} \vec{\nabla}^\xi \cdot \vec{\tilde{F}} = \vec{\nabla} \cdot \vec{F}, \tag{36}$$

Eq. (34) can be further expressed as

$$\frac{\partial Q_{i,j}}{\partial t} + \Pi_j \left(\vec{\nabla} \cdot \vec{F}(Q_i) \right) + \frac{2}{|J|_{i,j}} \sum_{f \in \partial V_i} \sum_l \alpha_{j,f,l} [F^n]_{f,l} S_{f,l} = 0, \tag{37}$$

where $S_{1,l} = |\vec{S}_\eta|_{1,l}$, $S_{2,l} = |\vec{S}_\xi + \vec{S}_\eta|_{2,l}$, $S_{3,l} = |\vec{S}_\xi|_{3,l}$.

The extension to quadrilateral element is straightforward as all the operations are one-dimensional using a tensor product basis. For 1D conservation laws, Eq. (37) reduces to

$$\frac{\partial Q_{i,j}}{\partial t} + \Pi_j \left(\frac{\partial F(Q_i)}{\partial x} \right) + \frac{1}{h_i} \left(\alpha_{L,j} [F^n]_L + \alpha_{R,j} [F^n]_R \right) = 0, \tag{38}$$

where h_i is the length of element i, which has two interfaces, the left one and right one, with unit face "areas" and unit face "normals" of -1 and 1 respectively, so that $[F^n]_L = -[F]_L$, $[F^n]_R = [F]_R$, $\alpha_{L,j}$ and $\alpha_{R,j}$ are constant lifting coefficients in 1D. Due to symmetry, we have $\alpha_{L,j} = \alpha_{R,k+2-j}$. For the 1D case, details can be found in the chapter by Huynh.[22]

For a quadrilateral element, two indices (j, m) are used to denote the solution point, and $\tilde{Q}_{i;j,m}$ denotes the DOFs. The CPR formulation is then

$$\begin{aligned}\frac{\partial \tilde{Q}_{i;j,m}}{\partial t} &+ \Pi_{j,m}\left(\vec{\nabla}^{\xi} \cdot \vec{\tilde{F}}(\tilde{Q}_i)\right) \\ &- [\tilde{F}_{com}(-1, \eta_{j,m}) - \tilde{F}_i(-1, \eta_{j,m})]\frac{\alpha_{L,j}}{2} \\ &+ [\tilde{F}_{com}(1, \eta_{j,m}) - \tilde{F}_i(1, \eta_{j,m})]\frac{\alpha_{R,j}}{2} \\ &- [\tilde{G}_{com}(\xi_{j,m}, -1) - \tilde{G}_i(\xi_{j,m}, -1)]\frac{\alpha_{L,m}}{2} \\ &+ [\tilde{G}_{com}(\xi_{j,m}, 1) - \tilde{G}_i(\xi_{j,m}, 1)]\frac{\alpha_{R,m}}{2} = 0.\end{aligned} \tag{39}$$

Note that the correction is done in a "one dimensional" manner. In other words, for quadrilateral cells, the operations are actually "one-dimensional", making the method more efficient per DOF than for triangular cells. The flux divergence projection in Eq. (39) can be performed using either the LP or the CR approach.

3. Treatment of Viscous Terms

3.1. *Basic framework*

The discretization of viscous term in the DG method has been studied extensively in the literature.[1,3,5,9,11,18,21,32,39] The extension of the CPR formulation to viscous flows follows existing compact approaches developed in Refs. 5,18,21 and 32. The Navier-Stokes equations can be written as

$$\frac{\partial Q}{\partial t} + \vec{\nabla} \cdot \vec{F}(Q) = \vec{\nabla} \cdot \vec{F}^{\nu}(Q, \vec{\nabla} Q), \tag{40}$$

where $\vec{F}^{\nu}(Q, \vec{\nabla} Q)$ denotes the viscous flux vector.

First, following Ref. 3, we introduce a new variable $\vec{R}$

$$\vec{R} = \vec{\nabla} Q. \tag{41}$$

Let $\vec{R}_i$ be an approximation of $\vec{R}$ on V_i, and $\vec{R}_i \in (P^k, P^k)$. Many studies have found that the obvious choice of $\vec{R}_i = \vec{\nabla} Q_i$ is not appropriate. Instead, the computation of $\vec{R}_i$ needs to involve data from neighboring cells. The CPR formulations of Eq. (40) and Eq. (41) on a linear triangle V_i can be expressed as

$$\frac{\partial Q_{i,j}}{\partial t} + \Pi_j \left(\nabla \cdot \vec{F}(Q_i)\right) - \Pi_j^\nu \left(\nabla \cdot \vec{F}^\nu(Q_i, \vec{R}_i)\right) + \frac{1}{|V_i|} \sum_{f \in \partial V_i} \sum_l \alpha_{j,f,l}([F^n]_{f,l} - [F^{\nu,n}]_{f,l}) S_f = 0, \tag{42}$$

$$\vec{R}_{i,j} = (\vec{\nabla} Q_i)_j + \frac{1}{|V_i|} \sum_{f \in \partial V_i} \sum_l \alpha_{j,f,l} [Q^{com} - Q_i]_{f,l} \vec{n}_f S_f, \tag{43}$$

where Π^ν is the projection operator for the divergence of the viscous flux vector to P^k, and

$$[F^{\nu,n}]_f \equiv \vec{F}^\nu(Q_f^{com}, \vec{\nabla} Q_f^{com}) \cdot \vec{n}_f - \vec{F}^\nu(Q_i, \vec{R}_i)\Big|_f \cdot \vec{n}_f, \tag{44}$$

with Q_f^{com} and $\vec{\nabla} Q_f^{com}$ the common solution and gradient on interface f respectively, and $Q_{i,f,l}$ is the solution within cell i on FP l of face f or the trace of Q_i on f. The computation of $\Pi^\nu \left(\nabla \cdot \vec{F}^\nu(Q_i, \vec{R}_i)\right)$ follows the LP approach. First, the viscous flux vector at each solution point is evaluated using

$$\vec{F}_{i,j}^\nu = \vec{F}^\nu(Q_{i,j}, \vec{R}_{i,j}). \tag{45}$$

After that, a Lagrange polynomial for the viscous flux vector is built with the values at all the solution points, i.e.,

$$\mathbf{I}_k(\vec{F}_i^\nu) = \sum_j \vec{F}_{i,j}^\nu L_j^{SP}. \tag{46}$$

Finally the divergence of this polynomial is used as the projection

$$\Pi_j^\nu \left(\vec{\nabla} \cdot \vec{F}^\nu(Q_i, \vec{R}_i)\right) \approx \vec{\nabla} \cdot \mathbf{I}_k(\vec{F}_i^\nu) = \sum_j \vec{F}_{i,j}^\nu \cdot \vec{\nabla} L_j^{SP}. \tag{47}$$

Various schemes for viscous fluxes differ in how the common solution Q_f^{com} and the common gradient $\vec{\nabla} Q_f^{com}$ are defined. In the following subsections, the BR2,[5] I-continuous,[21] interior penalty[11,18] and CDG[32] schemes are described. It is sometimes cleaner to use a face based notation, in which $Q_f^- \equiv Q_i$ and $Q_f^+ \equiv Q_{i+}$.

3.2. *Bassi-Rebay 2*

The common solution in BR2 is simply the average of the solutions at both sides of the face

$$Q_f^{com} = \frac{Q_f^- + Q_f^+}{2}. \tag{48}$$

The common gradient is computed with

$$\vec{\nabla} Q_f^{com} = \frac{1}{2}(\vec{\nabla} Q_f^- + \vec{r}_f^- + \vec{\nabla} Q_f^+ + \vec{r}_f^+), \tag{49}$$

where $\vec{\nabla} Q_{f,l}^-$ and $\vec{\nabla} Q_{f,l}^+$ are the gradients of the solution at the left and right cells without corrections, i.e.,

$$\vec{\nabla} Q_f^- = \vec{\nabla} Q_i\Big|_f, \ \vec{\nabla} Q_f^+ = \vec{\nabla} Q_{i+}\Big|_f, \tag{50}$$

while $\vec{r}_f^-$ and $\vec{r}_f^+$ are the corrections to the gradients due to the difference between the common solution and the solution at each side of face f. More specifically,

$$\begin{aligned} \vec{r}_{f,l}^- &= \frac{1}{|V^-|} \sum_{m=1}^{N_{fp}} \beta_{l,m} [Q^{com} - Q^-]_{f,m} \vec{n}_f S_f, \\ \vec{r}_{f,l}^+ &= \frac{1}{|V^+|} \sum_{m=1}^{N_{fp}} \beta_{l,m} [Q^{com} - Q^+]_{f,m} (-\vec{n}_f) S_f, \end{aligned} \tag{51}$$

where N_{fp} is the number of flux points on face f (which is $k+1$ in 2D), $\beta_{l.m}$ is the coefficient of correction due to face f. Note that the indices l and m vary on face f and, for our choice of solution points, $\beta_{l.m} = \alpha_{j,f,m}$, where index j is the solution point corresponding to flux point l on face f. For triangular elements, $\beta_{l,m}$ are identical for any face f with a fixed distribution of flux points.

For quadrilateral elements, because a tensor product basis is used, $\vec{r}_{f,l}^-$ and $\vec{r}_{f,l}^+$ are computed in a 1D manner, depending only on the difference between the common solution and the interior solution at the flux point.

3.3. *I-continuous*

The I-continuous approach was proposed by Huynh.[21] Its basic idea is the following: Instead of prescribing a common solution Q^{com} at the interface, Q^{com} is an unknown to be solved by the condition that the corrected derivatives are continuous across the interface f in the normal direction.

The corrected gradients on the left (−) and right (+) side can be expressed as

$$\begin{aligned}\vec{\nabla} Q_{f,l}^{com-} &= \vec{\nabla} Q_{f,l}^{-} + \frac{1}{|V^-|}\sum_{m=1}^{N_{fp}} \beta_{l,m}[Q^{com} - Q^{-}]_{f,m} S_f \vec{n}_f,\\ \vec{\nabla} Q_{f,l}^{com+} &= \vec{\nabla} Q_{f,l}^{+} + \frac{1}{|V^+|}\sum_{m=1}^{N_{fp}} \beta_{l,m}[Q^{com} - Q^{+}]_{f,m} S_f (-\vec{n}_f).\end{aligned} \tag{52}$$

Then we require the gradient to be continuous in the normal direction

$$\vec{\nabla} Q_{f,l}^{com-} \cdot \vec{n}_f = \vec{\nabla} Q_{f,l}^{com+} \cdot \vec{n}_f. \tag{53}$$

Substituting Eq. (52) in Eq. (53), we obtain

$$\begin{aligned}&\sum_{m=1}^{N_{fp}} \left(\frac{\beta_{l,m}}{|V^-|} + \frac{\beta_{l,m}}{|V^+|}\right) Q_{f,m}^{com} S_f\\ &= (\vec{\nabla} Q_{f,l}^{+} - \vec{\nabla} Q_{f,l}^{-}) \cdot \vec{n}_f + \sum_{m=1}^{N_{fp}} \beta_{l,m}\left(\frac{Q_{f,m}^{-}}{|V^-|} + \frac{Q_{f,m}^{+}}{|V^+|}\right) S_f.\end{aligned} \tag{54}$$

Equation (54) represents a linear system, from which $Q_{f,l}^{com}$ can be easily solved. Then, the common viscous flux $\vec{\nabla} Q_{f,l}^{com}$ is obtained by

$$\begin{aligned}\vec{\nabla} Q_{f,l}^{com} \cdot \vec{n}_f &= \vec{\nabla} Q^{com\pm} \cdot \vec{n}_f,\\ \vec{\nabla} Q_{f,l}^{com} \cdot \vec{t}_f &= \frac{\vec{\nabla} Q_{f,l}^{-} \cdot \vec{t}_f + \vec{\nabla} Q_{f,l}^{+} \cdot \vec{t}_f}{2},\end{aligned} \tag{55}$$

where $\vec{t}_f$ is a unit vector in the tangential direction of face f. Note that we need to solve a $k+1$ linear system for each face. The cost of this step is minimal: since the matrices are independent of the solution, they only need to be inverted once at the initialization stage. Therefore, the I-continuous approach can be made almost as efficient as the BR2 approach.

3.4. *Interior penalty*

The interior penalty approach is a simplified version of BR2 for triangular meshes, and is identical to BR2 for quadrilateral meshes with properly chosen coefficients. In BR2 the correction (or penalty) $\vec{r}_{f,l}^{+}$ and $\vec{r}_{f,l}^{-}$ at one face flux point is a linear combination of the solution differences at all flux points on the face. In interior penalty method, the penalty is only dependent on the solution difference at that point, i.e. the penalty is computed in a 1-D manner

$$\begin{aligned}\vec{r}^{-}_{f,l} &= [Q^{com} - Q^{-}]_{f,l}\beta_{l,l}\frac{S_f}{V^{-}}\vec{n}_f,\\ \vec{r}^{+}_{f,l} &= -[Q^{com} - Q^{+}]_{f,l}\beta_{l,l}\frac{S_f}{V^{+}}\vec{n}_f,\end{aligned} \tag{56}$$

where $\beta_{l,l}$ is a constant for any l.

3.5. *Compact discontinuous Galerkin*

The idea of CDG[32] is similar to the local DG approach.[8] The solution from one side of a face is used as the common solution, while the corrected gradient from the other side is used as the common gradient. CDG is compact for arbitrary unstructured meshes, while LDG may not be.

For example, if we use the right (+) side for the common solution and the left (−) side for the common gradient, we obtain,

$$Q^{com}_{f,l} = Q^{+}_{f,l}, \tag{57}$$

$$\vec{\nabla}Q^{com}_{f,l} = \vec{\nabla}Q^{-}_{f,l} + \vec{r}^{-}_{f,l}, \tag{58}$$

where

$$\vec{r}^{-}_{f,l} = \frac{1}{|V^{-}|}\sum_{m=1}^{N_{fp}}\beta_{l,m}[Q^{com} - Q^{-}]_{f,m}S_f\vec{n}_f. \tag{59}$$

Alternatively, we can also use the opposite sides for the common solution and common gradient.

4. Extension to 3D Elements

We focus on two element shapes, i.e., tetrahedron and triangular prism. The use of prismatic cells in addition to tetrahedral cells has the advantages in both accuracy and computational costs in resolving viscous boundary layers near solid walls. Again, all elements are transformed from the physical domain (x, y, z) into the corresponding standard elements in the computational domain (ξ, η, ζ) as shown in Fig. 5. Here we consider the transformations for high-order elements with curved sides (faces and edges). The discretization for the curved elements is conducted in the same way as the straight sided elements by applying the CPR formulation in the standard elements. Based on a set of nodes defining the shape of an element, a

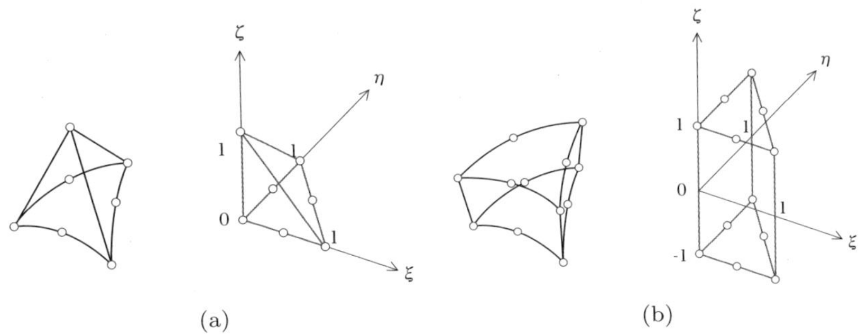

Fig. 5. Transformation of curve boundary tetrahedral and prismatic cells to the standard elements.

set of shape functions can be obtained.[45] The transformed equations in 3D can be obtained in a very similar manner to those in 2D. Let's assume a similar transformation to Eq. (28) exists between the physical domain and the computational domain. Denote J the Jacobian matrix of the transformation, and

$$\vec{S}_\xi = |J|\vec{\nabla}\xi, \; \vec{S}_\eta = |J|\vec{\nabla}\eta, \; \vec{S}_\zeta = |J|\vec{\nabla}\zeta. \tag{60}$$

Let the flux vector in the physical domain be $\vec{F} = (F, G, H)$. The transformed equations take the following form

$$\frac{\partial \tilde{Q}}{\partial t} + \vec{\nabla}^\xi \cdot \vec{\tilde{F}} = 0, \tag{61}$$

where

$$\tilde{Q} = |J|Q, \; \vec{\tilde{F}} = (\tilde{F}, \tilde{G}, \tilde{H}) \equiv (\vec{F}\cdot\vec{S}_\xi, \vec{F}\cdot\vec{S}_\eta, \vec{F}\cdot\vec{S}_\zeta). \tag{62}$$

Note that here we consider the Euler equations as the governing equations for the sake of simplicity. Extending the following discretization to the Navier-Stokes equations is straightforward following the approach in the last section.

4.1. *Discretization on a standard tetrahedron*

On a standard tetrahedron, the CPR formulation can be expressed as

$$\frac{\partial \tilde{Q}_{i,j}}{\partial t} + \Pi_j\left(\nabla^\xi \cdot \vec{\tilde{F}}(\tilde{Q}_i)\right) + \frac{1}{|V_i^\xi|}\sum_{f\in\partial V}\sum_l \alpha_{j,f,l}[\tilde{F}^n]^\xi_{f,l} S_f^\xi = 0. \tag{63}$$

For the standard tetrahedron, $|V_i^\xi| = 1/6$, the areas for the 4 faces are 1/2, 1/2, 1/2 and $\sqrt{3}/2$ respectively. For the face on the plane $\xi = 0$

(denoted as face 1), the outgoing unit normal in the computational domain is $\vec{n}_1^{\xi} = (-1, 0, 0)$

$$
\begin{aligned}
[\tilde{F}^n]^{\xi}_{1,l} S^{\xi}_1 &= -\frac{1}{2}(\tilde{F}_{com} - \tilde{F}(\tilde{Q}_i))_{1,l} \\
&= \frac{1}{2}(F^n_{com} - F^n(Q_i))_{1,l} |\vec{S}_{\xi}|_{1,l} \\
&= \frac{1}{2}[F^n]_{1,l} |\vec{S}_{\xi}|_{1,l} \equiv \frac{1}{2}[F^n]_{1,l} S_{1,l}.
\end{aligned} \tag{64}
$$

A similar expression can be obtained for the other faces, with a properly defined $S_{f,l}$. For the diagonal face,

$$
S_{f,l} = |\vec{S}_{\xi} + \vec{S}_{\eta} + \vec{S}_{\zeta}|_{f,l}. \tag{65}
$$

The final formulation can be written as

$$
\frac{\partial Q_{i,j}}{\partial t} + \Pi_j \left(\nabla \cdot \vec{F}(Q_i) \right) + \frac{3}{|J|_{i,j}} \sum_{f \in \partial V} \sum_{l} \alpha_{j,f,l} [F^n]_{f,l} S_{f,l} = 0. \tag{66}
$$

In 3D, to construct a complete polynomial of degree k, at least $k(k+1)(k+2)/3!$ SPs need to be specified. In order to achieve the most efficient implementation, SPs on edges are chosen to be the Legendre-Gauss Lobatto (LGL) points. For 4th- or higher order schemes, nodes inside the boundary triangle are chosen from Ref. 19. For 5th- or higher order schemes, nodes inside the tetrahedron are chosen from Ref. 44. The nodal set of the 4th-order CPR scheme is shown in Fig. 6(a). Note that the flux difference at a flux point corrects all solution points as shown in Eq. (66).

4.2. *Discretization on the standard prism*

For a standard triangular prism, the solution polynomial can be expressed as a tensor product of 1D and 2D Lagrange polynomials, i.e.,

$$
\tilde{Q}_i(\xi, \eta, \zeta) = \sum_{m} \sum_{j} \tilde{Q}_{i;j,m} L^{SP}_j(\xi, \eta) L^{SP}_m(\zeta), \tag{67}
$$

where $\tilde{Q}_{i;j,m}$ are the state variables at the solution point (j, m), with j the index on the $\xi - \eta$ plane and m the index in ζ direction, $L^{SP}_j(\xi, \eta)$ is a 2D Lagrange polynomial based on the solution points on the base triangle and $L^{SP}_m(\zeta)$ is a 1D Lagrange polynomial based on the solution points in the prism height direction. Figure 6(b) shows the locations of the solution points for $k = 3$. The nodal sets on the edge and the triangle are chosen in the same manner as the tetrahedral element.

The CPR formulation for a standard prism takes advantage of this tensor product basis, and is two dimensional in the $\xi - \eta$ plane and one dimensional in ζ direction

$$\begin{aligned}\frac{\partial \tilde{Q}_{i;j,m}}{\partial t} &+ \Pi_{j,m}\left(\nabla^{\xi}\cdot\vec{\tilde{F}}(\tilde{Q}_i)\right)\\ &+ \frac{1}{|V^{\xi}_{Tri}|}\sum_{f\in\partial V_{Tri}}\sum_{l}\alpha_{j,f,l}[\tilde{F}^n(\xi_{f,l},\eta_{f,l},\zeta_m)]S^{\xi}_f\\ &- [\tilde{H}_{com}(\xi_j,\eta_j,-1)-\tilde{H}(\xi_j,\eta_j,-1)]\frac{\alpha_{L,m}}{2}\\ &+ [\tilde{H}_{com}(\xi_j,\eta_j,1)-\tilde{H}(\xi_j,\eta_j,1)]\frac{\alpha_{R,m}}{2}=0,\end{aligned} \tag{68}$$

where the third term is the correction on the $\xi - \eta$ plane, which is computed with fixed $\zeta = \zeta_m$. This is nothing but the correction used in the 2D CPR method for a triangular element. In Eq. (68), $|V^{\xi}_{tri}|$ is the area of the base triangle, which is 1/2, S^{ξ}_f the length of the edge f of the base triangle, and l the index for flux points on f. Note that, $[\tilde{F}^n(\xi_{f,l},\eta_{f,l},\zeta_m)]$ corrects only the solution points on the triangle with fixed m instead of all solution points in the element as shown in Fig. 6(b). The last two terms denote the correction in the ζ direction, which is evaluated with the 1D CPR method.[20] The flux difference at an end point corrects only the solution points on the segment with fixed j as shown in Fig. 6(c). For prism cells, the number of solution points corrected by a flux point is smaller than that for tetrahedral cells due to the decoupled correction procedure. Hence, the method for prisms is more efficient per DOF than for tetrahedrons. This decoupled correction procedure also facilitates the implementation employing different degrees of polynomials in $\xi - \eta$ and ζ directions to adapt to flow features. An attempt employing higher order polynomials in the wall normal direction to resolve the boundary layer with coarser prism cells is presented in Ref. 17.

5. Numerical Results

5.1. *Accuracy study with vortex propagation problem*

This is an idealized problem with an exact solution for the Euler equations. The mean flow is $\{\rho, u, v, p\} = \{1, 1, 1, 1\}$. An isotropic vortex is then added to the mean flow, i.e., with perturbations in u, v, and temperature $T = p/\rho$, and no perturbation in entropy $S = p/\rho^\gamma$:

$$(\delta u, \delta v) = \frac{\epsilon}{2\pi} e^{0.5(1-r^2)}(-y, x),$$
$$\delta T = -\frac{(\gamma - 1)\epsilon^2}{8\gamma\pi^2} e^{1-r^2}, \delta S = 0, \tag{69}$$

where $r^2 = x^2 + y^2$, and the vortex strength $\epsilon = 5$. The exact solution is just the passive convection of the isotropic vortex with the mean velocity $(1, 1)$. In the numerical simulation, the computational domain is taken to be $[-5, 5] \times [-5, 5]$.

The numerical simulations are carried out until $t = 2$ on two different grids, one irregular triangular mesh and one mixed mesh as shown in Fig. 7. The finer irregular grids are generated recursively by cutting each coarser grid cell into four finer grid cells, while all mixed meshes are generated independently. On the irregular triangular mesh, we test both the LP and CR approaches in evaluating the interior flux vector divergence, while on the mixed mesh, we employ the CR approach. For time integration, a 3-stage Runge-Kutta explicit scheme[16] is used for time marching in all the cases.

In Table 1, the L_2 density errors at the solution points are presented on both set of meshes for $k = 1$ to 3. Note that the CR approach is more accurate than the LP approach on the triangular meshes for every polynomial degree and on every mesh. The CR approach not only produces the smaller errors, but also demonstrates more consistent numerical orders of

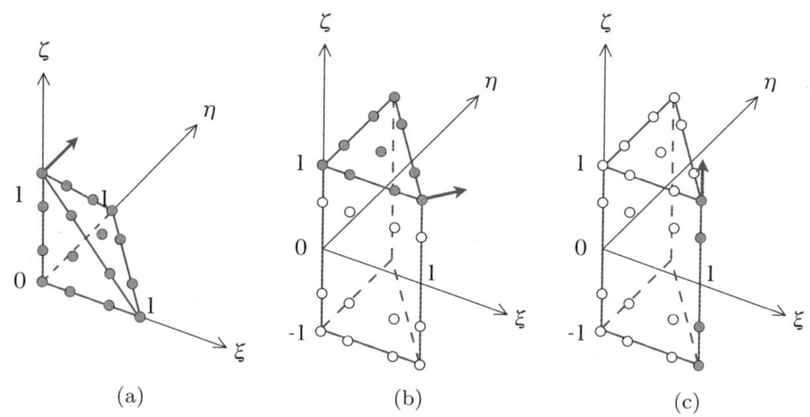

Fig. 6. Solution points in the standard tetrahedral and prism elements for $k = 3$ (only points on the visible faces are shown).

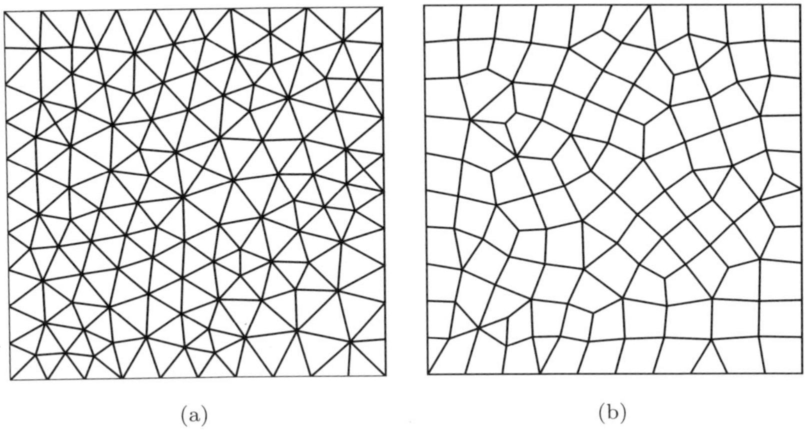

Fig. 7. Regular and irregular "10x10x2" triangular and mixed computational grids.

Table 1. Triangular mesh.

k	Grid size	Triangular mesh - LP		Triangular mesh - CR		Mixed mesh - CR	
		L_2 error	Order	L_2 error	Order	L_2 error	Order
1	10x10x2	2.01e-2	-	1.39e-2	-	1.58e-2	-
	20x20x2	6.67e-3	1.59	4.41e-3	1.66	5.32e-3	1.57
	40x40x2	1.73e-3	1.95	1.08e-3	2.03	1.50e-3	1.83
	80x80x2	4.84e-4	1.84	2.54e-4	2.09	3.54e-4	2.08
2	10x10x2	7.14e-3	-	4.41e-3	-	2.95e-3	-
	20x20x2	1.07e-3	2.74	5.19e-4	3.09	5.62e-4	2.39
	40x40x2	1.60e-4	2.74	5.84e-5	3.15	7.42e-5	2.92
	80x80x2	2.29e-5	2.80	6.94e-6	3.07	8.63e-6	3.10
3	10x10x2	1.79e-3	-	6.70e-4	-	5.79e-4	-
	20x20x2	1.40e-4	3.68	4.79e-5	3.81	5.05e-5	3.52
	40x40x2	9.75e-6	3.84	2.96e-6	4.02	3.51e-6	3.85
	80x80x2	6.96e-7	3.81	1.71e-7	4.11	1.89e-7	4.22

accuracy in the grid refinement study. In addition, the CPR method performs very well on the mixed grids, achieving the optimal order of accuracy on relatively poor quality meshes.

5.2. *Laminar flow around a NACA0012 airfoil*

Viscous laminar flow around a NACA 0012 airfoil is simulated with the CPR method, using the BR2, I-continuous, interior penalty and CDG schemes for the viscous flux. The flow conditions are Mach = 0.5 and Re = 5000, with an angle of attack of 1°. Under such conditions, steady laminar separations are expected on both upper and lower surfaces of the airfoil. Adiabatic

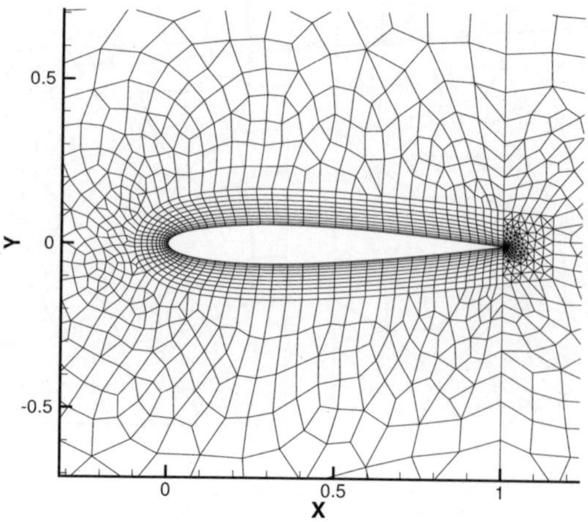

Fig. 8. Mixed mesh around a NACA0012 airfoil.

no-slip wall condition is prescribed at the airfoil surface. The curved wall boundary is represented by the same degree polynomial as the solution. The computational domain extends 20 chord lengths away from the center of the airfoil.

Figure 8 shows the computational mesh with 2,692 cells, which is composed of regular quadrilateral elements near the airfoil and irregular mixed elements elsewhere, with some refinement at the trailing edge. A block-preconditioned LU-SGS solver[6] was used for time integration and all cases converged to machine zero.

Figure 9 shows the computed Mach number contours of 2nd to 4th order schemes. Only the BR2 results are shown, since the results of other schemes are very similar. Because the mesh is coarse, the 2nd order results are not smooth, especially at the wake. Note that for the 3rd and 4th order cases, the contour lines are smooth across the interfaces between regular cells and irregular ones and also between triangular cells and quadrilateral ones. The 3rd and 4th order results are visibly similar.

Figure 10 shows the skin friction distribution near the separation point. For the four schemes, the 3rd and 4th order results are very close; this fact indicates convergence with p-refinement. For the 2nd order scheme, the I-continuous approach is the most accurate, while CDG is the least accurate and BR2 and IP are neary identical.

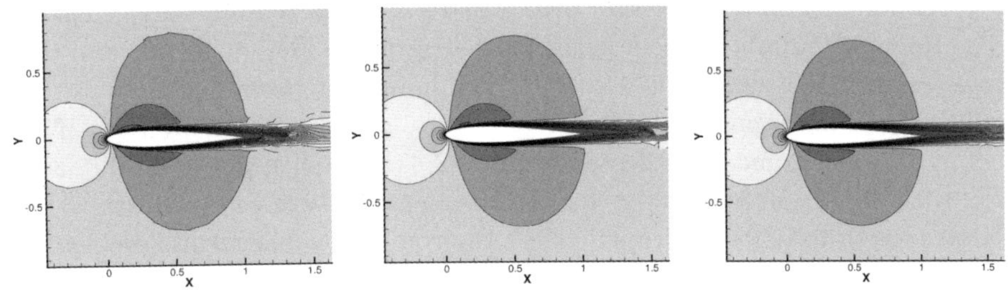

(a) 2nd order (b) 3rd order (c) 4th order

Fig. 9. Mach number contours of flow around an NACA 0012 airfoil.

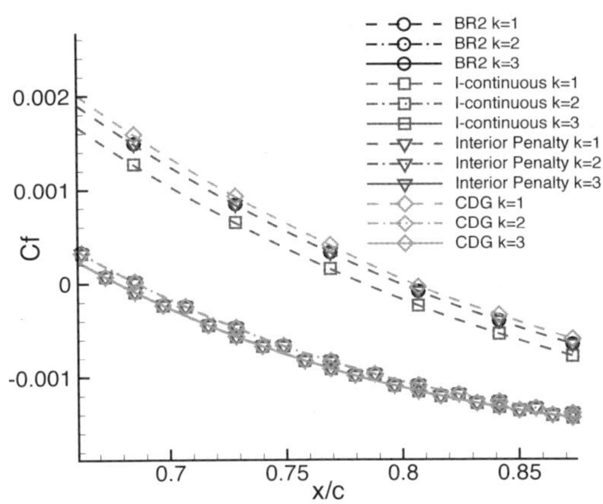

Fig. 10. Computed c_f distribution on the upper surface of NACA0012 airfoil with BR2, I-continuous, interior penalty and CDG.

5.3. *Laminar boundary layer on a flat plate*

One issue when we apply a CFD solver to engineering problems is the stiffness arising from using high aspect ratio cells near the solid wall to resolve the boundary layer especially for high Reynolds number flows. Here we attempt an approach to alleviate the stiffness by employing a small number of higher-order prism elements rather than having many lower order elements in the boundary layer. Since we use a tensor-product basis in prisms, we can use a higher order polynomial only in the normal direction to the wall while using a lower order one in the tangential direction of the wall so as to maximize the efficiency.

The laminar boundary layer over a plate is computed on a prism mesh. The Reynolds number based on the plate length L is $Re_L = 10,000$ and the freestream Mach number is $M = 0.2$. The boundary layer thickness at the trailing edge is estimated by the approximate relation $\delta = 5L/\sqrt{Re_L}$. The computational domain is selected to be $(-2 \le x \le 1, 0 \le y \le 100\delta, 0 \le z \le \delta)$, with $L = 1$. Note that the domain size in the y-direction is chosen to be large enough to not significantly affect the computational results especially in the v-velocity profiles. The prism mesh was produced from a Cartesian grid, with clustering at the wall and near the leading edge. In the spanwise z-direction, only one cell was generated.

Figure 11(a) shows the computed Mach number using polynomials of degree 5 in the y-direction and polynomials of degree 2 in x- and z- directions. The grid has only two cells in the boundary layer at $x = 1.0$ and 17 cells along the plate. The numbers of prism cells and DOFs are 728 and 26208 respectively. In comparison, a finer grid was generated by dividing each prism cell into two prism cells to have twice the number of cells in the y- direction. We employed degree 2 polynomials in all directions on this finer grid. Since each prism cell of degree 2 polynomials has 3 solution points, which is half of the solution points for the degree 5 polynomials, the total number of DOFs is the same as the case using degree 5 polynomials in the normal direction. The computed Mach number on this finer grid is shown in Fig. 11(b).

In Fig. 12, the computed v-velocity profiles in the boundary layer at $x = 5$ and skin friction profiles along the plate are shown. As we expected,

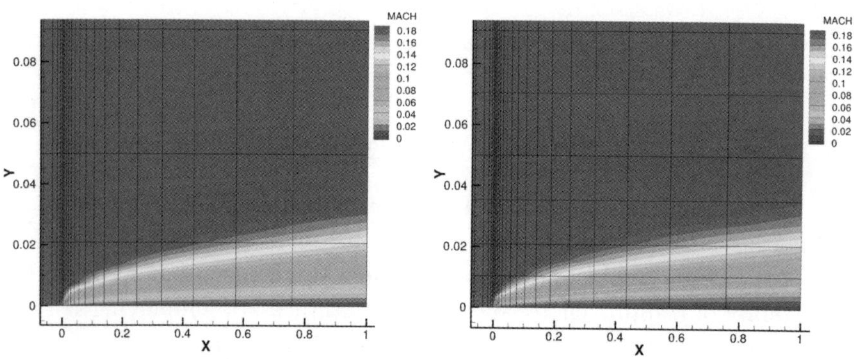

(a) $k = 5$ in the y-direction (b) $k = 2$ on a finer mesh in the y-direction

Fig. 11. Mach number contours of a laminar boundary layer on a flat plate (enlarged by a factor of 10 in y direction).

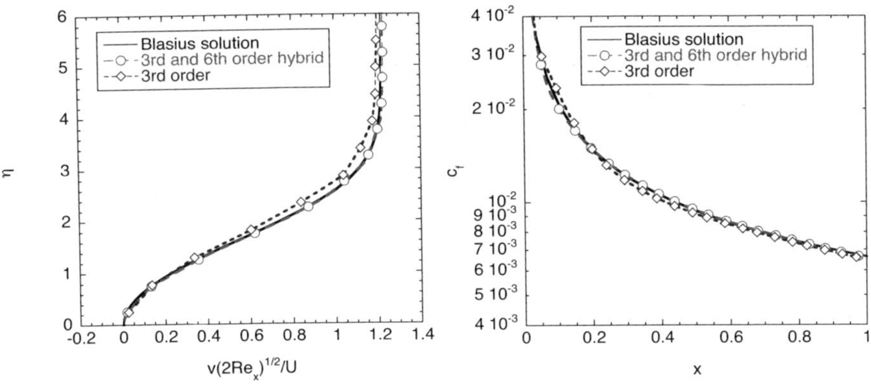

(a) v velocity profile at $x = 0.5$ (b) Skin friction profile

Fig. 12. Comparison of v-velocity and c_f profiles for the flat plate boundary layer problem.

the computed profiles using the higher order scheme agree better with the Blasius's solution. The convergence histories are compared in Fig. 13. The computations were performed using the block preconditioned LU-SGS scheme with several different time steps. It is shown that, a larger time step can be taken in the case employing the higher order elements with less grid cells and it takes fewer number of iterations to converge to the steady state.

5.4. *Unsteady subsonic flow over a sphere at Re=300*

Next, we consider an unsteady flow case over a sphere with a Reynolds number of 300 based on the diameter of the sphere. The inflow Mach number is 0.3. The hybrid prismatic and tetrahedral computational mesh is shown in Fig. 14. To resolve shedding vortices, finer elements are generated in the wake region. The total number of cells is 54,312. The local grid size around the sphere is $\sim 0.2r$ and the size in the wake region is $\sim 0.8r$ with r the radius of the sphere.

The computed Q isosurface colored by local Mach number using the 4th-order CPR scheme is shown in Fig. 15. The obtained plain symmetric wake vortex structure is comparable to the available experimental and computational results in Refs. 14 and 24 at least qualitatively. In Fig. 16 we plot the history of the drag coefficient C_d in terms of non-dimensional time t. The computed drag coefficient and the oscillating amplitude of drag and

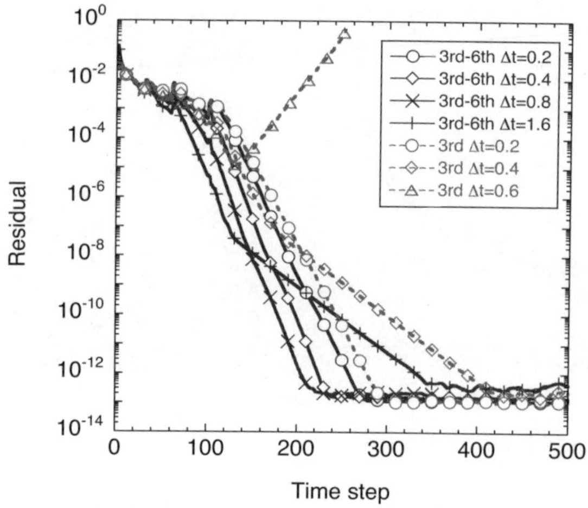

Fig. 13. Comparison of the convergence histories for the flat plate boundary layer problem.

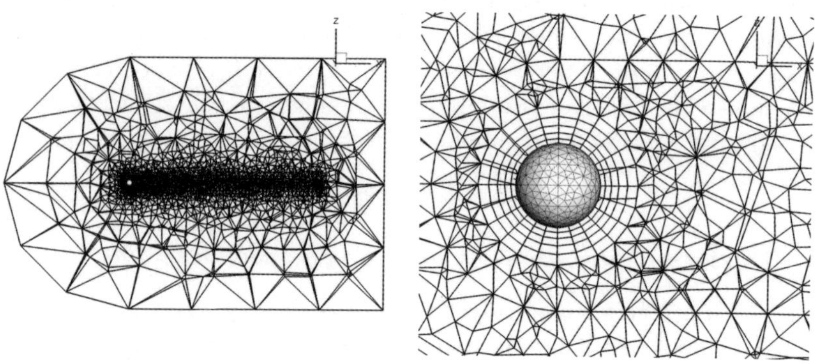

(a) Entire grid (b) Grid around the sphere

Fig. 14. Computational grid around a sphere for the unsteady viscous flow at Re = 300.

the Strouhal number St are shown in Table 2. For comparison, results from Gassner[14] using the 4th-order DG scheme on a tetrahedral grid and from Tomboulides[36] and Johnson and Patel[24] obtained with an incompressible simulation, are shown as well. The results computed with the CPR method agree reasonably well with those reference values.

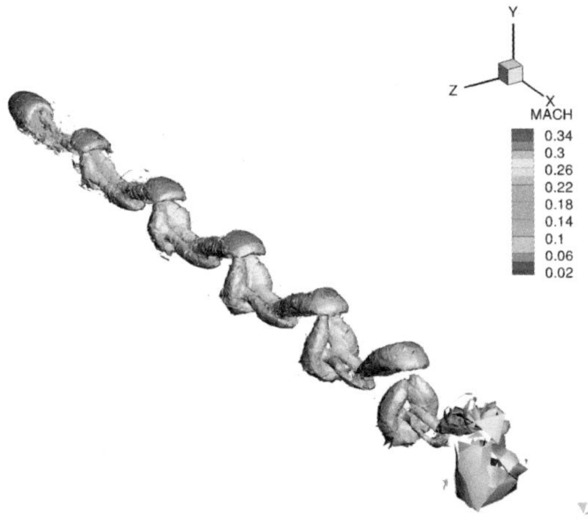

Fig. 15. Computed Q isosurfaces in the wake region of the viscous laminar flow over a sphere at Re=300.

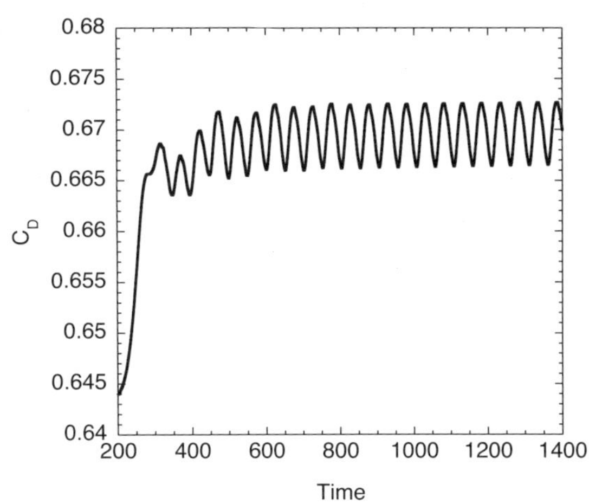

Fig. 16. Time history of the drag coefficient for unsteady flow over a sphere at Re = 300.

6. Conclusions

This chapter describes a discontinuous method named correction procedure via reconstruction or CPR for hybrid meshes. The CPR formulation unifies the discontinuous Galerkin, staggered grid, spectral volume and spectral

Table 2. Comparisons of the averaged drag coefficient, the amplitude of drag and the Strouhal number.

Method	C_d	ΔC_d	St
Present	0.670	0.0032	0.131
Gassner[14]	0.673	0.0031	0.135
Tomboulides[36]	0.671	0.0028	0.136
Johnson & Patel[24]	0.656	0.0035	0.137

difference methods into a single differential formulation, and is particular simple for high-order elements. The extensions to viscous flow, and to 3D mixed grids are also presented. Various accuracy studies have verified the CPR method is capable of obtaining the designed order of accuracy for both inviscid and viscous flow problems. Other benchmark and test cases have demonstrated the capability of the method. Future work includes the development of efficient, and low memory solvers, and solution based *hp*-adaptations.

Acknowledgments

The research on high-order methods has been funded by AFOSR grant FA9550-06-1-0146, and partially by DOE grant DE-FG02-05ER25677.

References

1. D. N. Arnold, F. Brezzi, B. Cockburn and L. D. Marini, Unified analysis of discontinuous Galerkin methods for elliptic problems, *SIAM J. Numer. Anal.* **19** (4), pp. 742–760, (2002).
2. T. J. Barth and P. O. Frederickson, High-order solution of the Euler equations on unstructured grids using quadratic reconstruction, *AIAA-90-0013*, (1990).
3. F. Bassi and S. Rebay, A high-order accurate discontinuous finite element method for the numerical solution of the compressible Navier-Stokes equations, *J. Comput. Phys.* **131**, pp. 267–279, (1997).
4. F. Bassi and S. Rebay, High-order accurate discontinuous finite element solution of the 2D Euler equations, *J. Comput. Phys.* **138** (2), pp. 251–285, (1997).
5. F. Bassi and S. Rebay, GMRES discontinuous Galerkin solution of the compressible Navier-Stokes equations, In eds. B. Cockburn, G.E. Karniadakis and C. W. Shu, *Discontinuous Galerkin methods: Theory, Computations and Applications.* pp. 197-208, Springer, Berlin, (2000).

6. R. F. Chen and Z. J. Wang, Fast, block lower-upper symmetric Gauss-Seidel scheme for arbitrary grids, *AIAA J.* **38** (12), pp. 2238–2245, (2000).
7. B. Cockburn and C. W. Shu, TVB Runge-Kutta local projection discontinuous Galerkin finite element method for conservation laws II: general framework, *Math. Comput.* **52**, pp. 411–435, (1989).
8. B. Cockburn and C. W. Shu, The Runge-Kutta discontinuous Galerkin method for conservation laws V: multidimensional systems, *J. Comput. Phys.* **141**, pp. 199–224, (1998).
9. B. Cockburn and C. W. Shu, The local discontinuous Galerkin methods for time-dependent convection diffusion systems, *SIAM J. Numer. Anal.* **35**, pp. 2440–2463, (1998).
10. M. Delanaye and Y. Liu, Quadratic reconstruction finite volume schemes on 3D arbitrary unstructured polyhedral grids, *AIAA-99-3259*, (1999).
11. V. Dolejší, On the discontinuous Galerkin method for numerical solution of the Navier-Stokes equations, *Int. J. Numer. Meth. Fluids.* **45**, pp. 1083–1106, (2004).
12. M. Dumbser, $P_N P_M$ schemes on unstructured meshes for time-dependent partial differential equations. In eds. Z. J. Wang, *Adaptive High-order Methods in Computational Fluid Dynamics.* pp. 233. World Scientific, Singapore, (2011).
13. J. A. Ekaterinaris, High-order accurate, low numerical diffusion methods for aerodynamics, *Progress in Aerospace Sciences.* **41**, pp. 192–300, (2005).
14. G. J. Gassner, J. F. Lorcher, C-D. Munz and J. S. Hesthaven, Polymorphic nodal elements and their application in discontinuous Galerkin methods, *J. Comput. Phys.* **228**, pp. 1573–1590, (2005).
15. S. K. Godunov, A finite-difference method for the numerical computation of discontinuos solutions of the equations of fluid dynamics, *Math. Sbornik.* **47**, pp. 271–306, (1959).
16. S. Gottlieb and C. W. Shu, Total variation diminishing Runge-Kutta schemes, *Math. Comput.* **67**, pp. 73–85, (1998).
17. T. Haga, H. Gao and Z. J. Wang, A high-order unifying discontinuous formulation for 3D mixed grids, *AIAA-2010-540*, (2010).
18. R. Hartmann and P. Houston, Symmetric interior penalty DG emthods for the compressible Navier-Stokes Equations I: Method formulation, *Int. J. Numer. Anal. Model..* **3** (1), pp. 1–20, (2006).
19. J. S. Hesthaven, From electrostatics to almost optimal nodal sets for polynomial interpolation in a simplex, *SIAM J. Numer. Anal.* **35** (2), pp. 655–676, (1998).
20. H. T. Huynh, A flux reconstruction approach to high-order schemes including discontinuous Galerkin methods, *AIAA-2007-4079*, (2007).
21. H. T. Huynh, A reconstruction approach to high-order schemes including discontinuous Galerkin for diffusion, *AIAA-2009-403*, (2009).
22. H. T. Huynh, High-order methods by correction procedures using reconstructions. In eds. Z. J. Wang, *Adaptive High-order Methods in Computational Fluid Dynamics.* pp. 422. World Scientific, Singapore, (2011).

23. A. Jameson, Analysis and design of numerical schemes for gas dynamics. I. Artificial diffusion, upwind biasing, limiters and their eefect on accuracy and multigrid convergence, *Int. J. Comput. Fluid Dyn.* **4**, pp. 171–218, (1994).
24. T. A. Johnson and V. C. Patel, Flow past a sphere up to a Reynolds number of 300, *J. Fluid Mech.* **378**, pp. 19–70, (1999).
25. G. E. Karniadakis and S. J. Sherwin, *Spectral-hp Element Methods.* Oxford University Press, Oxford, England, (1999).
26. D. A. Kopriva and J. H. Kolias, A conservative staggered-grid Chebyshev multidomain method for compressible flows, *J. Comput. Phys.* **125**, pp. 244–261, (1996).
27. M.-S. Liou, A sequel to AUSM, Part II: AUSM+-up for all speeds, *J. Comput. Phys.* **214**, pp. 137–170, (2006).
28. Y. Liu, M. Vinokur and Z. J. Wang, Spectral (finite) volume method for conservation laws on unstructured grids V: Extension to three-dimensional systems, *J. Comput. Phys.* **212**, pp. 454–472, (2006).
29. Y. Liu, M. Vinokur and Z. J. Wang, Discontinuous spectral difference method for conservation laws on unstructured grids, *J. Comput. Phys.*, **216**, pp. 780–801, (2006).
30. G. May and A. Jameson, A spectral difference method for the Euler and Navier-Stokes equations, *AIAA-2006-304*, (1996).
31. A. Nejata and C. Ollivier-Gooch, A high-order accurate unstrcutured finite voume Newton-Krylov algorithm for inviscid compressible flows, *J. Comput. Phys.* **227**, pp. 2582–2609, (2008).
32. J. Peraire and P.-O. Persson, The compact discontinuous Galerkin (CDG) method for elliptic problems, *SIAM J. Sci. Comput.* **30**, pp. 1806–1824, (2008).
33. W. H. Reed and T. R. Hill, Triangular mesh methods for the neutron transport equation, *Los Alamos Scientific Laboratory Report, LA-UR-73-479*, (1973).
34. P. L. Roe, Approximate Riemann solvers, parameter vectors, and difference schemes, *J. Comput. Phys.* **43**, pp. 357–372, (1981).
35. V. V. Rusanov, Calculation of interaction of non-steady shock waves with obtsacles, *SIAM J. Comput. Math. Phys.* **1**, pp. 261–279, (1961).
36. A. G. Tomboulides and S. A. Orzag, Numerical investigation of transitional and weak turbulent flow past a sphere, *J. Fluid Mech.* **416**, pp. 45–73, (2000).
37. K. Van der Abeele, C. Lacor and Z. J. Wang, On the stability and the accuracy of the spectral difference method, *J. Sci. Comput.* **37** (2), pp. 162–188, (2008).
38. B. Van Leer, Towards the ultimate conservative differences scheme V. a second order sequel to Godunov's method, *J. Comput. Phys.* **32**, pp. 101–136, (1979).
39. B. Van Leer and S. Nomura, Discontinuous Galerkin for diffusion, *AIAA-2005-5108*, (2005).
40. Z. J. Wang, Spectral (finite) volume method for conservation laws on unstructured grids: basic formulation, *J. Comput. Phys.* **178** (2), pp. 210–251, (2002).

41. Z. J. Wang, High-order methods for the Euler and Navier-Stokes equations on unstructured grids, *Progress in Aerospace Sciences.* **43**, pp. 1–47, (2007).
42. Z. J. Wang and H. Gao, A unifying lifting collocation penalty formulation including the discontinuous Galerkin, spectral volume/difference mthods for conservation laws on mixed grids, *J. Comput. Phys.* **228**, pp. 8161–8186, (2009).
43. Z. J. Wang and Y. Liu, Spectral (finite) volume method for conservation laws on unstructured grids II: extension to two-dimensional scalar equation, *J. Comput. Phys.* **179**, pp. 665–697, (2002).
44. T. Warburton, An explicit construction of interpolation nodes on the simplex, *J. Eng. Math.* **56** (2), pp. 247–262, (2006).
45. O. C. Zienkiewicz and R. L. Taylor, *The Finite Element Method The Basics, vol. 1.* Butterworth-Heinemann, Oxford, England, (2000).

Index